ENGINEERING MECHANICS OF COMPOSITE MATERIALS

ENGINEERING MECHANICS OF COMPOSITE MATERIALS

SECOND EDITION

Isaac M. Daniel
Departments of Civil and Mechanical Engineering
Northwestern University, Evanston, IL

Ori Ishai
Faculty of Mechanical Engineering
Technion-Israel Institute of Technology, Haifa, Israel

New York ■ Oxford
OXFORD UNIVERSITY PRESS
2006

Oxford University Press, Inc., publishes works that further Oxford University's
objective of excellence in research, scholarship, and education.

Oxford New York
Auckland Cape Town Dar es Salaam Hong Kong Karachi
Kuala Lumpur Madrid Melbourne Mexico City Nairobi
New Delhi Shanghai Taipei Toronto

With offices in
Argentina Austria Brazil Chile Czech Republic France Greece
Guatemala Hungary Italy Japan Poland Portugal Singapore
South Korea Switzerland Thailand Turkey Ukraine Vietnam

Published by Oxford University Press, Inc.
198 Madison Avenue, New York, New York 10016
http://www.oup.com

Oxford is a registered trademark of Oxford University Press

Library of Congress Cataloging-in-Publication Data

Daniel, Isaac M.
    Engineering mechanics of composite materials / Isaac M. Daniel, Ori Ishai.—2nd ed.
      p. cm.
    ISBN-13: 978-0-19-515097-1

      1. Composite materials—Mechanical properties.   2. Composite materials—Testing.
  I. Ishai, Ori.   II. Title.

TA418.9.C6D28 2005
620.1′183—dc22   2004065462

Printing number: 9 8 7 6 5 4

Printed in the United States of America
on acid-free paper

To my wife, Elaine
my children, Belinda, Rebecca, and Max
and the memory of my parents, Mordochai and Bella Daniel
Isaac M. Daniel

To my wife, Yael
and my children, Michal, Tami, Eran, and Yuval
Ori Ishai

Contents

Preface to the Second Edition

Writing this book and observing its widespread use and acceptance has been a very gratifying experience. We are very pleased and thankful for the many favorable comments we received over the years from colleagues around the world. Educators who used the book in their teaching of composite materials found it well organized, clear, and brief. Many of them offered valuable suggestions for corrections, revisions, and additions for this new edition.

After writing a textbook, one always thinks of ways in which it could be improved. Besides correcting some inevitable errors that appeared in the first edition, we wanted to revise, update, and expand the material in keeping with our teaching experience, the feedback we received, and the continuously expanding field of composite materials technology and applications.

This edition has been expanded to include two new chapters dealing with micromechanics. Materials in wide use today, such as textile-reinforced composites, are discussed in more detail. The database in the original Chapter 2 has been expanded to include more fabric composites, high-temperature composites, and three-dimensional properties, and it has been moved to an appendix for easier reference. A description has been added of processing methods since the quality and behavior of composite materials is intimately related to the fabrication process.

Chapter 3, a new chapter, gives a review of the micromechanics of elastic behavior, leading to the macromechanical elastic response of a composite lamina discussed in Chapter 4. Recognizing the current interest in three-dimensional effects, we included transformation relations for the three-dimensional case as well.

Chapter 5 describes the micromechanics of failure, including failure mechanisms and prediction of strength. Chapter 6 is a treatment and discussion of failure of a composite lamina from the macromechanical or phenomenological point of view. An updated review and description of macromechanical failure theories is given for the single lamina. Basic theories discussed in detail include maximum stress, maximum strain, phenomenological (interaction) theories (Tsai-Hill and Tsai-Wu), and mechanistic theories based on specific single or mixed failure modes (Hashin-Rotem). Their extension to three dimensions and their application to textile composites are described. Comparisons with experimental data have been added for the unidirectional lamina and the basic fabric lamina.

Chapter 7, which deals primarily with the classical lamination theory, has been expanded to include effects of transverse shear and application to sandwich plates. Except for some updating of data, few changes were made in Chapter 8, which describes hygrothermal effects. Chapter 9, which deals with stress and failure analysis of laminates has been revised extensively in view of the ongoing debate in technical circles on the applicability of the various failure theories. The discussion emphasizes progressive failure following first ply failure and evaluates the various theories based on their capability to predict ultimate laminate failure. Applications to textile composites are described, and comparisons between theoretical predictions and experimental results are discussed.

Chapter 10 has been revised primarily by adding test methods for fabric composites and for determination of three-dimensional properties. The book retains the same overall structure as the first edition. New problems have been added in the Problem sections.

We aimed to make this new edition more relevant by emphasizing topics related to current interests and technological trends. However, we believe that the uniqueness of this book lies primarily in its contribution to the continuously expanding educational activity in the field of composites.

We have tried to accomplish all of the above revisions and additions without expanding the size of the book significantly. We believe in placing more emphasis on the macromechanics of composite materials for structural applications.

We would like to acknowledge again, as with the first edition, the dedicated, expert, and enthusiastic help of Mrs. Yolande Mallian in typing and organizing the manuscript. We would like to thank Dr. Jyi-Jiin Luo for his assistance with the preparation of new illustrations, the evaluation of the various failure theories, and the writing of a new comprehensive and user-friendly computer program for predicting the failure of composite laminates and Drs. Jandro L. Abot, Patrick M. Schubel, and Asma Yasmin for their help with the preparation of new and revised illustrations. The valuable suggestions received from the following colleagues are greatly appreciated: John Botsis, Leif A. Carlsson, Kathleen Issen, Liviu Librescu, Ozden O. Ochoa, C. T. Sun, and George J. Weng.

Evanston, IL   I.M.D.
Haifa, Israel   O.I.

Preface to the First Edition

Although the underlying concepts of composite materials go back to antiquity, the technology was essentially developed and most of the progress occurred in the last three decades, and this development was accompanied by a proliferation of literature in the form of reports, conference proceedings, journals, and a few dozen books. Despite this plethora of literature, or because of it, we are constantly faced with a dilemma when asked to recommend a single introductory text for beginning students and engineers. This has convinced us that there is a definite need for a simple and up-to-date introductory textbook aimed at senior undergraduates, graduate students, and engineers entering the field of composite materials.

This book is designed to meet the above needs as a teaching textbook and as a self-study reference. It only requires knowledge of undergraduate mechanics of materials, although some knowledge of elasticity and especially anisotropic elasticity might be helpful.

The book starts with definitions and an overview of the current status of composites technology. The basic concepts and characteristics, including properties of constituents and typical composite materials of interest and in current use are discussed in Chapter 2. To keep the volume of material covered manageable, we omitted any extensive discussion of micromechanics. We felt that, although relevant, micromechanics is not essential in the analysis and design of composites. In Chapter 3 we deal with the elastic macromechanical response of the unidirectional lamina, including constitutive relations in terms of mathematical stiffnesses and compliances and in terms of engineering properties. We also deal with transformation relations for these mechanical properties. We conclude with a short discussion of micromechanical predictions of elastic properties. In Chapter 4 we begin with a discussion of microscopic failure mechanisms, which leads into the main treatment of failure from the macroscopic point of view. Four basic macroscopic failure theories are discussed in detail. Classical lamination theory, including hygrothermal effects, is developed in detail and then applied to stress and failure analyses of multidirectional laminates in Chapters 5, 6, and 7. We conclude Chapter 7 with a design methodology for structural composites, including a design example discussed in detail. Experimental methods for characterization and testing of the constituents and the composite material are described in Chapter 8.

Whenever applicable, in every chapter example problems are solved and a list of unsolved problems is given. Computational procedures are emphasized throughout, and flow charts for computations are presented.

The material in this book, which can be covered in one semester, is based on lecture notes that we have developed over the last fifteen years in teaching formal courses and condensed short courses at our respective institutions, and we have incorporated much of the feedback received from students. We hope this book is received as a useful and clear guide for introducing students and professionals to the field of composite materials.

We acknowledge with deep gratitude the outstanding, dedicated, and enthusiastic support provided by two people in the preparation of this work. Mrs. Yolande Mallian typed and proofread the entire manuscript, including equations and tables, with painstaking exactitude. Dr. Cho-Liang Tsai diligently and ably performed many computations and prepared all the illustrations.

Evanston, IL   I.M.D.
Haifa, Israel   O.I.
May 1993

ENGINEERING MECHANICS OF
COMPOSITE MATERIALS

1 Introduction

1.1 DEFINITION AND CHARACTERISTICS

A structural composite is a material system consisting of two or more phases on a macroscopic scale, whose mechanical performance and properties are designed to be superior to those of the constituent materials acting independently. One of the phases is usually discontinuous, stiffer, and stronger and is called the *reinforcement*, whereas the less stiff and weaker phase is continuous and is called the *matrix* (Fig. 1.1). Sometimes, because of chemical interactions or other processing effects, an additional distinct phase called an *interphase* exists between the reinforcement and the matrix. The properties of a composite material depend on the properties of the constituents, their geometry, and the distribution of the phases. One of the most important parameters is the volume (or weight) fraction of reinforcement or fiber volume ratio. The distribution of the reinforcement determines the homogeneity or uniformity of the material system. The more nonuniform the reinforcement distribution, the more heterogeneous the material, and the higher the

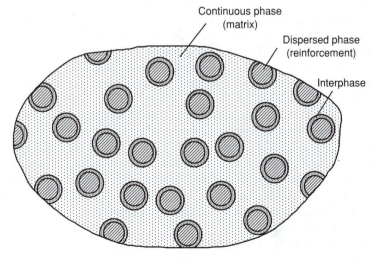

Continuous phase
(matrix)

Dispersed phase
(reinforcement)

Interphase

Fig. 1.1 Phases of a composite material.

scatter in properties and the probability of failure in the weakest areas. The geometry and orientation of the reinforcement affect the anisotropy of the system.

The phases of the composite system play different roles, which depend on the type and application of the composite material. In the case of low- to medium-performance composite materials, the reinforcement, usually in the form of short fibers or particles, may provide some stiffening but only limited strengthening of the material. The matrix, on the other hand, is the main load-bearing constituent governing the mechanical properties of the material. In the case of high-performance structural composites, the normally continuous fiber reinforcement is the backbone of the material, which determines its stiffness and strength in the fiber direction. The matrix phase provides protection for the sensitive fibers, bonding, support, and local stress transfer from one fiber to another. The interphase, although small in dimensions, can play an important role in controlling the failure mechanisms, failure propagation, fracture toughness and the overall stress-strain behavior to failure of the material.

1.2 HISTORICAL DEVELOPMENT

Conventional monolithic materials can be classified into three broad categories: metals, ceramics, and polymers. Composites are combinations of two or more materials from one or more of these categories. Human development and civilization are closely related to the utilization of materials. In the stone age primitive man relied primarily on ceramics (stone) for tools and weapons and on natural polymers and composites (wood). The use of metals started with gold and proceeded with copper, bronze, and iron. Metals, especially steel and aluminum, became dominant starting in the last century and continue to the present. A new trend is taking place presently where polymers, ceramics, and composites are regaining in relative importance. Whereas in the early years man used natural forms of these materials, the newer developments and applications emphasize man-made or engineered materials.[1]

Historically, the concept of fibrous reinforcement is very old, as quoted in biblical references to straw-reinforced clay bricks in ancient Egypt (Exodus 5:7). Achilles's shield is an example of composite laminate design as described in Homer's *Iliad* (verses 468–480). Iron rods were used to reinforce masonry in the nineteenth century, leading to the development of steel-reinforced concrete. Phenolic resin reinforced with asbestos fibers was introduced in the beginning of the last century. The first fiberglass boat was made in 1942, accompanied by the use of reinforced plastics in aircraft and electrical components. Filament winding was invented in 1946, followed by missile applications in the 1950s. The first boron and high-strength carbon fibers were introduced in the early 1960s, followed by applications of advanced composites to aircraft components in 1968. Metal-matrix composites such as boron/aluminum were introduced in 1970. Dupont developed Kevlar (or aramid) fibers in 1973. Starting in the late 1970s, applications of composites expanded widely to the aircraft, marine, automotive, sporting goods, and biomedical industries. The 1980s marked a significant increase in high-modulus fiber utilization. The 1990s marked a further expansion to infrastructure. Presently, a new frontier is opening, that of nanocomposites. The full potential of nanocomposites, having phases of dimensions on the order of nanometers, remains to be explored.

1.3 APPLICATIONS

Applications of composites abound and continue to expand. They include aerospace, aircraft, automotive, marine, energy, infrastructure, armor, biomedical, and recreational (sports) applications.

Aerospace structures, such as space antennae, mirrors, and optical instrumentation, make use of lightweight and extremely stiff graphite composites. A very high degree of dimensional stability under severe environmental conditions can be achieved because these composites can be designed to have nearly zero coefficients of thermal and hygric expansion.

The high-stiffness, high-strength, and low-density characteristics make composites highly desirable in primary and secondary structures of both military and civilian aircraft. The Boeing 777, for example, uses composites in fairings, floorbeams, wing trailing edge surfaces, and the empennage (Figs. 1.2 and 1.3).[2] The strongest sign of acceptance of composites in civil aviation is their use in the new Boeing 787 "Dreamliner" (Fig. 1.4) and the world's largest airliner, the Airbus A380 (Fig. 1.5). Composite materials, such as carbon/epoxy and graphite/titanium, account for approximately 50% of the weight of the Boeing 787, including most of the fuselage and wings. Besides the advantages of durability and reduced maintenance, composites afford the possibility of embedding sensors for on-board health monitoring. The Airbus A380 also uses a substantial amount of composites, including a hybrid glass/epoxy/aluminum laminate (GLARE), which combines the advantages and mitigates the disadvantages of metals and composites. Figure 1.6 shows a recently certified small aircraft with the primary structure made almost entirely of composite (composite sandwich with glass fabric/epoxy skins and PVC foam core). The stealth characteristics of

Fig. 1.2 Boeing 777 commercial aircraft. (Courtesy of Boeing Commercial Airplane Group.)

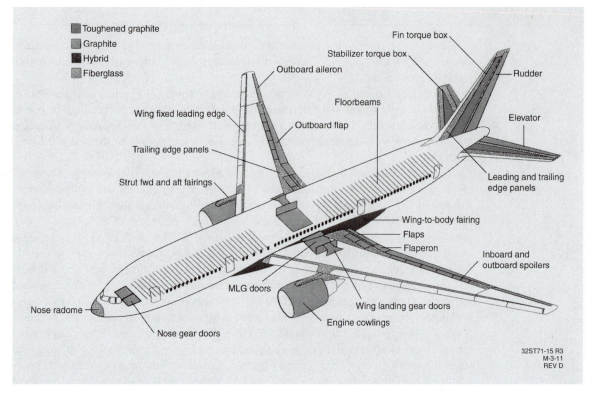

Fig. 1.3 Diagram illustrating usage of composite materials in various components of the Boeing 777 aircraft. (Courtesy of Boeing Commercial Airplane Group.)[2]

Fig. 1.4 Boeing 787 "Dreamliner" with most of the fuselage and wings made of composite materials. (Courtesy Boeing Commercial Airplane Group.)

Fig. 1.5 Airbus A380 containing a substantial amount of composite materials including glass/epoxy/ aluminum (GLARE). (Image by Navjot Singh Sandhu.)

Fig. 1.6 Small aircraft with primary structure made of composite materials. (Courtesy of Dr. Paul Brey, Cirrus Design Corporation.)

Fig. 1.7 B-2 stealth bomber made almost entirely of composite materials. (Courtesy of Dr. R. Ghetzler, Northrop Corporation.)

carbon/epoxy composites are highly desirable in military aircraft, such as the B-2 bomber (Fig. 1.7). Small unmanned air vehicles are also made almost entirely of composites (Fig. 1.8). The solar-powered flying wing Helios shown in Fig. 1.9, used by NASA for environmental research, was made of carbon and Kevlar fiber composites. It had a wing span of 75 m (246 ft) and weighed only 708 kg (1557 lb).

Composites are used in various forms in the transportation industry, including automotive parts and automobile, truck, and railcar frames. An example of a composite leaf spring is shown in Fig. 1.10, made of glass/epoxy composite and weighing one-fifth of the original steel spring. An example of an application to public transportation is the Cobra tram in Zurich (Fig. 1.11).

Ship structures incorporate composites in various forms, thick-section glass and carbon fiber composites and sandwich construction. The latter consists of thin composite facesheets bonded to a thicker lightweight core. Applications include minesweepers and

Fig. 1.8 Unmanned reconnaissance aircraft made of composite materials. (Courtesy of MALAT Division, Israel Aircraft Industries.)

Fig. 1.9 Solar-powered flying wing Helios. (Courtesy of Stuart Hindle, Sky Tower, Inc.; NASA Dryden Flight Center photograph.)[3]

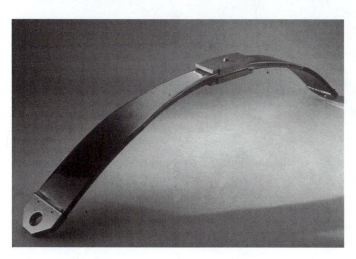

Fig. 1.10 Corvette rear leaf spring made of glass/epoxy composite weighing 3.6 kg (8 lb) compared to original steel spring weighing 18.6 kg (41 lb). (First production application of structural composite in automobiles; courtesy of Nancy Johnson, GM Research and Development Center, General Motors.)

Fig. 1.11 Cobra tram in Zurich, Switzerland, incorporating composite sandwich construction. (Courtesy of Alcan Airex AG, photograph by Bombardier Transportation.)

Fig. 1.12 Royal Danish Navy standard Flex 300 corvette: length, 55 m; displacement, 350 tons; materials, glass/polyester and PVC foam. (Courtesy of Professor Ole Thomsen, Aalborg University, Denmark.)

corvettes (Figs. 1.12 and 1.13). Composite ship structures have many advantages such as insulation, lower manufacturing cost, low maintenance, and lack of corrosion.

In the energy production field, carbon fiber composites have been used in the blades of wind turbine generators that significantly improve power output at a greatly reduced cost (Fig. 1.14). In offshore oil drilling installations, composites are used in drilling risers like the one installed in the field in 2001 and shown in Fig. 1.15.

Fig. 1.13 Royal Swedish Navy Visby class corvette: length, 72 m; displacement, 600 tons; materials, carbon/vinylester and PVC foam. (Courtesy of Professor Christian Berggreen, Technical University of Denmark and Kockums AB, Malmö, Sweden.)

Fig. 1.14 Composite wind turbine blade used for energy production. (Courtesy of Professor Ole Thomsen, Aalborg University, Denmark.)

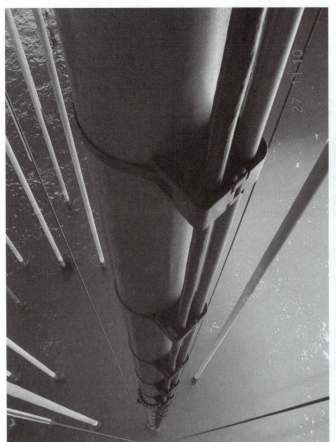

Fig. 1.15 Composite drilling riser for offshore oil drilling: 15 m (49 ft) long, 59 cm (22 in) inside diameter, 315 bar pressure; manufactured for Norske Conoco A/S and other oil companies. (Courtesy of Professor Ozden Ochoa, Texas A&M University, and Dr. Mamdouh M. Salama, ConocoPhillips.)

Biomedical applications include prosthetic devices and artificial limb parts (Figs. 1.16 and 1.17). Leisure products include tennis rackets, golf clubs, fishing poles, skis, and bicycles. An example of a composite bicycle frame is shown in Fig. 1.18.

Infrastructure applications are a more recent development. Composites are being used to reinforce structural members against earthquakes, to produce structural shapes for buildings and bridges, and to produce pipes for oil and water transport. An 80 cm (32 in) composite pipeline, made of glass/polyester composite is shown in Fig. 1.19. An example of a composite bridge is the 114 m (371 ft) long cable-stayed footbridge built in Aberfeldy, Scotland, in 1992 (Fig. 1.20). The deck structure rails and A-frame towers are made of glass/polyester, and the cables are Kevlar ropes.[1]

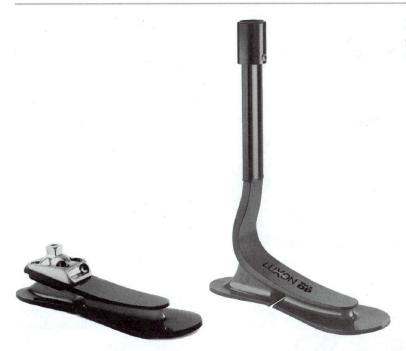

Fig. 1.16 Foot and leg prostheses incorporating carbon/epoxy components. (Courtesy of Otto Bock Health Care.)

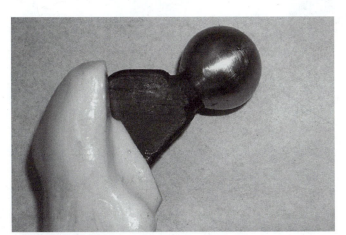

Fig. 1.17 Carbon/polysulfone hip prosthesis prototype (bottom, before implantation; top, implanted and ready for testing). (Courtesy of Professor Assa Rotem, Technion, Israel.)

Fig. 1.18 Bicycle frame made of carbon/epoxy composite and weighing 1.36 kg (3 lb), which is much less than the 5 kg (11 lb) weight of the corresponding steel frame.

Fig. 1.19 Composite pipe used for transport of drinking water: 80 cm (32 in) diameter glass/polyester pipe. (Courtesy of FIBERTEC Fiberglass Pipe Industry, Israel.)

Fig. 1.20 Footbridge in Aberfeldy, Scotland, using composite decking sections. (Courtesy of Maunsell Ltd., UK.)

1.4 OVERVIEW OF ADVANTAGES AND LIMITATIONS OF COMPOSITE MATERIALS

Composites have unique advantages over monolithic materials, such as high strength, high stiffness, long fatigue life, low density, and adaptability to the intended function of the structure. Additional improvements can be realized in corrosion resistance, wear resistance, appearance, temperature-dependent behavior, environmental stability, thermal insulation and conductivity, and acoustic insulation. The basis for the superior structural performance of composite materials lies in the high specific strength (strength to density ratio) and high specific stiffness (modulus to density ratio) and in the anisotropic and heterogeneous character of the material. The latter provides the composite with many degrees of freedom for optimum configuration of the material system.

Composites also have some limitations when compared with conventional monolithic materials. Below is a brief discussion of advantages and limitations of composites and conventional structural materials (mainly metals) when compared on the basis of various aspects, that is, micromechanics, macromechanics, material characterization, design and optimization, fabrication technology, maintenance and durability, and cost effectiveness.

1.4.1 Micromechanics

When viewed on the scale of fiber dimensions, composites have the advantage of high-stiffness and high-strength fibers. The usually low fracture toughness of the fiber is enhanced by the matrix ductility and the energy dissipation at the fiber/matrix interface. The stress transfer capability of the matrix enables the development of multiple-site and multiple-path failure mechanisms. On the other hand, the fibers exhibit a relatively high scatter in strength. Local stress concentrations around the fibers reduce the transverse tensile strength appreciably.

Conventional materials are more sensitive to their microstructure and local irregularities that influence the brittle or ductile behavior of the material. Their homogeneity makes them more susceptible to flaw growth under long-term cyclic loading.

1.4.2 Macromechanics

In macromechanical analysis, where the material is treated as quasi-homogeneous, its anisotropy can be used to advantage. The average material behavior can be controlled and predicted from the properties of the constituents. However, the anisotropic analysis is more complex and more dependent on the computational procedures. On the other hand, the analysis for conventional materials is much simpler due to their isotropy and homogeneity.

1.4.3 Mechanical Characterization

The analysis of composite structures requires the input of average material characteristics. These properties can be predicted on the basis of the properties and arrangement of the constituents. However, experimental verification of analysis or independent characterization requires a comprehensive test program for determination of a large number (more than ten) of basic material parameters. On the other hand, in the case of conventional isotropic materials, mechanical characterization is simple, as only two elastic constants and two strength parameters suffice.

1.4.4 Structural Design, Analysis, and Optimization

Composites afford the unique possibility of designing the material, the manufacturing procedure, and the structure in one unified and concurrent process. The large number of degrees of freedom available enables simultaneous material optimization for several given constraints, such as minimum weight, maximum dynamic stability, cost effectiveness, and so on. However, the entire process requires a reliable database of material properties, standardized structural analysis methods, modeling and simulation techniques, and models for materials processing. The numerous options available make the design and optimization process more involved and the analysis more complex. In the case of conventional materials, optimization is limited usually to one or two geometric parameters, due to the few degrees of freedom available.

1.4.5 Manufacturing Technology

The fabrication process is one of the most important steps in the application of composite materials. Structural parts, rather than generic material form, are fabricated with relatively simple tooling. A variety of fabrication methods suitable for various applications are available. They include autoclave molding, filament winding, pultrusion, fiber placement, and resin transfer molding (RTM). Structural components consisting of different materials, such as honeycomb sandwich structures, can be manufactured in one step by the so-called *cocuring process*. Thus, the number of parts to be assembled and joints required can be reduced significantly. On the negative side, composite fabrication is still dependent to some extent on skilled hand labor with limited automation and standardization. This requires more stringent, extensive, and costly quality control procedures.

In the case of conventional materials, material and structure fabrication are two separate processes. Structures usually necessitate complex tooling and elaborate assembly, with multiple elements and joints.

1.4.6 Maintainability, Serviceability, and Durability

Composites can operate in hostile environments for long periods of time. They have long fatigue lives and are easily maintained and repaired. However, composites and especially thermoset polymer composites suffer from sensitivity to hygrothermal environments. Service-induced damage growth may be internal, requiring sophisticated nondestructive techniques for its detection and monitoring. Sometimes it is necessary to apply protective coatings against erosion, surface damage, and lightning strike.

Conventional materials, usually metals, are susceptible to corrosion in hostile environments. Discrete flaws and cracks may be induced in service and may grow and propagate to catastrophic failure. Although detection of these defects may be easier, durable repair of conventional materials is not simple.

1.4.7 Cost Effectiveness

One of the important advantages of composites is reduction in acquisition and/or life cycle costs. This is effected through weight savings, lower tooling costs, reduced number of parts and joints, fewer assembly operations, and reduced maintenance. These advantages are somewhat diluted when one considers the high cost of raw materials, fibers, prepreg (resin preimpregnated fibers), and auxiliary materials used in fabrication and assembly of composite structures. Composite manufacturing processes are expensive, because they are not yet fully developed, automated, and optimized. They are labor intensive, may result in excessive waste, and require costly quality control and inspection. Affordability remains the biggest factor controlling further utilization of composites.[4]

In the case of conventional structural materials, the low cost of raw materials is more than offset by the high cost of tooling, machining, and assembly.

1.5 SIGNIFICANCE AND OBJECTIVES OF COMPOSITE MATERIALS SCIENCE AND TECHNOLOGY

The study of composites is a philosophy of material and process design that allows for the optimum material composition, along with structural design and optimization, in one concurrent and interactive process. It is a science and technology requiring close interaction of various disciplines such as structural design and analysis, materials science, mechanics of materials, and process engineering. The scope of composite materials research and technology consists of the following tasks:

1. investigation of basic characteristics of the constituent and composite materials
2. material optimization for given service conditions
3. development of effective and efficient fabrication procedures and evaluation of their effect on material properties
4. development of analytical procedures and numerical simulation models for determination of composite material properties and prediction of structural behavior
5. development of effective experimental methods for material characterization, stress analysis, and failure analysis
6. nondestructive evaluation of material integrity and structural reliability
7. assessment of durability, flaw criticality, and life prediction

1.6 CURRENT STATUS AND FUTURE PROSPECTS

The technology of composite materials has experienced a rapid development in the last four decades. Some of the underlying reasons and motivations for this development are

1. significant progress in materials science and technology in the area of fibers, polymers, and ceramics
2. requirements for high-performance materials in aircraft and aerospace structures
3. development of powerful and sophisticated numerical methods for structural analysis using modern computer technology
4. the availability of powerful desktop computers for the engineering community.

The initial driving force in the technology development, dominated by the aerospace industry, was performance through weight savings. Later, cost competitiveness with more conventional materials became equally important. In addition to these two requirements, today there is a need for quality assurance, reproducibility, and predictability of behavior over the lifetime of the structure.

New developments continue in all areas. For example, new types of carbon fibers have been introduced with higher strength and ultimate strain. Thermoplastic matrices are used under certain conditions because they are tough, have low sensitivity to moisture effects, and are more easily amenable to mass production and repair. Woven fabric and short-fiber reinforcements in conjunction with liquid molding processes are widely used. The design of structures and systems capable of operating in severe environments has spurred intensive research in high-temperature composites, including high-temperature polymer-matrix, metal-matrix, ceramic-matrix and carbon/carbon composites. Another area of interest is that of the so-called *smart composites* and structures incorporating active and passive sensors. A new area of growing interest is the utilization of nanocomposites and multiscale hybrid composites with multifunctional characteristics.

The utilization of conventional and new composite materials is intimately related to the development of fabrication methods. The manufacturing process is one of the most important stages in controlling the properties and assuring the quality of the finished product. A great deal of activity is devoted to intelligent processing of composites aimed at development of comprehensive and commercially viable approaches for fabrication of affordable, functional, and reliable composites. This includes the development and use of advanced hardware, software, and online sensing and controls.

The technology of composite materials, although still developing, has reached a stage of maturity. Prospects for the future are bright for a variety of reasons. The cost of the basic constituents is decreasing due to market expansion. The fabrication process is becoming less costly as more experience is accumulated, techniques are improved, and innovative methods are introduced. Newer high-volume applications, such as in the automotive industry and infrastructure, are expanding the use of composites greatly. The need for energy conservation motivates more uses of lightweight materials and products. The need for multifunctionality is presenting new challenges and opportunities for development of new material systems, such as nanocomposites with enhanced mechanical, electrical, and thermal properties. The availability of many good interactive computer programs and simulation methods makes structural design and analysis simpler and more manageable for engineers. Furthermore, the technology is vigorously enhanced by a younger generation of engineers and scientists well educated and trained in the field of composite materials.

REFERENCES

1. B. T. Åström, *Manufacturing of Polymer Composites*, Chapman and Hall, London, 1992.
2. G. E. Mabson, A. J. Fawcett, G. D. Oakes, "Composite Empennage Primary Structure Service Experience," *Proc. of Third Canadian Intern. Composites Conf.*, Montreal, Canada, August 2001.
3. "Solar-Powered Helios Completes Record-Breaking Flight," *High Performance Composites*, September/October 2001, p. 11.
4. C. W. Schneider, "Composite Materials at the Crossroads—Transition to Affordability," *Proc. of ICCM-11*, Gold Coast, Australia, 14–18 July 1997, pp. I-257–I-265.

2 Basic Concepts, Materials, Processes, and Characteristics

2.1 STRUCTURAL PERFORMANCE OF CONVENTIONAL MATERIALS

Conventional monolithic materials can be divided into three broad categories: metals, ceramics, and polymers. Although there is considerable variability in properties within each category, each group of materials has some characteristic properties that are more distinct for that group. In the case of ceramics one must make a distinction between two forms, bulk and fiber.

Table 2.1 presents a list of properties and a rating of the three groups of materials with regard to each property. The advantage or desirability ranking is marked as follows: superior (++), good (+), poor (−), and variable (v). For example, metals are superior with regard to stiffness and hygroscopic sensitivity (++), but they have high density and are subject to chemical corrosion (−). Ceramics in bulk form have low tensile strength and toughness (−) but good thermal stability, high hardness, low creep, and high erosion resistance (+). Ceramics in fibrous form behave very differently from those in bulk form and have some unique advantages. They rank highest with regard to tensile strength, stiffness, creep, and thermal stability (++). The biggest advantages that polymers have are their low density (++) and corrosion resistance (+), but they rank poorly with respect to stiffness, creep, hardness, thermal and dimensional stability, and erosion resistance (−). The observations above show that no single material possesses all the advantages for a given application (property) and that it would be highly desirable to combine materials in ways that utilize the best of each constituent in a synergistic way. A good combination, for example, would be ceramic fibers in a polymeric matrix.

2.2 GEOMETRIC AND PHYSICAL DEFINITIONS

2.2.1 Type of Material

Depending on the number of its constituents or phases, a material is called *single-phase* (or *monolithic*), *two-phase* (or *biphase*), *three-phase*, and *multiphase*. The different phases

TABLE 2.1 Structural Performance Ranking of Conventional Materials

Property	Metals	Ceramics		Polymers
		Bulk	Fibers	
Tensile strength	+	−	++	v
Stiffness	++	v	++	−
Fracture toughness	+	−	v	+
Impact strength	+	−	v	+
Fatigue endurance	+	v	+	+
Creep	v	v	++	−
Hardness	+	+	+	−
Density	−	+	+	++
Dimensional stability	+	v	+	−
Thermal stability	v	+	++	−
Hygroscopic sensitivity	++	v	+	v
Weatherability	v	v	v	+
Erosion resistance	+	+	+	−
Corrosion resistance	−	v	v	+

++ Superior
+ Good
− Poor
v Variable

of a structural composite have distinct physical and mechanical properties and characteristic dimensions much larger than molecular or grain dimensions.

2.2.2 Homogeneity

A material is called *homogeneous* if its properties are the same at every point or are independent of location. The concept of homogeneity is associated with a scale or characteristic volume and the definition of the properties involved. Depending on the scale or volume observed, the material can be more homogeneous or less homogeneous. If the variability from point to point on a macroscopic scale is low, the material is referred to as *quasi-homogeneous*.

2.2.3 Heterogeneity or Inhomogeneity

A material is heterogeneous or inhomogeneous if its properties vary from point to point, or depend on location. As in the case above, the concept of heterogeneity is associated with a scale or characteristic volume. As this scale decreases, the same material can be regarded as homogeneous, quasi-homogeneous, or heterogeneous.

2.2.4 Isotropy

Many material properties, such as stiffness, strength, thermal expansion, thermal conductivity, and permeability are associated with a direction or axis (vectorial or tensorial

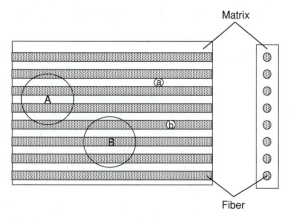

Fig. 2.1 Macroscopic (A, B) and microscopic (a, b) scales of observation in a unidirectional composite layer.

quantities). A material is *isotropic* when its properties are the same in all directions or are independent of the orientation of reference axes.

2.2.5 Anisotropy/Orthotropy

A material is *anisotropic* when its properties at a point vary with direction or depend on the orientation of reference axes. If the properties of the material along any direction are the same as those along a symmetric direction with respect to a plane, then that plane is defined as a *plane of material symmetry*. A material may have zero, one, two, three, or an infinite number of planes of material symmetry through a point. A material without any planes of symmetry is called *general anisotropic* (or *aeolotropic*). At the other extreme, an isotropic material has an infinite number of planes of symmetry.

Of special relevance to composite materials are *orthotropic* materials, that is, materials having at least three mutually perpendicular planes of symmetry. The intersections of these planes define three mutually perpendicular axes, called *principal axes of material symmetry* or simply *principal material axes*.

As in the case of homogeneity/heterogeneity discussed before, the concept of isotropy/anisotropy is also associated with a scale or characteristic volume. For example, the composite material in Fig. 2.1 is considered homogeneous and anisotropic on a macroscopic scale, because it has a similar composition at different locations (A and B) but properties varying with orientation. On a microscopic scale, the material is heterogeneous having different properties within characteristic volumes *a* and *b*. Within these characteristic volumes the material can be isotropic or anisotropic.

2.3 MATERIAL RESPONSE UNDER LOAD

Some of the intrinsic characteristics of the materials discussed before are revealed in their response to simple mechanical loading, for example, uniaxial normal stress and pure shear stress as illustrated in Fig. 2.2.

An isotropic material under uniaxial tensile loading undergoes an axial deformation (strain), ε_x, in the loading direction, a transverse deformation (strain), ε_y, and no shear deformation:

$$\varepsilon_x = \frac{\sigma_x}{E}$$

$$\varepsilon_y = -\frac{\nu\sigma_x}{E} \tag{2.1}$$

$$\gamma_{xy} = 0$$

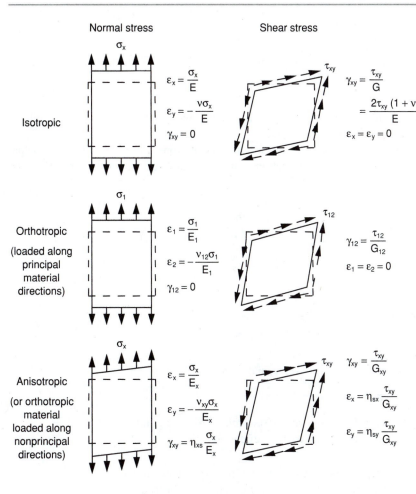

Fig. 2.2 Response of various types of materials under uniaxial normal and pure shear loadings.

where

$\varepsilon_x, \varepsilon_y, \gamma_{xy}$ = axial, transverse, and shear strains, respectively

σ_x = axial stress

E = Young's modulus

ν = Poisson's ratio

Under pure shear loading, τ_{xy}, the material undergoes a pure shear deformation, that is, a square element deforms into a diamond-shaped one with equal and unchanged side lengths. The shear strain (change of angle), γ_{xy}, and the normal strains, ε_x and ε_y, are

$$\gamma_{xy} = \frac{\tau_{xy}}{G} = \frac{2\tau_{xy}(1 + \nu)}{E}$$

$$\varepsilon_x = \varepsilon_y = 0$$

(2.2)

where

$$\tau_{xy} = \text{shear stress}$$

$$G = \text{shear modulus}$$

As indicated in Eq. (2.2) the shear modulus is not an independent constant, but is related to Young's modulus and Poisson's ratio.

An orthotropic material loaded in uniaxial tension along one of its principal material axes (1) undergoes deformations similar to those of an isotropic material and given by

$$\varepsilon_1 = \frac{\sigma_1}{E_1}$$

$$\varepsilon_2 = -\frac{\nu_{12}\sigma_1}{E_1} \tag{2.3}$$

$$\gamma_{12} = 0$$

where

$$\varepsilon_1, \varepsilon_2, \gamma_{12} = \text{axial, transverse, and shear strains, respectively}$$

$$\sigma_1 = \text{axial normal stress}$$

$$\nu_{12} = \text{Poisson's ratio associated with loading in the 1-direction}$$
$$\text{and strain in the 2-direction}$$

Under pure shear loading, τ_{12}, along the principal material axes, the material undergoes pure shear deformation, that is, a square element deforms into a diamond-shaped one with unchanged side lengths. The strains are

$$\gamma_{12} = \frac{\tau_{12}}{G_{12}}$$

$$\varepsilon_1 = \varepsilon_2 = 0 \tag{2.4}$$

Here, the shear modulus G_{12} is an independent material constant and is not directly related to the Young's moduli or Poisson's ratios.

In both cases discussed before, normal loading does not produce shear strain and pure shear loading does not produce normal strains. Thus, normal loading and shear deformation (as well as pure shear loading and normal strains) are independent or uncoupled.

A general anisotropic material under uniaxial tension, or an orthotropic material under uniaxial tension along a direction other than a principal material axis, undergoes axial, transverse, and shear deformations given by

$$\varepsilon_x = \frac{\sigma_x}{E_x}$$

$$\varepsilon_y = -\frac{\nu_{xy}\sigma_x}{E_x} \tag{2.5}$$

$$\gamma_{xy} = \eta_{xs}\frac{\sigma_x}{E_x}$$

where

$\varepsilon_x, \varepsilon_y, \gamma_{xy}$ = axial, transverse, and shear strains, respectively

σ_x = axial normal stress

E_x = Young's modulus in the x-direction

ν_{xy} = Poisson's ratio associated with loading in the x-direction and strain in the y-direction

η_{xs} = shear coupling coefficient (The first subscript denotes normal loading in the x-direction; the second subscript denotes shear strain.)

This mode of response characterized by η_{xs}, is called the *shear coupling effect* and will be discussed in detail in Chapter 4.

Under pure shear loading, τ_{xy}, along the same axes, the material undergoes both shear and normal deformations, that is, a square element deforms into a parallelogram with unequal sides. The shear and normal strains are given by

$$\gamma_{xy} = \frac{\tau_{xy}}{G_{xy}}$$

$$\varepsilon_x = \eta_{sx}\frac{\tau_{xy}}{G_{xy}} \tag{2.6}$$

$$\varepsilon_y = \eta_{sy}\frac{\tau_{xy}}{G_{xy}}$$

where

G_{xy} = shear modulus referred to the x- and y-axes

η_{sx}, η_{sy} = shear coupling coefficients (to be discussed in Chapter 4)

The above discussion illustrates the increasing complexity of material response with increasing anisotropy and the need to introduce additional material constants to describe this response.

2.4 TYPES AND CLASSIFICATION OF COMPOSITE MATERIALS

Two-phase composite materials are classified into three broad categories depending on the type, geometry, and orientation of the reinforcement phase, as illustrated in the chart of Fig. 2.3.

Particulate composites consist of particles of various sizes and shapes randomly dispersed within the matrix. Because of the usual randomness of particle distribution, these composites can be regarded as quasi-homogeneous and quasi-isotropic on a scale much larger than the particle size and spacing (macro scale). In some cases, such as polymer/clay nanocomposites, the silicate platelets can attain some degree of parallelism, thus making the material anisotropic. Particulate composites may consist of nonmetallic particles in a nonmetallic matrix (concrete, glass reinforced with mica flakes, brittle polymers reinforced with rubberlike particles, polymer/clay nanocomposites, ceramics reinforced with ceramic particles); metallic particles in nonmetallic matrices (aluminum particles in polyurethane rubber used in rocket propellants); metallic particles in metallic matrices (lead particles in copper alloys to improve machinability); nonmetallic particles in metallic matrices (silicon carbide particles in aluminum, SiC(p)/Al).

Discontinuous or *short-fiber composites* contain short fibers, nanotubes, or whiskers as the reinforcing phase. These short fibers, which can be fairly long compared to their diameter (i.e., have a high aspect ratio), can be either all oriented along one direction or randomly oriented. In the first instance the composite material tends to be markedly anisotropic, or more specifically orthotropic, whereas in the second, it can be regarded as quasi-isotropic. Nanocomposites reinforced with carbon nanotubes (approximately 1 nm in diameter and 1000 nm in length) are an example of this group of composites.

Continuous-fiber composites are reinforced by long continuous fibers and are the most efficient

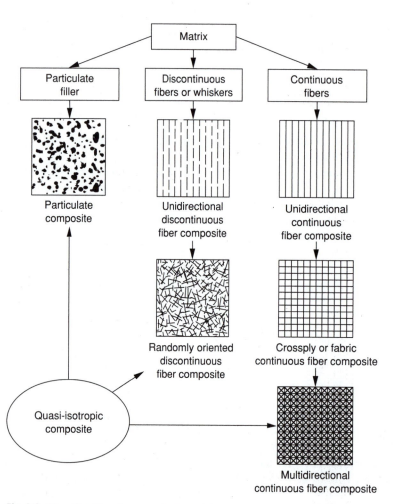

Fig. 2.3 Classification of composite material systems.

TABLE 2.2 Types of Fiber Composite Materials

Matrix Type	Fiber	Matrix
Polymer	E-glass	Epoxy
	S-glass	Phenolic
	Carbon (graphite)	Polyimide
	Aramid (Kevlar)	Bismaleimide
	Boron	Polyester
		Thermoplastics (PEEK, polysulfone, etc.)
Metal	Boron	Aluminum
	Borsic	Magnesium
	Carbon (graphite)	Titanium
	Silicon carbide	Copper
	Alumina	
Ceramic	Silicon carbide	Silicon carbide
	Alumina	Alumina
	Silicon nitride	Glass-ceramic
		Silicon nitride
Carbon	Carbon	Carbon

from the point of view of stiffness and strength. The continuous fibers can be all parallel (unidirectional continuous-fiber composite), can be oriented at right angles to each other (crossply or woven fabric continuous-fiber composite), or can be oriented along several directions (multidirectional continuous-fiber composite). In the latter case, for a certain number of fiber directions and distribution of fibers, the composite can be characterized as a quasi-isotropic material.

Fiber-reinforced composites can be classified into broad categories according to the matrix used: polymer-, metal-, ceramic-, and carbon-matrix composites (Table 2.2). *Polymer-matrix composites* include thermoset (epoxy, polyimide, polyester) or thermoplastic (poly-ether-ether-ketone, polysulfone) resins reinforced with glass, carbon (graphite), aramid (Kevlar), or boron fibers. They are used primarily in relatively low-temperature applications. *Metal-matrix composites* consist of metals or alloys (aluminum, magnesium, titanium, copper) reinforced with boron, carbon (graphite), or ceramic fibers. Their maximum use temperature is limited by the softening or melting temperature of the metal matrix. *Ceramic-matrix composites* consist of ceramic matrices (silicon carbide, aluminum oxide, glass-ceramic, silicon nitride) reinforced with ceramic fibers. They are best suited for very high-temperature applications. *Carbon/carbon composites* consist of carbon or graphite matrix reinforced with graphite yarn or fabric. They have unique properties of relatively high stiffness and moderate or low strength at high temperatures coupled with low thermal expansion and low density.

In addition to the types discussed above, there are *laminated composites* consisting of thin layers of different materials bonded together, such as bimetals, clad metals, plywood, Formica, and so on.

2.5 LAMINA AND LAMINATE—CHARACTERISTICS AND CONFIGURATIONS

A *lamina*, or ply, is a plane (or curved) layer of unidirectional fibers or woven fabric in a matrix. In the case of unidirectional fibers, it is also referred to as *unidirectional lamina* (UD). The lamina is an orthotropic material with principal material axes in the direction of the fibers (longitudinal), normal to the fibers in the plane of the lamina (in-plane transverse), and normal to the plane of the lamina (Fig. 2.4). These principal axes are designated as 1, 2, and 3, respectively. In the case of a woven fabric composite, the warp and the fill directions are the in-plane 1 and 2 principal directions, respectively (Fig. 2.4).

A *laminate* is made up of two or more unidirectional laminae or plies stacked together at various orientations (Fig. 2.5). The laminae (or plies, or layers) can be of various thicknesses and consist of different materials. Since the orientation of the principal material axes varies from ply to ply, it is more convenient to analyze laminates using a common fixed system or coordinates (*x*, *y*, *z*) as shown. The orientation of a given ply is given by the angle between the reference *x*-axis and the major principal material axis (fiber orientation or warp direction) of the ply, measured in a counterclockwise direction on the *x*-*y* plane.

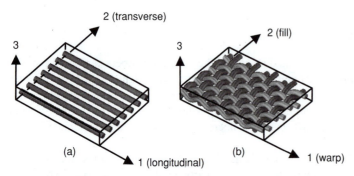

Fig. 2.4 Lamina and principal coordinate axes: (a) unidirectional reinforcement and (b) woven fabric reinforcement.

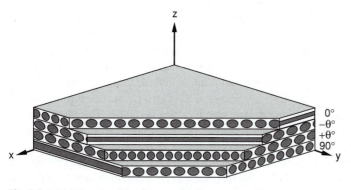

Fig. 2.5 Multidirectional laminate and reference coordinate system.

Composite laminates containing plies of two or more different types of materials are called *hybrid composites*, and more specifically *interply hybrid composites*. For example, a composite laminate may be made up of unidirectional glass/epoxy, carbon/epoxy and aramid/epoxy layers stacked together in a specified sequence. In some cases it may be advantageous to intermingle different types of fibers, such as glass and carbon or aramid and carbon, within the same unidirectional ply. Such composites are called *intraply hybrid composites*. Of course one may combine intraply hybrid layers with other layers to form an *intraply/interply hybrid composite*.

Composite laminates are designated in a manner indicating the number, type, orientation, and stacking sequence of the plies. The configuration of the laminate indicating its ply composition is called the *layup*. The configuration indicating, in addition to the ply composition, the exact location or sequence of the various plies, is called the *stacking sequence*. Following are some examples of laminate designations:

Unidirectional six-ply: $[0/0/0/0/0/0] = [0_6]$

Crossply symmetric: $[0/90/90/0] = [0/90]_s$

$[0/90/0/90/90/0/90/0] = [0/90]_{2s}$

$[0/90/0] = [0/\overline{90}]_s$

Angle-ply symmetric: $[+45/-45/-45/+45] = [\pm45]_s$

$[30/-30/30/-30/-30/30-30/30] = [\pm30]_{2s}$

Angle-ply asymmetric: $[30/-30/30/-30/30/-30/30/-30] = [\pm30]_4$

Multidirectional: $[0/45/-45/-45/45/0] = [0/\pm45]_s$

$[0/0/45/-45/0/0/0/0/-45/45/0/0] = [0_2/\pm45/0_2]_s$

$[0/15/-15/15/-15/0] = [0/\pm15/\pm15/0]_T = [0/(\pm15)_2/0]_T$

Hybrid: $[0^K/0^K/45^C/-45^C/90^G/-45^C/45^C/0^K/0^K]_T = [0_2^K/\pm45^C/\overline{90}^G]_s$

where subscripts and symbols signify the following:

number subscript = multiple of plies or group of plies

s = symmetric sequence

T = total number of plies

$^-$(overbar) = laminate is symmetric about the midplane of the ply

In the case of the hybrid laminate, superscripts K, C, and G denote Kevlar (aramid), carbon (graphite), and glass fibers, respectively.

2.6 SCALES OF ANALYSIS—MICROMECHANICS AND MACROMECHANICS

Composite materials can be viewed and analyzed at different levels and on different scales, depending on the particular characteristics and behavior under consideration. A schematic diagram of the various levels of consideration and the corresponding types of analysis is shown in Fig. 2.6.

At the constituent level the scale of observation is on the order of the fiber diameter, particle size, or matrix interstices between reinforcement. *Micromechanics* is the study of the interactions of the constituents on this microscopic level. It deals with the state of deformation and stress in the constituents and local failures, such as fiber failure (tensile, buckling, splitting), matrix failure (tensile, compressive, shear), and interface/interphase failure (debonding). The last two failure modes are referred to as *interfiber failure*. An example of the complex stress distributions on the transverse cross section of a transversely loaded unidirectional composite is illustrated in Fig. 2.7. Micromechanics is particularly important in the study of properties such as failure mechanisms and strength, fracture toughness, and fatigue life, which are strongly influenced by local characteristics

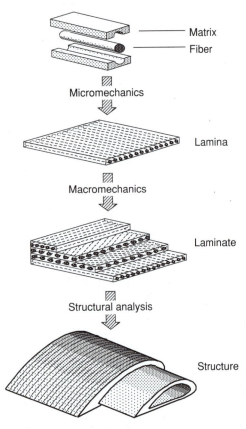

Fig. 2.6 Levels of observation and types of analysis for composite materials.

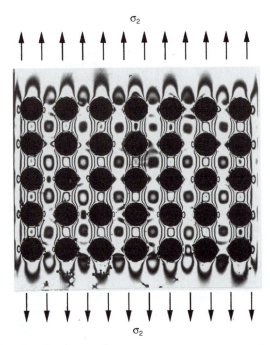

Fig. 2.7 Isochromatic fringe patterns in a model of transversely loaded unidirectional composite.

that cannot be integrated or averaged. Micromechanics also allows for the prediction of average behavior at the lamina level as a function of constituent properties and local conditions.

At the lamina level it is usually more expeditious to consider the material homogeneous, albeit anisotropic, and use average properties in the analysis. This type of analysis is called *macromechanics* and considers the unidirectional lamina as a quasi-homogeneous anisotropic material with its own average stiffness and strength properties. Failure criteria may be expressed in terms of average stresses and overall lamina strengths without reference to any particular local failure mechanisms. This approach, which assumes material continuity, is recommended in the study of the overall elastic, viscoelastic, or hygrothermal behavior of composite laminates and structures.

At the laminate level the macromechanical analysis is applied in the form of lamination theory dealing with overall behavior as a function of lamina properties and stacking sequence. Finally, at the component or structure level, methods such as finite element analysis coupled with lamination theory may predict the overall behavior of the structure as well as the state of stress in each lamina.

2.7 BASIC LAMINA PROPERTIES

The approach emphasized in this textbook is based on macromechanics. The unidirectional lamina or ply is considered the basic building block of any laminate or composite structure. In the case of textile composites, the basic ply contains the fabric layer. The basic material properties necessary for analysis and design are the average ply properties. With reference to Fig. 2.4 the unidirectional ply is characterized by the following properties:

E_1, E_2, E_3 = Young's moduli along the principal ply directions

G_{12}, G_{23}, G_{13} = shear moduli in 1-2, 2-3, and 1-3 planes, respectively (These are equal to G_{21}, G_{32}, and G_{31}, respectively.)

$\nu_{12}, \nu_{23}, \nu_{13}$ = Poisson's ratios (The first subscript denotes the loading direction, and the second subscript denotes the strain direction; these Poisson's ratios are different from ν_{21}, ν_{32}, and ν_{31}, that is, subscripts are not interchangeable.)

F_{1t}, F_{2t}, F_{3t} = tensile strengths along the principal ply directions

F_{1c}, F_{2c}, F_{3c} = compressive strengths along the principal ply directions

F_{12}, F_{23}, F_{13} = shear strengths in 1-2, 2-3, and 1-3 planes, respectively (These are equal to F_{21}, F_{32}, and F_{31}, respectively.)

$\alpha_1, \alpha_2, \alpha_3$ = coefficients of thermal expansion along principal ply directions

$\beta_1, \beta_2, \beta_3$ = coefficients of moisture expansion along principal ply directions

$\kappa_1, \kappa_2, \kappa_3$ = coefficients of thermal conductivity along principal ply directions

In addition to the above, the composite lamina is characterized by the following properties:

Fiber volume ratio: $V_f = \dfrac{\text{volume of fibers}}{\text{volume of composite}}$

Fiber weight ratio: $W_f = \dfrac{\text{weight of fibers}}{\text{weight of composite}}$

Matrix volume ratio: $V_m = \dfrac{\text{volume of matrix}}{\text{volume of composite}}$

Matrix weight ratio: $W_m = 1 - W_f = \dfrac{\text{weight of matrix}}{\text{weight of composite}}$

Void volume ratio: $V_v = 1 - V_f - V_m = \dfrac{\text{volume of voids}}{\text{volume of composite}}$

TABLE 2.3 Degrees of Anisotropy

	E_1/E_2	E_1/G_{12}	F_{1t}/F_{2t}	α_2/α_1
Silicon carbide/ceramic	1.09	2.75	17.8	1.04
Boron/aluminum	1.71	5.01	11.6	3.33
Silicon carbide/aluminum	1.73	5.02	17	1.96
S-glass/epoxy	4.1	10.0	35	4.3
E-glass/epoxy	4.0	9.5	29	3.7
Boron/epoxy	9.3	37.4	24.6	5.0
Carbon/epoxy	14.2	21.3	40	−30
Kevlar/epoxy	14.5	37	49	−30
Graphite (GY-70)/epoxy	46	60	34	−250

2.8 DEGREES OF ANISOTROPY

Some material properties, such as density, specific heat, absorptivity, and emittance, have no directionality associated with them and are described by one scalar quantity for both isotropic and anisotropic materials. On the other hand, properties such as stiffness, Poisson's ratio, strength, thermal expansion, moisture expansion, permeability, thermal conductivity, and electrical conductivity are associated with direction and are described by vector or tensor quantities. They are a function of orientation in anisotropic materials. Fiber composite materials can exhibit various degrees of anisotropy in the various properties. The largest differences occur between properties in the longitudinal (fiber) and transverse (normal to the fiber) directions in unidirectional composites. Ratios of some properties along these two directions for some typical composite materials are listed in Table 2.3.

2.9 CONSTITUENT MATERIALS

2.9.1 Reinforcement

The reinforcement phase of a composite may be in the form of continuous or short fibers, particles of various shapes, and whiskers. It contributes to or determines the composite stiffness and strength.

A large variety of fibers are available as reinforcement for composites. The desirable characteristics of most reinforcing fibers are high strength, high stiffness, and relatively low density. Each type of fiber has its own advantages and disadvantages as listed in Table 2.4. Table A.1 in Appendix A lists specific fibers with their properties of strength, modulus, and density. Additional properties of some fibers are given in Table A.2. More extensive discussions of fiber reinforcements for composite materials can be found elsewhere.[1–6]

Glass fibers are the most commonly used in low- to medium-performance composites because of their high tensile strength and low cost. They are limited in high-performance

TABLE 2.4 Advantages and Disadvantages of Reinforcing Fibers

Fiber	Advantages	Disadvantages
E-glass, S-glass	High strength Low cost	Low stiffness Short fatigue life High temperature sensitivity
Aramid (Kevlar)	High tensile strength Low density	Low compressive strength High moisture absorption
Boron	High stiffness High compressive strength	High cost
Carbon (AS4, T300, IM7)	High strength High stiffness	Moderately high cost
Graphite (GY-70, pitch)	Very high stiffness	Low strength High cost
Ceramic (Silicon carbide, alumina)	High stiffness High use temperature	Low strength High cost

composite applications because of their relatively low stiffness, low fatigue endurance, and rapid property degradation with exposure to severe hygrothermal conditions. Glass fibers are produced by extrusion of a molten mixture of silica (SiO_2) and other oxides through small holes of a platinum alloy bushing.[4] A coupling agent, or sizing, is applied to the fibers to protect their surface and ensure bonding to the resin matrix. An assembly of collimated glass fibers is called a *yarn* or *tow* and a group of collimated yarns is called a *roving*. Fiber diameters for composite applications are in the range of 10–20 μm (0.4×10^{-3}–0.8×10^{-3} in). Glass fibers are amorphous; therefore, they are considered isotropic.

Carbon fibers are the most widely used for advanced composites and come in many forms with a range of stiffnesses and strengths depending on the manufacturing process. Carbon fibers are manufactured from precursor organic fibers, such as rayon or polyacrylonitrile (PAN), or from petroleum pitch. In the former process (PAN), molten thermoplastic resin is spun into thin filaments in textile manufacturing equipment. The precursor fibers are initially drawn and oxidized under tension in air at temperatures between 200 °C (400 °F) and 315 °C (600 °F). Then, they are carbonized by pyrolysis at a temperature above 800 °C (1500 °F) in a nitrogen atmosphere. At this stage most fibers undergo surface treatment and sizing for use in composites manufacturing. This process yields high-strength and high-stiffness carbon fibers (AS4, T300, IM6, IM7). Graphite fibers, a subset of carbon fibers, are produced by further processing at temperatures above 2000 °C (3600 °F). This process, called *graphitization*, results in enhanced crystallinity and produces ultrahigh-stiffness graphite fibers with moduli over 410 GPa (60 Msi) and increased thermal conductivity in the axial direction. The increase in stiffness is achieved at the expense of strength as seen in Table A.1. Carbon fibers are anisotropic mechanically and thermally due to the nature of the manufacturing process. In the radial direction the stiffness is much lower and the coefficient of thermal expansion much higher than in the axial direction.

Aramid (or *Kevlar*) *fibers* are organic fibers manufactured by dissolving the polymer (aromatic polyamide) in sulfuric acid and extruding through small holes in a rotating device. The fiber diameter is typically 12 μm (0.5×10^{-3} in). Kevlar fibers have higher stiffness than glass fibers. Kevlar 49 and 149 are referred to as *high-* and *ultrahigh-modulus* grades. The former is the most commonly used for composites. Kevlar fibers have low density, about half that of glass, high tensile strength, and excellent toughness and impact resistance. However, Kevlar composites have very low longitudinal compressive and transverse tensile strengths and are sensitive to moisture absorption. Because of their high molecular orientation they are very anisotropic mechanically and thermally.

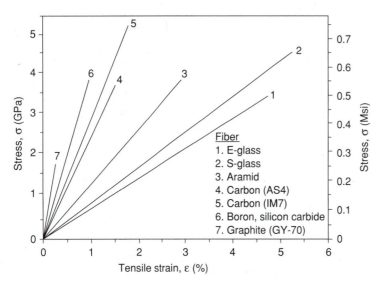

Fig. 2.8 Stress-strain curves of typical reinforcing fibers.

Boron and other ceramic fibers, such as silicon carbide (SiC) and alumina (Al_2O_3), are characterized by high stiffness, high use temperature, and reasonably high strength. They are not commonly used with polymeric matrices but rather with metal or ceramic matrices for high-temperature applications. Boron/polymer composites find limited applications in local stiffening and repair patching because of their high stiffness.

Most fibers behave linearly to failure as shown in Fig. 2.8. Carbon fibers, such as the AS4 fiber, however, display a nonlinear stiffening effect. One important property of the fiber related to strength and stiffness is the ultimate strain or strain to failure, because it influences greatly the strength of the composite laminate.

As mentioned previously, the basis of the superior performance of composites lies in the high specific strength (strength to density ratio) and high specific stiffness (modulus to density ratio). These two properties are controlled by the fibers. A two-dimensional comparative representation of some typical fibers from the point of view of specific strength and specific modulus is shown in Fig. 2.9. Here, these values are defined as the ratios of strength and modulus to specific weight of the material, respectively. Thus, they are expressed in units of length.

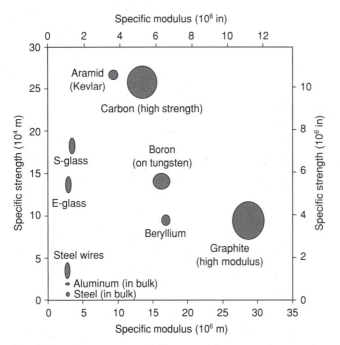

Fig. 2.9 Performance map of fibers used in structural composite materials.

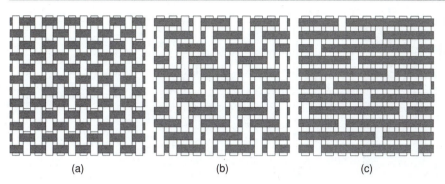

Fig. 2.10 Examples of fabric weave styles: (a) plane weave, (b) 2×2 twill, and (c) eight-harness satin weave.

(a)   (b)   (c)

The various fibers discussed before are not always used as straight yarns, but they are often used in the form of woven fabrics or textiles. An orthogonal woven fabric consists of two sets of interlaced yarns. The longitudinal direction of the fabric is called *warp* and the transverse direction *weft* or *fill*. The various types or styles of fabric are characterized by the repeat pattern of the interlaced regions as shown in Figs. 2.10 and 2.11. In the plain weave, for example, each yarn is interlaced over every other yarn in the other direction, that is, the smallest number of yarns involved in the repeat pattern in any direction is two ($n_g = 2$). In the twill fabric, each yarn is interlaced over every third yarn in the other direction ($n_g = 3$). In satin weaves each yarn is interlaced over every fourth, fifth, and so on, yarn in the other direction. These weaves are referred to as *four-harness* (4H), *five-harness* (5H), and so on, satin weaves. In addition to the fiber yarn type and weave style, the behavior of fabric reinforcement in a composite is characterized by the fabric crimp, which is a measure of the yarn waviness. The crimp fraction decreases and the drapeability of the fabric increases as we move from plain to twill and multiharness satin weaves. Although woven fabrics are usually two-dimensional and have warp and fill yarns normal to each other, it is possible to obtain fabrics with different yarn orientations and three-dimensional weaves. In addition to woven fabrics, other possible forms of reinforcement include knitted, braided, and nonwoven mats.

2.9.2 Matrices

As mentioned in Section 1.1, the main role of the matrix, especially in the case of high-performance composites, is to provide protection and support for the sensitive fibers and local stress transfer from one fiber to another. As shown in Table 2.2, the following four types of matrices are used in composites: polymeric, metallic, ceramic, and carbon. The most extensively used matrices are polymeric, which can be thermosets or thermoplastics. The other matrices are considered for high-temperature applications, with increasing use temperature from metallic to ceramic and carbon matrices.

Thermoset polymers are the most predominant types of matrix systems. Thermoset resins undergo polymerization and cross-linking during curing with the aid of a hardening agent and heating. They do not melt upon reheating, but they decompose thermally at high temperatures. The most commonly used thermosets are unsaturated polyesters, epoxies, polyimides, and vinylesters. Polyesters are used in large quantities with glass fiber reinforcement for quick-curing and room-temperature-curing systems in a variety of commercial products (automotive, boats, ships, structural components, storage tanks, and so

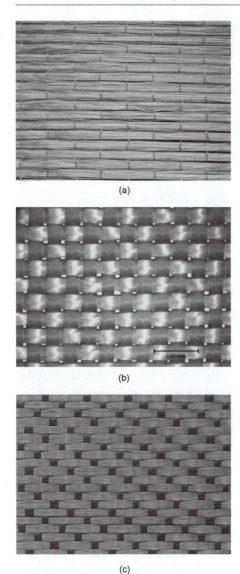

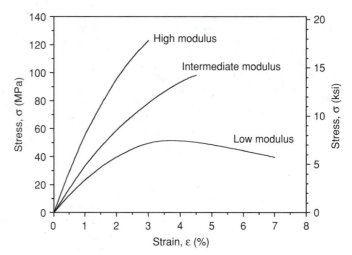

Fig. 2.12 Stress-strain curves of epoxy matrix resins of different moduli.

Fig. 2.11 Photographs of woven fabrics: (a) plain-weave UD (unidirectional) glass fabric, (b) plain-weave glass fabric, and (c) five-harness satin-weave carbon fabric.

on). Polyester-matrix composites have good mechanical properties and low cost, but they are sensitive to elevated temperatures.

The most highly developed of the thermoset polymers are epoxies of DGEBA type (diglycidyl ether of bisphenol A). They have better mechanical and thermal properties than polyesters. They can be formulated with a range of stiffnesses as shown in Fig. 2.12. Depending on the type of hardening agent, that is, amine or anhydride, epoxies can be cured at different temperatures, typically 120 °C (250 °F) or 175 °C (350 °F). The lower-temperature-curing epoxies are used in components exposed to low or moderate temperature variations (e.g., sporting goods). Those cured at a higher temperature are used in high-performance components exposed to high temperature and moisture variations (aircraft structures).

Vinylesters, being closely related to both polyesters and epoxies, combine some of the desirable properties of both, fast and simple curing with good mechanical and thermal properties. Vinylester-matrix composites are preferred in corrosive industrial and marine applications. Phenolics, originally used primarily with short-fiber reinforcement, are lately being used in higher-performance applications with continuous reinforcement because of improved processability. They are usually reinforced with glass fibers, and their composites are more heat and fire resistant.

Most thermosets are limited by their relatively low temperature resistance. Thermoset polyimides (PI) and bismaleimides (BMI) can be used in higher-temperature applications in excess of 300 °C (570 °F). However, these polymers have relatively lower strength and are more brittle at room temperature than epoxies. Table A.3 in Appendix A lists mechanical and physical properties of typical thermoset polymers.

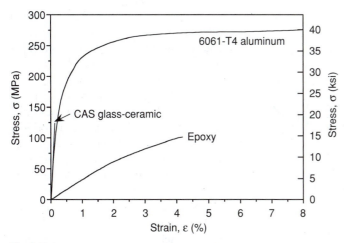

Fig. 2.13 Stress-strain curves of three typical matrices.

Thermoplastics are fully polymerized polymers and can be altered physically by softening or melting them with heat. Thermoplastics used as matrices for composites include polypropylene (PP), polyphenylene sulfide (PPS), polysulfone, poly-ether-ether-ketone (PEEK), and thermoplastic polyimides. They are more compatible with hot-forming and injection-molding fabrication methods. Compared to epoxies and thermoset polyimides, thermoplastics can be processed more quickly and have much higher glass transition and maximum use temperatures (up to 400 °C, 750 °F). They exhibit higher fracture toughness and are much less sensitive to moisture absorption. On the negative side, processing is not easily controlled, especially with crystalline or semicrystalline thermoplastics. They exhibit appreciable temperature-dependent behavior (viscoelasticity) and shorter fatigue life. Polypropylene is usually reinforced with glass fibers in mass-produced automotive and structural applications. Polyphenylene sulfide (PPS) is resistant to chemicals and fire and has reasonable mechanical properties. It is reinforced with glass or carbon fibers and used in some high-performance applications. Poly-ether-ether-ketone (PEEK) has high mechanical properties and high use temperature (see Table A.3). It is reinforced with glass or carbon fibers and used in some high-performance applications. Polysulfone has similar properties and used under similar conditions as PEEK. Thermoplastic polyimides have the highest temperature and environment resistance of the thermoplastics mentioned before. They have good mechanical properties but are costly. Polyimide composites are used in applications similar to those of PPS and PEEK.

Metal matrices are recommended for high-temperature applications up to approximately 800 °C (1500 °F). Commonly used metal matrices include aluminum, magnesium, and titanium alloys. Their use temperature is limited by the melting point.

Ceramic and carbon matrices are used for higher-temperature applications exceeding 1000 °C (1800 °F). They include glass, glass-ceramic, ceramic, and carbon matrices. Glass-ceramic matrices, such as lithium aluminosilicate (LAS) and calcium aluminosilicate (CAS), and ceramics such as reaction-bonded silicon nitride, are used with silicon carbide fibers. Carbon matrix is produced by vapor deposition of pyrolitic graphite onto a graphite fiber preform. The resulting composite can be used at temperatures up to 2600 °C (4700 °F). Stress-strain curves of three typical matrices, epoxy, aluminum, and glass-ceramic, are shown in Fig. 2.13.

2.10 MATERIAL FORMS—PREPREGS

Matrix and reinforcement can be combined in the so-called *prepreg* form and made ready for fabrication of composites. A prepreg tape consists of a layer of parallel or woven fibers

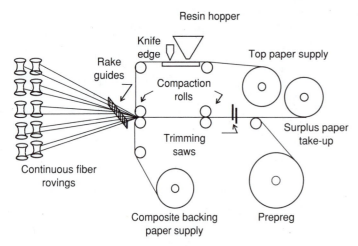

Fig. 2.14 Hot-melt system for prepreg manufacturing.

preimpregnated with resin partially cured (advanced) to a certain degree. Prepreg tapes are made to meet certain specifications, such as fiber volume ratio, ply thickness, and degree of partial cure (B-staging). The manufacturing of prepregs by the hot-melt impregnation system is illustrated in Fig. 2.14. It consists of fiber control, fiber collimation, resin impregnation, and tape production. Prepregs come in rolls of various widths with the resin-impregnated fibers supported on backing/release paper. They are characterized by the resin content (usually 32–42% by weight), tack (self-adhesion), drapeability or ability to conform to shapes, shelf life, out time, and gel time. The prepreg must be kept refrigerated at approximately −18 °C (0 °F) until final use.

2.11 MANUFACTURING METHODS FOR COMPOSITE MATERIALS

The manufacturing process is one of the most important steps in the application of composite materials. Ideally, the manufacturing method should be selected concurrently with material selection and structural design in a unified and interactive process. The manufacturing process is governed by the matrix used. This section deals only with polymer-matrix composites. The fabrication technology was originally developed in a semiempirical fashion, driven first by military applications and later by many and various civilian applications. The high cost of composite materials is the single most significant barrier to their more extensive utilization. Most of the expense is attributed to lack of cost-effective fabrication methods and the necessity for postprocess inspection to ensure quality of the material. The properties of the finished product are closely related to the manufacturing method.

The finished product must meet some general requirements. It must be free of defects (voids, cracks, fiber waviness), uniform in properties, fully cured (having expected properties, for example, stiffness, strength, fatigue endurance), and reproducible. Some of the specific goals of manufacturing are control of reinforcement location/orientation, ply thickness, fiber volume ratio, voids, residual stresses, and final dimensions. Regarding the process itself, the temperature must not exceed preset values, temperature distribution must be reasonably uniform throughout the part, and complete and uniform cure must be accomplished in the shortest possible time.

A large number of fabrication methods are in use today.[7] They include autoclave, vacuum bag and compression molding, filament winding, fiber placement, injection molding, pultrusion, and resin transfer molding (RTM). A brief description is given below of three of these methods.

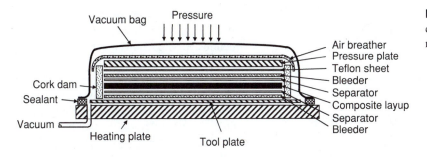

Fig. 2.15 Layup assembly for autoclave molding of composite laminates.

2.11.1 Autoclave Molding

The autoclave molding process is used for fabrication of high-performance advanced composites for military, aerospace, transportation, marine, and infrastructure applications. The method has few restrictions on size and shape and yields products with good dimensional tolerances. It is a low-volume process and labor intensive and therefore costly. Materials used are typically thermoset and thermoplastic resins reinforced with glass, carbon, and aramid fibers and fabrics. The autoclave process uses materials in prepreg form. Prepreg sheets are cut to size, oriented as desired, and stacked to form a layup. A bleeder/breather system consisting of dry glass fiber fabric or mat is used to absorb excess resin and allow the escape of volatiles during curing. The entire assembly of prepreg layup and auxiliary materials is sealed with a vacuum bag onto a tool plate (Fig. 2.15). Curing is effected by application of a prescribed temperature-pressure-vacuum-time cycle inside a chamber (autoclave). A typical curing cycle for a carbon/epoxy system is illustrated in Fig. 2.16. The layup is heated initially at a rate of 2–4 °C/min (3–7 °F/min) up to 110–125 °C (225–260 °F) under full vacuum in order to melt the resin and remove volatiles, but without causing excessive resin bleeding. Vacuum is maintained during a dwell period of approximately one hour at 110–125 °C (225–260 °F). At the end of this period, a pressure of 550–690 kPa (80–100 psi) is applied, followed by removal of the vacuum. Then, the temperature is raised at 2–4 °C/min (3–7 °F/min) up to 175 °C (350 °F) and maintained at that level for approximately two hours. The cooldown is gradual and may be controlled to minimize residual stresses and prevent microcracking.

2.11.2 Filament Winding

Filament winding consists of the winding under tension of preimpregnated or resin-coated reinforcement around a rotating mandrel (Fig. 2.17). By its nature, the process is best suited to products having surfaces of revolution, for example,

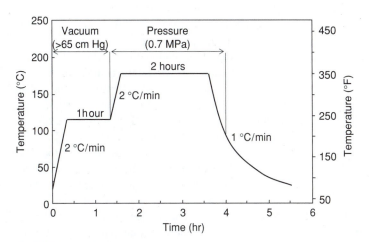

Fig. 2.16 Typical curing cycle for carbon/epoxy composites.

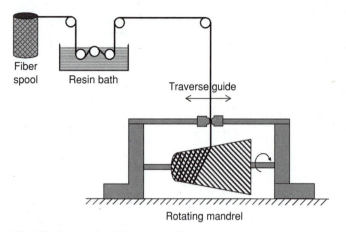

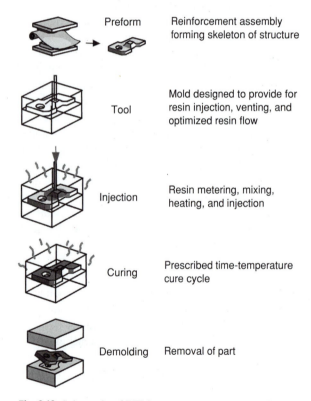

Fiber spool   Resin bath

Traverse guide

Rotating mandrel

Fig. 2.17 Schematic of filament winding process.

Preform — Reinforcement assembly forming skeleton of structure

Tool — Mold designed to provide for resin injection, venting, and optimized resin flow

Injection — Resin metering, mixing, heating, and injection

Curing — Prescribed time-temperature cure cycle

Demolding — Removal of part

Fig. 2.18 Schematic of RTM process.

pipes, containers, pressure vessels, rocket motor cases, and other tubular and conical structures. Compaction is achieved through fiber tension. The process yields products with high specific strength and well-controlled fiber orientation and uniformity. Fiber strands may be preimpregnated with partially cured resin (dry winding) or wetted by passing them through a resin bath (wet winding). In planar winding, the mandrel remains stationery while the fiber feeding arm rotates about it along the longitudinal axis. In helical winding, the mandrel rotates while the fiber feed carriage shuttles back and forth at a controlled speed to generate the desired helical angle. The method can also be applied to products not having surfaces of revolution. Helicopter rotor blades are made by a combination of filament winding (off-axis plies) and tape layup (longitudinal plies). Flat parts can be made by tape winding (dry winding). The wound mandrels may be cured in an oven or autoclave under a prescribed curing cycle.

2.11.3 Resin Transfer Molding

The resin transfer molding (RTM) process is suitable for high-volume production of complex or thick composite parts. It is a cost-effective process requiring low hardware and maintenance costs. The RTM process is used to produce military, aerospace, transportation, marine, and infrastructure products. Reinforcements used are glass, carbon and Kevlar fibers, fabrics, mats, and textiles; resins used include epoxies, unsaturated polyesters, and vinylesters.

The process consists of material selection (resin and reinforcement or preform), mold design, mold filling, curing, and controls. The dry fiber reinforcement assembly, or preform, forming the "skeleton" of the structure is prepared in the shape of the final part and placed into the mold (Fig. 2.18). The mold, or tool, is designed to provide for resin injection, venting, and optimized resin flow. During the injection stage, the resin, hardener, and accelerator (when applicable) are metered, mixed, heated, and injected at a specified pressure and/or flow rate through one or more gates into the closed mold. After completion of mold filling, the prescribed time-temperature curing cycle is applied until complete cure.

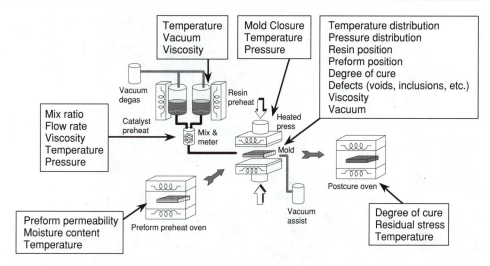

Fig. 2.19 RTM processing parameters.

The part is then removed from the mold and is evaluated nondestructively. The various relevant parameters at different stages of the RTM process are shown in Fig. 2.19.[8]

The process starts with product design, which takes into consideration design requirements, stress analysis, and materials selection. The selected materials are characterized by determining the preform permeability, resin viscosity as a function of temperature and degree of cure, and resin cure kinetics.

The mold is designed by selecting the location of injection gates, vents and vacuum ports, and heating and cooling lines. Process parameters, such as temperature, pressure, resin viscosity, and degree of cure can be monitored by in-mold sensors. Based on the material properties, that is, preform permeability, resin viscosity, and cure kinetics, a flow/cure model can be employed to predict the flow of the resin up to complete mold filling.[9,10] This type of modeling helps in determining the desired injection pressure and flow rate, injection resin temperature, and gate/vent opening sequence for optimum mold filling. Model predictions are checked with online monitoring sensors, which can also be used for control purposes. The cure cycle can be optimized, based on the kinetic model for the resin, for the purpose of achieving uniform and fast curing and minimizing residual stresses and dimensional distortions.[11] The cure progress can be monitored by temperature, degree of cure, and strain sensors, which can also be used for control purposes.

A state-of-the-art RTM system is shown in Fig. 2.20. It is capable of independent control of resin metering, mixing, heating, and injection. The system uses low-pressure syringe pumps and a holding tank for more effective mixing and degassing. Replaceable plastic tubing simplifies building, maintenance, and resin switchover. All functions of the system are computer controlled with a graphical user interface including supervisory control and data acquisition.

A variation of the RTM process is the vacuum-assisted resin transfer molding (VARTM). The preform is laid over an open mold surface, covered by a peel ply and/or a resin distribution fabric. The stack is covered with a vacuum bag sealed to the mold periphery. The resin is injected at one point while drawing vacuum at another point. The

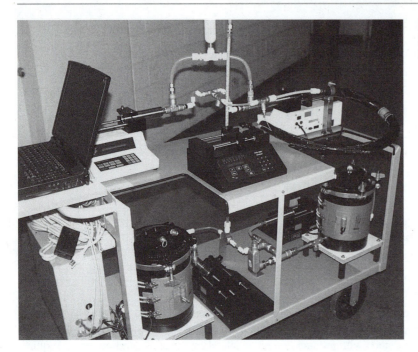

Fig. 2.20 RTM system. (Center for Intelligent Processing of Composites, Northwestern University.)

resin flow through the preform (and resin distribution blanket) is assisted by the vacuum. The VARTM process requires less expensive tooling and is suitable for fabrication of large components. However, only one side of the composite has a mold finish. The resins used must have very low viscosity, thus compromising on mechanical properties.

2.12 PROPERTIES OF TYPICAL COMPOSITE MATERIALS

Composite materials incorporating the various constituents discussed before display a wide range of characteristics. As mentioned before, the quality of performance of composite materials can be ranked on the basis of specific strength and specific modulus. A comparative representation of the performance of typical structural composites from the point of view of these properties is shown in Fig. 2.21. The range shown for the composites corresponds to the variation between quasi-isotropic and unidirectional laminates. As can be seen in the figure, most composites have higher specific modulus and specific strength than metals. Among the various composites, carbon/epoxy in its unidirectional form seems to provide the best combination of high specific modulus and strength.

The behavior of unidirectional composites in the fiber direction, especially the stiffness, is usually dominated by the fiber properties. Stress-strain curves of typical unidirectional composites in the fiber direction are shown in Fig. 2.22 and compared with that of aluminum. Some general trends can be observed. As the stiffness increases, the ultimate strain decreases. For a certain group of materials (identified as materials 4, 5, 6, and 7 in

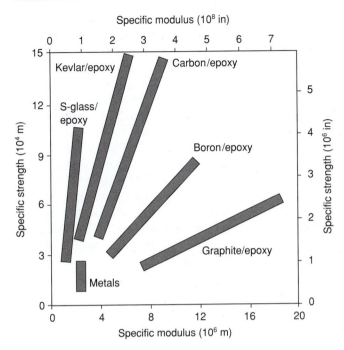

Fig. 2.21 Performance map of structural composites.

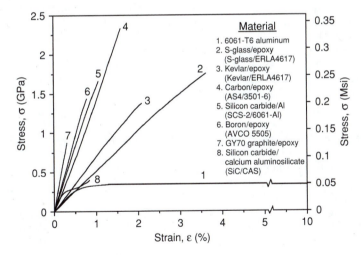

Fig. 2.22 Stress-strain curves of typical unidirectional composites in the fiber direction.

the figure) the increase in stiffness is accompanied by a drastic decrease in strength and ultimate strain. The behavior of unidirectional composites in the transverse to the fiber direction, especially the strength, is dominated by the matrix and interfacial properties. Stress-strain curves of typical unidirectional composites in the transverse to the fiber direction are shown in Fig. 2.23. All of these materials exhibit quasi-linear behavior with relatively low ultimate strains and strengths. In particular, the four polymer-matrix composites depicted in this figure show almost the same transverse strength.

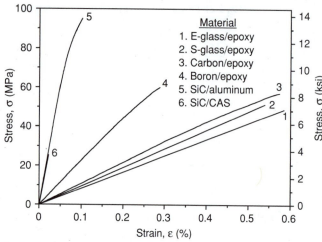

Fig. 2.23 Stress-strain curves of typical unidirectional composites in the transverse to the fiber direction.

A more comprehensive list of properties of typical composite materials is given in Tables A.4–A.7 in Appendix A. The composite properties listed are at ambient temperature (24 °C; 75 °F) and dry conditions. The values listed are typical for the material systems but can vary from batch to batch of the same material. They can be used for instructional and preliminary design purposes. For a final design of a component, it is recommended that the designer obtain by direct testing more exact properties for the particular batch of material used. Table A.8 shows for comparison corresponding properties of three structural metals, aluminum, steel, and titanium.

REFERENCES

1. E. Fitzer, *Carbon Fibres and Their Composites*, Springer-Verlag, Berlin, 1985.
2. A. R. Bunsell, Ed., *Fiber Reinforcements for Composite Materials*, Elsevier, Amsterdam, 1988.
3. K. K. Chawla, *Ceramic Matrix Composites*, Chapman and Hall, London, 1993.
4. J. C. Watson and N. Raghupati, "Glass Fibers," in *Engineered Materials Handbook, Volume 1, Composites*, ASM International, Metals Park, OH, 1987, pp. 107–111.
5. R. J. Diefendorf, "Carbon/Graphite Fibers," in *Engineered Materials Handbook, Volume 1, Composites*, ASM International, Metals Park, OH, 1987, pp. 49–53.
6. J. W. S. Hearle, P. Grossberg, S. Backer, *Structural Mechanics of Fibers, Yarns and Fabrics*, Wiley-Interscience, New York, 1969.
7. B. T. Åström, *Manufacturing of Polymer Composites*, Chapman and Hall, London, 1997.
8. J. M. Fildes and I. M. Daniel, "Intelligent RTM Processing of Composites," *Proc. of SAMPE '99*, Long Beach, CA, 1999.
9. I. M. Daniel, M. K. Um, B. W. Childs, and D. H. Kim, "Permeability and Resin Flow Measurements and Simulations in Composite Preforms," *Proc. of 45th International SAMPE Symposium*, May 21–25, 2000, pp. 761–775.
10. M. K. Um and I. M. Daniel, "A New Kinetic Model for Degree of Cure and Viscosity in Liquid Molding Applications," *Proc. of 45th International SAMPE Symposium*, May 21–25, 2000, pp. 1598–1610.
11. Y. K. Kim and I. M. Daniel, "Cure Cycle Effect on Composite Structures Manufactured by Resin Transfer Molding," *J. Composite Materials*, Vol. 36, 2002, pp. 1725–1743.

3 Elastic Behavior of Composite Lamina—Micromechanics

3.1 SCOPE AND APPROACHES

As mentioned before, composite materials can be viewed and analyzed on different scales. On the micromechanical scale cognizance is taken of the local states of deformation and stress of the constituents (reinforcement and matrix) and their interaction. Such local states of stress may be complex, as illustrated by the photoelastic fringe pattern in Fig. 2.7. The behavior of a composite lamina, which forms the basic building block of composite laminates and structures, is a function of the constituent properties and geometric characteristics, such as fiber volume ratio and geometric parameters.

Typical transverse cross sections of unidirectional composites are shown in Fig. 3.1. It is shown that composites with low fiber volume ratio tend to have a random fiber distribution, whereas fibers in composites with high fiber volume ratio tend to nest in nearly hexagonal packing. One objective of micromechanics is to obtain functional relationships for average elastic properties of the composite, such as stiffness, in the form

$$C^* = f(C_f, C_m, V_f, S, A) \tag{3.1}$$

where

$$C^* = \text{average composite stiffness}$$

$$C_f, C_m = \text{fiber and matrix stiffnesses, respectively}$$

$$V_f = \text{fiber volume ratio}$$

$$S, A = \text{geometric parameters describing the shape and array of the reinforcement, respectively}$$

A variety of methods have been used to predict properties of composite materials.[1,2] The approaches used fall into the following general categories:

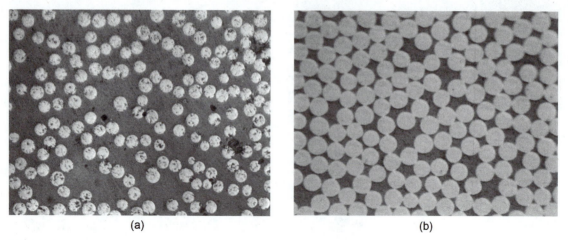

Fig. 3.1 Photomicrographs of typical transverse cross sections of unidirectional composites: (a) silicon carbide/ glass-ceramic (SiC/CAS), average fiber diameter 15 μm, fiber volume ratio $V_f = 0.40$ and (b) carbon/epoxy (AS4/3501- 6), fiber diameter 8 μm, fiber volume ratio $V_f = 0.70$.

1. mechanics of materials
2. numerical
3. self-consistent field
4. bounding (variational approach)
5. semiempirical
6. experimental

The mechanics of materials approach is based on simplifying assumptions of either uniform strain or uniform stress in the constituents.[3] The mechanics of materials predictions are adequate for longitudinal properties such as Young's modulus E_1 and major Poisson's ratio v_{12} of a unidirectional continuous-fiber composite. These properties are not sensitive to fiber shape and distribution. On the other hand, the mechanics of materials approach underestimates the transverse and shear properties, that is, transverse modulus E_2 and shear modulus G_{12} of such unidirectional materials.

Numerical approaches using finite difference, finite element, periodic cell, or boundary element methods yield the best predictions; however, they are time consuming and do not yield closed-form expressions. Results are usually presented in the form of families of curves.[4,5] In the periodic cell model described by Aboudi, a two-phase characteristic volume element is analyzed, consisting of a fiber cell and surrounding matrix cells.[6] The model satisfies equilibrium and continuity conditions between cells and neighboring volume elements. The model is capable of predicting the overall behavior of a composite material with elastic or nonelastic constituents.

In the self-consistent field approach, a simplified composite model is considered, consisting of a typical fiber surrounded by a cylindrical matrix phase (Fig. 3.2). This composite element is considered embedded in a larger (infinite), homogeneous medium whose properties are identical to the average properties of the composite material.

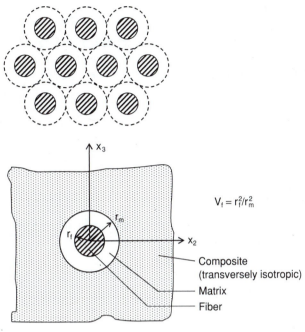

$$V_f = r_f^2/r_m^2$$

Composite (transversely isotropic)

Matrix

Fiber

Fig. 3.2 Self-consistent field model.

Classical elasticity theory has been used to obtain closed-form solutions for the various elastic constants of the composite.[7,8] This approach, because of the gross geometric simplifications involved, neglects interaction effects between fibers and as a result tends to underestimate composite properties for higher fiber volume ratios.

Variational methods based on energy principles have been developed to establish bounds on effective properties.[9,10] The bounds are close to each other in the case of longitudinal properties (E_1, ν_{12}), but they can be far apart in the case of transverse and shear properties (E_2, G_{12}).

Semiempirical relationships have been developed to circumvent the difficulties with the theoretical approaches above and to facilitate computation.[11] The so-called *Halpin-Tsai relationships* have a consistent form for all properties and represent an attempt at judicious interpolation between the series and parallel models used in the mechanics of materials approach or between the upper and lower bounds of the variational approach. This is expressed in terms of a parameter ξ, which is a measure of the reinforcing efficiency (or load transfer) and can be determined with the aid of experiment.

The micromechanics of load transfer and the correlation between constituent properties and average composite properties must be ultimately verified experimentally. Photoelastic models have proven useful in illustrating stress transfer and determining local stress distributions and stress concentrations for composites of various constituents and various geometric parameters.[12]

3.2 MICROMECHANICS METHODS

The objective of these methods is to characterize the elastic response of a representative volume element of the lamina as a function of the material and geometric properties of the constituents. Specifically, it is required to determine average properties that characterize the response of the representative volume element under simple loadings, such as longitudinal, transverse, in-plane shear, and transverse shear. The relevant engineering elastic properties are

Young's moduli:	E_1, E_2, E_3
Shear moduli:	G_{12}, G_{23}, G_{13}
Poisson's ratios:	$\nu_{12}, \nu_{23}, \nu_{13}$

3.2.1 Mechanics of Materials Methods

In the mechanics of materials approach, the composite can be described either by a parallel (Voigt) model or a series (Reuss) model. In the former it is assumed that the reinforcement and matrix are under equal uniform strain. This leads to the following expression for the stiffness:

$$C^* = V_f C_f + V_m C_m \tag{3.2}$$

where

$$C^*, C_f, C_m = \text{composite, fiber, and matrix stiffnesses, respectively}$$

In the series model it is assumed that the phases are under equal and uniform stress. The resulting expression for the compliance is of the form

$$S^* = V_f S_f + V_m S_m \tag{3.3}$$

where

$$S^*, S_f, S_m = \text{composite, fiber, and matrix compliances, respectively}$$

Assuming the compliance to be the inverse of the stiffness, $S^* = 1/C^*$, Eq. (3.3) gives the following relation for the stiffness:

$$C^* = \frac{1}{V_f/C_f + V_m/C_m} \tag{3.4}$$

In reality the state of stress or strain in the composite is not uniform and the relations above represent limits bounding the true value of the stiffness,

$$\frac{1}{V_f/C_f + V_m/C_m} \le C^* \le V_f C_f + V_m C_m \tag{3.5}$$

3.2.2 Bounding Methods

The bounding methods aimed at improving (narrowing) the above primitive bounds are based on energy principles. The actual stored energy in a representative volume of the material is

$$U = \frac{1}{2} \int_V \sigma_i \varepsilon_i dV \tag{3.6}$$

or, in terms of the strain field,

$$U(\varepsilon) = \frac{1}{2}\int_V (C_{ij}\varepsilon_j)\varepsilon_i dV \tag{3.7}$$

and in terms of the stress field,

$$U(\sigma) = \frac{1}{2}\int_V \sigma_i(S_{ij}\sigma_j)dV \tag{3.8}$$

(Here, a contracted notation, to be discussed later (Eqs. (4.5)–(4.7) in Section 4.1.1), has been used for the stress, strain, stiffness, and compliance tensors.)

Bounds for the stiffness or the compliance are obtained by using the following energy theorems.

Theorem of Least Work (Complementary Energy)

For any state of stress that satisfies prescribed traction boundary conditions and equilibrium throughout the body, but not necessarily compatibility, the stored elastic energy is greater than or equal to the actual energy

$$U \leq U(\sigma) \tag{3.9}$$

or

$$\int_V \sigma_i(S_{ij}^*\sigma_j)dV \leq \int_V \sigma_i(S_{ij}^\sigma\sigma_j)dV \tag{3.10}$$

where S_{ij}^* is the actual compliance and S_{ij}^σ is the compliance obtained by assuming a state of stress. Then,

$$S_{ij}^* \leq S_{ij}^\sigma \tag{3.11}$$

or

$$C_{ij}^* \geq C^\sigma \tag{3.12}$$

This result indicates that the stiffness obtained by assuming a state of stress (e.g., series model) is a lower bound.

Theorem of Minimum Potential Energy

For any state of strain that satisfies prescribed displacement boundary conditions and strain compatibility, but not necessarily equilibrium, the stored elastic energy is greater than or equal to the actual energy:

$$U \leq U(\varepsilon) \tag{3.13}$$

or

$$\int_V (C_{ij}^*\varepsilon_j)\varepsilon_i dV \leq \int_V (C_{ij}^\varepsilon\varepsilon_j)\varepsilon_i dV \tag{3.14}$$

where C_{ij}^* is the actual stiffness and C_{ij}^ε is the stiffness obtained by assuming a state of strain. Then,

$$C_{ij}^* \leq C_{ij}^\varepsilon \tag{3.15}$$

This result indicates that the stiffness obtained by assuming a state of strain (e.g., parallel model) is an upper bound. Combining Eqs. (3.12) and (3.15) we obtain

$$C_{ij}^\sigma \leq C_{ij}^* \leq C_{ij}^\varepsilon \tag{3.16}$$

Specific values for the bounds are obtained by assuming trial stress and strain fields. Simple forms of stress or strain fields lead to primitive bounds (too far apart). More realistic fields lead to tighter bounds.

3.2.3 Semiempirical Methods

Semiempirical methods, exemplified by the Halpin-Tsai relations, represent an attempt to make a judicious interpolation between the upper and lower bounds. The general composite property $P*$ is obtained as

$$P* = \frac{P_m(1 + \xi \eta V_f)}{1 - \eta V_f} \tag{3.17}$$

where

$$\eta = \frac{P_f - P_m}{P_f + \xi P_m}$$

ξ = estimated parameter or "reinforcing efficiency"

P_f, P_m = fiber and matrix properties, respectively

The relations above can be written as

$$P* = \frac{P_m[P_f + \xi P_m + \xi V_f(P_f - P_m)]}{P_f + \xi P_m - V_f(P_f - P_m)} \tag{3.18}$$

For $\xi \to \infty$ we obtain the parallel (Voigt) model

$$P* = V_f P_f + V_m P_m \tag{3.19}$$

For $\xi = 0$ we obtain the series (Reuss) model

$$P* = \left(\frac{V_f}{P_f} + \frac{V_m}{P_m} \right)^{-1} \tag{3.20}$$

The appropriate value of parameter ξ can be determined by experiment. For an experimental value of a composite property P^* for a given fiber volume ratio V_f, the parameter ξ is given by

$$\xi = \frac{P_f(P^* - P_m) - V_f P^*(P_f - P_m)}{P_m[(P_f - P^*) - V_m(P_f - P_m)]} \tag{3.21}$$

3.3 GEOMETRIC ASPECTS AND ELASTIC SYMMETRY

Most micromechanical analyses deal with the simplest type of composite, one consisting of continuous parallel fibers in a matrix. The properties of the unidirectional lamina, as discussed previously, depend not only on the fiber volume ratio, but also on the packing geometry of the fibers. The three idealized packing geometries, rectangular, square, and hexagonal, are illustrated in Fig. 3.3. The fiber volume ratios for the three fiber packing geometries are related to the fiber radius and fiber spacing as follows:

$$V_f = \frac{\pi}{4}\left(\frac{r^2}{R_2 R_3}\right) \qquad \text{(rectangular packing)}$$

$$V_f = \frac{\pi}{4}\left(\frac{r}{R}\right)^2 \qquad \text{(square packing)} \tag{3.22}$$

$$V_f = \frac{\pi}{2\sqrt{3}}\left(\frac{r}{R}\right)^2 \qquad \text{(hexagonal packing)}$$

The maximum values of the fiber volume ratios for the three cases above are 0.785, 0.785, and 0.907, respectively. Laminae with rectangular, square, and hexagonal fiber packing are characterized by nine, six, and five independent elastic constants, respectively (Fig. 3.3).

3.4 LONGITUDINAL ELASTIC PROPERTIES—CONTINUOUS FIBERS

Longitudinal properties associated with loading in the fiber direction are dominated by the fibers that are usually stronger, stiffer, and have a lower ultimate strain (Fig. 3.4). Assuming a perfect bond between matrix and fibers, longitudinal strains are uniform throughout and equal for the matrix and fibers. This leads to the so-called *rule of mixtures*, or parallel model, for the longitudinal modulus:

$$E_1 = V_f E_{1f} + V_m E_m \tag{3.23}$$

where E_{1f} and E_m are longitudinal fiber and matrix moduli, respectively, and V_f and V_m are the fiber and matrix volume ratios, respectively. In the relation above it is assumed that the fiber can be anisotropic with different properties in the axial and transverse (radial) directions and that the matrix is isotropic.

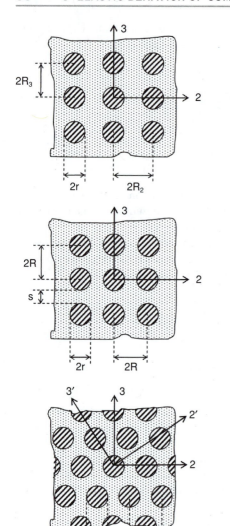

Rectangular

Nine independent constants

E_1, E_2, E_3

G_{12}, G_{23}, G_{13}

$\nu_{12}, \nu_{23}, \nu_{13}$

Square

Six independent constants

$E_1, E_2 = E_3$

$G_{12} = G_{13}, G_{23}$

$\nu_{12} = \nu_{13}, \nu_{23}$

Hexagonal

Five independent constants

$E_1, E_2 = E_3 = E_{2'} = E_{3'}$

$G_{12} = G_{13} = G_{12'} = G_{13'}$

$$G_{23} = \frac{E_{23}}{2(1+\nu_{23})}$$

$\nu_{12} = \nu_{13} = \nu_{12'} = \nu_{13'}, \nu_{23}$

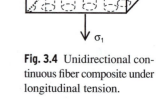

Fig. 3.4 Unidirectional continuous fiber composite under longitudinal tension.

Fig. 3.3 Idealized fiber packing geometries and elastic symmetry.

The same assumptions lead to a similar relation for the major (longitudinal) Poisson's ratio:

$$\nu_{12} = V_f \nu_{12f} + V_m \nu_m \tag{3.24}$$

where ν_{12f} is the longitudinal Poisson's ratio of the fiber and ν_m is Poisson's ratio of the matrix. In the case of isotropic fibers (e.g., glass fibers), E_{1f} and ν_{12f} in the above equations are replaced by E_f and ν_f, respectively.

The self-consistent field method in this case (Fig. 3.2) uses strain-displacement and constitutive relations, continuity conditions at the fiber/matrix interface, and regularity conditions at the center of the fiber. The solution of the elasticity problem yields more refined expressions for the longitudinal modulus and Poisson's ratio:

$$E_1 = V_f E_{1f} + V_m E_m + \frac{4(\nu_m - \nu_{12f})^2 K_f K_m G_m V_m V_f}{K_f K_m + G_m(V_f K_f + V_m K_m)} \tag{3.25}$$

$$\nu_{12} = V_f \nu_{12f} + V_m \nu_m + \frac{(\nu_m - \nu_{12f})(K_m - K_f)G_m V_m V_f}{K_f K_m + G_m(V_f K_f + V_m K_m)} \tag{3.26}$$

where K_f and K_m are the plane strain bulk moduli of the fiber and matrix, respectively. The plane strain bulk modulus for an isotropic material of modulus E and Poisson's ratio ν is

$$K_{(\varepsilon)} = K + \frac{G}{3} = \frac{E}{3(1-2\nu)} + \frac{E}{6(1+\nu)} = \frac{E}{2(1+\nu)(1-2\nu)} \tag{3.27}$$

The last terms in Eqs. (3.25) and (3.26) are not significant for most composites; therefore, the rule of mixtures relations in Eqs. (3.23) and (3.24) are very good approximations.

The variational approach described by Hashin gives the following improved bounds for the longitudinal modulus:[1]

$$V_f E_{1f} + V_m E_m + \frac{4(\nu_m - \nu_{12f})^2 K_f K_m G_m V_m V_f}{K_f K_m + G_m(V_f K_f + V_m K_m)} \leq E_1$$

$$\leq V_f E_{1f} + V_m E_m + \frac{4(\nu_{12f} - \nu_m)^2 K_f K_m G_f V_m V_f}{K_f K_m + G_f(V_f K_f + V_m K_m)} \tag{3.28}$$

In this case the lower bound is identical to the solution of the self-consistent field method, Eq. (3.25). The upper bound is obtained from the lower one by interchanging the "fiber" and "matrix" designations in the expression for the lower bound.

The semiempirical approach using the Halpin-Tsai relation, Eq. (3.17), yields exactly the rule of mixtures predictions of Eqs. (3.23) and (3.24) when the parameter $\xi \to \infty$.

In the case of longitudinal properties, predictions by all methods are very close to each other and in very good agreement with experimental results.

3.5 TRANSVERSE ELASTIC PROPERTIES—CONTINUOUS FIBERS

In the case of transverse normal loading, the state of stress in the matrix surrounding the fibers is very nonuniform and is influenced by the interaction with neighboring fibers, as illustrated in the photoelastic fringe pattern of Fig. 2.7. The transverse modulus is a matrix-dominated property and sensitive to the local state of stress. Approaches based on assumptions of simplified stress states do not yield accurate results.

In the mechanics of materials approach, the unidirectional lamina can be idealized as a plate consisting of rectangular cross section fiber and matrix strips in series (Fig. 3.5).

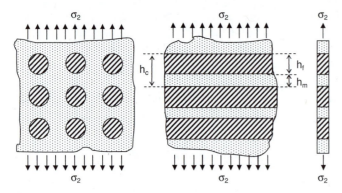

Fig. 3.5 Mechanics of materials idealization of transversely loaded unidirectional composite.

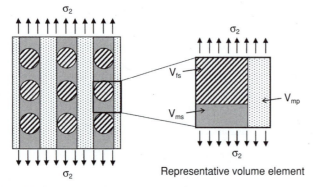

Representative volume element

Fig. 3.6 Geometric idealization for combined series and parallel model (combination model).

Thus, it is assumed that both matrix and fibers are under equal and uniform stress. This leads to the following relation for the transverse modulus:

$$\frac{1}{E_2} = \frac{V_f}{E_{2f}} + \frac{V_m}{E_m} \qquad (3.29)$$

or

$$E_2 = \frac{E_{2f} E_m}{V_f E_m + V_m E_{2f}} \qquad (3.30)$$

where E_{2f} is the transverse modulus of the fiber. The matrix modulus E_m in Eqs. (3.29) and (3.30) is usually replaced by

$$E'_m = \frac{E_m}{1 - v_m^2} \qquad (3.31)$$

where v_m is Poisson's ratio of the matrix. This accounts for the constraint imposed on the matrix by the fibers in the fiber direction. Thus, Eq. (3.30) is modified as follows:

$$E_2 = \frac{E_{2f} E'_m}{V_f E'_m + V_m E_{2f}} \qquad (3.32)$$

The mechanics of materials prediction above tends to underestimate the transverse modulus.

A refinement of the above approach was proposed by Shaffer in the form of a combined series and parallel model.[3] The transverse cross-sectional slice of the material is assumed to consist of columns of matrix and fiber elements in series acting in parallel with columns of pure matrix between them (Fig. 3.6). The resulting modulus for the composite is obtained as

$$E_2 = E_m V_{mp} + E_{2s}(1 - V_{mp}) \qquad (3.33)$$

where E_{2s} is the modulus of the series portion of the element given by

$$E_{2s} = \frac{E_{2f} E_m}{E_m V_{fs} + E_{2f} V_{ms}} \qquad (3.34)$$

and

V_{mp} = overall volume ratio of parallel matrix columns

V_{fs}, V_{ms} = fiber and matrix volume ratios of series portion, respectively

The Halpin-Tsai semiempirical relation for the transverse modulus is[11]

$$E_2 = \frac{E_m(1 + \xi\eta V_f)}{1 - \eta V_f} \tag{3.35}$$

where

$$\eta = \frac{E_{2f} - E_m}{E_{2f} + \xi E_m} \tag{3.36}$$

The parameter ξ can be treated as a curve-fitting parameter and can be obtained from an experimental value of E_2. Experimental results fall within a band of $1 < \xi < 2$. Usually, it is assumed that $\xi = 1$ for hexagonal arrays (glass and carbon composites with high fiber volume ratios) and $\xi = 2$ for square arrays (boron composite). For $\xi = 1$ the above equations yield

$$E_2 = E_m \frac{(1 + V_f)E_{2f} + V_m E_m}{V_m E_{2f} + (1 + V_f)E_m} \tag{3.37}$$

In the case of isotropic fibers, E_{2f} is replaced by E_f. The variation of transverse modulus as a function of fiber volume ratio for several composite materials obtained by the Halpin-Tsai relation is shown in Fig. 3.7. Similar results have been obtained by Adams and Doner using numerical methods.[4]

The self-consistent field model yields a complex expression for the transverse composite modulus in terms of the bulk and transverse shear moduli,[1]

$$E_2 = \frac{1}{\dfrac{1}{4K_2} + \dfrac{1}{4G_{23}} + \dfrac{v_{12}^2}{E_1}} \tag{3.38}$$

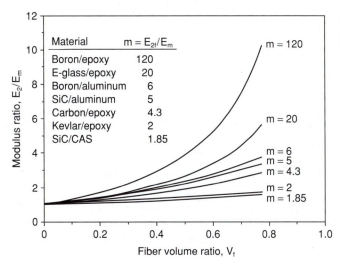

Fig. 3.7 Transverse modulus of unidirectional composites as a function of fiber volume ratio (Halpin-Tsai equations[11]).

where

$$K_2 = \frac{(K_{2f} + G_m)K_m + (K_{2f} - K_m)G_mV_f}{(K_{2f} + G_m) - (K_{2f} - K_m)V_f} \tag{3.39}$$

and

$$G_{23} = \frac{G_m[K_m(G_m + G_{23f}) + 2G_{23f}G_m + K_m(G_{23f} - G_m)V_f]}{K_m(G_m + G_{23f}) + 2G_{23f}G_m - (K_m + 2G_m)(G_{23f} - G_m)V_f} \tag{3.40}$$

The transverse plane strain bulk modulus, K_2, is obtained from the relationship

$$\bar{\sigma} = 2K_2\bar{\varepsilon}$$

where

$$\bar{\sigma} = \tfrac{1}{2}(\sigma_2 + \sigma_3) = p$$

$$\bar{\varepsilon} = \tfrac{1}{2}(\varepsilon_2 + \varepsilon_3)$$

when the material is under a loading of $\sigma_2 = \sigma_3 = p$ and an axial stress σ_1 resulting in $\varepsilon_1 = 0$.

The transverse Poisson's ratio, ν_{23}, is also related to the other moduli and longitudinal (major) Poisson's ratio as follows:

$$\nu_{23} = 1 - \frac{E_2}{2K_2} - 2\nu_{12}^2\frac{E_2}{E_1} \tag{3.41}$$

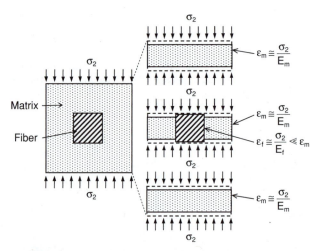

For a transversely isotropic material (with 2-3 as the plane of isotropy), it has been proven that the transverse shear modulus is related to the transverse Young's modulus and Poisson's ratio by the familiar isotropic relation[1]

$$G_{23} = \frac{E_2}{2(1 + \nu_{23})} \tag{3.42}$$

The energy principles used in the bounding method can be clearly illustrated in the case of the transverse modulus. Figure 3.8 shows a representative unit volume element containing a fiber subelement embedded in the matrix. If a uniform state of stress is assumed throughout (satisfying equilibrium and traction boundary conditions) it is seen that, by considering the free body diagrams in Fig. 3.8, the deformation

Fig. 3.8 Loading and deformation of representative unit volume element of composite under uniform stress (compatibility not satisfied).

of the middle block containing the fiber does not satisfy compatibility, that is, after deformation, the three slices do not fit together. The associated strain energy is

$$U^\sigma = \frac{1}{2} \int_V \sigma_2 \varepsilon_2 dV = \frac{\sigma_2^2}{2} \int_V \frac{dV}{E} = \frac{\sigma_2^2}{2} \left(\frac{V_f}{E_f} + \frac{V_m}{E_m} \right) \tag{3.43}$$

which should be greater than or equal to the strain energy U of a corresponding homogeneous unit volume element of modulus E_2,

$$U = \frac{\sigma_2^2}{2E_2}$$

Therefore,

$$\frac{\sigma_2^2}{2E_2} \leq \frac{\sigma_2^2}{2} \left(\frac{V_f}{E_f} + \frac{V_m}{E_m} \right)$$

or

$$E_2 \geq \frac{1}{\dfrac{V_f}{E_f} + \dfrac{V_m}{E_m}} \tag{3.44}$$

The above lower bound is identical to the mechanics of materials prediction, Eq. (3.30).

An upper bound is found by assuming a uniform state of strain throughout (Fig. 3.9). In order to maintain the state of constant uniform strain, different stresses must be applied to the matrix and fiber elements of the middle block. The stresses at the interfaces between the fiber element and the top and bottom blocks are $\sigma_f \cong E_f \varepsilon_2$, and the stresses on all other matrix elements are $\sigma_m \cong E_m \varepsilon_2$. Therefore, the top and bottom blocks are not in equilibrium. The associated strain energy

$$U^\varepsilon = \frac{1}{2} \int_V \sigma_2 \varepsilon_2 dV = \frac{\varepsilon_2^2}{2} \int_V E dV$$
$$= \frac{\varepsilon_2^2}{2} (E_f V_f + E_m V_m) \tag{3.45}$$

is greater than or equal to the strain energy U of an equivalent homogeneous unit volume element of modulus E_2,

$$U = \frac{\varepsilon_2^2 E_2}{2}$$

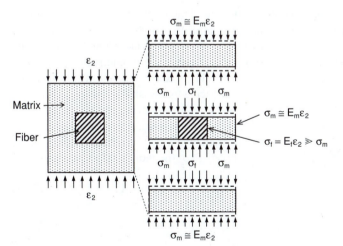

Fig. 3.9 Deformation and loading of representative unit volume element of composite under uniform strain (equilibrium not satisfied).

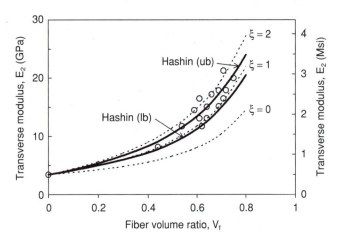

Fig. 3.10 Comparison of predicted and experimental results for transverse modulus of unidirectional glass/epoxy composite. (Predictions are based on Hashin's[13] improved upper and lower bounds and Halpin-Tsai relation for given values of ξ; experimental results were obtained by Tsai.[14])

Therefore,

$$\frac{\varepsilon_2^2 E_2}{2} \le \frac{\varepsilon_2^2}{2}(E_f V_f + E_m V_m)$$

or

$$E_2 \le E_f V_f + E_m V_m \qquad (3.46)$$

The above upper bound is identical to the parallel model prediction of the longitudinal modulus E_1, Eq. (3.23). The bounds obtained above by assuming oversimplified states of stress or strain are "primitive" bounds and may be far apart. These bounds can be improved (narrowed) by assuming more realistic states of stress or strain based on elastic stress analysis.

Improved bounds obtained by Hashin are in very good agreement with experimental results, with the upper bound being a closer fit, as shown in Fig. 3.10.[13] The experimental results in this figure were obtained by Tsai.[14] For comparison, results from the Halpin-Tsai predictions for $\xi = 1$ and $\xi = 2$ are plotted in the same figure and shown to bracket the experimental results. The prediction for $\xi = 0$, which corresponds to the series model of the mechanics of materials approach, falls far below the experimental data.

3.6 IN-PLANE SHEAR MODULUS

The behavior of a unidirectional composite under in-plane (longitudinal) shear loading is also dominated by the matrix properties and the local stress distributions. In the mechanics of materials approach, the response of a unidirectional composite layer under shear parallel to the fibers can be idealized as that of a series model consisting of alternating matrix and fiber layers under constant shear stress, τ_{12}, (Fig. 3.11). Each matrix and fiber element is subjected to the same shear stress, but their shear deformations, γ_m and γ_{12f}, are different. The average shear deformation of the representative volume element is the volume average of the individual shear strains:

$$\gamma_{12} = \gamma_{12f} V_f + \gamma_m V_m \qquad (3.47)$$

or

$$\frac{\tau_{12}}{G_{12}} = \frac{\tau_{12}}{G_{12f}} V_f + \frac{\tau_{12}}{G_m} V_m$$

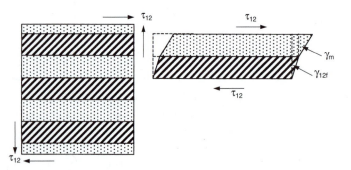

Fig. 3.11 Unidirectional composite element under in-plane shear.

This leads to the series model relation

$$\frac{1}{G_{12}} = \frac{V_f}{G_{12f}} + \frac{V_m}{G_m} \tag{3.48}$$

or

$$G_{12} = \frac{G_{12f}G_m}{V_f G_m + V_m G_{12f}} \tag{3.49}$$

where G_{12f} and G_m are the shear moduli of the fiber and matrix, respectively. In the case of isotropic fibers, G_{12f} is replaced with G_f. As in the case of transverse modulus, this approach tends to underestimate the in-plane shear modulus.

The Halpin-Tsai semiempirical relation in this case is

$$G_{12} = G_m \frac{1 + \xi\eta V_f}{1 - \eta V_f} \tag{3.50}$$

where

$$\eta = \frac{G_{12f} - G_m}{G_{12f} + \xi G_m}$$

Best agreement with experimental results has been found for $\xi = 1$, when the relation in Eq. (3.50) becomes

$$G_{12} = G_m \frac{(G_{12f} + G_m) + V_f(G_{12f} - G_m)}{(G_{12f} + G_m) - V_f(G_{12f} - G_m)} = G_m \frac{(1 + V_f)G_{12f} + V_m G_m}{V_m G_{12f} + (1 + V_f)G_m} \tag{3.51}$$

The self-consistent field model gives the following relation for the shear modulus:

$$G_{12} = G_m \frac{(1 + V_f)G_{12f} + V_m G_m}{V_m G_{12f} + (1 + V_f)G_m} \tag{3.52}$$

And the bounding method gives the following improved bounds:[1]

$$G_m \frac{(1 + V_f)G_{12f} + V_m G_m}{V_m G_{12f} + (1 + V_f)G_m} \leq G_{12} \leq G_{12f} \frac{(1 + V_m)G_m + V_f G_{12f}}{V_f G_m + (1 + V_m)G_{12f}} \tag{3.53}$$

The upper bound is obtained from the lower one by interchanging matrix and fiber designations.

It is noteworthy that the predictions of G_{12} by the Halpin-Tsai relation for $\xi = 1$, the self-consistent field method, and the lower bound above are identical. Despite this agreement, all predictions of G_{12} tend to underestimate the experimentally measured values. Figure 3.12 shows the variation of the in-plane shear modulus with fiber volume ratio for several composite materials obtained by numerical analysis by Adams and Doner.[5]

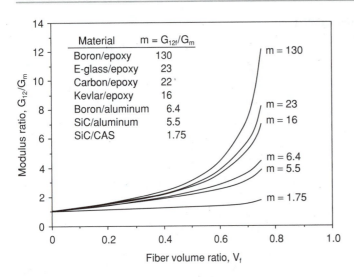

Fig. 3.12 In-plane shear modulus of unidirectional composites as a function of fiber volume ratio. (From Adams and Doner.[5])

3.7 LONGITUDINAL PROPERTIES—DISCONTINUOUS (SHORT) FIBERS

3.7.1 Elastic Stress Transfer Model—Shear Lag Analysis (Cox[15])

The easiest case to analyze is that of short aligned fibers of equal length (Fig. 3.13). One classical approach is the elastic stress transfer model developed by Cox.[15] It is assumed that fibers and matrix behave elastically and that the interface transfers stress from the fiber to the matrix without yielding or slippage.

Figure 3.13 shows a composite with short aligned fibers under uniform stress σ_1 producing an average strain ε_1. Figure 3.14 shows a representative cylindrical volume element of radius r_0 before and after deformation. The radius r_0 is related to the fiber radius and the fiber volume ratio as follows:

$$V_f = \frac{\pi r^2}{\pi r_0^2}$$

or

$$r_0 = \frac{r}{\sqrt{V_f}} \tag{3.54}$$

It is assumed that the strain at the outer surface of the cylindrical volume element is equal to the average composite strain ε_1. The shear lag analysis described by Cox leads to the following solution for the axial stress in the fiber:

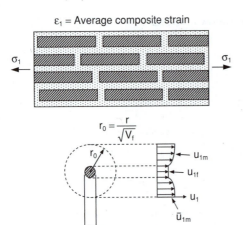

Fig. 3.13 Composite with short aligned fibers under uniform longitudinal loading and displacement variation on cross section of characteristic volume element.

$$\sigma_{1f} = E_{1f}\varepsilon_1 \left[1 - \frac{\cosh \beta x}{\cosh (\beta l/2)} \right] = E_{1f}\varepsilon_1 \left[1 - \frac{\cosh (nx/r)}{\cosh (ns)} \right] \tag{3.55}$$

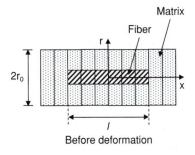

Matrix

Fiber

r

$2r_0$

x

l

Before deformation

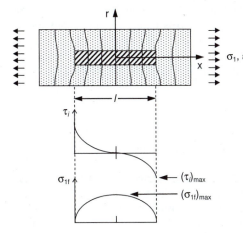

σ_1, ε_1

After deformation

Fig. 3.14 Representative volume element before and after deformation.

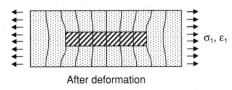

r

x

σ_1, ε_1

l

τ_i

σ_{1f}

$(\tau_i)_{max}$

$(\sigma_{1f})_{max}$

Fig. 3.15 Representative volume element after deformation and distributions of interfacial shear and axial fiber stresses.

where

$$\beta^2 = \frac{2G_m}{E_{1f} r^2 \log(r_0/r)} = \frac{n^2}{r^2}$$

$$n^2 = \frac{2G_m}{E_{1f} \log(r_0/r)}$$

$$s = \frac{l}{2r} = \frac{l}{d} \quad \text{(fiber aspect ratio)}$$

r, d = fiber radius and diameter, respectively

The interfacial shear stress is

$$\tau_i = \frac{\beta r}{2} E_{1f} \varepsilon_1 \frac{\sinh \beta x}{\cosh (\beta l/2)} = \frac{n E_{1f} \varepsilon_1}{2} \frac{\sinh (nx/r)}{\cosh (ns)} \quad (3.56)$$

The distribution of these stresses along the fiber is illustrated in Fig. 3.15.

The average stress in the composite volume element is obtained by a rule of mixtures of the average fiber and matrix stresses, $\bar{\sigma}_{1f}$ and $\bar{\sigma}_m$, as

$$\sigma_1 = V_f \bar{\sigma}_{1f} + V_m \bar{\sigma}_m$$

$$= \left[V_f E_{1f} \left(1 - \frac{\tanh (\beta l/2)}{\beta l/2} \right) + V_m E_m \right] \varepsilon_1 \quad (3.57)$$

from which we obtain the following expression for the longitudinal composite modulus:

$$E_1 = \frac{\sigma_1}{\varepsilon_1} = E_{1f} V_f \eta_l + V_m E_m \quad (3.58)$$

where

$$\eta_l = 1 - \frac{\tanh (\beta l/2)}{\beta l/2} = 1 - \frac{\tanh (ns)}{ns} \quad (3.59)$$

The term η_l is a "length correction" factor for the rule of mixtures. For very stiff fibers compared to the matrix, $E_{1f} \gg E_m$, and sufficiently high V_f, the above expression leads to the relation

$$\eta_l \cong \frac{E_1(short \; fibers)}{E_1(continuous \; fibers)} \quad (3.60)$$

As ns increases, $\tanh(ns)$ and η_l approach unity, that is, as the aspect ratio $s = l/2r$ increases, the behavior of the discontinuous-fiber composite approaches that of a continuous-fiber composite.

3.7.2 Semiempirical Relation (Halpin[16])

The longitudinal modulus of a short-fiber composite can also be estimated by a semi-empirical approach using the Halpin-Tsai relation. By setting the parameter ξ as

$$\xi = \frac{l}{r} = 2s$$

we obtain from Eq. (3.18)

$$E_1 = \frac{E_m[E_{1f} + 2sE_m + 2sV_f(E_{1f} - E_m)]}{E_{1f} + 2sE_m - V_f(E_{1f} - E_m)}$$

or

$$E_1 = E_m \frac{E_{1f}(1 + 2sV_f) + 2sE_mV_m}{E_{1f}V_m + E_m(2s + V_f)} \tag{3.61}$$

As the fiber aspect ratio increases ($s \rightarrow \infty$), the material becomes a continuous-fiber composite, and the above equation reduces to the parallel model prediction of Eq. (3.23).

As the fibers get shorter (i.e., as $s \rightarrow 0$), Eq. (3.60) is reduced to the series model prediction of Eq. (3.30). This would be the predicted modulus of a composite reinforced with thin parallel platelets in the direction normal to the platelets, for example, the transverse modulus of a polymer/clay nanocomposite with exfoliated and parallel clay platelets.

REFERENCES

1. Z. Hashin, "Analysis of Composite Materials—A Survey," *J. Appl. Mech.,* Vol. 50, 1983, pp. 481–505.
2. J. M. Whitney and R. L. McCullough, *Micromechanical Materials Modeling,* Delaware Composites Design Encyclopedia, Vol. 2, Technomic Publishing Co., Lancaster, 1990.
3. B. W. Shaffer, "Stress-Strain Relations of Reinforced Plastics Parallel and Normal to the Internal Filaments," *AIAA Journal,* Vol. 2, 1964, p. 348.
4. D. F. Adams and D. R. Doner, "Transverse Normal Loading of a Unidirectional Composite," *J. Composite Materials,* Vol. 1, 1967, pp. 152–164.
5. D. F. Adams and D. R. Doner, "Longitudinal Shear Loading of a Unidirectional Composite," *J. Composite Materials,* Vol. 1, 1967, pp. 4–17.
6. J. Aboudi, *Mechanics of Composite Materials—A Unified Micromechanical Approach,* Elsevier, Amsterdam, 1991.
7. R. Hill, "Theory of Mechanical Properties of Fibre-Strengthened Materials: III, Self-Consistent Model," *J. Mech. Phys. Solids,* Vol. 13, 1965, p. 189.
8. J. M. Whitney and M. B. Riley, "Elastic Properties of Fiber Reinforced Composite Materials," *AIAA J.,* Vol. 4, 1966, p. 1537.
9. B. Paul, "Prediction of Elastic Constants of Multiphase Materials," *Trans. Metal., Soc. AIME,* Vol. 218, 1960, pp. 36–41.
10. Z. Hashin and B. W. Rosen, "The Elastic Moduli of Fiber-Reinforced Materials," *J. Appl. Mech.,* Vol. 21, 1964, pp. 233–242.

11. J. C. Halpin and S. W. Tsai, *Effects of Environmental Factors on Composite Materials*, Air Force Technical Report AFML-TR-67-423, Wright Aeronautical Labs, Dayton, OH, 1967.

12. I. M. Daniel, "Photoelastic Investigation of Composites," in *Composite Materials*, Vol. 2, *Mechanics of Composite Materials*, G. P. Sendeckyj, Vol. Ed., L. J. Broutman and R. H. Krock, Series Eds., Academic Press, New York, 1974.

13. Z. Hashin, *Theory of Fiber Reinforced Materials*, NASA Technical Report, Contract No. NAS1-8818, 1970.

14. S. W. Tsai, *Structural Behavior of Composite Materials*, NASA Report CR-71, 1964.

15. H. L. Cox, "The Elasticity and Strength of Paper and Other Fibrous Materials," *Brit. J. Appl. Physics*, Vol. 3, 1952, pp. 72–79.

16. J. C. Halpin, "Stiffness and Expansion Estimates for Oriented Short Fiber Composites," *J. Comp. Mat.*, Vol. 3, 1969, pp. 732–734.

PROBLEMS

3.1 Derive the expressions in Eqs. (3.22) for fiber volume ratio and determine its maximum value in each case.

3.2 Plot clear fiber spacing, s/r, versus fiber volume ratio for the square and hexagonal packings.

3.3 The fibers of a carbon/epoxy composite are 8 μm in diameter and are coated with a 1 μm thick epoxy coating. Determine the maximum fiber volume ratio, V_f, that can be achieved by bonding them with a similar epoxy matrix.

3.4 Arrange carbon fibers with half as many glass fibers in the closest possible packing in an epoxy matrix and determine the maximum longitudinal modulus E_1 that can be achieved if E_{1f} (glass) = 72.5 GPa, E_{1f} (carbon) = 235 GPa, E_m = 3.45 GPa, d_f (glass) = 13 μm, and d_f (carbon) = 8 μm.

3.5 In the general Halpin-Tsai expression for composite properties, prove that the value of parameter $\xi = 0$ corresponds to the series model and $\xi \to \infty$ corresponds to the parallel model.

3.6 In the general Halpin-Tsai expression, find the limiting values of parameter η and corresponding composite properties for the cases of rigid inclusions (fibers), homogeneous material (fibers of the same material as the matrix), and voids (or totally debonded fibers).

3.7 Derive the relations in Eqs. (3.23) and (3.24).

3.8 Derive the relation in Eq. (3.29).

3.9 Derive the expression for the transverse modulus E_2 using the combination model for a unidirectional composite with square array, in terms of constituent properties E_m, E_{2f}, and V_f.

3.10 Derive the expression for the transverse modulus E_2 using the combination model for a unidirectional composite with hexagonal array, in terms of constituent properties E_m, E_{2f}, and V_f.

3.11 Determine the transverse modulus E_2 of a unidirectional carbon/epoxy composite with the properties

$$E_{2f} = 14.8 \text{ GPa (2.15 Msi)}$$
$$E_m = 3.45 \text{ GPa (0.5 Msi)}$$
$$\nu_m = 0.36$$
$$V_f = 0.65$$

using the mechanics of materials approach and the Halpin-Tsai relationship with $\xi = 1$.

3.12 Determine the transverse modulus E_2 of a unidirectional silicon carbide/aluminum (SiC/Al) composite with properties

$$E_{2f} = 366 \text{ GPa (53 Msi)}$$
$$E_m = 69 \text{ GPa (10 Msi)}$$
$$\nu_m = 0.33$$
$$V_f = 0.40$$

using the mechanics of materials approach and the Halpin-Tsai relationship with $\xi = 2$.

3.13 Plot curves of E_2/E_m versus V_f for a unidirectional carbon/epoxy material with the constituent properties

$$E_{2f} = 14.8 \text{ GPa (2.15 Msi)}$$
$$E_m = 3.45 \text{ GPa (0.5 Msi)}$$
$$\nu_m = 0.36$$

using

(a) the mechanics of materials approach, Eq. (3.32), and

(b) the Halpin-Tsai relation, Eq. (3.37), with $\xi = 1$

3.14 The measured transverse modulus E_2 of a unidirectional carbon/epoxy composite is $E_2 = 10.3$ GPa

(1.49 Msi). Given that $E_m = 3.45$ GPa (0.5 Msi) and $V_f = 0.65$, determine the transverse fiber modulus E_{2f} of the fiber using the Halpin-Tsai relation with $\xi = 1.5$.

3.15 A unidirectional glass/epoxy composite with properties $E_m = 3.45$ GPa (0.5 Msi), $E_{2f} = 69$ GPa (10 Msi), and $V_f = 0.55$ has a transverse modulus of $E_2 = 8.56$ GPa (1.24 Msi). Determine the appropriate value of ξ in the Halpin-Tsai relation and subsequently determine the transverse modulus of a composite of the same constituents but with a different fiber volume ratio of $V_f = 0.65$.

3.16 Derive the relation for shear modulus in Eq. (3.48).

3.17 Determine the in-plane shear modulus G_{12} of a glass/epoxy composite with the properties

$$G_f = 28.3 \text{ GPa (4.10 Msi)}$$
$$G_m = 1270 \text{ MPa (184 ksi)}$$
$$V_f = 0.55$$

using the mechanics of materials approach and the Halpin-Tsai relationship with $\xi = 1$.

3.18 Derive the expression for the in-plane shear modulus G_{12} using the combination model for a unidirectional composite with square array in terms of G_{12f}, G_m, and V_f. Determine the value of G_{12} for the properties given in the previous problem.

3.19 Plot curves of G_{12}/G_m versus V_f for a unidirectional glass/epoxy composite with the properties

$$G_f = 28.3 \text{ GPa (4.10 Msi)}$$
$$G_m = 1.27 \text{ GPa (184 ksi)}$$

using

(a) the mechanics of materials approach, Eq. (3.49), and

(b) the Hashin upper and lower bounds, Eq. (3.53)

3.20 The in-plane shear modulus of a unidirectional carbon/epoxy composite was measured to be $G_{12} = 6.9$ GPa (1.0 Msi). Determine the value of parameter ξ in the Halpin-Tsai relation for the following properties:

$$G_m = 1.27 \text{ GPa (0.18 Msi)}$$
$$G_{12f} = 13.1 \text{ GPa (1.9 Msi)}$$
$$V_f = 0.60$$

3.21 The in-plane shear modulus of a unidirectional carbon/epoxy composite was measured to be $G_{12} = 6.9$ GPa (1.0 Msi). Using the Halpin-Tsai relation for $\xi = 10$ and the properties

$$G_m = 1.27 \text{ GPa (184 ksi)}$$
$$V_f = 0.60$$

determine the in-plane shear modulus of the fiber G_{12f}.

3.22 The measured shear modulus of a boron/epoxy composite with a fiber volume ratio of $V_f = 0.50$ is $G_{12} = 5.4$ GPa (0.78 Msi). Using the Halpin-Tsai relation, determine the appropriate value of ξ and shear modulus G_{12} for a similar material with $V_f = 0.70$ and the following constituent properties:

$$G_{12f} = 165 \text{ GPa (23.9 Msi)}$$
$$G_m = 1.27 \text{ GPa (184 ksi)}$$

3.23 The longitudinal modulus of a glass/epoxy composite containing short aligned fibers of length l is $E_1 = 40$ GPa (5.8 Msi). Using Halpin's semiempirical relation, determine the length l of the fibers for the properties

$$V_f = 0.60$$
$$d_f = 10 \text{ μm } (4 \times 10^{-4} \text{ in})$$
$$E_f = 70 \text{ GPa (10.1 Msi)}$$
$$E_m = 3.5 \text{ GPa (0.51 Msi)}$$

3.24 Given a glass/epoxy composite containing short aligned fibers of length l and radius r, with $E_f = 69$ GPa (10 Msi), $E_m = 3.45$ GPa (0.5 Msi), $\nu_m = 0.36$, and $V_f = 0.50$, determine l/r by Cox's and Halpin's approaches so that E_1 (discont.)/E_1 (cont.) = 0.9.

3.25 The longitudinal modulus of a unidirectional glass/epoxy composite containing short aligned fibers, is $E_1 = 39.1$ GPa (5.7 Msi). Using Cox's theory, determine the length of the fibers for the properties

$$V_f = 0.60$$
$$d_f = 10 \text{ μm } (4 \times 10^{-4} \text{ in})$$
$$E_f = 70 \text{ GPa (10.1 Msi)}$$
$$E_m = 3.5 \text{ GPa (0.51 Msi)}$$
$$\nu_m = 0.36$$

3.26 Determine the necessary fiber volume ratios V_f of two different composites having the same matrix and the same longitudinal modulus E_1 equal to ten times the matrix modulus. The first is a nanocomposite consisting of parallel carbon nanotubes dispersed in the matrix; the second is a continuous-fiber Kevlar/epoxy composite. The following properties are given:

Nanotubes
  Modulus: $E_{1f} = 1000$ GPa (145 Msi)
  Length: $l = 1$ μm (40 μ in)
  Diameter: $d = 1$ nm (0.04 μ in)
Kevlar Fibers
  Modulus: $E_{1f} = 130$ GPa (18.8 Msi)
Matrix
  Modulus: $E_m = 3.33$ GPa (480 ksi)

4 Elastic Behavior of Composite Lamina—Macromechanics

4.1 STRESS-STRAIN RELATIONS

4.1.1 General Anisotropic Material

The state of stress at a point in a general continuum can be represented by nine stress components σ_{ij} (where $i, j = 1, 2, 3$) acting on the sides of an elemental cube with sides parallel to the 1-, 2-, and 3-axes of a reference coordinate system (Fig. 4.1). Similarly, the state of deformation is represented by nine strain components, ε_{ij}. In the most general case the stress and strain components are related by the generalized Hooke's law as follows:

$$
\begin{bmatrix} \sigma_{11} \\ \sigma_{22} \\ \sigma_{33} \\ \sigma_{23} \\ \sigma_{31} \\ \sigma_{12} \\ \sigma_{32} \\ \sigma_{13} \\ \sigma_{21} \end{bmatrix} =
\begin{bmatrix}
C_{1111} & C_{1122} & C_{1133} & C_{1123} & C_{1131} & C_{1112} & C_{1132} & C_{1113} & C_{1121} \\
C_{2211} & C_{2222} & C_{2233} & C_{2223} & C_{2231} & C_{2212} & C_{2232} & C_{2213} & C_{2221} \\
C_{3311} & C_{3322} & C_{3333} & C_{3323} & C_{3331} & C_{3312} & C_{3332} & C_{3313} & C_{3321} \\
C_{2311} & C_{2322} & C_{2333} & C_{2323} & C_{2331} & C_{2312} & C_{2332} & C_{2313} & C_{2321} \\
C_{3111} & C_{3122} & C_{3133} & C_{3123} & C_{3131} & C_{3112} & C_{3132} & C_{3113} & C_{3121} \\
C_{1211} & C_{1222} & C_{1233} & C_{1223} & C_{1231} & C_{1212} & C_{1232} & C_{1213} & C_{1221} \\
C_{3211} & C_{3222} & C_{3233} & C_{3223} & C_{3231} & C_{3212} & C_{3232} & C_{3213} & C_{3221} \\
C_{1311} & C_{1322} & C_{1333} & C_{1323} & C_{1331} & C_{1312} & C_{1332} & C_{1313} & C_{1321} \\
C_{2111} & C_{2122} & C_{2133} & C_{2123} & C_{2131} & C_{2112} & C_{2132} & C_{2113} & C_{2121}
\end{bmatrix}
\begin{bmatrix} \varepsilon_{11} \\ \varepsilon_{22} \\ \varepsilon_{33} \\ \varepsilon_{23} \\ \varepsilon_{31} \\ \varepsilon_{12} \\ \varepsilon_{32} \\ \varepsilon_{13} \\ \varepsilon_{21} \end{bmatrix}
\quad (4.1)
$$

and

$$
\begin{bmatrix} \varepsilon_{11} \\ \varepsilon_{22} \\ \varepsilon_{33} \\ \varepsilon_{23} \\ \varepsilon_{31} \\ \varepsilon_{12} \\ \varepsilon_{32} \\ \varepsilon_{13} \\ \varepsilon_{21} \end{bmatrix} =
\begin{bmatrix}
S_{1111} & S_{1122} & S_{1133} & S_{1123} & S_{1131} & S_{1112} & S_{1132} & S_{1113} & S_{1121} \\
S_{2211} & S_{2222} & S_{2233} & S_{2223} & S_{2231} & S_{2212} & S_{2232} & S_{2213} & S_{2221} \\
S_{3311} & S_{3322} & S_{3333} & S_{3323} & S_{3331} & S_{3312} & S_{3332} & S_{3313} & S_{3321} \\
S_{2311} & S_{2322} & S_{2333} & S_{2323} & S_{2331} & S_{2312} & S_{2332} & S_{2313} & S_{2321} \\
S_{3111} & S_{3122} & S_{3133} & S_{3123} & S_{3131} & S_{3112} & S_{3132} & S_{3113} & S_{3121} \\
S_{1211} & S_{1222} & S_{1233} & S_{1223} & S_{1231} & S_{1212} & S_{1232} & S_{1213} & S_{1221} \\
S_{3211} & S_{3222} & S_{3233} & S_{3223} & S_{3231} & S_{3212} & S_{3232} & S_{3213} & S_{3221} \\
S_{1311} & S_{1322} & S_{1333} & S_{1323} & S_{1331} & S_{1312} & S_{1332} & S_{1313} & S_{1321} \\
S_{2111} & S_{2122} & S_{2133} & S_{2123} & S_{2131} & S_{2112} & S_{2132} & S_{2113} & S_{2121}
\end{bmatrix}
\begin{bmatrix} \sigma_{11} \\ \sigma_{22} \\ \sigma_{33} \\ \sigma_{23} \\ \sigma_{31} \\ \sigma_{12} \\ \sigma_{32} \\ \sigma_{13} \\ \sigma_{21} \end{bmatrix}
\quad (4.2)
$$

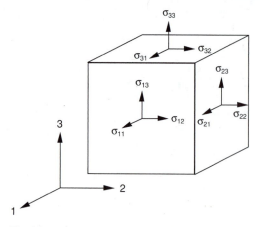

Fig. 4.1 State of stress at a point of a continuum.

or, in indicial notation

$$\sigma_{ij} = C_{ijkl}\varepsilon_{kl}$$
$$\varepsilon_{ij} = S_{ijkl}\sigma_{kl}$$

$$(i, j, k, l = 1, 2, 3) \qquad (4.3)$$

where

$$C_{ijkl} = \text{stiffness components}$$
$$S_{ijkl} = \text{compliance components}$$

Repeated subscripts in the relations above imply summation for all values of those subscripts. The compliance matrix $[S_{ijkl}]$ is the inverse of the stiffness matrix $[C_{ijkl}]$.

Thus, in general, it would require 81 elastic constants to characterize a material fully. However, the symmetry of the stress and strain tensors

$$\sigma_{ij} = \sigma_{ji}$$
$$\varepsilon_{ij} = \varepsilon_{ji}$$

$$(4.4)$$

reduces the number of independent elastic constants to 36.

It is customary in mechanics of composites to use a contracted notation for the stress, strain, stiffness, and compliance tensors as follows:

$$\sigma_{11} = \sigma_1$$
$$\sigma_{22} = \sigma_2$$
$$\sigma_{33} = \sigma_3$$
$$\sigma_{23} = \tau_{23} = \sigma_4 = \tau_4$$
$$\sigma_{31} = \tau_{31} = \sigma_5 = \tau_5$$
$$\sigma_{12} = \tau_{12} = \sigma_6 = \tau_6$$

$$(4.5)$$

$$\varepsilon_{11} = \varepsilon_1$$
$$\varepsilon_{22} = \varepsilon_2$$
$$\varepsilon_{33} = \varepsilon_3$$
$$2\varepsilon_{23} = \gamma_{23} = \varepsilon_4 = \gamma_4$$
$$2\varepsilon_{31} = \gamma_{31} = \varepsilon_5 = \gamma_5$$
$$2\varepsilon_{12} = \gamma_{12} = \varepsilon_6 = \gamma_6$$

$$(4.6)$$

$$C_{1111} = C_{11}, \quad C_{1122} = C_{12}, \quad C_{1133} = C_{13}, \quad C_{1123} = C_{14}, \quad C_{1131} = C_{15}, \quad C_{1112} = C_{16}$$

$$C_{2211} = C_{21}, \quad C_{2222} = C_{22}, \quad C_{2233} = C_{23}, \quad C_{2223} = C_{24}, \quad C_{2231} = C_{25}, \quad C_{2212} = C_{26}$$

$$C_{3311} = C_{31}, \quad C_{3322} = C_{32}, \quad C_{3333} = C_{33}, \quad C_{3323} = C_{34}, \quad C_{3331} = C_{35}, \quad C_{3312} = C_{36}$$

$$C_{2311} = C_{41}, \quad C_{2322} = C_{42}, \quad C_{2333} = C_{43}, \quad C_{2323} = C_{44}, \quad C_{2231} = C_{45}, \quad C_{2312} = C_{46}$$

$$C_{3111} = C_{51}, \quad C_{3122} = C_{52}, \quad C_{3133} = C_{53}, \quad C_{3123} = C_{54}, \quad C_{3131} = C_{55}, \quad C_{3112} = C_{56}$$

$$C_{1211} = C_{61}, \quad C_{1222} = C_{62}, \quad C_{1233} = C_{63}, \quad C_{1223} = C_{64}, \quad C_{1231} = C_{65}, \quad C_{1212} = C_{66}$$

(4.7)

Thus, the stress-strain relations for an anisotropic body can be written in the contracted notation as

$$
\begin{bmatrix} \sigma_1 \\ \sigma_2 \\ \sigma_3 \\ \tau_4 \\ \tau_5 \\ \tau_6 \end{bmatrix}
=
\begin{bmatrix}
C_{11} & C_{12} & C_{13} & C_{14} & C_{15} & C_{16} \\
C_{21} & C_{22} & C_{23} & C_{24} & C_{25} & C_{26} \\
C_{31} & C_{32} & C_{33} & C_{34} & C_{35} & C_{36} \\
C_{41} & C_{42} & C_{43} & C_{44} & C_{45} & C_{46} \\
C_{51} & C_{52} & C_{53} & C_{54} & C_{55} & C_{56} \\
C_{61} & C_{62} & C_{63} & C_{64} & C_{65} & C_{66}
\end{bmatrix}
\begin{bmatrix} \varepsilon_1 \\ \varepsilon_2 \\ \varepsilon_3 \\ \gamma_4 \\ \gamma_5 \\ \gamma_6 \end{bmatrix}
$$

(4.8)

$$
\begin{bmatrix} \varepsilon_1 \\ \varepsilon_2 \\ \varepsilon_3 \\ \gamma_4 \\ \gamma_5 \\ \gamma_6 \end{bmatrix}
=
\begin{bmatrix}
S_{11} & S_{12} & S_{13} & S_{14} & S_{15} & S_{16} \\
S_{21} & S_{22} & S_{23} & S_{24} & S_{25} & S_{26} \\
S_{31} & S_{32} & S_{33} & S_{34} & S_{35} & S_{36} \\
S_{41} & S_{42} & S_{43} & S_{44} & S_{45} & S_{46} \\
S_{51} & S_{52} & S_{53} & S_{54} & S_{55} & S_{56} \\
S_{61} & S_{62} & S_{63} & S_{64} & S_{65} & S_{66}
\end{bmatrix}
\begin{bmatrix} \sigma_1 \\ \sigma_2 \\ \sigma_3 \\ \tau_4 \\ \tau_5 \\ \tau_6 \end{bmatrix}
$$

(4.9)

or, in indicial notation,

$$\sigma_i = C_{ij}\varepsilon_j$$
$$\varepsilon_i = S_{ij}\sigma_j$$
$$(i, j = 1, 2, 3, \ldots, 6)$$

(4.10)

Energy considerations require additional symmetries. The work per unit volume is expressed as

$$W = \frac{1}{2} C_{ij}\varepsilon_i\varepsilon_j$$

(4.11)

The stress-strain relation, Eq. (4.10), can be obtained by differentiating Eq. (4.11):

$$\sigma_i = \frac{\partial W}{\partial \varepsilon_i} = C_{ij}\varepsilon_j$$

(4.12)

By differentiating again we obtain

$$C_{ij} = \frac{\partial^2 W}{\partial \varepsilon_i \partial \varepsilon_j}$$

(4.13)

In a similar manner, by reversing the order of differentiation, we obtain

$$C_{ji} = \frac{\partial^2 W}{\partial \varepsilon_j \partial \varepsilon_i}$$

(4.14)

Since the order of differentiation of W is immaterial, Eqs. (4.13) and (4.14) yield

$$C_{ij} = C_{ji}$$

(4.15)

In a similar manner we can show that

$$S_{ij} = S_{ji}$$

(4.16)

that is, the stiffness and compliance matrices are symmetric. Thus, the state of stress (or strain) at a point can be described by six components of stress (or strain), and the stress-strain Eqs. (4.8) and (4.9) are expressed in terms of 21 independent stiffness (or compliance) constants.

4.1.2 Specially Orthotropic Material

In the case of an orthotropic material (which has three mutually perpendicular planes of material symmetry) the stress-strain relations in general have the same form as Eqs. (4.8) and (4.9). However, the number of independent elastic constants is reduced to nine, as various stiffness and compliance terms are interrelated. This is clearly seen when the reference system of coordinates is selected along principal planes of material symmetry, that is, in the case of a *specially orthotropic* material. Then,

$$
\begin{bmatrix} \sigma_1 \\ \sigma_2 \\ \sigma_3 \\ \tau_4 \\ \tau_5 \\ \tau_6 \end{bmatrix}
=
\begin{bmatrix}
C_{11} & C_{12} & C_{13} & 0 & 0 & 0 \\
C_{21} & C_{22} & C_{23} & 0 & 0 & 0 \\
C_{31} & C_{32} & C_{33} & 0 & 0 & 0 \\
0 & 0 & 0 & C_{44} & 0 & 0 \\
0 & 0 & 0 & 0 & C_{55} & 0 \\
0 & 0 & 0 & 0 & 0 & C_{66}
\end{bmatrix}
\begin{bmatrix} \varepsilon_1 \\ \varepsilon_2 \\ \varepsilon_3 \\ \gamma_4 \\ \gamma_5 \\ \gamma_6 \end{bmatrix}
$$

(4.17)

and

$$
\begin{bmatrix} \varepsilon_1 \\ \varepsilon_2 \\ \varepsilon_3 \\ \gamma_4 \\ \gamma_5 \\ \gamma_6 \end{bmatrix}
=
\begin{bmatrix}
S_{11} & S_{12} & S_{13} & 0 & 0 & 0 \\
S_{21} & S_{22} & S_{23} & 0 & 0 & 0 \\
S_{31} & S_{32} & S_{33} & 0 & 0 & 0 \\
0 & 0 & 0 & S_{44} & 0 & 0 \\
0 & 0 & 0 & 0 & S_{55} & 0 \\
0 & 0 & 0 & 0 & 0 & S_{66}
\end{bmatrix}
\begin{bmatrix} \sigma_1 \\ \sigma_2 \\ \sigma_3 \\ \tau_4 \\ \tau_5 \\ \tau_6 \end{bmatrix}
$$

(4.18)

It is clearly shown that an orthotropic material can be characterized by nine independent elastic constants. This number does not change by changing the reference system of coordinates to one in which the stiffness and compliance matrices in Eqs. (4.17) and (4.18) are fully populated. The terms of either the stiffness or compliance matrix can be obtained by inversion of the other. Thus, relationships can be obtained between C_{ij} and S_{ij}.

Three important observations can be made with respect to the stress-strain relations in Eqs. (4.17) and (4.18):

1. No coupling exists between normal stresses σ_1, σ_2, σ_3 and shear strains γ_4, γ_5, γ_6; that is, normal stresses acting along principal material directions produce only normal strains.
2. No coupling exists between shear stresses τ_4, τ_5, τ_6 and normal strains ε_1, ε_2, ε_3; that is, shear stresses acting on principal material planes produce only shear strains.
3. No coupling exists between a shear stress acting on one plane and a shear strain on a different plane; that is, a shear stress acting on a principal plane produces a shear strain only on that plane.

4.1.3 Transversely Isotropic Material

An orthotropic material is called *transversely isotropic* when one of its principal planes is a *plane of isotropy*, that is, at every point there is a plane on which the mechanical properties are the same in all directions. Many unidirectional composites with fibers packed in a hexagonal array, or close to it, can be considered transversely isotropic, with the 2-3 plane (normal to the fibers) as the plane of isotropy (Fig. 4.2). This is the case with unidirectional carbon/epoxy, aramid/epoxy, and glass/epoxy composites with relatively high fiber volume ratios.

The stress-strain relations for a transversely isotropic material are simplified by noting that subscripts 2 and 3 (for a 2-3 plane of isotropy) in the material constants are interchangeable in Eqs. (4.17) and (4.18), that is,

$$C_{12} = C_{13}$$
$$C_{22} = C_{33} \quad (4.19)$$

and

$$S_{12} = S_{13}$$
$$S_{22} = S_{33}$$

Also, subscripts 5 and 6 are interchangeable; thus,

$$C_{55} = C_{66}$$
$$S_{55} = S_{66} \quad (4.20)$$

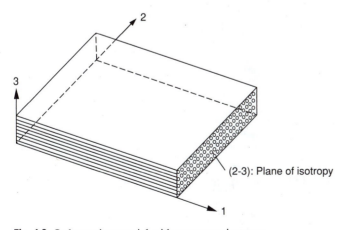

Fig. 4.2 Orthotropic material with transverse isotropy.

(2-3): Plane of isotropy

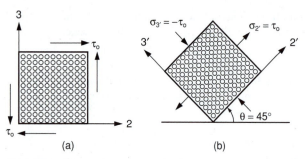

Fig. 4.3 Stress transformations on plane of isotropy of transversely isotropic material.

Furthermore, the simple stress transformation illustrated in Fig. 4.3 shows that stiffness C_{44} (or compliance S_{44}) is not independent. Considering an element with sides parallel to the 2- and 3-axes (Fig. 4.3a) under pure shear stress $\tau_0 = \tau_{23}$ and resulting shear strain $\gamma_0 = \gamma_{23}$, we have from Eq. (4.17),

$$\tau_4 = \tau_{23} = C_{44}\gamma_{23} = C_{44}\gamma_4 = \tau_0 \qquad (4.21)$$

The state of stress shown in Fig. 4.3a is equivalent to that of an element rotated by 45° and subjected to equal tensile and compressive normal stresses (Fig. 4.3b),

$$\sigma_{2'} = \tau_0$$
$$\sigma_{3'} = -\tau_0 \qquad (4.22)$$

resulting in normal strains

$$\varepsilon_{2'} = -\varepsilon_{3'} = \frac{\gamma_{23}}{2}$$
$$\varepsilon_1 = 0 \qquad (4.23)$$

Then, from Eq. (4.17),

$$\sigma_{2'} = C_{2'2'}\varepsilon_{2'} + C_{2'3'}\varepsilon_{3'} = C_{22}\varepsilon_{2'} - C_{23}\varepsilon_{2'}$$

or

$$\sigma_{2'} = \varepsilon_{2'}(C_{22} - C_{23}) = \frac{\gamma_{23}}{2}(C_{22} - C_{23}) \qquad (4.24)$$

since

$$C_{2'2'} = C_{22}$$
$$C_{2'3'} = C_{23}$$

due to transverse isotropy.

From Eqs. (4.21), (4.22), and (4.24), we obtain

$$C_{44} = \frac{C_{22} - C_{23}}{2} \qquad (4.25)$$

Thus, the stress-strain relations for a transversely isotropic material are reduced to

$$
\begin{bmatrix} \sigma_1 \\ \sigma_2 \\ \sigma_3 \\ \tau_4 \\ \tau_5 \\ \tau_6 \end{bmatrix} = \begin{bmatrix} C_{11} & C_{12} & C_{12} & 0 & 0 & 0 \\ C_{12} & C_{22} & C_{23} & 0 & 0 & 0 \\ C_{12} & C_{23} & C_{22} & 0 & 0 & 0 \\ 0 & 0 & 0 & (C_{22} - C_{23})/2 & 0 & 0 \\ 0 & 0 & 0 & 0 & C_{55} & 0 \\ 0 & 0 & 0 & 0 & 0 & C_{55} \end{bmatrix} \begin{bmatrix} \varepsilon_1 \\ \varepsilon_2 \\ \varepsilon_3 \\ \gamma_4 \\ \gamma_5 \\ \gamma_6 \end{bmatrix}
\tag{4.26}
$$

and the inverse relations are reduced to

$$
\begin{bmatrix} \varepsilon_1 \\ \varepsilon_2 \\ \varepsilon_3 \\ \gamma_4 \\ \gamma_5 \\ \gamma_6 \end{bmatrix} = \begin{bmatrix} S_{11} & S_{12} & S_{12} & 0 & 0 & 0 \\ S_{12} & S_{22} & S_{23} & 0 & 0 & 0 \\ S_{12} & S_{23} & S_{22} & 0 & 0 & 0 \\ 0 & 0 & 0 & 2(S_{22} - S_{23}) & 0 & 0 \\ 0 & 0 & 0 & 0 & S_{55} & 0 \\ 0 & 0 & 0 & 0 & 0 & S_{55} \end{bmatrix} \begin{bmatrix} \sigma_1 \\ \sigma_2 \\ \sigma_3 \\ \tau_4 \\ \tau_5 \\ \tau_6 \end{bmatrix}
\tag{4.27}
$$

The relations above show that an orthotropic material with transverse isotropy is characterized by only five independent elastic constants.

4.1.4 Orthotropic Material Under Plane Stress

In most structural applications, composite materials are used in the form of thin laminates loaded in the plane of the laminate. Thus, composite laminae (and laminates) can be considered to be under a condition of plane stress, with all stress components in the out-of-plane direction (3-direction) being zero, that is,

$$
\sigma_3 = 0
$$
$$
\tau_{23} = \tau_4 = 0
\tag{4.28}
$$
$$
\tau_{13} = \tau_5 = 0
$$

The orthotropic stress-strain relations, Eq. (4.17), are reduced to

$$
\begin{bmatrix} \sigma_1 \\ \sigma_2 \\ 0 \\ 0 \\ 0 \\ \tau_6 \end{bmatrix} = \begin{bmatrix} C_{11} & C_{12} & C_{13} & 0 & 0 & 0 \\ C_{12} & C_{22} & C_{23} & 0 & 0 & 0 \\ C_{13} & C_{23} & C_{33} & 0 & 0 & 0 \\ 0 & 0 & 0 & C_{44} & 0 & 0 \\ 0 & 0 & 0 & 0 & C_{55} & 0 \\ 0 & 0 & 0 & 0 & 0 & C_{66} \end{bmatrix} \begin{bmatrix} \varepsilon_1 \\ \varepsilon_2 \\ \varepsilon_3 \\ \gamma_4 \\ \gamma_5 \\ \gamma_6 \end{bmatrix}
\tag{4.29}
$$

which, in expanded form, are

$$\sigma_1 = C_{11}\varepsilon_1 + C_{12}\varepsilon_2 + C_{13}\varepsilon_3$$
$$\sigma_2 = C_{12}\varepsilon_1 + C_{22}\varepsilon_2 + C_{23}\varepsilon_3$$
$$0 = C_{13}\varepsilon_1 + C_{23}\varepsilon_2 + C_{33}\varepsilon_3 \qquad (4.30)$$
$$\gamma_4 = \gamma_5 = 0$$
$$\tau_6 = C_{66}\gamma_6$$

Eliminating strain ε_3 from Eq. (4.30), we obtain

$$\sigma_1 = \left(C_{11} - \frac{C_{13}C_{13}}{C_{33}}\right)\varepsilon_1 + \left(C_{12} - \frac{C_{13}C_{23}}{C_{33}}\right)\varepsilon_2 = Q_{11}\varepsilon_1 + Q_{12}\varepsilon_2$$

$$\sigma_2 = \left(C_{12} - \frac{C_{23}C_{13}}{C_{33}}\right)\varepsilon_1 + \left(C_{22} - \frac{C_{23}C_{23}}{C_{33}}\right)\varepsilon_2 = Q_{12}\varepsilon_1 + Q_{22}\varepsilon_2$$

$$\tau_6 = C_{66}\gamma_6 = Q_{66}\gamma_6$$

or

$$\begin{bmatrix} \sigma_1 \\ \sigma_2 \\ \tau_6 \end{bmatrix} = \begin{bmatrix} Q_{11} & Q_{12} & 0 \\ Q_{12} & Q_{22} & 0 \\ 0 & 0 & Q_{66} \end{bmatrix} \begin{bmatrix} \varepsilon_1 \\ \varepsilon_2 \\ \gamma_6 \end{bmatrix} \qquad (4.31)$$

or, in brief,

$$[\sigma]_{1,2} = [Q]_{1,2}\,[\varepsilon]_{1,2}$$

where the reduced stiffness matrix components are

$$Q_{ij} = C_{ij} - \frac{C_{i3}C_{j3}}{C_{33}} \qquad (i, j = 1, 2, 6) \qquad (4.32)$$

The inverse relation is written as

$$\begin{bmatrix} \varepsilon_1 \\ \varepsilon_2 \\ \gamma_6 \end{bmatrix} = \begin{bmatrix} S_{11} & S_{12} & 0 \\ S_{12} & S_{22} & 0 \\ 0 & 0 & S_{66} \end{bmatrix} \begin{bmatrix} \sigma_1 \\ \sigma_2 \\ \tau_6 \end{bmatrix} \qquad (4.33)$$

or, in brief,

$$[\varepsilon]_{1,2} = [S]_{1,2}[\sigma]_{1,2}$$

TABLE 4.1 Independent Elastic Constants for Various Types of Materials

Material	No. of Independent Elastic Constants
General anisotropic material	81
Anisotropic material considering symmetry of stress and strain tensors ($\sigma_{ij} = \sigma_{ji}$, $\varepsilon_{ij} = \varepsilon_{ji}$)	36
Anisotropic material with elastic energy considerations	21
General orthotropic material	9
Orthotropic material with transverse isotropy	5
Isotropic material	2

Thus, the in-plane stress-strain relations for an orthotropic layer under plane stress can be expressed in terms of only four independent elastic parameters, that is, the reduced stiffnesses Q_{11}, Q_{12}, Q_{22}, and Q_{66} or the compliances S_{11}, S_{12}, S_{22}, and S_{66}. It should be noted that, under plane stress, the nonzero out-of-plane strain ε_3 (or ε_z) is related to the in-plane stresses σ_1 and σ_2 through the compliances S_{13} and S_{23}. This requires two additional independent elastic parameters over and above the four needed for the in-plane stress-strain relations.

4.1.5 Isotropic Material

An isotropic material is characterized by an infinite number of planes of material symmetry through a point. For such a material, subscripts 1, 2, and 3 in the material constants are interchangeable. Then, the stress-strain relations in Eq. (4.17) are reduced to

$$
\begin{bmatrix} \sigma_1 \\ \sigma_2 \\ \sigma_3 \\ \tau_4 \\ \tau_5 \\ \tau_6 \end{bmatrix}
=
\begin{bmatrix}
C_{11} & C_{12} & C_{12} & 0 & 0 & 0 \\
C_{12} & C_{11} & C_{12} & 0 & 0 & 0 \\
C_{12} & C_{12} & C_{11} & 0 & 0 & 0 \\
0 & 0 & 0 & (C_{11} - C_{12})/2 & 0 & 0 \\
0 & 0 & 0 & 0 & (C_{11} - C_{12})/2 & 0 \\
0 & 0 & 0 & 0 & 0 & (C_{11} - C_{12})/2
\end{bmatrix}
\begin{bmatrix} \varepsilon_1 \\ \varepsilon_2 \\ \varepsilon_3 \\ \gamma_4 \\ \gamma_5 \\ \gamma_6 \end{bmatrix}
\tag{4.34}
$$

Thus, an isotropic material is fully characterized by only two independent constants, for example, the stiffnesses C_{11} and C_{12}.

The conclusions discussed before regarding the required number of independent elastic constants for the various types of materials are summarized in Table 4.1.

4.2 RELATIONS BETWEEN MATHEMATICAL AND ENGINEERING CONSTANTS

The stress-strain relations discussed before acquire more physical meaning when expressed in terms of the familiar engineering constants, that is, moduli and Poisson's ratios. Relations between mathematical and engineering constants are obtained by conducting imaginary elementary experiments as illustrated in Figs. 4.4 and 4.5.

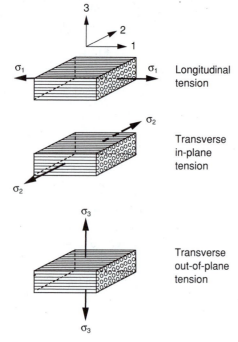

Fig. 4.4 Elementary experiments for obtaining relations between mathematical and engineering constants (normal stress loading).

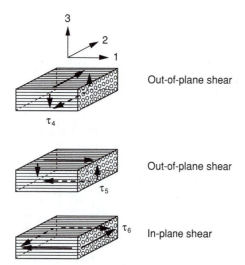

Fig. 4.5 Elementary experiments for obtaining relations between mathematical and engineering constants (shear loading).

If an orthotropic material element is subjected to uniaxial tensile loading in the longitudinal direction, σ_1, then from Eq. (4.18) we have

$$\varepsilon_1 = S_{11}\sigma_1$$
$$\varepsilon_2 = S_{12}\sigma_1$$
$$\varepsilon_3 = S_{13}\sigma_1 \qquad (4.35)$$
$$\gamma_4 = \gamma_5 = \gamma_6 = 0$$

From engineering considerations we have

$$\varepsilon_1 = \frac{\sigma_1}{E_1}$$
$$\varepsilon_2 = -\frac{\nu_{12}}{E_1}\sigma_1$$
$$\varepsilon_3 = -\frac{\nu_{13}}{E_1}\sigma_1 \qquad (4.36)$$
$$\gamma_4 = \gamma_5 = \gamma_6 = 0$$

Recall that the first and second subscripts in Poisson's ratio denote stress and strain directions, respectively.

From Eqs. (4.35) and (4.36) we obtain the relations

$$S_{11} = \frac{1}{E_1} \qquad S_{12} = -\frac{\nu_{12}}{E_1} \qquad S_{13} = -\frac{\nu_{13}}{E_1} \qquad (4.37)$$

If a material element is subjected to uniaxial tensile loading σ_2 in the in-plane transverse direction, we have in a similar fashion

$$\varepsilon_1 = S_{12}\sigma_2 = -\frac{\nu_{21}}{E_2}\sigma_2$$
$$\varepsilon_2 = S_{22}\sigma_2 = \frac{\sigma_2}{E_2}$$
$$\varepsilon_3 = S_{23}\sigma_2 = -\frac{\nu_{23}}{E_2}\sigma_2 \qquad (4.38)$$
$$\gamma_4 = \gamma_5 = \gamma_6 = 0$$

from which we obtain the relations

$$S_{12} = -\frac{\nu_{21}}{E_2} \qquad S_{22} = \frac{1}{E_2} \qquad S_{23} = -\frac{\nu_{23}}{E_2} \qquad (4.39)$$

Uniaxial normal loading σ_3 in the out-of-plane transverse direction yields

$$\varepsilon_1 = S_{13}\sigma_3 = -\frac{\nu_{31}}{E_3}\sigma_3$$

$$\varepsilon_2 = S_{23}\sigma_3 = -\frac{\nu_{32}}{E_3}\sigma_3 \qquad (4.40)$$

$$\varepsilon_3 = S_{33}\sigma_3 = \frac{\sigma_3}{E_3}$$

from which we obtain the relations

$$S_{13} = -\frac{\nu_{31}}{E_3} \qquad S_{23} = -\frac{\nu_{32}}{E_3} \qquad S_{33} = \frac{1}{E_3} \qquad (4.41)$$

In-plane pure shear loading, τ_6, yields

$$\varepsilon_1 = \varepsilon_2 = \varepsilon_3 = \gamma_4 = \gamma_5 = 0$$

$$\gamma_6 = S_{66}\tau_6 = \frac{\tau_6}{G_{12}} \qquad (4.42)$$

from which we obtain

$$S_{66} = \frac{1}{G_{12}} \qquad (4.43)$$

Out-of-plane pure shear loading τ_4 on the 2-3 plane yields

$$\varepsilon_1 = \varepsilon_2 = \varepsilon_3 = \gamma_5 = \gamma_6 = 0$$

$$\gamma_4 = S_{44}\tau_4 = \frac{\tau_4}{G_{23}} \qquad (4.44)$$

from which we obtain

$$S_{44} = \frac{1}{G_{23}} \qquad (4.45)$$

Finally, out-of-plane pure shear loading τ_5 on the 1-3 plane yields

$$\varepsilon_1 = \varepsilon_2 = \varepsilon_3 = \gamma_4 = \gamma_6 = 0$$

$$\gamma_5 = S_{55}\tau_5 = \frac{\tau_5}{G_{13}} \qquad (4.46)$$

from which we obtain

$$S_{55} = \frac{1}{G_{13}} \tag{4.47}$$

The stress-strain relations in Eq. (4.18) can then be expressed in terms of engineering constants as follows:

$$
\begin{bmatrix} \varepsilon_1 \\ \varepsilon_2 \\ \varepsilon_3 \\ \gamma_4 \\ \gamma_5 \\ \gamma_6 \end{bmatrix}
=
\begin{bmatrix}
\dfrac{1}{E_1} & -\dfrac{\nu_{21}}{E_2} & -\dfrac{\nu_{31}}{E_3} & 0 & 0 & 0 \\[2mm]
-\dfrac{\nu_{12}}{E_1} & \dfrac{1}{E_2} & -\dfrac{\nu_{32}}{E_3} & 0 & 0 & 0 \\[2mm]
-\dfrac{\nu_{13}}{E_1} & -\dfrac{\nu_{23}}{E_2} & \dfrac{1}{E_3} & 0 & 0 & 0 \\[2mm]
0 & 0 & 0 & \dfrac{1}{G_{23}} & 0 & 0 \\[2mm]
0 & 0 & 0 & 0 & \dfrac{1}{G_{13}} & 0 \\[2mm]
0 & 0 & 0 & 0 & 0 & \dfrac{1}{G_{12}}
\end{bmatrix}
\begin{bmatrix} \sigma_1 \\ \sigma_2 \\ \sigma_3 \\ \tau_4 \\ \tau_5 \\ \tau_6 \end{bmatrix}
\tag{4.48}
$$

From the symmetry of the compliance matrix $[S_{ij}]$ and the above we conclude that

$$\frac{\nu_{12}}{E_1} = \frac{\nu_{21}}{E_2}$$

$$\frac{\nu_{13}}{E_1} = \frac{\nu_{31}}{E_3} \tag{4.49}$$

$$\frac{\nu_{23}}{E_2} = \frac{\nu_{32}}{E_3}$$

and in general

$$\frac{\nu_{ij}}{E_i} = \frac{\nu_{ji}}{E_j} \quad \text{or} \quad \frac{\nu_{ij}}{\nu_{ji}} = \frac{E_i}{E_j} \quad (i, j = 1, 2, 3)$$

(Note: The above can also be deduced from Betti's reciprocal law, according to which transverse deformation due to a stress applied in the longitudinal direction is equal to the longitudinal deformation due to an equal stress applied in the transverse direction.[1])

As seen above, the relations between compliances S_{ij} and engineering constants are fairly simple. This, however, is not the case for the relations between stiffnesses C_{ij} and engineering constants. To obtain such relationships, we need first to invert the compliance matrix $[S_{ij}]$ and express the stiffnesses C_{ij} as a function of the compliances S_{ij} as follows:

$$C_{11} = \frac{S_{22}S_{33} - S_{23}^2}{S}$$

$$C_{22} = \frac{S_{33}S_{11} - S_{13}^2}{S}$$

$$C_{33} = \frac{S_{11}S_{22} - S_{12}^2}{S}$$

$$C_{12} = \frac{S_{13}S_{23} - S_{12}S_{33}}{S} \tag{4.50}$$

$$C_{23} = \frac{S_{12}S_{13} - S_{23}S_{11}}{S}$$

$$C_{13} = \frac{S_{12}S_{23} - S_{13}S_{22}}{S}$$

$$C_{44} = \frac{1}{S_{44}} \qquad C_{55} = \frac{1}{S_{55}} \qquad C_{66} = \frac{1}{S_{66}}$$

where

$$S = \begin{vmatrix} S_{11} & S_{12} & S_{13} \\ S_{12} & S_{22} & S_{23} \\ S_{13} & S_{23} & S_{33} \end{vmatrix} \tag{4.51}$$

Substituting the relations between S_{ij} and engineering constants in the above, we obtain[2]

$$C_{11} = \frac{1 - \nu_{23}\nu_{32}}{E_2 E_3 \Delta}$$

$$C_{22} = \frac{1 - \nu_{13}\nu_{31}}{E_1 E_3 \Delta}$$

$$C_{33} = \frac{1 - \nu_{12}\nu_{21}}{E_1 E_2 \Delta}$$

$$C_{12} = \frac{\nu_{21} + \nu_{31}\nu_{23}}{E_2 E_3 \Delta} = \frac{\nu_{12} + \nu_{13}\nu_{32}}{E_1 E_3 \Delta} \tag{4.52}$$

$$C_{23} = \frac{\nu_{32} + \nu_{12}\nu_{31}}{E_1 E_3 \Delta} = \frac{\nu_{23} + \nu_{21}\nu_{13}}{E_1 E_2 \Delta}$$

$$C_{13} = \frac{\nu_{13} + \nu_{12}\nu_{23}}{E_1 E_2 \Delta} = \frac{\nu_{31} + \nu_{21}\nu_{32}}{E_2 E_3 \Delta}$$

$$C_{44} = G_{23} \qquad C_{55} = G_{13} \qquad C_{66} = G_{12}$$

where

$$\Delta = \frac{1}{E_1 E_2 E_3} \begin{vmatrix} 1 & -v_{21} & -v_{31} \\ -v_{12} & 1 & -v_{32} \\ -v_{13} & -v_{23} & 1 \end{vmatrix} \tag{4.53}$$

It should be noted that in the case of a transversely isotropic material with the 2-3 plane as the plane of isotropy,

$$E_2 = E_3$$

$$G_{12} = G_{13} \tag{4.54}$$

$$v_{12} = v_{13}$$

4.3 STRESS-STRAIN RELATIONS FOR A THIN LAMINA (TWO-DIMENSIONAL)

A thin, unidirectional lamina is assumed to be under a state of plane stress; therefore, the stress-strain relations in Eqs. (4.31) and (4.33) are applicable. They relate the in-plane stress components with the in-plane strain components along the principal material axes:

$$\begin{bmatrix} \sigma_1 \\ \sigma_2 \\ \tau_6 \end{bmatrix} = \begin{bmatrix} Q_{11} & Q_{12} & 0 \\ Q_{12} & Q_{22} & 0 \\ 0 & 0 & Q_{66} \end{bmatrix} \begin{bmatrix} \varepsilon_1 \\ \varepsilon_2 \\ \gamma_6 \end{bmatrix} \tag{4.31 bis}$$

and

$$\begin{bmatrix} \varepsilon_1 \\ \varepsilon_2 \\ \gamma_6 \end{bmatrix} = \begin{bmatrix} S_{11} & S_{12} & 0 \\ S_{12} & S_{22} & 0 \\ 0 & 0 & S_{66} \end{bmatrix} \begin{bmatrix} \sigma_1 \\ \sigma_2 \\ \tau_6 \end{bmatrix} \tag{4.33 bis}$$

The relations above can be expressed in terms of engineering constants by noting that

$$S_{11} = \frac{1}{E_1}$$

$$S_{22} = \frac{1}{E_2}$$

$$S_{12} = S_{21} = -\frac{v_{12}}{E_1} = -\frac{v_{21}}{E_2} \tag{4.55}$$

$$S_{66} = \frac{1}{G_{12}}$$

and

$$Q_{11} = \frac{E_1}{1 - \nu_{12}\nu_{21}}$$

$$Q_{22} = \frac{E_2}{1 - \nu_{12}\nu_{21}}$$

$$Q_{12} = Q_{21} = \frac{\nu_{21}E_1}{1 - \nu_{12}\nu_{21}} = \frac{\nu_{12}E_2}{1 - \nu_{12}\nu_{21}}$$

$$Q_{66} = G_{12}$$

(4.56)

Thus, as far as the in-plane stress-strain relations are concerned, a single orthotropic lamina (e.g., unidirectional lamina) can be fully characterized by four independent constants—the four reduced stiffnesses Q_{11}, Q_{22}, Q_{12}, and Q_{66}; or the four compliances S_{11}, S_{22}, S_{12}, and S_{66}; or four engineering constants E_1, E_2, G_{12}, and ν_{12}. Poisson's ratio ν_{21} is not independent, as it is related to ν_{12}, E_1, and E_2 by Eq. (4.49).

4.4 TRANSFORMATION OF STRESS AND STRAIN (TWO-DIMENSIONAL)

Normally, the lamina principal axes (1, 2) do not coincide with the loading or reference axes (x, y) (Fig. 4.6). Then, the stress and strain components referred to the principal material axes (1, 2) can be expressed in terms of those referred to the loading axes (x, y) by the following transformation relations:

$$\begin{bmatrix} \sigma_1 \\ \sigma_2 \\ \tau_6 \end{bmatrix} = [T] \begin{bmatrix} \sigma_x \\ \sigma_y \\ \tau_s \end{bmatrix}$$

(4.57)

or, in brief,

$$[\sigma]_{1,2} = [T][\sigma]_{x,y}$$

and

$$\begin{bmatrix} \varepsilon_1 \\ \varepsilon_2 \\ \frac{1}{2}\gamma_6 \end{bmatrix} = [T] \begin{bmatrix} \varepsilon_x \\ \varepsilon_y \\ \frac{1}{2}\gamma_s \end{bmatrix}$$

(4.58)

or, in brief,

$$[\varepsilon]_{1,2} = [T][\varepsilon]_{x,y}$$

where the transformation matrix $[T]$ is given by

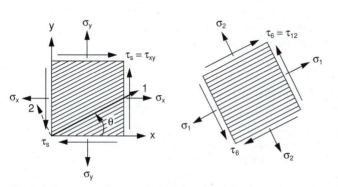

Fig. 4.6 Stress components in unidirectional lamina referred to loading and material axes.

$$[T] = \begin{bmatrix} m^2 & n^2 & 2mn \\ n^2 & m^2 & -2mn \\ -mn & mn & m^2 - n^2 \end{bmatrix} \tag{4.59}$$

and

$$m = \cos \theta$$

$$n = \sin \theta$$

The angle θ is measured positive counterclockwise from the x-axis to the 1-axis.

By inversion of the relations above we obtain

$$\begin{bmatrix} \sigma_x \\ \sigma_y \\ \tau_s \end{bmatrix} = [T^{-1}] \begin{bmatrix} \sigma_1 \\ \sigma_2 \\ \tau_6 \end{bmatrix} \tag{4.60}$$

and

$$\begin{bmatrix} \varepsilon_x \\ \varepsilon_y \\ \frac{1}{2}\gamma_s \end{bmatrix} = [T^{-1}] \begin{bmatrix} \varepsilon_1 \\ \varepsilon_2 \\ \frac{1}{2}\gamma_6 \end{bmatrix} \tag{4.61}$$

where

$$[T^{-1}] = [T(-\theta)] = \begin{bmatrix} m^2 & n^2 & -2mn \\ n^2 & m^2 & 2mn \\ mn & -mn & m^2 - n^2 \end{bmatrix} \tag{4.62}$$

In the contracted notation used here, the subscript s in the above equations corresponds to shear stress or strain components referred to the x-y system of coordinates, that is, $\tau_s = \tau_{xy}$ and $\gamma_s = \gamma_{xy}$. The subscript 6, as mentioned before, is a contraction of the subscripts 12.

The laws of stress and strain transformation are independent of material properties, that is, they are the same for isotropic and anisotropic materials.

4.5 TRANSFORMATION OF ELASTIC PARAMETERS (TWO-DIMENSIONAL)

The stress-strain relations in Eqs. (4.31) and (4.33) show that, when the lamina is loaded only in tension or compression along the principal material axes, there is no shear strain. Similarly, when the lamina is loaded under pure shear, τ_6, on the principal plane (1-2) only a shear strain, γ_6, is produced on the 1-2 plane. Thus, there is no coupling between normal stresses and shear deformation and between shear stress and normal strains. This is not the case when the lamina is loaded along arbitrary axes x and y. Then, the stress-strain relations take the form

$$\begin{bmatrix} \sigma_x \\ \sigma_y \\ \tau_s \end{bmatrix} = \begin{bmatrix} Q_{xx} & Q_{xy} & Q_{xs} \\ Q_{yx} & Q_{yy} & Q_{ys} \\ Q_{sx} & Q_{sy} & Q_{ss} \end{bmatrix} \begin{bmatrix} \varepsilon_x \\ \varepsilon_y \\ \gamma_s \end{bmatrix} \tag{4.63}$$

or, in brief,

$$[\sigma]_{x,y} = [Q]_{x,y}[\varepsilon]_{x,y}$$

with the reduced stiffness matrix fully populated. However, the number of independent constants is still four, as in the case of Eq. (4.31). What is needed, then, is the relationship between the transformed stiffnesses $[Q]_{x,y}$ and principal stiffnesses $[Q]_{1,2}$.

Equation (4.63) can be rewritten in the form

$$\begin{bmatrix} \sigma_x \\ \sigma_y \\ \tau_s \end{bmatrix} = \begin{bmatrix} Q_{xx} & Q_{xy} & 2Q_{xs} \\ Q_{yx} & Q_{yy} & 2Q_{ys} \\ Q_{sx} & Q_{sy} & 2Q_{ss} \end{bmatrix} \begin{bmatrix} \varepsilon_x \\ \varepsilon_y \\ \frac{1}{2}\gamma_s \end{bmatrix} \tag{4.64}$$

Introducing the stress-strain relation, Eq. (4.31), into the transformation relation, Eq. (4.60), we have

$$\begin{bmatrix} \sigma_x \\ \sigma_y \\ \tau_s \end{bmatrix} = [T^{-1}] \begin{bmatrix} \sigma_1 \\ \sigma_2 \\ \tau_6 \end{bmatrix} = [T^{-1}] \begin{bmatrix} Q_{11} & Q_{12} & 0 \\ Q_{21} & Q_{22} & 0 \\ 0 & 0 & Q_{66} \end{bmatrix} \begin{bmatrix} \varepsilon_1 \\ \varepsilon_2 \\ \gamma_6 \end{bmatrix}$$

$$= [T^{-1}] \begin{bmatrix} Q_{11} & Q_{12} & 0 \\ Q_{21} & Q_{22} & 0 \\ 0 & 0 & 2Q_{66} \end{bmatrix} \begin{bmatrix} \varepsilon_1 \\ \varepsilon_2 \\ \frac{1}{2}\gamma_6 \end{bmatrix} = [T^{-1}] \begin{bmatrix} Q_{11} & Q_{12} & 0 \\ Q_{21} & Q_{22} & 0 \\ 0 & 0 & 2Q_{66} \end{bmatrix} [T] \begin{bmatrix} \varepsilon_x \\ \varepsilon_y \\ \frac{1}{2}\gamma_s \end{bmatrix} \tag{4.65}$$

Comparison of Eqs. (4.64) and (4.65) leads to the following transformation relation for the stiffness matrix:

$$\begin{bmatrix} Q_{xx} & Q_{xy} & 2Q_{xs} \\ Q_{yx} & Q_{yy} & 2Q_{ys} \\ Q_{sx} & Q_{sy} & 2Q_{ss} \end{bmatrix} = [T^{-1}] \begin{bmatrix} Q_{11} & Q_{12} & 0 \\ Q_{21} & Q_{22} & 0 \\ 0 & 0 & 2Q_{66} \end{bmatrix} [T] \tag{4.66}$$

From the relation above we obtain the transformed reduced stiffnesses as a function of the principal lamina stiffnesses:

$$\begin{aligned}
Q_{xx} &= m^4 Q_{11} + n^4 Q_{22} + 2m^2 n^2 Q_{12} + 4m^2 n^2 Q_{66} \\
Q_{yy} &= n^4 Q_{11} + m^4 Q_{22} + 2m^2 n^2 Q_{12} + 4m^2 n^2 Q_{66} \\
Q_{xy} &= m^2 n^2 Q_{11} + m^2 n^2 Q_{22} + (m^4 + n^4) Q_{12} - 4m^2 n^2 Q_{66} \\
Q_{xs} &= m^3 n Q_{11} - mn^3 Q_{22} - mn(m^2 - n^2) Q_{12} - 2mn(m^2 - n^2) Q_{66} \\
Q_{ys} &= mn^3 Q_{11} - m^3 n Q_{22} + mn(m^2 - n^2) Q_{12} + 2mn(m^2 - n^2) Q_{66} \\
Q_{ss} &= m^2 n^2 Q_{11} + m^2 n^2 Q_{22} - 2m^2 n^2 Q_{12} + (m^2 - n^2)^2 Q_{66}
\end{aligned} \tag{4.67}$$

The transformed strain-stress relations can be obtained either by direct inversion of the stress-strain relations in Eq. (4.63) or by transformation of the strain-stress relations in Eq. (4.33) referred to the principal material axes. The transformed strain-stress relations are

$$\begin{bmatrix} \varepsilon_x \\ \varepsilon_y \\ \gamma_s \end{bmatrix} = \begin{bmatrix} S_{xx} & S_{xy} & S_{xs} \\ S_{yx} & S_{yy} & S_{ys} \\ S_{sx} & S_{sy} & S_{ss} \end{bmatrix} \begin{bmatrix} \sigma_x \\ \sigma_y \\ \tau_s \end{bmatrix} \tag{4.68}$$

or, in brief,

$$[\varepsilon]_{x,y} = [S]_{x,y}[\sigma]_{x,y}$$

which can be rewritten in the form

$$\begin{bmatrix} \varepsilon_x \\ \varepsilon_y \\ \frac{1}{2}\gamma_s \end{bmatrix} = \begin{bmatrix} S_{xx} & S_{xy} & S_{xs} \\ S_{yx} & S_{yy} & S_{ys} \\ \frac{1}{2}S_{sx} & \frac{1}{2}S_{sy} & \frac{1}{2}S_{ss} \end{bmatrix} \begin{bmatrix} \sigma_x \\ \sigma_y \\ \tau_s \end{bmatrix} \tag{4.69}$$

A series of transformations similar to those of Eq. (4.65) gives

$$\begin{bmatrix} \varepsilon_x \\ \varepsilon_y \\ \frac{1}{2}\gamma_s \end{bmatrix} = [T^{-1}] \begin{bmatrix} \varepsilon_1 \\ \varepsilon_2 \\ \frac{1}{2}\gamma_6 \end{bmatrix} = [T^{-1}] \begin{bmatrix} S_{11} & S_{12} & 0 \\ S_{21} & S_{22} & 0 \\ 0 & 0 & \frac{1}{2}S_{66} \end{bmatrix} \begin{bmatrix} \sigma_1 \\ \sigma_2 \\ \tau_6 \end{bmatrix} = [T^{-1}] \begin{bmatrix} S_{11} & S_{12} & 0 \\ S_{21} & S_{22} & 0 \\ 0 & 0 & \frac{1}{2}S_{66} \end{bmatrix} [T] \begin{bmatrix} \sigma_x \\ \sigma_y \\ \tau_s \end{bmatrix} \tag{4.70}$$

Comparison of Eqs. (4.69) and (4.70) leads to the following transformation relation for the compliance matrix:

$$\begin{bmatrix} S_{xx} & S_{xy} & S_{xs} \\ S_{yx} & S_{yy} & S_{ys} \\ \frac{1}{2}S_{sx} & \frac{1}{2}S_{sy} & \frac{1}{2}S_{ss} \end{bmatrix} = [T^{-1}] \begin{bmatrix} S_{11} & S_{12} & 0 \\ S_{21} & S_{22} & 0 \\ 0 & 0 & \frac{1}{2}S_{66} \end{bmatrix} [T] \tag{4.71}$$

This relation leads to the following ones for the transformed compliances as a function of the principal lamina compliances:

$$\begin{aligned}
S_{xx} &= m^4 S_{11} + n^4 S_{22} + 2m^2 n^2 S_{12} + m^2 n^2 S_{66} \\
S_{yy} &= n^4 S_{11} + m^4 S_{22} + 2m^2 n^2 S_{12} + m^2 n^2 S_{66} \\
S_{xy} &= m^2 n^2 S_{11} + m^2 n^2 S_{22} + (m^4 + n^4)S_{12} - m^2 n^2 S_{66} \\
S_{xs} &= 2m^3 n S_{11} - 2mn^3 S_{22} - 2mn(m^2 - n^2)S_{12} - mn(m^2 - n^2)S_{66} \\
S_{ys} &= 2mn^3 S_{11} - 2m^3 n S_{22} + 2mn(m^2 - n^2)S_{12} + mn(m^2 - n^2)S_{66} \\
S_{ss} &= 4m^2 n^2 S_{11} + 4m^2 n^2 S_{22} - 8m^2 n^2 S_{12} + (m^2 - n^2)^2 S_{66}
\end{aligned} \tag{4.72}$$

The transformation relations are presented in tabular form for easy reference in Table 4.2.

TABLE 4.2 Relations for Stiffness and Compliance Transformation

	$S_{11}(Q_{11})$	$S_{22}(Q_{22})$	$S_{12}(Q_{12})$	$S_{66}(4Q_{66})$
$S_{xx}(Q_{xx})$	m^4	n^4	$2m^2n^2$	m^2n^2
$S_{yy}(Q_{yy})$	n^4	m^4	$2m^2n^2$	m^2n^2
$S_{xy}(Q_{xy})$	m^2n^2	m^2n^2	$(m^4 + n^4)$	$-m^2n^2$
$S_{ss}(4Q_{ss})$	$4m^2n^2$	$4m^2n^2$	$-8m^2n^2$	$(m^2 - n^2)^2$
$S_{xs}(2Q_{xs})$	$2m^3n$	$-2mn^3$	$-2mn(m^2 - n^2)$	$-mn(m^2 - n^2)$
$S_{ys}(2Q_{ys})$	$2mn^3$	$-2m^3n$	$2mn(m^2 - n^2)$	$mn(m^2 - n^2)$

$m = \cos\theta, n = \sin\theta$

4.6 TRANSFORMATION OF STRESS-STRAIN RELATIONS IN TERMS OF ENGINEERING CONSTANTS (TWO-DIMENSIONAL)

The strain-stress relations referred to the principal material axes as given by Eqs. (4.33) are expressed in terms of engineering constants by using Eqs. (4.55)

$$\begin{bmatrix} \varepsilon_1 \\ \varepsilon_2 \\ \gamma_6 \end{bmatrix} = \begin{bmatrix} S_{11} & S_{12} & 0 \\ S_{21} & S_{22} & 0 \\ 0 & 0 & S_{66} \end{bmatrix} \begin{bmatrix} \sigma_1 \\ \sigma_2 \\ \tau_6 \end{bmatrix} = \begin{bmatrix} \dfrac{1}{E_1} & -\dfrac{v_{21}}{E_2} & 0 \\ -\dfrac{v_{12}}{E_1} & \dfrac{1}{E_2} & 0 \\ 0 & 0 & \dfrac{1}{G_{12}} \end{bmatrix} \begin{bmatrix} \sigma_1 \\ \sigma_2 \\ \tau_6 \end{bmatrix} \tag{4.73}$$

These relations, when transformed to the x-y coordinate system, are expressed by Eq. (4.68) in terms of mathematical compliance constants, S_{xx}, S_{yy}, S_{xy}, and so on. To obtain the relationships between these constants and engineering parameters, we conduct simple imaginary experiments on an element with sides parallel to the x- and y-axes. For example, the terms of the first column of the compliance matrix in Eq. (4.68) are the strain components, ε_x, ε_y, and γ_s produced by a unit normal stress $\sigma_x = 1$.

A uniaxial stress σ_x produces the following strains:

$$\varepsilon_x = \frac{\sigma_x}{E_x}$$

$$\varepsilon_y = -\frac{v_{xy}}{E_x}\sigma_x \tag{4.74}$$

$$\gamma_s = \frac{\eta_{xs}}{E_x}\sigma_x$$

In the above, Poisson's ratio v_{xy}, corresponding to stress in the x-direction and strain in the y-direction, is the negative ratio of the transverse strain ε_y to the axial strain ε_x. The shear coupling coefficient η_{xs}, corresponding to normal stress in the x-direction and shear strain in the x-y plane, is the ratio of the shear strain γ_s (γ_{xy}) to the axial strain ε_x.

In a similar manner, a uniaxial stress σ_y produces the strains

$$\varepsilon_x = -\frac{v_{yx}}{E_y}\sigma_y$$

$$\varepsilon_y = \frac{\sigma_y}{E_y} \qquad (4.75)$$

$$\gamma_s = \frac{\eta_{ys}}{E_y}\sigma_y$$

with Poisson's ratio v_{yx} and shear coupling coefficient η_{ys} defined as before.

A pure shear stress τ_s (τ_{xy}) produces the following strains:

$$\varepsilon_x = \frac{\eta_{sx}}{G_{xy}}\tau_s$$

$$\varepsilon_y = \frac{\eta_{sy}}{G_{xy}}\tau_s \qquad (4.76)$$

$$\gamma_s = \frac{\tau_s}{G_{xy}}$$

where the shear coupling coefficients η_{sx} and η_{sy} are the ratios of the normal strains ε_x and ε_y to the shear strain γ_s, respectively, for the applied pure shear loading. Here, the first subscript s denotes shear stress in the x-y plane and the second subscripts x and y denote normal strains in the x- and y-directions, respectively.

By superposition of the three loadings discussed above, we obtain the following strain-stress relations in terms of engineering constants:

$$\begin{bmatrix} \varepsilon_x \\ \varepsilon_y \\ \gamma_s \end{bmatrix} = \begin{bmatrix} \dfrac{1}{E_x} & -\dfrac{v_{yx}}{E_y} & \dfrac{\eta_{sx}}{G_{xy}} \\ -\dfrac{v_{xy}}{E_x} & \dfrac{1}{E_y} & \dfrac{\eta_{sy}}{G_{xy}} \\ \dfrac{\eta_{xs}}{E_x} & \dfrac{\eta_{ys}}{E_y} & \dfrac{1}{G_{xy}} \end{bmatrix} \begin{bmatrix} \sigma_x \\ \sigma_y \\ \tau_s \end{bmatrix} \qquad (4.77)$$

From symmetry considerations of the compliance matrix we obtain

$$\frac{v_{xy}}{E_x} = \frac{v_{yx}}{E_y} \qquad\qquad \frac{v_{xy}}{v_{yx}} = \frac{E_x}{E_y}$$

$$\frac{\eta_{xs}}{E_x} = \frac{\eta_{sx}}{G_{xy}} \qquad \text{or} \qquad \frac{\eta_{xs}}{\eta_{sx}} = \frac{E_x}{G_{xy}} \qquad (4.78)$$

$$\frac{\eta_{ys}}{E_y} = \frac{\eta_{sy}}{G_{xy}} \qquad\qquad \frac{\eta_{ys}}{\eta_{sy}} = \frac{E_y}{G_{xy}}$$

Comparison of equivalent strain-stress relations in Eqs. (4.68) and (4.77) yields the following relationships:

$$S_{xx} = \frac{1}{E_x}$$

$$S_{yy} = \frac{1}{E_y}$$

$$S_{ss} = \frac{1}{G_{xy}}$$

$$S_{xy} = S_{yx} = -\frac{\nu_{xy}}{E_x} = -\frac{\nu_{yx}}{E_y}$$

$$S_{xs} = S_{sx} = \frac{\eta_{xs}}{E_x} = \frac{\eta_{sx}}{G_{xy}}$$

$$S_{ys} = S_{sy} = \frac{\eta_{ys}}{E_y} = \frac{\eta_{sy}}{G_{xy}}$$

(4.79)

or

$$E_x = \frac{1}{S_{xx}}$$

$$E_y = \frac{1}{S_{yy}}$$

$$G_{xy} = \frac{1}{S_{ss}}$$

$$\nu_{xy} = -\frac{S_{yx}}{S_{xx}} \qquad \nu_{yx} = -\frac{S_{xy}}{S_{yy}}$$

$$\eta_{xs} = \frac{S_{sx}}{S_{xx}} \qquad \eta_{sx} = \frac{S_{xs}}{S_{ss}}$$

$$\eta_{ys} = \frac{S_{sy}}{S_{yy}} \qquad \eta_{sy} = \frac{S_{ys}}{S_{ss}}$$

(4.80)

The relations in Eq. (4.77) can be inverted to yield stress-strain relations in terms of engineering constants. These relations would be more complex than the strain-stress relations in Eq. (4.77).

4.7 TRANSFORMATION RELATIONS FOR ENGINEERING CONSTANTS (TWO-DIMENSIONAL)

Using the relations between engineering constants and compliances, Eqs. (4.79) and (4.80), in the compliance transformation relations, Eq. (4.72), we obtain the following transformation relations for the engineering constants:

$$\frac{1}{E_x} = \frac{m^2}{E_1}(m^2 - n^2 v_{12}) + \frac{n^2}{E_2}(n^2 - m^2 v_{21}) + \frac{m^2 n^2}{G_{12}}$$

$$\frac{1}{E_y} = \frac{n^2}{E_1}(n^2 - m^2 v_{12}) + \frac{m^2}{E_2}(m^2 - n^2 v_{21}) + \frac{m^2 n^2}{G_{12}}$$

$$\frac{1}{G_{xy}} = \frac{4m^2 n^2}{E_1}(1 + v_{12}) + \frac{4m^2 n^2}{E_2}(1 + v_{21}) + \frac{(m^2 - n^2)^2}{G_{12}}$$

$$\frac{v_{xy}}{E_x} = \frac{v_{yx}}{E_y} = \frac{m^2}{E_1}(m^2 v_{12} - n^2) + \frac{n^2}{E_2}(n^2 v_{21} - m^2) + \frac{m^2 n^2}{G_{12}}$$

$$\frac{\eta_{xs}}{E_x} = \frac{\eta_{sx}}{G_{xy}} = \frac{2m^3 n}{E_1}(1 + v_{12}) - \frac{2mn^3}{E_2}(1 + v_{21}) - \frac{mn(m^2 - n^2)}{G_{12}}$$

$$\frac{\eta_{ys}}{E_y} = \frac{\eta_{sy}}{G_{xy}} = \frac{2mn^3}{E_1}(1 + v_{12}) - \frac{2m^3 n}{E_2}(1 + v_{21}) + \frac{mn(m^2 - n^2)}{G_{12}}$$

$$(4.81)$$

A computational procedure for calculation of transformed elastic constants is illustrated by the flowchart in Fig. 4.7. It is assumed that the input consists of the basic engineering constants E_1, E_2, G_{12}, and v_{12} referred to the principal material axes of the lamina and obtained from characterization tests. Then, relations in Eqs. (4.55) and (4.56) are used to obtain the reduced principal compliances and stiffnesses $[S]_{1,2}$ and $[Q]_{1,2}$. The transformation relations in Eqs. (4.67) and (4.72) are used to obtain the transformed lamina stiffnesses $[Q]_{x,y}$ and compliances $[S]_{x,y}$. Finally, relations in Eq. (4.80) are used to obtain the transformed engineering constants (E_x, E_y, G_{xy}, v_{xy}, v_{yx}, η_{xs}, η_{ys}, η_{sx}, η_{sy}) referred to the x-y system of coordinates. Alternatively, relations in Eq. (4.81) can be used to obtain the transformed engineering constants directly from the given engineering constants referred to the principal material axes.

The variation of the transformed engineering constants with fiber orientation is illustrated in Figs. 4.8 and 4.9 for a typical unidirectional carbon/epoxy (AS4/3501-6) material. Young's modulus decreases monotonically from its maximum value E_1 at $\theta = 0°$ to its minimum E_2 at $\theta = 90°$. The shear modulus G_{xy} peaks at $\theta = 45°$ and reaches its minimum value at $\theta = 0°$ and $\theta = 90°$. Poisson's ratio v_{xy} decreases monotonically from its maximum value v_{12} at

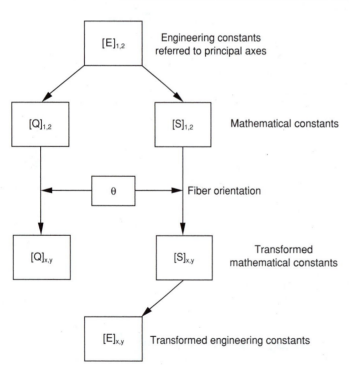

Fig. 4.7 Flowchart for determination of transformed elastic constants of unidirectional lamina (two-dimensional).

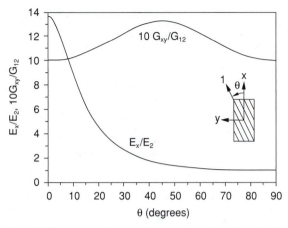

Fig. 4.8 Young's modulus and shear modulus of unidirectional composite as a function of fiber orientation (AS4/3501-6 carbon/epoxy).

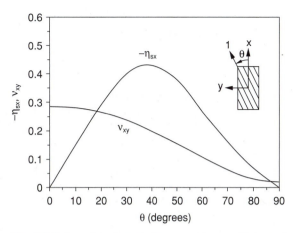

Fig. 4.9 Poisson's ratio and shear coupling coefficient of unidirectional composite as a function of fiber orientation (AS4/3501-6 carbon/epoxy).

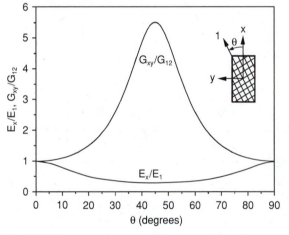

Fig. 4.10 Young's modulus and shear modulus of carbon fabric/epoxy material as a function of warp fiber orientation (AGP370-5H/3501-6S).

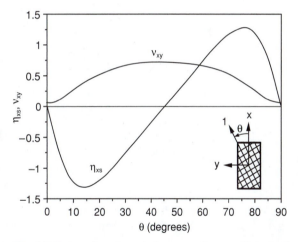

Fig. 4.11 Poisson's ratio and shear coupling coefficient of carbon fabric/epoxy material as a function of warp fiber orientation (AGP370-5H/3501-6S).

$\theta = 0°$ to its minimum value v_{21} at $\theta = 90°$. The shear coupling coefficient η_{sx} is negative throughout and peaks at around $\theta = 38°$. The curves in Figs. 4.8 and 4.9 are typical of high-stiffness, highly anisotropic composites. The form of the variation of elastic constants depends on the relative magnitudes of the basic constants referred to the principal material axes.

An example of a carbon fabric/epoxy material (AGP370-5H/3501-6S) is illustrated in Figs. 4.10 and 4.11. Here, Young's modulus decreases from a maximum value at $\theta = 0°$ to an absolute minimum at $\theta = 45°$, and then it increases to a local maximum at $\theta = 90°$. The shear modulus behaves similarly as in the case of unidirectional carbon/epoxy, but it attains a maximum value at $\theta = 45°$ that is higher than the Young's modulus at $\theta = 45°$. Poisson's ratio has minimum values at $\theta = 0°$ and $\theta = 90°$, and it peaks at $\theta \cong 45°$. The shear coupling coefficient changes signs as shown.

SAMPLE PROBLEM 4.1
Transformation of Young's Modulus

Given the basic properties E_1, E_2, G_{12}, and ν_{12} of a unidirectional lamina, it is required to determine Young's modulus E_x at an angle $\theta = 45°$ to the fiber direction (Fig. 4.12). From Eqs. (4.81) we obtain the following exact relation:

$$\left(\frac{1}{E_x}\right)_{\theta=45°} = \frac{1-\nu_{12}}{4E_1} + \frac{1-\nu_{21}}{4E_2} + \frac{1}{4G_{12}} \qquad (4.82)$$

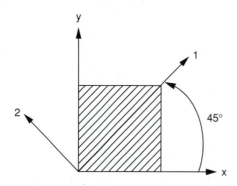

Fig. 4.12 Unidirectional lamina with fiber orientation at 45° to the reference axis.

The above relation can be simplified for the case of a high-stiffness composite for which $E_1 \gg E_2$ and $\nu_{21} \ll 1$. In that case we obtain the following approximate expression:

$$(E_x)_{\theta=45°} \cong \frac{4G_{12}E_2}{G_{12} + E_2} \qquad (4.83)$$

This means that Young's modulus at 45° to the fiber direction is a matrix-dominated property, since it depends primarily on E_2 and G_{12}, which are also matrix-dominated properties.

SAMPLE PROBLEM 4.2
Transformation of Shear Modulus

Given the basic properties E_1, E_2, G_{12}, and ν_{12} of a unidirectional lamina, it is required to determine the shear modulus G_{xy} at 45° to the fiber direction (see Fig. 4.12). From Eqs. (4.81) we obtain the following exact relation:

$$\left(\frac{1}{G_{xy}}\right)_{\theta=45°} = \frac{1+\nu_{12}}{E_1} + \frac{1+\nu_{21}}{E_2} \qquad (4.84)$$

For a high-stiffness composite, for which $E_1 \gg E_2$ and $\nu_{21} \ll 1$, we obtain

$$(G_{xy})_{\theta=45°} \cong E_2 \qquad (4.85)$$

which means that in this case the above shear modulus is a matrix-dominated property.

SAMPLE PROBLEM 4.3
Transformation of Poisson's Ratio

Given the basic properties E_1, E_2, G_{12}, and v_{12} of a unidirectional lamina, it is required to determine Poisson's ratio v_{xy} at 45° to the fiber direction (see Fig. 4.12). From Eqs. (4.81) we obtain the following exact relation:

$$\left(\frac{v_{xy}}{E_x}\right)_{\theta=45°} = \frac{1}{4E_1}(v_{12}-1) + \frac{1}{4E_2}(v_{21}-1) + \frac{1}{4G_{12}} \tag{4.86}$$

For a high-stiffness composite, for which $E_1 \gg E_2$ and $v_{21} \ll 1$, taking into consideration the approximate relation in Eq. (4.83), we obtain

$$(v_{xy})_{\theta=45°} \cong \frac{E_2 - G_{12}}{E_2 + G_{12}} \tag{4.87}$$

which means that in this case Poisson's ratio is a matrix-dominated property.

SAMPLE PROBLEM 4.4
Transformation of Shear Coupling Coefficient

Given the basic properties E_1, E_2, G_{12}, and v_{12} of a unidirectional lamina, it is required to determine the shear coupling coefficient η_{sx} at $\theta = 45°$ to the fiber direction (see Fig. 4.12). From Eqs. (4.81) we obtain the following exact relation:

$$\left(\frac{\eta_{sx}}{G_{xy}}\right)_{\theta=45°} = \frac{1}{2E_1}(1+v_{12}) - \frac{1}{2E_2}(1+v_{21}) = \frac{1}{2}\left(\frac{1}{E_1} - \frac{1}{E_2}\right) \tag{4.88}$$

This relation can be simplified as follows for the case of a high-stiffness composite:

$$\left(\frac{\eta_{sx}}{G_{xy}}\right)_{\theta=45°} \cong -\frac{1}{2E_2} \tag{4.89}$$

Recalling from Eq. (4.85) that in this case $(G_{xy})_{\theta=45°} \cong E_2$, we obtain

$$(\eta_{sx})_{\theta=45°} \cong -\frac{1}{2} \tag{4.90}$$

which is independent of the lamina properties.

4.8 TRANSFORMATION OF STRESS AND STRAIN (THREE-DIMENSIONAL)

4.8.1 General Transformation

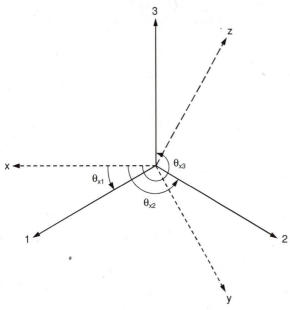

Fig. 4.13 Illustration of coordinate systems referred to in three-dimensional transformation relations.

Many structural applications of composites involve thick components subjected to three-dimensional states of stress. Assuming the material to be orthotropic, deformation and failure analysis is facilitated by referring stress and strain components and their relations to the principal system of material coordinates (1, 2, 3). These stress and strain components are related to those referred to an arbitrary coordinate system (x, y, z) by the following transformation relations (Fig. 4.13):

$$
\begin{bmatrix} \sigma_1 \\ \sigma_2 \\ \sigma_3 \\ \tau_4 \\ \tau_5 \\ \tau_6 \end{bmatrix} = [T_{ij}] \begin{bmatrix} \sigma_x \\ \sigma_y \\ \sigma_z \\ \tau_{yz} \\ \tau_{zx} \\ \tau_{xy} \end{bmatrix} = [T_{ij}] \begin{bmatrix} \sigma_x \\ \sigma_y \\ \sigma_z \\ \tau_q \\ \tau_r \\ \tau_s \end{bmatrix} \quad (4.91)
$$

or, in brief,

$$
[\sigma]_{1,2,3} = [T][\sigma]_{x,y,z}
$$

and

$$
\begin{bmatrix} \varepsilon_1 \\ \varepsilon_2 \\ \varepsilon_3 \\ \frac{1}{2}\gamma_4 \\ \frac{1}{2}\gamma_5 \\ \frac{1}{2}\gamma_6 \end{bmatrix} = [T_{ij}] \begin{bmatrix} \varepsilon_x \\ \varepsilon_y \\ \varepsilon_z \\ \frac{1}{2}\gamma_{yz} \\ \frac{1}{2}\gamma_{zx} \\ \frac{1}{2}\gamma_{xy} \end{bmatrix} = [T_{ij}] \begin{bmatrix} \varepsilon_x \\ \varepsilon_y \\ \varepsilon_z \\ \frac{1}{2}\gamma_q \\ \frac{1}{2}\gamma_r \\ \frac{1}{2}\gamma_s \end{bmatrix} \quad (4.92)
$$

or, in brief,

$$
[\varepsilon]_{1,2,3} = [T][\varepsilon]_{x,y,z}
$$

Subscripts q, r, and s in the above equations are contractions of subscripts yz, zx, and xy, respectively.

The transformation matrix $[T_{ij}]$ is given by

$$[T_{ij}] = \begin{bmatrix} m_1^2 & n_1^2 & p_1^2 & 2n_1p_1 & 2p_1m_1 & 2m_1n_1 \\ m_2^2 & n_2^2 & p_2^2 & 2n_2p_2 & 2p_2m_2 & 2m_2n_2 \\ m_3^2 & n_3^2 & p_3^2 & 2n_3p_3 & 2p_3m_3 & 2m_3n_3 \\ m_2m_3 & n_2n_3 & p_2p_3 & n_2p_3 + n_3p_2 & p_2m_3 + p_3m_2 & m_2n_3 + m_3n_2 \\ m_3m_1 & n_3n_1 & p_3p_1 & n_3p_1 + n_1p_3 & p_3m_1 + p_1m_3 & m_3n_1 + m_1n_3 \\ m_1m_2 & n_1n_2 & p_1p_2 & n_1p_2 + n_2p_1 & p_1m_2 + p_2m_1 & m_1n_2 + m_2n_1 \end{bmatrix} \quad (4.93)$$

where m_i, n_i, and p_i are the direction cosines of axis i, that is,

$$m_1 = \cos \theta_{x1} \qquad n_1 = \cos \theta_{y1} \qquad p_1 = \cos \theta_{z1}$$

$$m_2 = \cos \theta_{x2} \qquad n_2 = \cos \theta_{y2} \qquad p_2 = \cos \theta_{z2}$$

$$m_3 = \cos \theta_{x3} \qquad n_3 = \cos \theta_{y3} \qquad p_3 = \cos \theta_{z3}$$

The angles θ_{ij} are measured from axis i to axis j as shown in Fig. 4.13.

By inverting Eqs. (4.91) and (4.92) we obtain

$$\begin{bmatrix} \sigma_x \\ \sigma_y \\ \sigma_z \\ \tau_q \\ \tau_r \\ \tau_s \end{bmatrix} = [T_{ij}^{-1}] \begin{bmatrix} \sigma_1 \\ \sigma_2 \\ \sigma_3 \\ \tau_4 \\ \tau_5 \\ \tau_6 \end{bmatrix} \quad (4.94)$$

and

$$\begin{bmatrix} \varepsilon_x \\ \varepsilon_y \\ \varepsilon_z \\ \frac{1}{2}\gamma_q \\ \frac{1}{2}\gamma_r \\ \frac{1}{2}\gamma_s \end{bmatrix} = [T_{ij}^{-1}] \begin{bmatrix} \varepsilon_1 \\ \varepsilon_2 \\ \varepsilon_3 \\ \frac{1}{2}\gamma_4 \\ \frac{1}{2}\gamma_5 \\ \frac{1}{2}\gamma_6 \end{bmatrix} \quad (4.95)$$

where $[T_{ij}^{-1}]$ is the inverse of matrix $[T_{ij}]$ in Eq. (4.93).

4.8.2 Rotation About 3-Axis

In the case where the loading direction z coincides with the through-the-thickness principal direction 3, the above relations are simplified by noting that

$$p_1 = p_2 = 0 \qquad p_3 = 1$$

$$m_3 = n_3 = 0 \quad (4.96)$$

and, referring to the definitions of m and n in the two-dimensional case,

$$m_1 = m \qquad n_1 = n$$

$$m_2 = -n \qquad n_2 = m$$

Then, the transformation matrix, $[T'_{ij}]$, takes the form

$$[T'_{ij}] = \begin{bmatrix} m^2 & n^2 & 0 & 0 & 0 & 2mn \\ n^2 & m^2 & 0 & 0 & 0 & -2mn \\ 0 & 0 & 1 & 0 & 0 & 0 \\ 0 & 0 & 0 & m & -n & 0 \\ 0 & 0 & 0 & n & m & 0 \\ -mn & mn & 0 & 0 & 0 & m^2 - n^2 \end{bmatrix} \tag{4.97}$$

The inverse of the transformation matrix can be obtained by changing the sign of the rotation angle about the z- or 3-axis, that is, by replacing n with $-n$.

$$[T'^{-1}_{ij}] = \begin{bmatrix} m^2 & n^2 & 0 & 0 & 0 & -2mn \\ n^2 & m^2 & 0 & 0 & 0 & 2mn \\ 0 & 0 & 1 & 0 & 0 & 0 \\ 0 & 0 & 0 & m & n & 0 \\ 0 & 0 & 0 & -n & m & 0 \\ mn & -mn & 0 & 0 & 0 & m^2 - n^2 \end{bmatrix} \tag{4.98}$$

The above transformation matrices can be decomposed into two parts. By considering only the first, second, and sixth rows and columns, they are reduced to the transformation matrices $[T]$ and $[T^{-1}]$ defined in Eqs. (4.59) and (4.62) for the two-dimensional case. The remaining parts describe the transformation of the out-of-plane shear components.

$$\begin{bmatrix} \tau_{23} \\ \tau_{31} \end{bmatrix} = \begin{bmatrix} m & -n \\ n & m \end{bmatrix} \begin{bmatrix} \tau_{yz} \\ \tau_{zx} \end{bmatrix} \quad \text{or} \quad \begin{bmatrix} \tau_4 \\ \tau_5 \end{bmatrix} = \begin{bmatrix} m & -n \\ n & m \end{bmatrix} \begin{bmatrix} \tau_q \\ \tau_r \end{bmatrix} \tag{4.99}$$

and

$$\begin{bmatrix} \tau_{yz} \\ \tau_{zx} \end{bmatrix} = \begin{bmatrix} m & n \\ -n & m \end{bmatrix} \begin{bmatrix} \tau_{23} \\ \tau_{31} \end{bmatrix} \quad \text{or} \quad \begin{bmatrix} \tau_q \\ \tau_r \end{bmatrix} = \begin{bmatrix} m & n \\ -n & m \end{bmatrix} \begin{bmatrix} \tau_4 \\ \tau_5 \end{bmatrix} \tag{4.100}$$

4.9 TRANSFORMATION OF ELASTIC PARAMETERS (THREE-DIMENSIONAL)

The stress-strain relations of Eqs. (4.17) for a specially orthotropic material take the following form when referred to an arbitrary system of coordinates (x, y, z):

$$
\begin{bmatrix} \sigma_x \\ \sigma_y \\ \sigma_z \\ \tau_q \\ \tau_r \\ \tau_s \end{bmatrix} = \begin{bmatrix} C_{xx} & C_{xy} & C_{xz} & 2C_{xq} & 2C_{xr} & 2C_{xs} \\ C_{yx} & C_{yy} & C_{yz} & 2C_{yq} & 2C_{yr} & 2C_{ys} \\ C_{zx} & C_{zy} & C_{zz} & 2C_{zq} & 2C_{zr} & 2C_{zs} \\ C_{qx} & C_{qy} & C_{qz} & 2C_{qq} & 2C_{qr} & 2C_{qs} \\ C_{rx} & C_{ry} & C_{rz} & 2C_{rq} & 2C_{rr} & 2C_{rs} \\ C_{sx} & C_{sy} & C_{sz} & 2C_{sq} & 2C_{sr} & 2C_{ss} \end{bmatrix} \begin{bmatrix} \varepsilon_x \\ \varepsilon_y \\ \varepsilon_z \\ \tfrac{1}{2}\gamma_q \\ \tfrac{1}{2}\gamma_r \\ \tfrac{1}{2}\gamma_s \end{bmatrix} \tag{4.101}
$$

Introducing the stress-strain relations, Eqs. (4.17), into the transformation relations, Eqs. (4.94), we obtain

$$
\begin{bmatrix} \sigma_x \\ \sigma_y \\ \sigma_z \\ \tau_q \\ \tau_r \\ \tau_s \end{bmatrix} = [T_{ij}^{-1}] \begin{bmatrix} \sigma_1 \\ \sigma_2 \\ \sigma_3 \\ \tau_4 \\ \tau_5 \\ \tau_6 \end{bmatrix} = [T_{ij}^{-1}] \begin{bmatrix} C_{11} & C_{12} & C_{13} & 0 & 0 & 0 \\ C_{21} & C_{22} & C_{23} & 0 & 0 & 0 \\ C_{31} & C_{32} & C_{33} & 0 & 0 & 0 \\ 0 & 0 & 0 & 2C_{44} & 0 & 0 \\ 0 & 0 & 0 & 0 & 2C_{55} & 0 \\ 0 & 0 & 0 & 0 & 0 & 2C_{66} \end{bmatrix} \begin{bmatrix} \varepsilon_1 \\ \varepsilon_2 \\ \varepsilon_3 \\ \tfrac{1}{2}\gamma_4 \\ \tfrac{1}{2}\gamma_5 \\ \tfrac{1}{2}\gamma_6 \end{bmatrix}
$$

$$
= [T_{ij}^{-1}] \begin{bmatrix} C_{11} & C_{12} & C_{13} & 0 & 0 & 0 \\ C_{21} & C_{22} & C_{23} & 0 & 0 & 0 \\ C_{31} & C_{32} & C_{33} & 0 & 0 & 0 \\ 0 & 0 & 0 & 2C_{44} & 0 & 0 \\ 0 & 0 & 0 & 0 & 2C_{55} & 0 \\ 0 & 0 & 0 & 0 & 0 & 2C_{66} \end{bmatrix} [T_{ij}] \begin{bmatrix} \varepsilon_x \\ \varepsilon_y \\ \varepsilon_z \\ \tfrac{1}{2}\gamma_q \\ \tfrac{1}{2}\gamma_r \\ \tfrac{1}{2}\gamma_s \end{bmatrix} \tag{4.102}
$$

The following transformation relation for the stiffness matrix is obtained by comparison of Eqs. (4.101) and (4.102):

$$
\begin{bmatrix} C_{xx} & C_{xy} & C_{xz} & 2C_{xq} & 2C_{xr} & 2C_{xs} \\ C_{yx} & C_{yy} & C_{yz} & 2C_{yq} & 2C_{yr} & 2C_{ys} \\ C_{zx} & C_{zy} & C_{zz} & 2C_{zq} & 2C_{zr} & 2C_{zs} \\ C_{qx} & C_{qy} & C_{qz} & 2C_{qq} & 2C_{qr} & 2C_{qs} \\ C_{rx} & C_{ry} & C_{rz} & 2C_{rq} & 2C_{rr} & 2C_{rs} \\ C_{sx} & C_{sy} & C_{sz} & 2C_{sq} & 2C_{sr} & 2C_{ss} \end{bmatrix} = [T_{ij}^{-1}] \begin{bmatrix} C_{11} & C_{12} & C_{13} & 0 & 0 & 0 \\ C_{21} & C_{22} & C_{23} & 0 & 0 & 0 \\ C_{31} & C_{32} & C_{33} & 0 & 0 & 0 \\ 0 & 0 & 0 & 2C_{44} & 0 & 0 \\ 0 & 0 & 0 & 0 & 2C_{55} & 0 \\ 0 & 0 & 0 & 0 & 0 & 2C_{66} \end{bmatrix} [T_{ij}] \tag{4.103}
$$

The above yields transformation relations for C_{ij} ($i, j = x, y, z, q, r, s$) in terms of $C_{\alpha\beta}$ ($\alpha, \beta = 1, 2, 3, 4, 5, 6$) similar to Eqs. (4.67) for two dimensions. Similar relations are obtained for the transformation of compliances.

The complete transformation relations for stiffnesses and compliances in three dimensions are given in Table B.1 of Appendix B. If the z-axis coincides with the 3-direction, the transformation matrix $[T_{ij}]$ and its inverse $[T_{ij}^{-1}]$ are reduced to the forms $[T_{ij}']$ and

TABLE 4.3 Three-Dimensional Transformation Relations for Elastic Parameters of Composite Lamina*

	$S_{11}(C_{11})$	$S_{12}(C_{12})$	$S_{13}(C_{13})$	$S_{22}(C_{22})$	$S_{23}(C_{23})$	$S_{33}(C_{33})$	$S_{44}(4C_{44})$	$S_{55}(4C_{55})$	$S_{66}(4C_{66})$
$S_{xx}(C_{xx})$	m^4	$2m^2n^2$	0	n^4	0	0	0	0	m^2n^2
$S_{xy}(C_{xy})$	m^2n^2	m^4+n^4	0	m^2n^2	0	0	0	0	$-m^2n^2$
$S_{xz}(C_{xz})$	0	0	m^2	0	n^2	0	0	0	0
$S_{xs}(2C_{xs})$	$2m^3n$	$-2mn(m^2-n^2)$	0	$-2mn^3$	0	0	0	0	$-mn(m^2-n^2)$
$S_{yy}(C_{yy})$	n^4	$2m^2n^2$	0	m^4	0	0	0	0	m^2n^2
$S_{yz}(C_{yz})$	0	0	n^2	0	m^2	0	0	0	0
$S_{ys}(2C_{ys})$	$2mn^3$	$2mn(m^2-n^2)$	0	$-2m^3n$	0	0	0	0	$mn(m^2-n^2)$
$S_{zz}(C_{zz})$	0	0	0	0	0	1	0	0	0
$S_{zs}(2C_{zs})$	0	0	$2mn$	0	$-2mn$	0	0	0	0
$S_{qq}(4C_{qq})$	0	0	0	0	0	0	m^2	n^2	0
$S_{qr}(4C_{qr})$	0	0	0	0	0	0	$-mn$	mn	0
$S_{rr}(4C_{rr})$	0	0	0	0	0	0	n^2	m^2	0
$S_{ss}(4C_{ss})$	$4m^2n^2$	$-8m^2n^2$	0	$4m^2n^2$	0	0	0	0	$(m^2-n^2)^2$

* Simplified for the case when the z-direction and the 3-direction coincide.
$m_1 = m, n_1 = n, p_1 = 0$
$m_2 = -n, n_2 = m, p_2 = 0$
$m_3 = 0, n_3 = 0, p_3 = 1$

$[T_{ij}'^{-1}]$ in Eqs. (4.97) and (4.98). The simplified transformation relations are presented in tabular form in Table 4.3.

Transformation relations for engineering constants can be obtained as in the two-dimensional case discussed before. The strain-stress relations are expressed in terms of transformed engineering constants, and relations are obtained between transformed compliances S_{ij} ($i, j = x, y, z, q, r, s$) and transformed engineering constants. Then, the transformation relations for engineering constants are obtained by making use of the compliance transformation relations (Table 4.3 and Table B.1 in Appendix B).

REFERENCES

1. E. Betti, *Il Nuovo Cimento*, Ser. 2, Vols. 7 and 8, 1872.
2. R. M. Jones, *Mechanics of Composite Materials*, Second Edition, Taylor and Francis, Philadelphia, 1999.

PROBLEMS

4.1 Derive Eqs. (4.52) and (4.53) from Eqs. (4.18), (4.48), (4.50), and (4.51).

4.2 Assuming that the stiffness and compliance matrices are positive definite, that is, they have positive principal values ($C_{11}, C_{22}, C_{33} > 0$), and using Eqs. (4.48) to (4.53), prove that Poisson's ratios of an orthotropic material must satisfy the condition

$$\nu_{ij} < \left(\frac{E_i}{E_j}\right)^{1/2} \quad (i, j = 1, 2, 3)$$

4.3 Derive the transformation relations, Eqs. (4.67), for Q_{xx}, Q_{xy}, Q_{xs}, and Q_{ss}.

4.4 Using the equations of stiffness transformation prove that the quantity $Q_{11} + Q_{22} + 2Q_{12}$ is an invariant (i.e., it is independent of axes orientation) by showing that

$$Q_{xx} + Q_{yy} + 2Q_{xy} = Q_{11} + Q_{22} + 2Q_{12}$$

for any angle θ between the fiber direction and the reference axes.

4.5 Show that the transformed stiffness Q_{xx} can be expressed in the form

$$Q_{xx} = U_1 + U_2 \cos 2\theta + U_3 \cos 4\theta$$

Determine U_1, U_2, and U_3, and show that they are invariants (i.e., independent of orientation of coordinate axes).

4.6 Show that an orthotropic material whose principal stiffnesses satisfy the following relations is isotropic:

$$Q_{11} = Q_{22}$$
$$Q_{66} = \tfrac{1}{2}(Q_{11} - Q_{12})$$

(In other words, show that Q_{xx}, Q_{xy}, and Q_{ss} are independent of orientation, and that $Q_{xs} = Q_{ys} = 0$.)

4.7 Derive the transformation relation for E_x in Eqs. (4.81).

4.8 Derive the transformation relation for G_{xy} in Eqs. (4.81).

4.9 Derive the transformation relation for ν_{xy} in Eqs. (4.81).

4.10 Derive the transformation relation for η_{xs} in Eqs. (4.81).

4.11 Write the stress-strain relations for a unidirectional lamina in terms of engineering constants referred to an arbitrary coordinate system (x, y).

4.12 A specimen is made by joining along one edge two unidirectional laminae at $0°$ and $45°$ to the loading axis as shown in Fig. P4.12. How—(a), (b), or (c)—

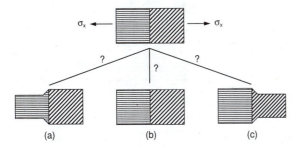

(a)   (b)   (c)

Fig. P4.12

is the specimen going to deform in the transverse direction for a material of the following properties? Prove your answer by numerical computation.

$$E_1 = 145 \text{ GPa (21 Msi)}$$
$$E_2 = 10.4 \text{ GPa (1.5 Msi)}$$
$$G_{12} = 6.9 \text{ GPa (1.0 Msi)}$$
$$\nu_{12} = 0.28$$

4.13 Show that a material whose properties satisfy the following relations is isotropic:

$$E_1 = E_2$$
$$G_{12} = \frac{E_1}{2(1 + \nu_{12})}$$

(i.e., show that E_x is independent of orientation).

4.14 Compare exact and approximate values of Young's modulus at $45°$ to the fiber direction for a carbon/epoxy material having the following properties:

$$E_1 = 145 \text{ GPa (21 Msi)}$$
$$E_2 = 10.45 \text{ GPa (1.5 Msi)}$$
$$G_{12} = 6.9 \text{ GPa (1.0 Msi)}$$
$$\nu_{12} = 0.28$$

4.15 Knowing E_1, E_2, ν_{12}, and $(E_x)_{\theta=45°}$ (modulus at $45°$ with the fiber direction), determine G_{12}, first exactly and then using the approximation for high-stiffness composites. Compare numerical values for the carbon/epoxy material of Problem 4.14.

4.16 Knowing E_1, E_2, G_{12}, and ν_{12} for a unidirectional lamina, determine $(\nu_{xy})_{\theta=45°}$, first exactly and then using the approximation for high-stiffness composites. Compare numerical values for the carbon/epoxy material of Problem 4.14.

4.17 Compare exact and approximate values of the shear modulus $(G_{xy})_{\theta=45°}$ for the material of Problem 4.14.

4.18 Compare exact and approximate values of $(\eta_{sx})_{\theta=45°}$ for the material of Problem 4.14.

4.19 Find a general expression for the coupling coefficient η_{xs} for a unidirectional composite with fiber orientation $\theta = 45°$ in terms of constituent properties. Then, obtain an approximate expression for a high-stiffness composite. Is η_{xs} a fiber-dominated or matrix-dominated property?

4.20 Determine Poisson's ratio at $45°$ to the principal material axes of a woven glass/epoxy composite with the following properties:

$$E_1/E_2 = 1.25$$
$$G_{12}/E_2 = 0.25$$
$$\nu_{12} = 0.16$$

4.21 Determine Poisson's ratio v_{xy} at angle $\theta = 30°$ to the fiber direction for the two materials below with the given properties:

(a) $E_1/E_2 = 3$
$G_{12}/E_2 = 0.5$
$v_{12} = 0.25$

(b) $E_1/E_2 = 15$
$G_{12}/E_2 = 0.625$
$v_{12} = 0.30$

4.22 Obtain an approximate value for Poisson's ratio v_{xy} of an off-axis unidirectional lamina at an angle $\theta = 60°$ to the fiber direction for a high-stiffness composite material with $E_1 \gg E_2$ and $E_2 = 1.5G_{12}$.

4.23 Obtain Poisson's ratio for an off-axis unidirectional composite at an angle $\theta = 60°$ to the fiber direction for a material with $E_1 = 15E_2 = 22.5G_{12}$ and $v_{12} = 0.30$.

4.24 A unidirectional lamina is characterized by determining E_1 and E_2. Subsequently, it is loaded at an angle $\theta = 30°$ to the fibers, and the modulus $(E_x)_{\theta=30°}$ is determined. Obtain a relationship for the modulus $(E_x)_{\theta=45°}$ in terms of E_1, E_2, and $(E_x)_{\theta=30°}$.

4.25 A unidirectional lamina is loaded at angles $\theta = 30°$ and $60°$ to the fiber direction, and the corresponding moduli $(E_x)_{\theta=30°}$ and $(E_x)_{\theta=60°}$ are obtained. Determine a relationship between these two moduli and E_1 and E_2. Find an approximate expression for E_2 in terms of $(E_x)_{\theta=30°}$ and $(E_x)_{\theta=60°}$ for a high-stiffness composite $(E_1 \gg E_2)$.

4.26 For a composite material with $E_1 = E_2 = E$ and $v_{12} = v_{21} \cong 0$ (woven composite), determine $(v_{xy})_{max}$ in terms of E and G_{12} and the corresponding angle θ where this maximum occurs.

4.27 For a composite material with $E_1 = 2E_2$ and $v_{12} \cong v_{21} \cong 0$ (woven composite), determine $(v_{xy})_{max}$ in terms of E_2 and G_{12} and the corresponding angle θ where this maximum occurs.

4.28 A woven carbon/epoxy composite has identical moduli and Poisson's ratios in the warp and fill directions, $E_1 = E_2 = E$ and $v_{12} \cong v_{21} = v$. Determine the maximum values of Poisson's ratio $(v_{xy})_{max}$, shear coupling coefficient $(\eta_{sx})_{max}$, and the corresponding angles θ to the principal material directions.

4.29 Determine the extreme values of the shear coupling coefficient η_{sx} and corresponding angles to the principal material directions for a woven glass/epoxy composite with the following properties:

$E_1 = 15.8$ GPa (2.3 Msi)
$E_2 = 12.9$ GPa (1.9 Msi)
$G_{12} = 2.7$ GPa (0.4 Msi)
$v_{12} = 0.16$

4.30 Using the transformation relation for E_x in Eqs. (4.81), determine its maximum and minimum values. Prove that E_x can have a maximum value for some value of θ in the range $0 < \theta < 90°$ when

$$G_{12} > \frac{E_1}{2(1 + v_{12})}$$

(This means that for some orthotropic materials, the off-axis modulus E_x can be higher than E_1.)

4.31 Using the same procedure as in the preceding problem, prove that E_x can have a minimum value for some value of θ in the range $0 < \theta < 90°$ when

$$G_{12} < \frac{E_1}{2[(E_1/E_2) + v_{12}]}$$

(This means that for some orthotropic materials, the off-axis modulus E_x can be lower than E_2.)

4.32 Using the same procedure as in Problem 4.30, prove that E_x attains its maximum value E_1 at $\theta = 0°$ and its minimum value E_2 at $\theta = 90°$ when

$$\frac{E_1}{2[(E_1/E_2) + v_{12}]} < G_{12} < \frac{E_1}{2(1 + v_{12})}$$

4.33 A unidirectional lamina is loaded under a uniaxial stress $\sigma_1 = \sigma_0$, and principal strains ε_1 and ε_2 are measured (Fig. P4.33). Compute the transverse

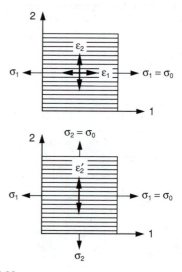

Fig. P4.33

strain ε_2' of the same lamina loaded under equal biaxial normal stresses $\sigma_1 = \sigma_2 = \sigma_0$ as a function of the strains ε_1 and ε_2 obtained before and the modulus ratio $k_E = E_1/E_2$.

4.34 A unidirectional lamina is loaded under uniaxial tension σ_y at 45° to the fiber direction as shown in Fig. P4.34. Determine the three strain components ε_x, ε_y, and $\gamma_s(\gamma_{xy})$ as a function of σ_y and the basic lamina properties $(E_1, E_2, G_{12}, \nu_{12})$ referred to the principal material axes. Obtain first the strains in terms of the transformed engineering constants $(E_x, E_y, \nu_{xy}, \eta_{xs}$, and so on) and then use approximate relations for these constants for a high-stiffness composite.

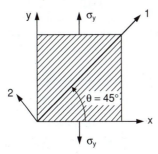

Fig. P4.34

4.35 A 30° off-axis unidirectional lamina is under biaxial normal loading along the x- and y-axes as shown in Fig. P4.35.
(a) Express the ratio $k = \sigma_y/\sigma_x$ in terms of transformed engineering constants $(E_x, E_y, G_{xy}, \eta_{xs}$, and so on), such that there is no shear deformation in the lamina.
(b) Obtain an expression for k in terms of principal engineering constants $(E_1, E_2, G_{12}, \nu_{12})$, and then obtain an approximate value of k for a high-stiffness composite $(E_1 \gg E_2)$ and $E_2 = 1.5G_{12}$.

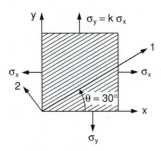

Fig. P4.35

4.36 An off-axis unidirectional lamina is under biaxial normal loading along the x- and y-axes (Fig. P4.36). Find a general expression in terms of angle θ for the ratio of the two normal stresses, $k = \sigma_y/\sigma_x$, such that there is no shear deformation in the lamina. Obtain first an exact expression, and then an approximate one for a high-stiffness composite, with $E_2 = 2G_{12}$.

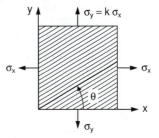

Fig. P4.36

4.37 For a unidirectional lamina loaded in pure shear τ_0 at 45° with the fiber direction, obtain expressions for the three strain components ε_x, ε_y, and γ_s as a function of the basic engineering properties E_1, E_2, G_{12}, ν_{12} (and/or ν_{21}), and the shear stress τ_0 (Fig. P4.37). Obtain first the exact relations and then the approximate ones for a high-stiffness composite.

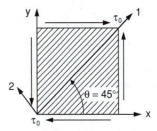

Fig. P4.37

4.38 An off-axis unidirectional lamina is loaded as shown in Fig. P4.38, and the strain ε_x in the x-direction is measured with a strain gage.

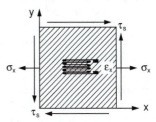

Fig. P4.38

(a) Find an expression for the shear coupling coefficient η_{sx} in terms of E_x, G_{xy}, σ_x, τ_s, and the measured strain ε_x.

(b) Determine the shear strain γ_s for the following values:

$$\varepsilon_x = 2 \times 10^{-3}$$
$$E_x = 58.7 \text{ GPa } (8.5 \text{ Msi})$$
$$G_{xy} = 9.7 \text{ GPa } (1.4 \text{ Msi})$$
$$\sigma_x = 193 \text{ MPa } (28 \text{ ksi})$$
$$\tau_s = 48.3 \text{ MPa } (7 \text{ ksi})$$

4.39 An off-axis unidirectional lamina is loaded as shown in Fig. P4.39 under uniaxial tension and in-plane shear at an angle to the fiber direction. Express the normal strains ε_x and ε_y in terms of the engineering properties (E_x, E_y, G_{xy}, ν_{xy}, ν_{yx}, η_{xs}, η_{ys}, η_{sx}, η_{sy}). What relation must these engineering properties satisfy, so that for a certain ratio of τ_s/σ_x the material behaves as an infinitely rigid one, that is, $\varepsilon_x = \varepsilon_y = 0$? What is the stress ratio $k = \tau_s/\sigma_x$ in that case?

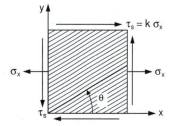

Fig. P4.39

4.40 For the same lamina and loading of Problem 4.39:

(a) Determine the ratio $k = \tau_s/\sigma_x$ in terms of the off-axis engineering constants (E_x, E_y, G_{xy}, ν_{xy}, ν_{yx}, η_{xs}, η_{ys}, η_{sx}, η_{sy}) for which the shear strain $\gamma_s = 0$.

(b) What relationship must the engineering constants satisfy in order that $\varepsilon_y = 0$ for the above loading ratio?

(c) Under what condition is $\varepsilon_x = 0$ for the same loading ratio?

4.41 An off-axis unidirectional lamina is loaded under uniaxial tension and in-plane shear of equal magnitude ($\sigma_x = \tau_s = \sigma_0$) (Fig. P4.41). Determine the fiber orientation θ for which there is no shear deformation along the principal material axes.

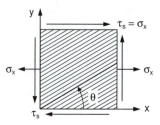

Fig. P4.41

4.42 A unidirectional lamina is loaded as shown in Fig. P4.42, first under biaxial normal loading σ_x and σ_y and then under pure shear τ_s, both at the same angle θ to the fiber direction. The strains produced by the first (biaxial) loading are ε_x^A, ε_y^A, and γ_s^A; the strains produced by the second (shear) loading are ε_x^S, ε_y^S, and γ_s^S. Develop expressions for ε_y^A and γ_s^A in terms of σ_x, σ_y, ε_x^A, τ_s, ε_x^S, ε_y^S, E_x, and E_y.

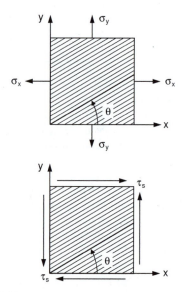

Fig. P4.42

4.43 Two cylindrical pressure vessels made of the same unidirectional lamina but with different fiber orientations were loaded as shown in Fig. P4.43 and gave the following strain readings:

Cylinder A
    Diameter: $D = 100$ mm
    Thickness: $h = 5$ mm
    Hoop wound (i.e., fibers in the circumferential direction)

Internal pressure: $p = 20$ MPa
Measured strains: $\varepsilon_x = 8 \times 10^{-3}$,
                           $\varepsilon_y = \varepsilon_\theta = 2.75 \times 10^{-3}$

Cylinder B
   Diameter: $D = 100$ mm
   Thickness: $h = 5$ mm
   Helically wound at $\theta = 45°$
   Axial load: $P = 147$ kN
   Measured strain: $\varepsilon_x = 7 \times 10^{-3}$

Assuming $\nu_{12} = 0.3$, determine lamina moduli E_1, E_2, and G_{12}.

Cylinder A

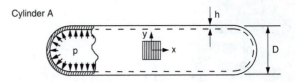

Cylinder B

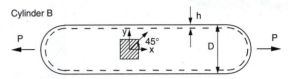

Fig. P4.43

5 Strength of Unidirectional Lamina—Micromechanics

5.1 INTRODUCTION

In the preceding chapters, the elastic behavior of a lamina was discussed from the microscopic and macroscopic points of view. In the case of failure phenomena and strength of a lamina, it is helpful and important to understand the underlying failure mechanisms and processes within and between the constituents of the composite and their effect on the ultimate macroscopic behavior (see Section 2.6). The failure mechanisms and processes on a micromechanical scale vary with the type of loading and are intimately related to the properties of the constituents, that is, fiber, matrix, and interface/interphase. These micromechanical processes and the macroscopic strength predictions based on them are discussed below for various types of loading.

5.2 LONGITUDINAL TENSION—FAILURE MECHANISMS AND STRENGTH

Under longitudinal tension, the phase with the lower ultimate strain will fail first. For perfectly bonded fibers, the average longitudinal stress in the composite, σ_1, is given by the rule of mixtures as

$$\sigma_1 = \sigma_f V_f + \sigma_m V_m \tag{5.1}$$

where

σ_f, σ_m = average longitudinal stresses in the fiber and matrix, respectively

V_f, V_m = fiber and matrix volume ratios, respectively

Under the simple deterministic assumption of uniform strengths, two cases are distinguished depending on the relative magnitudes of the ultimate tensile strains of the constituents.

In the case in which the ultimate tensile strain of the fiber is lower than that of the matrix, that is, when

$$\varepsilon_{ft}^{u} < \varepsilon_{mt}^{u} \tag{5.2}$$

the composite will fail when its longitudinal strain reaches the ultimate tensile strain in the fiber (Fig. 5.1). Then, the longitudinal tensile strength of the composite can be approximated by the relation

$$F_{1t} \cong F_{ft}V_f + \sigma_m'V_m \tag{5.3}$$

where

F_{1t} = longitudinal composite tensile strength

F_{ft} = longitudinal fiber tensile strength

σ_m' = average longitudinal matrix stress when ultimate fiber strain is reached

Assuming linear elastic behavior for the constituents, Eq. (5.3) is written as

$$F_{1t} \cong F_{ft}V_f + E_m\varepsilon_{ft}^{u}V_m = F_{ft}\left(V_f + V_m\frac{E_m}{E_f}\right) \tag{5.4}$$

For composites with very stiff fibers, that is, when $E_f \gg E_m$, and for reasonable values of V_f, the above relation can be further simplified as

$$F_{1t} \cong F_{ft}V_f \tag{5.5}$$

When the ultimate tensile strain of the matrix is lower than that of the fiber, that is, when

$$\varepsilon_{mt}^{u} < \varepsilon_{ft}^{u} \tag{5.6}$$

the composite fails when its longitudinal strain reaches the ultimate tensile strain of the matrix (Fig. 5.2). Then, the longitudinal tensile strength of the composite can be approximated by the relation

$$F_{1t} \cong \sigma_f'V_f + F_{mt}V_m \tag{5.7}$$

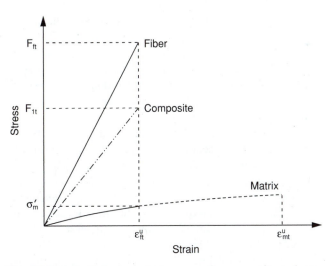

Fig. 5.1 Longitudinal stress-strain curves for composite and constituents for the case of fiber-dominated strength ($\varepsilon_{ft}^{u} < \varepsilon_{mt}^{u}$).

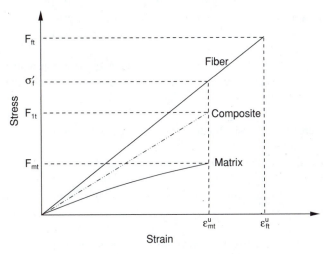

Fig. 5.2 Longitudinal stress-strain curves of composite and constituents for the case of matrix-dominated strength ($\varepsilon_{mt}^{u} < \varepsilon_{ft}^{u}$).

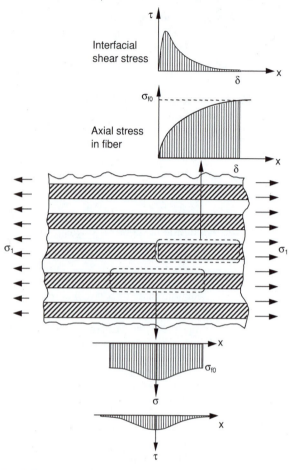

Fig. 5.3 Local stress distributions around a fiber break in a unidirectional composite under longitudinal tension.

which can be further approximated as

$$F_{1t} \cong F_{mt}\left(V_f \frac{E_f}{E_m} + V_m\right) \qquad (5.8)$$

where

F_{mt} = matrix tensile strength

σ'_f = longitudinal fiber stress when ultimate matrix strain is reached

The results above do not take into consideration the statistical distribution of fiber and matrix strengths. In the case of fiber-dominated strength, for example, fiber strength varies from point to point and from fiber to fiber. Not all fibers fail simultaneously, but isolated single fiber breaks (singlets) occur at weak points. A nonuniform state of stress is developed around the fiber break (Fig. 5.3).[1] An interfacial shear stress results with a high peak near the fiber break and helps transfer the stress to the broken fiber. The stress transmitted by the fiber is zero at the break but increases gradually to the far-field value at some characteristic distance, δ, from the break. This is roughly the same distance at which the interfacial shear stress drops to zero. The effect of the fiber break on adjacent fibers is a local increase in both fiber stress and interfacial shear stress. The net effect of a single fiber break is to reduce the load-carrying fiber length by the ineffective length 2δ.

Depending on the properties of the constituents, these initial fiber breaks produce different types of failure in their vicinity (Fig. 5.4).[2] These failure mechanisms take the following forms:

1. matrix cracking transverse to the tensile loading in composites with a brittle matrix and a relatively strong interface
2. fiber/matrix debonding in the case of a relatively weak interface and/or relatively high ultimate fiber strain
3. conical shear fractures in the matrix in the case of a relatively ductile matrix and a strong interface

In most cases the damage is localized and arrested by the adjacent fibers. The net effect of this localized damage is to increase the ineffective length of the fiber.

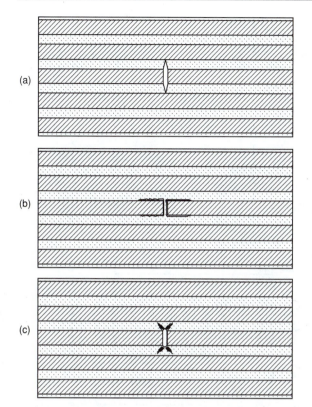

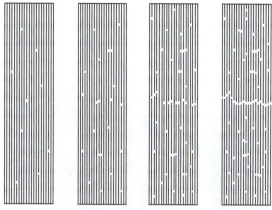

Fig. 5.5 Failure sequence in a unidirectional composite with fiber-dominated strength under longitudinal tensile loading.

Fig. 5.4 Failure mechanisms around a single fiber break in a unidirectional composite under longitudinal tension: (a) transverse matrix cracking for brittle matrix and relatively strong interface, (b) fiber/matrix debonding for relatively weak interface and/or relatively high fiber ultimate strain, and (c) conical shear fractures in relatively ductile matrix and strong interface.

As the load increases, the single fiber breaks increase in density and interact to produce adjacent fiber breaks (e.g., doublets, triplets) (Fig. 5.5). These localized failures interact and eventually coalesce to produce catastrophic failure. The exact sequence of events and the final failure pattern vary with constituent properties and the fiber volume ratio.[1,3] Typical failure patterns under longitudinal tension are shown in Fig. 5.6 for two composite materials, boron/epoxy and S-glass/epoxy. The boron/epoxy shows more brittle failure and limited fiber/matrix debonding, whereas the glass/epoxy shows extensive interfacial debonding associated with the relatively high ultimate strain of the glass fiber. These results can be incorporated into a statistical analysis to yield the overall composite strength in terms of the fiber volume ratio and the parameters of the statistical distribution of fiber strengths.[1]

In the case of brittle-matrix composites, such as ceramic-matrix composites, the failure strain of the matrix is usually lower than that of the fibers, and damage initiates with the development of multiple matrix cracks analogous to the fiber breaks discussed before (Fig. 5.7). These cracks produce local stress distributions with high interfacial shear stresses and increased tensile stress in adjacent fibers. These cracks are accompanied or followed by fiber/matrix debonding and fiber breaks. A typical fractograph of such a failure in a silicon carbide/glass-ceramic composite is shown in Fig. 5.8. It shows clearly the transverse matrix crack, fiber breaks, and fiber pullout. Analysis of stresses and failure mechanisms of such composites provides predictions of longitudinal tensile strength as a function of material and geometric parameters.[4]

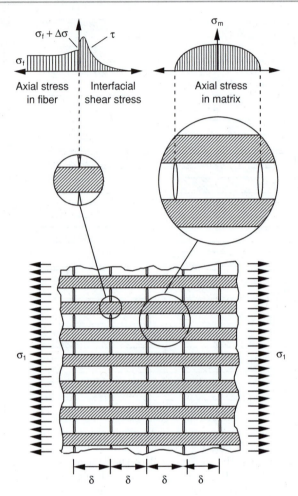

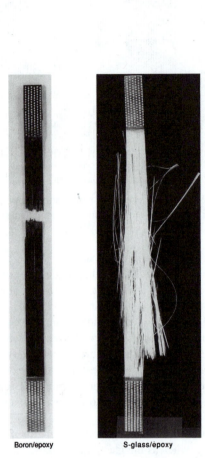

Boron/epoxy   S-glass/epoxy

Fig. 5.6 Typical failure patterns of uni-directional composites under longitudinal tension.

Fig. 5.7 Matrix cracking and local stress distributions in a unidirectional brittle-matrix composite under longitudinal tension.

5.3 LONGITUDINAL TENSION—INEFFECTIVE FIBER LENGTH

In the case of a composite with a ductile matrix discussed before, a fiber breaks at a weak point and the fiber stress σ_f increases exponentially from zero at the break to the far-field value σ_{fo} at some distance δ, called *ineffective* or *transfer* length. This distance can be defined as that at which the fiber stress reaches a certain fraction k of the far-field fiber stress:

$$\frac{\sigma_f}{\sigma_{fo}} = k \tag{5.9}$$

This fraction k is usually taken to be 0.9 (Fig. 5.9).

One simple way of calculating δ for a ductile matrix is by assuming the interfacial shear stress to be uniform over the length δ. Then, from equilibrium it follows that

$$\frac{\delta}{d} = \frac{1}{4}\left(\frac{\sigma_{fo}}{\tau_o}\right)$$
(5.10)

where

$$d = \text{fiber diameter}$$

$$\tau_o = \text{average shear stress}$$

This relation gives an upper bound for the transfer length if τ_o is taken as the average interfacial shear strength.

According to Rosen et al.,[5] the fiber stress variation near the fiber break is given by

$$\sigma_f(x) = \sigma_{fo}(1 - e^{-\gamma x})$$
(5.11)

where

$$x = \text{distance from the fiber break}$$

$$\gamma^2 = \frac{G_m}{E_f}\left(\frac{V_f^{1/2}}{1 - V_f^{1/2}}\right)\frac{1}{r^2}$$
(5.12)

$$r = \text{fiber radius}$$

By setting $x = \delta$ for $k = 0.9$, we obtain

$$\frac{\delta}{d} = -\frac{\log(1 - k)}{\gamma d} = 1.15\left[\frac{E_f}{G_m}\left(\frac{1 - V_f^{1/2}}{V_f^{1/2}}\right)\right]^{1/2}$$
(5.13)

The interfacial shear stress τ_i is obtained from Eq. (5.11) using the equilibrium relation

$$\frac{d\sigma_f}{dx} = -\frac{2\tau_i}{r}$$
(5.14)

Thus,

$$\frac{\tau_i(x)}{\sigma_{fo}} = -\frac{1}{2}\gamma r e^{-\gamma x} = -\frac{1}{2}\left[\frac{G_m}{E_f}\left(\frac{V_f^{1/2}}{1 - V_f^{1/2}}\right)\right]^{1/2} e^{-\gamma x}$$
(5.15)

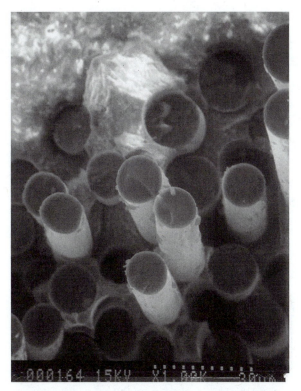

Fig. 5.8 Scanning electron microscope (SEM) fractograph of longitudinal tensile failure in silicon carbide/glass-ceramic (SiC/CAS) composite.

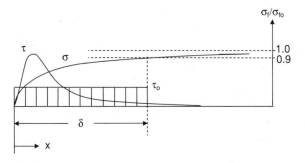

Fig. 5.9 Fiber stress and interfacial shear stress distribution near fiber break.

For a typical carbon/epoxy material with $G_m = 1.27$ GPa (185 ksi), $E_f = 235$ GPa (34 Msi), and $V_f = 0.65$,

$$\frac{|\tau_i(0)|}{\sigma_{fo}} = 0.075$$

For a fiber stress at failure of $\sigma_{fo} = F_{ft} = 3100$ MPa (450 ksi), the maximum interfacial shear stress would be $(\tau_i)_{max} = 232$ MPa (33.7 ksi), a value much higher than the matrix or interfacial shear strength. Therefore, interface debonding will inevitably follow a fiber break.

The transfer length from a fiber break or the end of a short fiber can also be obtained from Cox's shear lag analysis discussed in Chapter 3.[6] The variation of fiber stress along a fiber of length l given by Eq. (3.54) is

$$\sigma_f(x) = E_f\varepsilon_1\left[1 - \frac{\cosh\beta x}{\cosh\frac{\beta l}{2}}\right] \cong \sigma_{fo}\left[1 - \frac{\cosh\beta x}{\cosh\frac{\beta l}{2}}\right] \tag{5.16}$$

where

$$\beta = \frac{1}{r}\left[\frac{2G_m}{E_f\log\frac{r_0}{r}}\right]^{1/2}$$

ε_1 = average longitudinal strain in composite and fiber

and the origin of the x-axis is at the center of the fiber as shown in Fig. 3.15. The ratio k of the fiber stress at a distance δ from the end of the fiber to the far-field fiber stress for a fiber of length l is

$$k = \frac{\sigma_{f(x=l/2-\delta)}}{\sigma_{fo}} = 1 - \frac{\cosh\beta(\frac{l}{2} - \delta)}{\cosh\frac{\beta l}{2}} \tag{5.17}$$

For $k = 0.9$ and for large l, when $\tanh(\beta l/2) \cong 1$, the transfer length is obtained as

$$\frac{\delta}{d} \cong 1.15\left[\frac{E_f}{2G_m}\log\frac{r_0}{r}\right]^{1/2} \tag{5.18}$$

The radius r_0 in the above equation is the radius of the representative cylindrical element, $r_0 = r/\sqrt{V_f}$, as shown in Fig. 3.13.

The maximum fiber stress, which occurs at the center of the fiber ($x = 0$), is

$$(\sigma_f)_{max} = E_f\varepsilon_1\left[1 - \frac{1}{\cosh\frac{\beta l}{2}}\right] \tag{5.19}$$

The maximum interfacial shear stress occurs at the end of the fiber ($x = l/2$):

$$(\tau_i)_{max} = \tau_{(x=l/2)} = \frac{\beta r}{2} E_f \varepsilon_1 \tanh \frac{\beta l}{2} \qquad (5.20)$$

It is seen from Eqs. (5.13), (5.15), (5.18), and (5.20) that as the modulus ratio E_f/G_m increases, the ineffective length δ increases and the maximum interfacial shear stress decreases. The reinforcing efficiency of a composite depends to a large extent on the strength of the interface. High interfacial shear stresses lead to several failure mechanisms, namely interfacial (shear) debonding, cohesive failure of the matrix, cohesive failure of the fiber, or shear yielding of the matrix.

5.4 LONGITUDINAL COMPRESSION

Failure mechanisms and strength of continuous-fiber composites under longitudinal compression have been studied by many investigators over the last four decades. Rosen[7] in 1965 recognized that longitudinal compressive failure is associated with microbuckling of the fibers within the matrix (Fig. 5.10). He analyzed two idealized microbuckling patterns and gave predictions for the longitudinal compressive strength. For low values of fiber volume ratio, the extensional or out-of-phase mode of microbuckling is predicted with a corresponding compressive strength[7]

$$F_{1c} \cong 2V_f \left[\frac{E_m E_f V_f}{3(1 - V_f)} \right]^{1/2} \qquad (5.21)$$

For higher values of V_f, the shear or in-phase mode is predicted with a compressive strength[7]

$$F_{1c} \cong \frac{G_m}{1 - V_f} \qquad (5.22)$$

The fiber microbuckling failure mechanism was further studied by Greszczuk,[8] Wang,[9] and Hahn and Williams.[10]

Flexural stresses in a fiber due to in-phase buckling lead to the formation of kink zones, which can cause pronounced deformation in ductile fibers, such as aramid, or fracture planes in brittle fibers, such as carbon (Fig. 5.11). The mechanism of kink band formation was studied by Argon,[11] Berg and Salama,[12] Weaver and Williams,[13] and Evans and Adler.[14] Budiansky,[15] and Budiansky and Fleck[16] considered plastic deformation during kinking. More recently the combined buckling/kinking failure mechanism was analyzed by Steif,[17] Chung and Weitsman,[18] and Schapery.[19] Figure 5.12 shows kink band formation in a carbon/epoxy composite loaded in longitudinal

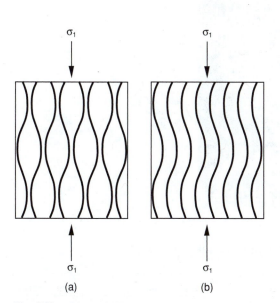

Fig. 5.10 Microbuckling modes in a unidirectional composite under longitudinal compression: (a) out-of-phase or extensional mode and (b) in-phase or shear mode.

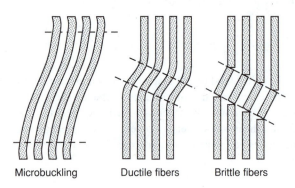

Microbuckling   Ductile fibers   Brittle fibers

Fig. 5.11 Microbuckling leading to formation of kink zones with excessive deformation or fracture planes for ductile or brittle fibers, respectively.

compression. Figure 5.13 shows fractographs of a carbon/epoxy composite after failure under longitudinal compression.[20] Figure 5.13a shows the stepped nature of the macroscopic failure surface resulting from fiber microbuckling, kinking, and fiber fractures. Figure 5.13b shows an enlarged view of the same fracture surface showing cracking of the fibers in the direction normal to the axis of microbuckling.

Compressive strength predictions from Eqs. (5.21) and (5.22) are much higher than experimentally measured values, as illustrated in Table 5.1 for three typical composite materials.

The difference between theoretical and experimental results is attributed to preexisting fiber misalignment, which reduces the strength appreciably.

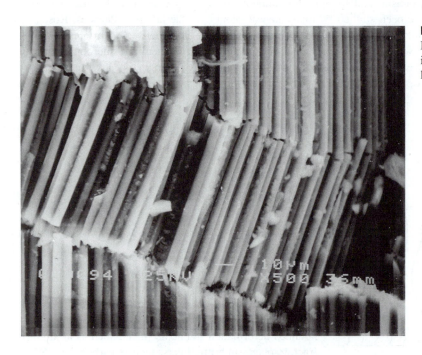

Fig. 5.12 Kink band in unidirectional IM6/3501-6 carbon/epoxy composite under longitudinal compressive loading.

TABLE 5.1 Predicted and Measured Longitudinal Compressive Strengths, F_{1c}, MPa (ksi)

Material	Extensional Mode	Shear Mode	Experimental
Glass/epoxy	8700 (1260)	2200 (320)	690 (100)
Carbon/epoxy	22,800 (3300)	2900 (420)	1730 (250)
Kevlar/epoxy	13,200 (1900)	2900 (420)	340 (50)

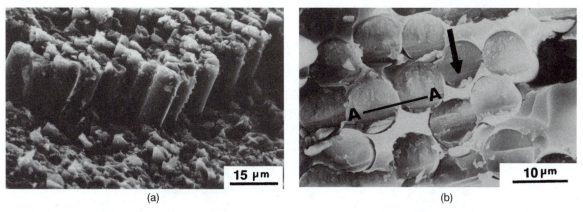

Fig. 5.13 Fractographs of carbon/epoxy composite under longitudinal compression:[20] (a) stepped microscopic fracture surface resulting from fiber microbuckling and kinking and (b) arrow indicating direction of crack propagation normal to microbuckling axis A-A.

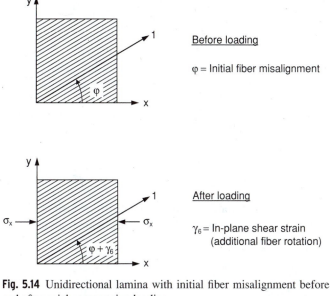

Before loading

φ = Initial fiber misalignment

After loading

γ_6 = In-plane shear strain (additional fiber rotation)

Fig. 5.14 Unidirectional lamina with initial fiber misalignment before and after axial compressive loading.

A simple model for predicting longitudinal compressive strength, taking into consideration the initial fiber misalignment, is described below.[21,22] Consider a unidirectional lamina element with an initial fiber misalignment φ, loaded in compression in the x-direction (Fig. 5.14). Because of the fiber misalignment, a two-dimensional state of stress is developed with respect to the principal material axes. The shear stress component produces a shear strain γ_6, which effectively increases the fiber misalignment to $(\varphi + \gamma_6)$. Under the applied axial stress σ_x, the stresses referred to the principal material axes obtained by stress transformation are

$$\sigma_1 = \sigma_x \cos^2(\varphi + \gamma_6)$$
$$\sigma_2 = \sigma_x \sin^2(\varphi + \gamma_6) \qquad (5.23)$$
$$\tau_6 = -\sigma_x \sin(\varphi + \gamma_6)\cos(\varphi + \gamma_6)$$

For small values of φ and γ_6

$$\sigma_1 \cong \sigma_x$$
$$\sigma_2 \cong 0 \qquad (5.24)$$
$$\tau_6 \cong -\sigma_x(\varphi + \gamma_6)$$

Then,

$$\sigma_x \cong \frac{-\tau_6}{\varphi + \gamma_6} \qquad (5.25)$$

The limiting value of $\sigma_x \cong \sigma_1$ is the longitudinal compressive strength F_{1c}. The necessary condition for a maximum σ_x is

$$\frac{\partial \sigma_x}{\partial \gamma_6} = 0$$

or, from Eq. (5.25)

$$\left[-\frac{\partial \tau_6}{\partial \gamma_6}(\varphi + \gamma_6) + \tau_6 \right] \frac{1}{(\varphi + \gamma_6)^2} = 0$$

or

$$\frac{\partial \tau_6}{\partial \gamma_6} = \frac{\tau_6}{\varphi + \gamma_6} \qquad (5.26)$$

The shear stress τ_6 is a nonlinear function of the shear strain γ_6 as obtained from experimental characterization of the material (Fig. 5.15). The solution of Eq. (5.26) corresponds to the point (τ^*, γ^*) on the τ_6 versus γ_6 curve where the tangent is equal to $\tau^*/(\varphi + \gamma^*)$, that is,

$$\frac{\partial \tau_6}{\partial \gamma_6} = \frac{\tau_6}{\varphi + \gamma_6} = \frac{\tau^*}{\varphi + \gamma^*} \qquad (5.27)$$

The corresponding value of $|\sigma_x|$, which is a maximum, is equal to the longitudinal compressive strength

$$F_{1c} = |\sigma_x|_{max} = \frac{\tau^*}{\varphi + \gamma^*} \qquad (5.28)$$

According to this prediction, the longitudinal compressive strength depends only on the in-plane shear stress-strain curve and the initial fiber misalignment. Both of these characteristics need to be obtained experimentally for a given material system. Figure 5.16 shows the variation of the predicted compressive strength F_{1c} with fiber misalignment for a carbon/epoxy composite

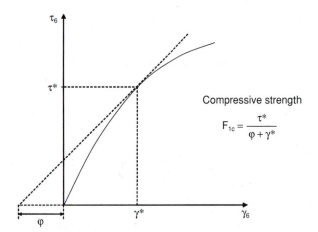

Compressive strength

$$F_{1c} = \frac{\tau^*}{\varphi + \gamma^*}$$

Fig. 5.15 Illustration of graphical calculation of longitudinal compressive strength.

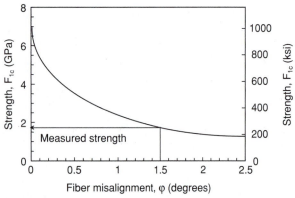

Fig. 5.16 Predicted longitudinal compressive strength of carbon/epoxy composite (IM6G/3501-6) as a function of initial fiber misalignment.

(IM6G/3501-6). The experimentally measured compressive strength of this material, $F_{1c} = 1725$ MPa (250 ksi), corresponds to a fiber misalignment of $\varphi \cong 1.5°$.

If the initial misalignment is much larger than the shear strain, $\varphi \gg \gamma$, Eq. (5.28) is reduced to

$$F_{1c} \cong \frac{(\tau_6)_{max}}{\varphi} = \frac{F_6}{\varphi} \qquad (5.29)$$

where F_6 is the in-plane shear strength of the material. The above solution is identical to Argon's prediction.[11] The same result is obtained by using the maximum stress criterion to be discussed in Chapter 6.

When the initial misalignment is very small compared to the induced shear strain, $\varphi \ll \gamma_6$, Eq. (5.28) is reduced to

$$F_{1c} \cong \frac{(\tau_6)_{max}}{(\gamma_6)_{max}} \cong (G_{12})_{sec} \qquad (5.30)$$

where $(G_{12})_{sec}$ is the secant shear modulus of the τ_6 versus γ_6 curve at failure. If this curve is assumed to be linear, then

$$F_{1c} \cong G_{12} \qquad (5.31)$$

which is Rosen's prediction. Using the series model for G_{12} and assuming very stiff fibers, $G_{12f} \gg G_m$, we can rewrite Eq. (5.31) as

$$F_{1c} \cong \frac{G_m}{1 - V_f} \qquad (5.32)$$

which is identical to Eq. (5.22).

As the fiber volume ratio increases, debonding precedes in-phase microbuckling, and failure results from the three-dimensional state of stress. At the highest values of V_f for well-aligned fibers, one may encounter pure compressive failure, which can be related to shear failure of the fibers. For the case of high fiber volume ratios, the shear failure mode governed by the shear strength of the fiber is illustrated in Fig. 5.17. The predicted strength based on this mode and a maximum stress criterion for the fibers is

$$F_{1c} = 2F_{fs} \left[V_f + (1 - V_f) \frac{E_m}{E_f} \right] \qquad (5.33)$$

where F_{fs} is the shear strength of the fiber.

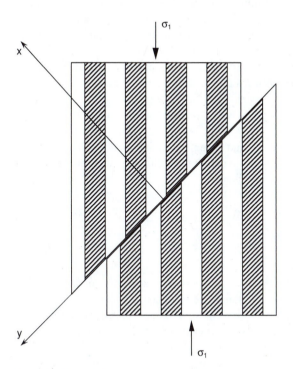

Fig. 5.17 Shear failure mode of a unidirectional composite with high fiber volume ratio under longitudinal compression.

5.5 TRANSVERSE TENSION

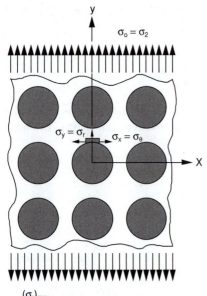

$$k_\sigma = \frac{(\sigma_r)_{max}}{\sigma_2} = \text{Stress concentration factor}$$

Fig. 5.18 Local stresses in transversely loaded unidirectional composite.

The most critical loading of a unidirectional composite is transverse tensile loading. This type of loading results in high stress and strain concentrations in the matrix and interface/interphase. Stress distributions around the fiber can be obtained analytically by finite element, finite difference, complex variable, or boundary element methods and experimentally by means of two- and three-dimensional photoelastic models. The critical stresses and strains usually occur at the fiber/matrix interface.

Superimposed on the load-induced stresses are stresses due to matrix shrinkage during cure, and hygrothermal stresses due to variations in temperature and moisture concentration of the material. Local failure mechanisms include matrix cracking, interface debonding, interphase failure, and fiber splitting.

The peak stress in the matrix for a square array is the normal stress at the interface along the loading direction (Fig. 5.18). The stress concentration factor, defined as the ratio of this peak stress to the applied average stress, is shown in Fig. 5.19 as a function of fiber volume ratio for three typical composites. Results shown were obtained by finite difference[23] and photoelastic[24] methods. The boron/epoxy material with a fiber to matrix modulus ratio of 120 represents one of the most severe cases. In most other cases, the values of stress concentration are lower.

The stress concentration factor can also be calculated approximately by considering a thin slice through the centers of the fibers in the loading direction as a series model with an average strain equal to the composite average strain ε_2. Then, for a square array and using the series model relation for E_2, we obtain the following stress concentration factor:

$$k_\sigma = \frac{(\sigma_r)_{max}}{\sigma_2} = \frac{1 - V_f(1 - E_m/E_f)}{1 - (4V_f/\pi)^{1/2}(1 - E_m/E_f)}$$
(5.34)

A more characteristic quantity for a transversely loaded composite is the strain concentration factor that is related to the stress concentration factor as follows:[24]

$$k_\varepsilon = \frac{\varepsilon_{max}}{\varepsilon_2} \cong k_\sigma \left(\frac{E_2}{E_m}\right)\frac{(1 + v_m)(1 - 2v_m)}{1 - v_m}$$
(5.35)

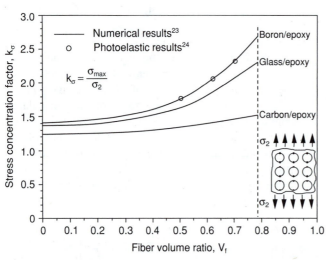

Fig. 5.19 Stress concentration factor in matrix of unidirectional composites with square fiber array under transverse tension.

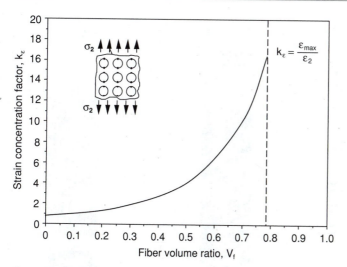

Fig. 5.20 Strain concentration factor in matrix of unidirectional boron/epoxy composite under transverse tension.

where ε_{max} and ε_2 are the maximum and average strains, respectively, and v_m the matrix Poisson's ratio. In the expression above, it was assumed that the fibers are much stiffer than, and perfectly bonded to, the matrix. The variation of the strain concentration factor with fiber volume ratio for a boron/epoxy composite is shown in Fig. 5.20. It is shown how it increases sharply for fiber volume ratios above 0.5.

A strain concentration factor can also be obtained as follows from the relation of Eq. (5.34) and is the same as that given by Kies[25] based on a simple deformation analysis of a unit cell in a square array:

$$k_\varepsilon = \frac{1}{1 - (4V_f/\pi)^{1/2}(1 - E_m/E_f)} \quad (5.36)$$

In predicting failure of a transversely loaded composite, the residual stresses and strains due to curing of the matrix, or thermal stresses and strains due to thermal expansion mismatch, must be taken into account. Assuming a maximum tensile stress or strain failure criterion, linear elastic behavior to failure for the matrix, and very stiff perfectly bonded fibers, one can predict the transverse tensile strength for a unidirectional composite,

$$F_{2t} = \frac{1}{k_\sigma}(F_{mt} - \sigma_{rm}) \quad (5.37)$$

for the maximum tensile stress criterion, and

$$F_{2t} = \frac{1 - v_m}{k_\sigma(1 + v_m)(1 - 2v_m)}(F_{mt} - \varepsilon_{rm}E_m) \quad (5.38)$$

for the maximum tensile strain criterion, where σ_{rm} and ε_{rm} are the radial (maximum) residual stress and residual strain, respectively.

The above prediction was corroborated experimentally by conducting transverse tensile tests on a glass/epoxy composite, for which the following properties were measured:[26]

$$E_2 = 12.6 \text{ GPa (1.83 Msi)}$$

$$V_f = 0.50$$

$$k_\sigma = 2.0 \text{ (measured photoelastically)}$$

$$F_{2t} = 47.3 \text{ MPa (6850 psi)}$$

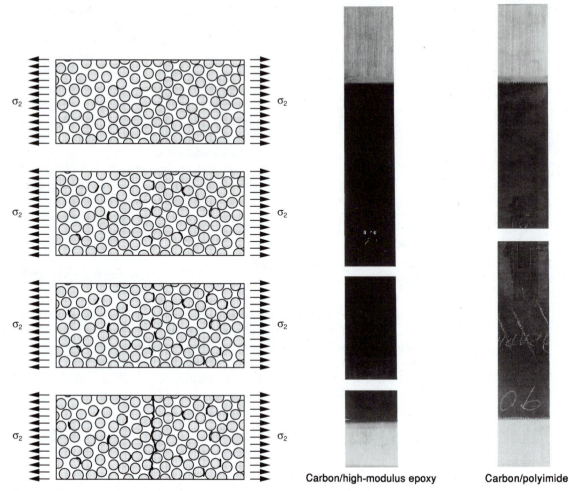

Fig. 5.21 Progressive microcracking leading to ultimate failure in unidirectional composite under transverse tension.

Fig. 5.22 Typical failure patterns of unidirectional composites under transverse tension.

Carbon/high-modulus epoxy   Carbon/polyimide

The strength predicted by Eq. (5.38) is

$$F_{2t} = 41.7 \text{ MPa (6050 psi)}$$

and is lower than the measured one. This is due to the fact that the predicted value is for local failure initiation, whereas the measured value is the ultimate strength, which must be higher.

In general, the micromechanical predictions above are based on a local deterministic failure criterion at a point. In reality, as in the case of longitudinal tensile loading, failure takes the form of isolated interfacial microcracks increasing in number with loading and finally coalescing into a catastrophic macrocrack (Fig. 5.21). Typical failure patterns under transverse tension are shown in Fig. 5.22 for two composite materials, carbon/

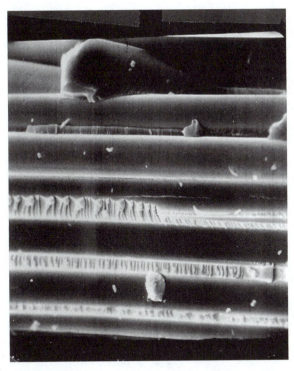

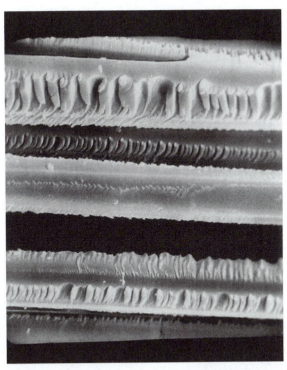

Fig. 5.23 Scanning electron microscope (SEM) fractograph of transverse tensile failure in E-glass/epoxy composite (brittle matrix).[27]

Fig. 5.24 Scanning electron microscope (SEM) fractograph of transverse tensile failure in E-glass/epoxy composite (ductile matrix).[27]

high-modulus epoxy and carbon/polyimide. Typical fractographs of transverse tensile failures are shown in Figs. 5.23, 5.24, and 5.25.

5.6 TRANSVERSE COMPRESSION

Under transverse compression a unidirectional composite may fail under a number of failure mechanisms. The high compressive stress at the interface may cause compressive or shear failure in the matrix and/or fiber crushing. The predicted composite strength for this failure mechanism is

$$F_{2c} = \frac{F_{mc} + \sigma_{rm}}{k_\sigma} \tag{5.39}$$

where F_{mc} is the compressive strength of the matrix and σ_{rm} is the maximum residual radial stress at the interface. High interfacial shear stresses may cause matrix shear failure and/or debonding, leading to an overall shear failure mode, as illustrated schematically in Fig. 5.26.

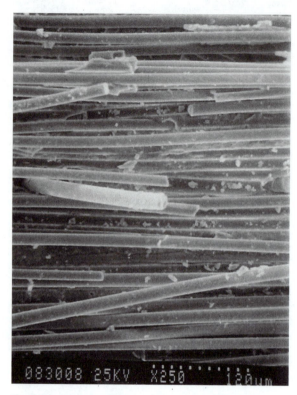

Fig. 5.25 Scanning electron microscope (SEM) fractograph of transverse tensile failure in silicon carbide/glass-ceramic (SiC/CAS) composite.

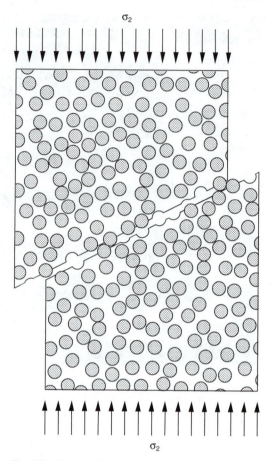

Fig. 5.26 Shear failure mode in unidirectional composite under transverse compression.

5.7 IN-PLANE SHEAR

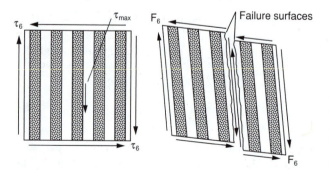

Fig. 5.27 Failure mode of unidirectional composite under in-plane shear.

Under in-plane shear, as illustrated in Fig. 5.27, a high shear stress concentration develops at the fiber/matrix interface. The variation of the shear stress concentration factor with material and fiber volume ratio has been obtained by a finite difference procedure.[28] The high shear stress at the interface can cause shear failure in the matrix and/or fiber/matrix debonding (Fig. 5.27). Figure 5.28 shows the failure pattern of a carbon/epoxy composite under in-plane shear. The in-plane shear strength of the composite based on matrix shear failure can be predicted as

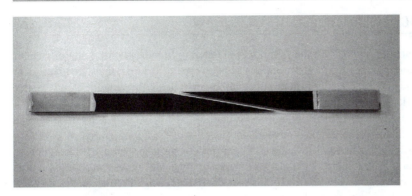

Fig. 5.28 Failure pattern of unidirectional carbon/epoxy composite under in-plane shear (10° off-axis specimen loaded in axial tension).

$$F_6 = \frac{F_{ms}}{k_\tau} \tag{5.40}$$

where F_{ms} is the matrix shear strength and k_τ the shear stress concentration factor. An approximate expression for the shear stress concentration factor analogous to Eq. (5.34) is

$$k_\tau = \frac{1 - V_f(1 - G_m/G_{12f})}{1 - (4V_f/\pi)^{1/2}(1 - G_m/G_{12f})} \tag{5.41}$$

5.8 OUT-OF-PLANE LOADING

Out-of-plane loading includes normal loading σ_3, which can be tensile or compressive, and shear loadings τ_4 (τ_{23}) and τ_5 (τ_{13}). For transversely isotropic materials with the 2-3 plane as the plane of isotropy, the out-of-plane normal loading σ_3 is equivalent to the transverse in-plane loading σ_2. The behavior of the material is the same as that discussed before in Sections 5.5 and 5.6 for transverse tension and transverse compression, respectively. Similarly, the behavior under out-of-plane shear loading τ_5 (τ_{13}) is equivalent to that under in-plane shear τ_6 (τ_{12}), discussed in Section 5.7.

The out-of-plane shear loading τ_4 (τ_{23}) is equivalent to equal biaxial tension/compression $\sigma_{2'} = -\sigma_{3'} = \tau_4$ along axes 2' and 3', which are inclined at 45° to the 2- and 3-axes, respectively (Fig. 5.29). The most likely failure mode is matrix cracking normal to the 2'-direction because the material is usually much weaker in transverse tension than in transverse compression. The stress $\sigma_{2'} = \tau_4$ at failure would be close, but not necessarily equal, to the failure stress under uniaxial transverse tension σ_2, because of tension/compression ($\sigma_{2'}$, $\sigma_{3'}$) interaction effects.

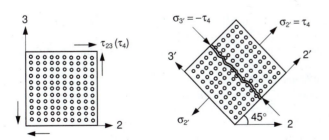

Fig. 5.29 Failure mode of unidirectional composite under out-of-plane shear, τ_4 (τ_{23}).

5.9 GENERAL MICROMECHANICS APPROACH

In this approach the overall lamina failure is predicted on the basis of the micromechanical state of stress in the constituents. For any general type of loading, the state of stress in a characteristic volume element is determined either analytically by dividing it into matrix and fiber subelements (Aboudi[29]) or numerically by finite element analysis (Gotsis et al.[30]). Appropriate failure criteria are applied for the matrix and fiber constituents to determine failure of the characteristic volume element and thereby of the entire lamina.

The preceding discussion shows how the various constituent characteristics influence failure in a composite lamina. The micromechanics analysis of failure can serve as a useful guide in selecting constituent materials and designing composites from the point of view of failure behavior. It can be seen, however, that the micromechanics approach to failure prediction is complex, based on assumptions and approximations, and not always reliable. At least, local failure initiation can be predicted, but the results cannot be easily or reliably extended to global failure prediction.

REFERENCES

1. B. W. Rosen, "Tensile Failure of Fibrous Composites," *AIAA J.*, Vol. 2, 1964, pp. 1985–1991.
2. J. Mullin, J. M. Berry, and A. Gatti, "Some Fundamental Fracture Mechanisms Applicable to Advanced Filament Reinforced Composites," *J. Composite Materials*, Vol. 2, No. 1, 1968, pp. 82–103.
3. M. G. Bader, "Tensile Strength of Uniaxial Composites," *Sci. and Eng. of Composite Materials*, Vol. 1, No. 1, 1988, pp. 1–11.
4. J.-W. Lee and I. M. Daniel, "Deformation and Failure of Longitudinally Loaded Brittle-Matrix Composites," in *Composite Materials: Testing and Design* (Tenth Volume), ASTM STP 1120, Glen C. Grimes, Ed., American Society for Testing and Materials, West Conshohocken, PA, 1992, pp. 204–221.
5. B. W. Rosen, N. F. Dow, and Z. Hashin, "Mechanical Properties of Fibrous Composites," NASA Report, CR-31, April 1964.
6. H. L. Cox, "The Elasticity and Strength of Paper and Other Fibrous Materials," *Brit. J. Appl. Phys.*, Vol. 3, 1952, pp. 72–79.
7. B. W. Rosen, "Mechanics of Composite Strengthening," Ch. 3, in *Fiber Composite Materials*, ASM, Metals Park, OH, 1965.
8. L. B. Greszczuk, "Microbuckling Failure of Circular Fiber-Reinforced Composites," *AIAA J.*, Vol. 13, 1975, pp. 1311–1318.
9. A. S. D. Wang, "A Non-linear Microbuckling Model Predicting the Compressive Strength of Unidirectional Composites," ASME Paper 78-WA/Aero-1, 1978.
10. H. T. Hahn and J. G. Williams, "Compression Failure Mechanisms in Unidirectional Composites," *Composite Materials: Testing and Design* (Seventh Conference), ASTM STP 893, American Soc. for Testing and Materials, West Conshohocken, PA, 1986, pp. 115–139.
11. A. S. Argon, "Fracture of Composites," *Treatise on Materials Science and Technology*, Vol. 1, H. Herman, Ed., Academic Press, New York, 1972, pp. 79–114.
12. C. A. Berg and M. Salama, "Fatigue of Graphite Fiber-Reinforced Epoxy in Compression," *Fibre Sci. and Tech.*, Vol. 6, 1973, pp. 79–118.
13. C. W. Weaver and J. G. Williams, "Deformation of a Carbon/Epoxy Composite Under Hydrostatic Pressure," *J. Mater. Sci.*, Vol. 10, 1975, pp. 1323–1333.

14. A. G. Evans and W. G. Adler, "Kinking as a Mode of Structural Degradation in Carbon Fiber Composites," *Acta Metall.*, Vol. 26, 1978, pp. 725–738.

15. B. Budiansky, "Micromechanics," *Computers and Structures*, Vol. 16, 1983, pp. 3–12.

16. B. Budiansky and N. A. Fleck, "Compressive Failure of Fibre Composites," *J. Mech. Phys. Solids*, Vol. 41, No. 1, 1993, pp. 183–211.

17. P. S. Steif, "A Model for Kinking in Fiber Composites; I. Fiber Breakage via Microbuckling; II. Kink Band Formation," *Int. J. Solids Struct.*, Vol. 26, 1990, pp. 549–558.

18. I. Chung and Y. J. Weitsman, "Model for the Micro-buckling/Micro-kinking Compressive Response of Fiber-reinforced Composites," *Mechanics USA 1994, Proc. of the Twelfth U.S. Nat. Congress of Applied Mechanics*, A. S. Kobayashi, Ed., ASME, 1994, pp. S256–S261.

19. R. A. Schapery, "Prediction of Compressive Strength and Kink Bands in Composites Using a Work Potential," *Int. J. Solids Struct.*, Vol. 32, (6/7), 1995, pp. 739–765.

20. L. Shikhmanter, I. Eldror, and B. Cina, "Fractography of Unidirectional CFRP Composites," *J. Materials Sci.*, Vol. 24, 1989, pp. 167–172.

21. J. G. Häberle and F. L. Matthews, "A Micromechanics Model for Compressive Failure of Unidirectional Fiber-Reinforced Plastics," *J. Composite Materials*, Vol. 28, No. 17, 1994, pp. 1618–1639.

22. I. M. Daniel, H.-M. Hsiao, and S.-C. Wooh, "Failure Mechanisms in Thick Composites Under Compressive Loading," *Composites, Part B*, Vol. 27B, 1996, pp. 543–552.

23. D. F. Adams and D. R. Doner, "Transverse Normal Loading of a Unidirectional Composite," *J. Composite Materials*, Vol. 1, 1967, pp. 152–164.

24. I. M. Daniel, "Photoelastic Investigation of Composites," in *Composite Materials*, Vol. 2, *Mechanics of Composite Materials*, G. P. Sendeckyj, Vol. Ed., L. J. Broutman and R. H. Krock, Series Eds., Academic Press, 1974, pp. 433–489.

25. J. A. Kies, "Maximum Strains in the Resin of Fiberglass Composites," Naval Research Laboratory, NRL Report 5752, 1962.

26. I. M. Daniel, G. M. Koller, and T. Niiro, "Photoelastic Studies of Internal Stress Distributions of Unidirectional Composites," Army Materials and Mechanics Research Center, Report AMMRC TR 80-56, 1980.

27. O. Ishai, "Failure of Unidirectional Composites in Tension," *J. Eng. Mech. Div.*, Proc. of the ASCE, Vol. 97, No. EM2, 1971, pp. 205–221.

28. D. F. Adams and D. R. Doner, "Longitudinal Shear Loading of a Unidirectional Composite," *J. Composite Materials*, Vol. 1, No. 1, 1967, pp. 4–17.

29. J. Aboudi, *Mechanics of Composite Materials—A Unified Micromechanical Approach*, Elsevier, Amsterdam, 1991.

30. P. K. Gotsis, C. C. Chamis, and L. Minetyan, "Prediction of Composite Laminate Fracture: Micromechanics and Progressive Fracture," *Composites Science and Technology*, Vol. 58, 1998, pp. 1137–1149.

PROBLEMS

5.1 Determine the longitudinal modulus E_1 and the longitudinal tensile strength F_{1t} of a unidirectional E-glass/epoxy composite with the constituent properties

Fiber volume ratio:	$V_f = 0.65$	
Fiber modulus:	$E_f = 69$ GPa (10 Msi)	
Matrix modulus:	$E_m = 3.45$ GPa (0.5 Msi)	
Fiber tensile strength:	$F_{ft} = 3450$ MPa (500 ksi)	

Matrix tensile strength: $F_{mt} = 104$ MPa (15 ksi)

Assume linear elastic behavior to failure for both fiber and matrix. Everything else being equal, how does the strength F_{1t} vary with E_f?

5.2 Determine the longitudinal modulus E_1 and the longitudinal tensile strength F_{1t} of a unidirectional carbon/epoxy composite with the properties

$$V_f = 0.65$$
$$E_{1f} = 235 \text{ GPa (34 Msi)}$$
$$E_m = 4.14 \text{ GPa (0.6 Msi)}$$
$$F_{ft} = 3450 \text{ MPa (500 ksi)}$$
$$F_{mt} = 104 \text{ MPa (15 ksi)}$$

5.3 Determine the longitudinal modulus E_1 and longitudinal tensile strength F_{1t} of a unidirectional carbon/glass composite with the properties

$$V_f = 0.45$$
$$E_{1f} = 235 \text{ GPa (34 Msi)}$$
$$E_m = 70 \text{ GPa (10 Msi)}$$
$$F_{ft} = 3500 \text{ MPa (510 ksi)}$$
$$F_{mt} = 140 \text{ MPa (20 ksi)}$$

(Note: Strength is defined as the composite stress at failure initiation in one of the phases.)

5.4 Determine the longitudinal modulus E_1 and longitudinal tensile strength F_{1t} of a unidirectional silicon carbide/ceramic composite with the properties

$$V_f = 0.40$$
$$E_f = 172 \text{ GPa (25 Msi)}$$
$$E_m = 97 \text{ GPa (14 Msi)}$$
$$F_{ft} = 1930 \text{ MPa (280 ksi)}$$
$$F_{mt} = 138 \text{ MPa (20 ksi)}$$

Assume linear elastic behavior for both fiber and matrix. Everything else being equal, how does the strength F_{1t} vary with fiber modulus E_{1f}? (Note: Strength is defined here as the composite stress at failure initiation of one of the phases.)

5.5 The longitudinal strength of a unidirectional composite, F_{1t}, must be higher than the net matrix strength, $\sigma_m = V_m F_{mt}$. In order for this to happen, the fiber volume ratio must be higher than a certain critical value $(V_f)_{crit}$. Determine $(V_f)_{crit}$ as a function of constituent properties, E_f, E_m, F_{ft}, and F_{mt}. Consider both cases,

$$\varepsilon_{ft}^u < \varepsilon_{mt}^u \quad \text{and} \quad \varepsilon_{mt}^u < \varepsilon_{ft}^u$$

5.6 A glass matrix is reinforced with unidirectional silicon carbide fibers and loaded in longitudinal tension until matrix failure. Determine the minimum fiber volume ratio, so that the composite does not fail catastrophically (i.e., the unbroken fibers can support the load) immediately after matrix failure, for the constituent properties

$$E_f = 400 \text{ GPa (58 Msi)}$$
$$E_m = 69 \text{ GPa (10 Msi)}$$
$$F_{ft} = 3175 \text{ MPa (460 ksi)}$$
$$F_{mt} = 125 \text{ MPa (18 ksi)}$$

5.7 Using Cox's and Rosen's analyses determine the ineffective fiber length δ/d at a fiber break for a carbon/epoxy composite with the properties

$$E_{1f} = 235 \text{ GPa (34 Msi)}$$
$$G_m = 1.27 \text{ GPa (184 ksi)}$$
$$d = 7 \text{ μm}$$
$$V_f = 0.65$$

5.8 A unidirectional continuous-fiber composite is loaded in longitudinal tension. Assume that at some point all fibers have fractured in fragments of length equal to 2δ prior to ultimate failure, as shown in Fig. P5.8. Determine the initial longitudinal modulus, E_1^0, and the one after fiber fragmentation, E_1' for the properties

$$E_m = 3.45 \text{ GPa (0.50 Msi)}$$
$$G_m = 1.27 \text{ GPa (184 ksi)}$$
$$E_f = 235 \text{ GPa (34 Msi)}$$
$$V_f = 0.60$$
$$r = 4 \text{ μm} (1.6 \times 10^{-4} \text{ in})$$

(Hint: Use the Rosen theory to calculate δ and Halpin's formula to determine the modulus E_1' after fragmentation.)

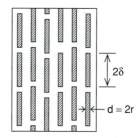

Fig. P5.8

5.9 A unidirectional continuous-fiber composite is loaded in longitudinal tension. Determine the maximum fiber volume ratio V_f allowed to prevent interfacial debonding at the fiber breaks for the material properties

$$E_m = 3.45 \text{ GPa (500 ksi)}$$
$$E_{1f} = 235 \text{ GPa (34 Msi)}$$
$$G_m = 1.27 \text{ GPa (184 ksi)}$$
$$F_{ft} = 3.5 \text{ GPa (500 ksi)}$$
$$F_{is} = 200 \text{ MPa (29 ksi)} \quad \text{(interfacial shear strength)}$$

(Hint: Use the Rosen model.)

5.10 Using the Rosen model, determine the maximum interfacial shear stress developed at a fiber break for a composite with the properties

$$G_m = 1.27 \text{ GPa (184 ksi)}$$
$$E_{1f} = 235 \text{ GPa (34 Msi)}$$
$$V_f = 0.65$$
$$F_{ft} = 3450 \text{ MPa (500 ksi)}$$

5.11 Determine the maximum interfacial shear stress at a fiber break by Cox's shear lag analysis given a carbon/epoxy composite with the properties

$$E_{1f} = 235 \text{ GPa (34 Msi)}$$
$$F_{ft} = 3.45 \text{ GPa (500 ksi)}$$
$$G_m = 1.27 \text{ GPa (184 ksi)}$$
$$V_f = 0.65$$

5.12 A composite consists of epoxy resin with aligned carbon nanotubes. Determine the longitudinal modulus of the composite and the average applied longitudinal tensile stress at which either the matrix or the interface at the end of the nanotube fails (assuming the nanotubes do not fail) given the properties

Matrix
$$E_m = 3.45 \text{ GPa (0.5 Msi)}$$
$$G_m = 1.27 \text{ GPa (184 ksi)}$$
$$F_{mt} = 100 \text{ MPa (14.5 ksi)}$$
Fiber (nanotubes)
$$l = 1 \text{ μm } (40 \times 10^{-6} \text{ in})$$
$$d = 1 \text{ nm } (40 \times 10^{-9} \text{ in})$$
$$E_f = 1000 \text{ GPa (145 Msi)}$$
$$V_f = 0.01$$
Interfacial shear strength
$$F_{is} = \tau_{max} = 200 \text{ MPa (29 ksi)}$$

(Hint: Use the Rosen model.)

5.13 The nonlinear shear stress-strain relation of a unidirectional carbon/epoxy composite is given as

$$\tau_6 = G_{12}(\gamma_6 - 20\gamma_6^2)$$

Determine the longitudinal compressive strength of the composite (F_{1c}) for an initial fiber misalignment of 2° and $G_{12} = 7$ GPa (1 Msi).

5.14 Derive the expressions in Eqs. (5.34) and (5.36) for stress and strain concentration factors in a unidirectional composite under transverse normal loading.

5.15 A unidirectional E-glass/epoxy composite is loaded in transverse tension. Obtain the stress concentration factor from Fig. 5.19 and calculate the trans-

verse tensile strength based on the maximum stress and maximum strain criteria, for the constituent and composite properties

$$V_f = 0.65$$
$$E_f = 69 \text{ GPa (10 Msi)}$$
$$E_m = 3.45 \text{ GPa (0.5 Msi)}$$
$$v_m = 0.36$$
$$F_{mt} = 104 \text{ MPa (15 ksi)}$$

(Hint: Neglect residual stresses.)

5.16 Determine the transverse tensile strength, F_{2t}, of a glass/epoxy composite with the properties

$$V_f = 0.50$$
$$E_f = 70 \text{ GPa (10 Msi)}$$
$$E_m = 3.45 \text{ GPa (0.5 Msi)}$$
$$F_{mt} = 104 \text{ MPa (15 ksi)}$$

Use the maximum stress criterion, neglecting residual stresses, and Eq. (5.34) for stress concentration.

5.17 A 10° off-axis specimen is loaded as shown in Fig. P5.17, so that it fails in in-plane (parallel to the fibers) shear. Using the theoretical prediction for shear stress concentration factor in the matrix (Eq. 5.41) and assuming a maximum shear stress failure criterion, determine the axial failure stress F_{xt} for a material with the properties

$$V_f = 0.65$$
$$G_{12f} = 13 \text{ GPa (1.9 Msi)}$$
$$G_m = 1.24 \text{ GPa (180 ksi)}$$
$$F_{ms} = 170 \text{ MPa (25 ksi)} \quad \text{(matrix shear strength)}$$

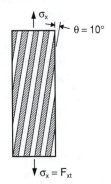

Fig. P5.17

6 Strength of Composite Lamina—Macromechanics

6.1 INTRODUCTION

As discussed in the preceding chapter, the failure mechanisms vary greatly with material properties and type of loading. Strength predictions were based on micromechanical analyses and point failure criteria. Even when the predictions are accurate with regard to failure initiation at critical points, they are only approximate as far as global failure of the lamina is concerned. Furthermore, the possible interaction of failure mechanisms makes it difficult to obtain reliable strength predictions under a general type of loading. For these reasons, a macromechanical or phenomenological approach to failure analysis may be preferable.

From the macromechanical point of view, the strength of a lamina is an anisotropic property, that is, it varies with orientation. It is desirable, for example, to correlate the strength along an arbitrary direction to some basic strength parameters. A lamina may be characterized by a number of basic strength parameters referred to its principal material directions in a manner analogous to the stiffness parameters defined before. For in-plane loading, a lamina may be characterized by five strength parameters as listed in Fig. 6.1. These are the longitudinal tensile and compressive strengths, F_{1t} and F_{1c}, the transverse tensile and compressive strengths, F_{2t} and F_{2c}, and the in-plane shear strength, $F_6(F_{12})$. Four additional lamina strength parameters, which are relevant in three-dimensional analysis, are the out-of-plane or interlaminar tensile, compressive, and shear strengths, F_{3t}, F_{3c}, $F_4(F_{23})$, and $F_5(F_{13})$. For transversely isotropic composites, with the 2-3 plane as the plane of isotropy, $F_{3t} \cong F_{2t}$, $F_{3c} \cong F_{2c}$, and $F_5 \cong F_6$. All strength parameters are measured by their absolute numerical values.

This characterization recognizes the fact that most composite materials have different strengths in tension and compression. No distinction is necessary between positive and negative shear strength as long as it is referred to the principal material directions. This is illustrated in Fig. 6.2, where a unidirectional lamina is subjected to positive and negative shear stress according to the sign convention used in mechanics of materials. As can be

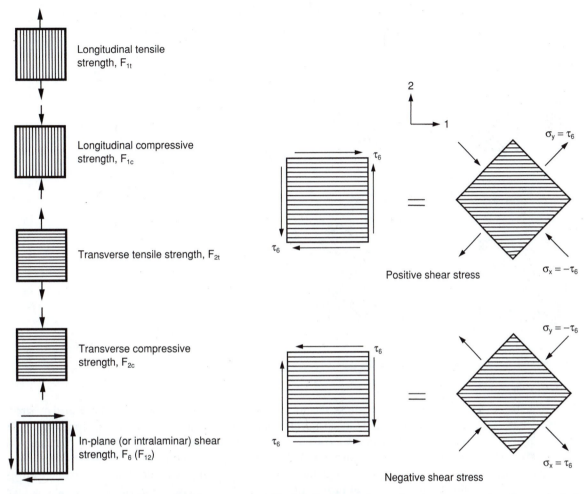

Fig. 6.1 Basic strength parameters of unidirectional lamina for in-plane loading.

Fig. 6.2 Positive and negative shear stress acting along principal material directions.

observed, both cases are equivalent to equal tensile and compressive normal loading at 45° to the fiber direction. Thus, the sign of the shear stress is immaterial. The shear strength referred to the principal material directions does not depend on the difference between tensile and compressive strengths of the material.

However, this is not the case when the shear stress is applied at an angle to the principal material directions. Figure 6.3 shows an example of a lamina loaded in shear at 45° to the fiber direction. As can be observed, positive shear stress corresponds to a tensile stress in the fiber direction and an equal compressive stress in the transverse to the fiber direction, whereas negative shear stress corresponds to a compressive stress in the fiber direction and an equal tensile stress in the transverse direction. Since most composites have different tensile and compressive strengths and they are weakest in transverse tension, it follows that in this case the lamina would be stronger under positive shear.

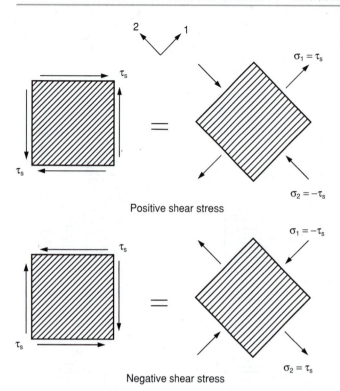

Fig. 6.3 Positive and negative shear stress acting at 45° to principal material directions.

Given a state of stress, the principal stresses and their directions are obtained by stress transformation that is independent of material properties. The principal strains and their directions are obtained by using the appropriate anisotropic stress-strain relations and strain transformation. In general, the principal stress, principal strain, and material symmetry directions do not coincide. Since strength varies with orientation, maximum stress alone is not the critical factor in failure. Anisotropic failure theories are needed that take into account both the stress and strength variation with orientation.

6.2 FAILURE THEORIES

Failure criteria for homogeneous isotropic materials, such as maximum normal stress (Rankine), maximum shear stress (Tresca), maximum distortional energy (von Mises), and so forth, are well established. Macromechanical failure theories for composites have been proposed by extending and adapting isotropic failure theories to account for the anisotropy in stiffness and strength of the composite. Surveys of anisotropic failure theories have been published by Sandhu,[1] Owen and Rice,[2] Tsai,[3] Chen and Matthews,[4] Sun,[5] Christensen,[6] and Hinton et al.[7]

In 1965 Azzi and Tsai[8] adapted Hill's[9] theory, which was originally developed for homogeneous anisotropic ductile materials, to anisotropic heterogeneous and brittle composites and introduced the so-called *Tsai-Hill theory*. They verified their theory experimentally with test results from unidirectional off-axis specimens. An early adaptation of the maximum stress theory to composites was described by Kelly[10] in 1966, who used it to predict the off-axis strength of unidirectional composites. In 1971 Ishai[11] presented experimental results for off-axis strength of composites with ductile and brittle matrices. The results for both types of composites were in agreement with predictions of the Tsai-Hill theory. In 1971 Tsai and Wu[12] introduced the fully interactive Tsai-Wu theory, which is widely used today. In 1973 Hashin and Rotem[13] introduced a partially interactive theory based on actual failure modes, which also enjoys wide popularity.

Over the last four decades numerous other theories were proposed, which combined features of the ones mentioned before for isotropic materials, that is, the interactive von Mises approach and the maximum stress criterion. The discussion below deals with failure theories for a single isolated lamina and not one within a multidirectional laminate. The application of these theories to multidirectional laminates will be discussed later in Chapter 9. Most theories are based on assumptions of homogeneity and linear stress-strain behavior to failure. Failure criteria are usually expressed in terms of the basic strength parameters referred to the principal material axes (Fig. 6.1). A few phenomenological theories, such as Puck's,[14] rely on some constituent properties in addition to lamina properties and account for nonlinear behavior. Most theories can identify the mode of failure, such as fiber or interfiber, but some of them postulate such failure modes as failure criteria. Some theories do not account for interaction of stress components, while others do so to varying degrees. Some interaction theories require additional lamina strength properties obtained by multiaxial testing.

Lamina failure theories can be classified in the following three groups:

1. *Limit or noninteractive theories*, in which specific failure modes are predicted by comparing individual lamina stresses or strains with corresponding strengths or ultimate strains, for example, maximum stress and maximum strain theories. No interaction among different stress components on failure is considered.
2. *Interactive theories* (e.g., the Tsai-Hill and Tsai-Wu theories), in which all stress components are included in one expression (failure criterion). Overall failure is predicted without reference to particular failure modes.
3. *Partially interactive or failure-mode-based theories* (e.g., the Hashin-Rotem and Puck theories), where separate criteria are given for fiber and interfiber (matrix or interface) failures.

The most representative and widely used of the above theories are discussed in the following sections.

6.3 MAXIMUM STRESS THEORY

As mentioned before, Kelly adapted the maximum stress theory to composites under plane stress conditions.[10] He used it to predict the off-axis strength of a unidirectional lamina as a function of fiber orientation by three different curves corresponding to three different failure modes. According to the maximum stress theory, failure occurs when at least one stress component along one of the principal material axes exceeds the corresponding strength in that direction. The stresses acting on a unidirectional composite are resolved along the principal material axes and planes (σ_1, σ_2, σ_3, τ_4, τ_5, τ_6), and the failure condition is expressed in the form of the following subcriteria:

$$\sigma_1 = \begin{cases} F_{1t} & \text{when} \quad \sigma_1 > 0 \\ -F_{1c} & \text{when} \quad \sigma_1 < 0 \end{cases} \tag{6.1a}$$

$$\sigma_2 = \begin{cases} F_{2t} & \text{when} \quad \sigma_2 > 0 \\ -F_{2c} & \text{when} \quad \sigma_2 < 0 \end{cases} \tag{6.1b}$$

$$\sigma_3 = \begin{cases} F_{3t} & \text{when} \quad \sigma_3 > 0 \\ -F_{3c} & \text{when} \quad \sigma_3 < 0 \end{cases} \tag{6.1c}$$

$$|\tau_4| = F_4 \tag{6.1d}$$

$$|\tau_5| = F_5 \tag{6.1e}$$

$$|\tau_6| = F_6 \tag{6.1f}$$

For a material with transverse isotropy on the 2-3 plane,

$$F_{2t} = F_{3t}$$
$$F_{2c} = F_{3c} \tag{6.2}$$
$$F_5 = F_6$$

For a two-dimensional state of stress, Eqs. (6.1c), (6.1d), and (6.1e) are not relevant. For a two-dimensional state of stress with $\tau_6 = 0$, the failure envelope takes the form of a rectangle as shown in Fig. 6.4.

In the more general case, the stresses are transformed along the principal material axes, and each stress component is related to the corresponding strength parameter. Consider, for example, the case of uniaxial loading of an off-axis lamina (Fig. 6.5). The stress components along the principal material axes are

$$\sigma_1 = \sigma_x \cos^2\theta$$
$$\sigma_2 = \sigma_x \sin^2\theta \tag{6.3}$$
$$\tau_6 = -\sigma_x \sin\theta \cos\theta$$

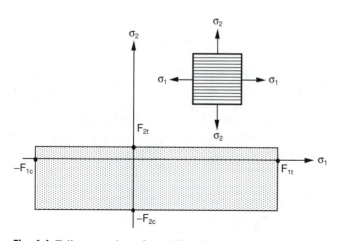

Fig. 6.4 Failure envelope for unidirectional lamina under biaxial normal loading (maximum stress theory).

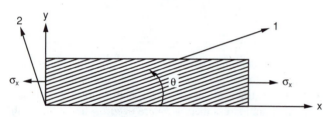

Fig. 6.5 Uniaxial loading of off-axis unidirectional lamina.

By equating the stress components in Eq. (6.3) to the corresponding strengths, we obtain the following ultimate values of σ_x, that is, the off-axis strength F_x: when $\sigma_x > 0$,

$$F_{xt} = \frac{F_{1t}}{\cos^2\theta} \tag{6.4a}$$

$$F_{xt} = \frac{F_{2t}}{\sin^2\theta} \tag{6.4b}$$

$$F_{xt} = \frac{F_6}{\sin\theta\cos\theta} \tag{6.4c}$$

and when $\sigma_x < 0$,

$$F_{xc} = \frac{F_{1c}}{\cos^2\theta} \tag{6.5a}$$

$$F_{xc} = \frac{F_{2c}}{\sin^2\theta} \tag{6.5b}$$

$$F_{xc} = \frac{F_6}{\sin\theta\cos\theta} \tag{6.5c}$$

It should be noted that in the case of shear stress and strength referred to the principal material axes, the sign of the shear stress is immaterial and only absolute values need be used.

Using the strength properties of a material such as E-glass/epoxy from Table A.4 in Appendix A, one can obtain the variation of lamina strength as a function of fiber orientation (Fig. 6.6). By taking the lowest values of the predicted strength, we obtain a failure envelope for F_x as a function of θ. This envelope is characterized by cusps at the intersections of the curves for the various subcriteria. Three regions can be identified, corresponding to three different modes of failure:

1. fiber failure (tensile and compressive)
2. in-plane shear interfiber failure
3. transverse normal stress interfiber failure (tensile and compressive)

The maximum stress theory is more applicable for the brittle modes of failure of the material, closer to transverse and longitudinal tension, and does not take into account any stress interaction under a general biaxial state of stress.

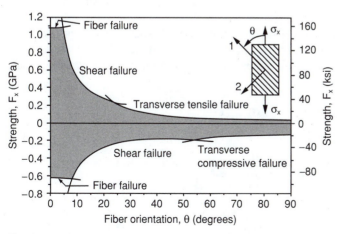

Fig. 6.6 Uniaxial strength of off-axis E-glass/epoxy unidirectional lamina as a function of fiber orientation (maximum stress theory).

6.4 MAXIMUM STRAIN THEORY

According to the maximum strain theory, failure occurs when at least one of the strain components along the principal material axes exceeds the corresponding ultimate strain in that direction. This theory allows for some interaction of stress components due to Poisson's ratio effects. It is expressed in the form of the following subcriteria:

$$\varepsilon_1 = \begin{cases} \varepsilon_{1t}^u & \text{when} \quad \varepsilon_1 > 0 \\ \varepsilon_{1c}^u & \text{when} \quad \varepsilon_1 < 0 \end{cases} \tag{6.6a}$$

$$\varepsilon_2 = \begin{cases} \varepsilon_{2t}^u & \text{when} \quad \varepsilon_2 > 0 \\ \varepsilon_{2c}^u & \text{when} \quad \varepsilon_2 < 0 \end{cases} \tag{6.6b}$$

$$\varepsilon_3 = \begin{cases} \varepsilon_{3t}^u & \text{when} \quad \varepsilon_3 > 0 \\ \varepsilon_{3c}^u & \text{when} \quad \varepsilon_3 < 0 \end{cases} \tag{6.6c}$$

$$|\gamma_4| = \gamma_4^u \tag{6.6d}$$

$$|\gamma_5| = \gamma_5^u \tag{6.6e}$$

$$|\gamma_6| = \gamma_6^u \tag{6.6f}$$

where ε_1, ε_2, ε_3, γ_4, γ_5, and γ_6 are the strain components referred to the principal material axes and

ε_{1t}^u, ε_{2t}^u, ε_{3t}^u = ultimate tensile strains along the 1-, 2-, and 3-directions, respectively

ε_{1c}^u, ε_{2c}^u, ε_{3c}^u = ultimate compressive strains along the 1-, 2-, and 3-directions, respectively

γ_4^u, γ_5^u, γ_6^u = ultimate shear strains on the 2-3, 3-1, and 1-2 planes, respectively

To apply the theory for a given general state of stress, the stress components along the principal material axes and planes (σ_1, σ_2, σ_3, τ_4, τ_5, τ_6) are first obtained by stress transformation, and then the corresponding strains are obtained from the stress-strain relations, Eq. (4.48).

$$\varepsilon_1 = \frac{\sigma_1}{E_1} - \nu_{21}\frac{\sigma_2}{E_2} - \nu_{31}\frac{\sigma_3}{E_3} = \frac{1}{E_1}(\sigma_1 - \nu_{12}\sigma_2 - \nu_{13}\sigma_3)$$

$$\varepsilon_2 = \frac{\sigma_2}{E_2} - \nu_{12}\frac{\sigma_1}{E_1} - \nu_{32}\frac{\sigma_3}{E_3} = \frac{1}{E_2}(\sigma_2 - \nu_{21}\sigma_1 - \nu_{23}\sigma_3)$$

$$\varepsilon_3 = \frac{\sigma_3}{E_3} - \nu_{13}\frac{\sigma_1}{E_1} - \nu_{23}\frac{\sigma_2}{E_2} = \frac{1}{E_3}(\sigma_3 - \nu_{31}\sigma_1 - \nu_{32}\sigma_2)$$

$$\gamma_4 = \frac{\tau_4}{G_{23}} \qquad \gamma_5 = \frac{\tau_5}{G_{13}} \qquad \gamma_6 = \frac{\tau_6}{G_{12}}$$

$$\tag{6.7}$$

The ultimate strains obtained by uniaxial or pure shear testing of the unidirectional composite, assuming linear behavior to failure, are related to the basic strength parameters of the material as follows:

$$\varepsilon_{1t}^u = \frac{F_{1t}}{E_1}, \qquad \varepsilon_{1c}^u = -\frac{F_{1c}}{E_1}$$

$$\varepsilon_{2t}^u = \frac{F_{2t}}{E_2}, \qquad \varepsilon_{2c}^u = -\frac{F_{2c}}{E_2}$$

$$\varepsilon_{3t}^u = \frac{F_{3t}}{E_3}, \qquad \varepsilon_{3c}^u = -\frac{F_{3c}}{E_3}$$

$$\gamma_4^u = \frac{F_4}{G_{23}} \qquad \gamma_5^u = \frac{F_5}{G_{13}} \qquad \gamma_6^u = \frac{F_6}{G_{12}}$$

(6.8)

In view of relations in Eqs. (6.7) and (6.8), the failure subcriteria of Eqs. (6.6) can be expressed in terms of stresses as follows:

$$\sigma_1 - v_{12}\sigma_2 - v_{13}\sigma_3 = \begin{cases} F_{1t} & \text{when} \quad \varepsilon_1 > 0 \\ -F_{1c} & \text{when} \quad \varepsilon_1 < 0 \end{cases}$$

$$\sigma_2 - v_{21}\sigma_1 - v_{23}\sigma_3 = \begin{cases} F_{2t} & \text{when} \quad \varepsilon_2 > 0 \\ -F_{2c} & \text{when} \quad \varepsilon_2 < 0 \end{cases}$$

(6.9)

$$\sigma_3 - v_{31}\sigma_1 - v_{32}\sigma_2 = \begin{cases} F_{3t} & \text{when} \quad \varepsilon_3 > 0 \\ -F_{3c} & \text{when} \quad \varepsilon_3 < 0 \end{cases}$$

$$|\tau_4| = F_4 \qquad |\tau_5| = F_5 \qquad |\tau_6| = F_6$$

For a two-dimensional state of stress ($\sigma_3 = \tau_4 = \tau_5 = 0$) and $\tau_6 = 0$, the failure envelope on the σ_1-σ_2 space takes the form of a parallelogram with its center off the origin of the σ_1-σ_2 coordinate system (Fig. 6.7).

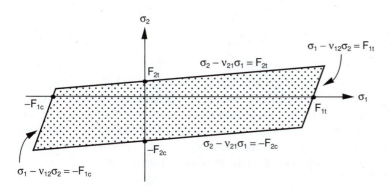

Fig. 6.7 Failure envelope for unidirectional lamina under biaxial normal loading (maximum strain theory).

6.5 ENERGY-BASED INTERACTION THEORY (TSAI-HILL)

The deviatoric or distortional energy has been proposed by many investigators (e.g., von Mises, Hencky, Nadai, Novozhilov) in various forms as a failure criterion for isotropic ductile metals. For a two-dimensional state of stress referred to the principal stress directions, the von Mises yield criterion has the form

$$\sigma_1^2 + \sigma_2^2 - \sigma_1\sigma_2 = \sigma_{yp}^2 \tag{6.10}$$

where σ_{yp} is the yield stress.

Hill[9] modified this criterion for the case of ductile metals with anisotropy and proposed the following form:

$$A\sigma_1^2 + B\sigma_2^2 + C\sigma_1\sigma_2 + D\tau_6^2 = 1 \tag{6.11}$$

where A, B, C, and D are material parameters characteristic of the current state of anisotropy. The above form cannot be technically referred to as distortional energy criterion, since distortion cannot be separated from dilatation in anisotropic materials.

Azzi and Tsai[8] adapted this criterion to orthotropic composite materials, such as a unidirectional lamina with transverse isotropy. The parameters of Eq. (6.11) can be related to the basic strength parameters of the lamina by conducting real or imaginary elementary experiments as discussed previously in Section 4.2 (Figs. 4.4 and 4.5).

For uniaxial longitudinal loading to failure, $\sigma_1^u = F_1$, $\sigma_2 = \tau_6 = 0$, and Eq. (6.11) yields

$$A = \frac{1}{F_1^2} \tag{6.12}$$

For uniaxial transverse loading to failure, $\sigma_2^u = F_2$, $\sigma_1 = \tau_6 = 0$, and Eq. (6.11) yields

$$B = \frac{1}{F_2^2} \tag{6.13}$$

For in-plane shear loading to failure, $\sigma_1 = \sigma_2 = 0$, $\tau_6^u = F_6$, and Eq. (6.11) yields

$$D = \frac{1}{F_6^2} \tag{6.14}$$

The superscript u in the above denotes the ultimate value of stress at failure.

The remaining parameter C, accounting for interaction between normal stresses σ_1 and σ_2, must be determined by means of a biaxial test. Under equal biaxial normal loading, $\sigma_1 = \sigma_2 \neq 0$, $\tau_6 = 0$, it can be assumed that the material follows the maximum stress criterion, that is, failure will occur when the transverse stress σ_2 reaches the transverse strength value, F_2, which is much lower than the longitudinal strength F_1. Equation (6.11) then yields

$$C = -\frac{1}{F_1^2} \tag{6.15}$$

Substituting the values of the parameters A, B, C, and D into Eq. (6.11), we obtain the Tsai-Hill criterion for a two-dimensional state of stress:

$$\frac{\sigma_1^2}{F_1^2} + \frac{\sigma_2^2}{F_2^2} + \frac{\tau_6^2}{F_6^2} - \frac{\sigma_1\sigma_2}{F_1^2} = 1 \tag{6.16}$$

In the above, no distinction is made between tensile and compressive strengths. However, the appropriate strength values can be used in Eq. (6.16) according to the signs of the normal stresses σ_1 and σ_2. Thus,

$$F_1 = \begin{cases} F_{1t} & \text{when} \quad \sigma_1 > 0 \\ F_{1c} & \text{when} \quad \sigma_1 < 0 \end{cases} \tag{6.17a}$$

$$F_2 = \begin{cases} F_{2t} & \text{when} \quad \sigma_2 > 0 \\ F_{2c} & \text{when} \quad \sigma_2 < 0 \end{cases} \tag{6.17b}$$

The failure envelope described by the Tsai-Hill criterion in Eq. (6.16) is a closed surface in the σ_1-σ_2-τ_6 space. Failure envelopes for constant values of $k = \tau_6/F_6$ have the form

$$\frac{\sigma_1^2}{F_1^2} + \frac{\sigma_2^2}{F_2^2} - \frac{\sigma_1\sigma_2}{F_1^2} = 1 - k^2 \tag{6.18}$$

The above form represents four different elliptical arcs joined at the σ_1- and σ_2-axes.

Consider, for example, the case of uniaxial off-axis loading shown in Fig. 6.5. By transforming the applied stress σ_x along the principal material axes, Eq. (6.3), and substituting into Eq. (6.16), we obtain the following equation for the axial strength $F_x = \sigma_x^u$:

$$\frac{1}{F_x^2} = \frac{m^4}{F_1^2} + \frac{n^4}{F_2^2} + \left[\frac{1}{F_6^2} - \frac{1}{F_1^2} \right] m^2 n^2 \tag{6.19}$$

where

$$m = \cos \theta$$

$$n = \sin \theta$$

In the case of advanced high-strength composites, the longitudinal strength is much higher than the shear strength, that is, $F_1 \gg F_6$. Then, Eq. (6.19) can be approximated by the following:

$$\frac{1}{F_x^2} \cong \frac{m^4}{F_1^2} + \frac{n^4}{F_2^2} + \frac{m^2 n^2}{F_6^2} \tag{6.20}$$

In the case of a three-dimensional state of stress, the failure criterion is obtained by similar reasoning, and assuming $F_2 \cong F_3$ and $F_5 \cong F_6$, it takes the form

$$\frac{\sigma_1^2}{F_1^2} + \frac{\sigma_2^2 + \sigma_3^2}{F_2^2} + \frac{\tau_4^2}{F_4^2} + \frac{\tau_5^2 + \tau_6^2}{F_6^2} - \frac{\sigma_1\sigma_2}{F_1^2} - \frac{\sigma_2\sigma_3}{F_2^2} - \frac{\sigma_1\sigma_3}{F_1^2} = 1$$

or

$$\frac{\sigma_1^2 - \sigma_1\sigma_2 - \sigma_1\sigma_3}{F_1^2} + \frac{\sigma_2^2 + \sigma_3^2 - \sigma_2\sigma_3}{F_2^2} + \frac{\tau_4^2}{F_4^2} + \frac{\tau_5^2 + \tau_6^2}{F_6^2} = 1 \qquad (6.21)$$

The interactive quadratic nature of the Tsai-Hill theory has been criticized because it is based on Hill's theory, which is suitable for homogeneous, anisotropic, and ductile materials, whereas most composites are strongly heterogeneous and brittle. Azzi and Tsai verified the theory by off-axis tensile testing of unidirectional composites.[8]

The Tsai-Hill failure theory is expressed in terms of a single criterion instead of the multiple subcriteria required in the maximum stress and maximum strain theories. The theory allows for considerable interaction among the stress components. One disadvantage, however, is that it does not distinguish directly between tensile and compressive strengths. The strength parameters in Eq. (6.16) must be specified and used according to the given state of stress, Eq. (6.17).

6.6 INTERACTIVE TENSOR POLYNOMIAL THEORY (TSAI-WU)

One of the first attempts to develop a general failure theory for anisotropic materials without the limitations of previous theories was discussed by Gol'denblat and Kopnov.[15] This theory is capable of predicting strength under general states of stress for which no experimental data are available. It uses the concept of strength tensors, which allows for transformation from one coordinate system to another. It has the form of an invariant formed from stress and strain tensor components, and, most important, it has the capability to account for the difference between tensile and compressive strengths. The proposed original form of the criterion (in contracted notation) is

$$(f_i\sigma_i)^\alpha + (f_{ij}\sigma_i\sigma_j)^\beta + (f_{ijk}\sigma_i\sigma_j\sigma_k)^\gamma + \ldots = 1 \qquad (6.22)$$

where repeated subscripts in a term imply summation, with $i, j, k = 1, 2, \ldots, 6$. The coefficients $f_i, f_{ij}, f_{ijk} \ldots$ are strength tensors of second, fourth, sixth, and higher orders and can be related to the basic strength constants of the material. To make the criterion homogeneous, the exponents are taken as $\alpha = 1$, $\beta = 1/2$, and $\gamma = 1/3$. In its simplest form the criterion takes the form

$$f_i\sigma_i + [f_{ij}\sigma_i\sigma_j]^{1/2} = 1 \qquad (6.23)$$

Tsai and Wu[12] proposed a modified tensor polynomial theory by assuming the existence of a failure surface in the stress space. In contracted notation it takes the form

$$f_i\sigma_i + f_{ij}\sigma_i\sigma_j = 1 \qquad (6.24)$$

where f_i and f_{ij} are second- and fourth-order strength tensors, and $i, j = 1, 2, \ldots, 6$.

In expanded form the Tsai-Wu criterion (6.24) is expressed as

$$
\begin{aligned}
&f_1\sigma_1 + f_2\sigma_2 + f_3\sigma_3 + f_4\tau_4 + f_5\tau_5 + f_6\tau_6 + f_{11}\sigma_1^2 + f_{22}\sigma_2^2 + f_{33}\sigma_3^2 + f_{44}\tau_4^2 \\
&+ f_{55}\tau_5^2 + f_{66}\tau_6^2 + 2f_{12}\sigma_1\sigma_2 + 2f_{13}\sigma_1\sigma_3 + 2f_{14}\sigma_1\tau_4 + 2f_{15}\sigma_1\tau_5 + 2f_{16}\sigma_1\tau_6 \\
&+ 2f_{23}\sigma_2\sigma_3 + 2f_{24}\sigma_2\tau_4 + 2f_{25}\sigma_2\tau_5 + 2f_{26}\sigma_2\tau_6 + 2f_{34}\sigma_3\tau_4 + 2f_{35}\sigma_3\tau_5 \\
&+ 2f_{36}\sigma_3\tau_6 + 2f_{45}\tau_4\tau_5 + 2f_{46}\tau_4\tau_6 + 2f_{56}\tau_5\tau_6 = 1
\end{aligned}
\qquad (6.25)
$$

The linear terms in normal stress in this expression allow for the distinction between tensile and compressive strengths. The coefficients $f_{12}, f_{23},$ and f_{13} allow for interaction among normal stresses $\sigma_1, \sigma_2,$ and σ_3.

Since the shear strengths of a unidirectional composite, $F_4, F_5,$ and F_6, referred to the principal material planes (specially orthotropic material), are independent of the sign of the shear stress (τ_4, τ_5, τ_6), all linear terms in the shear stresses and all terms associated with interaction of normal and shear stresses and between shear stresses acting on different planes must vanish. Thus,

$$f_4 = f_5 = f_6 = f_{14} = f_{15} = f_{16} = f_{24} = f_{25} = f_{26} = f_{34} = f_{35} = f_{36} = f_{45} = f_{46} = f_{56} = 0 \quad (6.26)$$

The number of coefficients is reduced further by assuming transverse isotropy, with the 2-3 plane as the plane of isotropy, and noting that

$$f_2 = f_3 \qquad f_{22} = f_{33} \qquad f_{55} = f_{66} \qquad f_{12} = f_{13} \qquad (6.27)$$

Then, the criterion is written as

$$
\begin{aligned}
&f_1\sigma_1 + f_2(\sigma_2 + \sigma_3) + f_{11}\sigma_1^2 + f_{22}(\sigma_2^2 + \sigma_3^2) + f_{44}\tau_4^2 \\
&+ f_{66}(\tau_5^2 + \tau_6^2) + 2f_{12}(\sigma_1\sigma_2 + \sigma_1\sigma_3) + 2f_{23}\sigma_2\sigma_3 = 1
\end{aligned}
\qquad (6.28)
$$

For a two-dimensional state of stress ($\sigma_1, \sigma_2, \sigma_6$) the above criterion is reduced to the most familiar form

$$f_1\sigma_1 + f_2\sigma_2 + f_{11}\sigma_1^2 + f_{22}\sigma_2^2 + f_{66}\tau_6^2 + 2f_{12}\sigma_1\sigma_2 = 1 \qquad (6.29)$$

The coefficients of the general quadratic Tsai-Wu criterion are obtained by applying elementary loadings to the lamina. Thus, for longitudinal tensile loading to failure $\sigma_1^u = F_{1t}$, $\sigma_2 = \sigma_3 = \tau_4 = \tau_5 = \tau_6 = 0$, and

$$f_1 F_{1t} + f_{11} F_{1t}^2 = 1 \qquad (6.30)$$

and for longitudinal compressive loading $\sigma_1^u = -F_{1c}$, $\sigma_2 = \sigma_3 = \tau_4 = \tau_5 = \tau_6 = 0$, and

$$-f_1 F_{1c} + f_{11} F_{1c}^2 = 1 \qquad (6.31)$$

Equations (6.30) and (6.31) yield the values of coefficients f_1 and f_{11} as

$$f_1 = \frac{1}{F_{1t}} - \frac{1}{F_{1c}} \qquad f_{11} = \frac{1}{F_{1t}F_{1c}} \tag{6.32}$$

Similarly, for transverse uniaxial tensile and compressive loadings along the 2- and 3-directions we obtain

$$f_2 = \frac{1}{F_{2t}} - \frac{1}{F_{2c}} \qquad f_{22} = \frac{1}{F_{2t}F_{2c}} \tag{6.33}$$

$$f_3 = \frac{1}{F_{3t}} - \frac{1}{F_{3c}} \qquad f_{33} = \frac{1}{F_{3t}F_{3c}} \tag{6.34}$$

For pure shear loading to failure on each one of the three principal planes of material symmetry we obtain

$$f_{44} = \frac{1}{F_4^2} \qquad f_{55} = \frac{1}{F_5^2} \qquad f_{66} = \frac{1}{F_6^2} \tag{6.35}$$

The interaction coefficients f_{12} (f_{13}) must be obtained by some form of biaxial testing. Under equal biaxial normal loading $\sigma_1^u = \sigma_2^u = F_{(12)}$, $\sigma_3 = \tau_4 = \tau_5 = \tau_6 = 0$, we obtain

$$(f_1 + f_2)F_{(12)} + (f_{11} + f_{22} + 2f_{12})F_{(12)}^2 = 1 \tag{6.36}$$

where $F_{(12)}$ is the experimentally determined strength under equal biaxial tensile loading $(\sigma_1 = \sigma_2)$. Equation (6.36) is then solved for f_{12}. Thus, f_{12} is a function of the basic strength parameters plus the equal biaxial strength $F_{(12)}$.

Direct biaxial testing is not easy or practical to perform. An easier test producing a biaxial state of stress is the off-axis tensile test, that is, uniaxial loading σ_x at an angle θ to the fiber direction on the 1-2 plane. For $\theta = 45°$, Eq. (6.3) yields

$$\sigma_1 = \sigma_2 = |\tau_6| = \frac{1}{2}F_{45t} \tag{6.37}$$

where F_{45t} is the off-axis tensile strength of the lamina at 45° to the fiber direction. Substituting the above in Eq. (6.29) we obtain

$$2F_{45t}(f_1 + f_2) + F_{45t}^2(f_{11} + f_{22} + f_{66} + 2f_{12}) = 4 \tag{6.38}$$

which can be solved for f_{12} as a function of the other coefficients and the off-axis strength F_{45t}.

In general it is difficult, or not accurate enough, to determine the interaction coefficient experimentally. It was shown that the coefficient f_{12} can be bounded by imposing certain restrictions to the failure envelope.[16] It was shown that for typical carbon and glass fiber composites the coefficient f_{12} is bounded as follows:[16]

$$-0.9(f_{11}f_{12})^{1/2} \le f_{12} \le 0 \tag{6.39}$$

A reasonable approximation of f_{12} is

$$f_{12} \cong -\frac{1}{2}(f_{11}f_{22})^{1/2} \tag{6.40}$$

This relation is compatible with the von Mises yield criterion in Eq. (6.10) for isotropic materials where

$$f_{11} = f_{22} = \frac{1}{\sigma_{yp}^2} \tag{6.41}$$

For a transversely isotropic material, the interaction coefficient f_{13} is equal to f_{12}. The remaining interaction coefficient f_{23} can be expressed in terms of basic strength parameters. Consider a lamina under uniaxial tensile loading $\sigma_{2'}$ on the 2-3 plane of isotropy along the $2'$-axis at 45° to the 2-axis (Fig. 6.8). Then, at failure,

$$\sigma_{2'}^u = \sigma_2^u = F_{2t} \tag{6.42}$$

and, by transformation to the 2- and 3-axes,

$$\sigma_2 = \sigma_3 = \tau_4 = \frac{\sigma_{2'}^u}{2} = \frac{F_{2t}}{2} \tag{6.43}$$

Substituting this state of stress, with $\sigma_1 = \tau_5 = \tau_6 = 0$, in the Tsai-Wu criterion, Eq. (6.28), we obtain

$$f_2 F_{2t} + \frac{1}{2}f_{22}F_{2t}^2 + \frac{1}{4}f_{44}F_{2t}^2 + \frac{1}{2}f_{23}F_{2t}^2 = 1$$

and, using relations in Eqs. (6.33) and (6.35), we obtain

$$f_{23} = \frac{1}{F_{2t}F_{2c}} - \frac{1}{2F_4^2} \tag{6.44}$$

or

$$f_{23} = f_{22} - \frac{f_{44}}{2}$$

This is exactly analogous to the relationship between compliances S_{44}, S_{22}, and S_{23} in Eq. (4.27).

In general all interaction terms must be constrained by the stability condition[12]

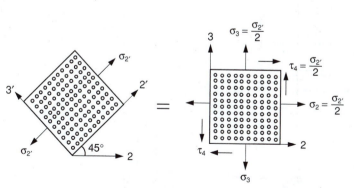

Fig. 6.8 Equivalent states of stress on transverse plane of isotropy (2-3).

$$f_{ii}f_{jj} - f_{ij}^2 \geq 0 \tag{6.45}$$

where repeated subscripts do not signify summation in this case.

By extending the assumption of transverse isotropy to the 1-2 and 3-1 planes as well, we obtain for an isotropic material (with equal tensile and compressive strengths)

$$f_1 = f_2 = f_3 = 0$$

$$f_{11} = f_{22} = f_{33}$$

$$f_{12} = f_{23} = f_{13} = -\frac{f_{11}}{2} \tag{6.46}$$

$$f_{44} = f_{55} = f_{66} = 3f_{11}$$

and the Tsai-Wu criterion, Eq. (6.28), is reduced to

$$\sigma_1^2 + \sigma_2^2 + \sigma_3^2 - \sigma_1\sigma_2 - \sigma_2\sigma_3 - \sigma_3\sigma_1 + 3(\tau_4^2 + \tau_5^2 + \tau_6^2) = F^2 \tag{6.47}$$

where F is the uniaxial strength (or yield point). The above is the well-established distortional energy failure criterion for isotropic materials (Huber-Hencky-von Mises).

The off-axis strength of unidirectional E-glass/epoxy (Table A.4) was calculated by means of the Tsai-Wu criterion and plotted as a function of fiber orientation in Fig. 6.9, using the approximate relation in Eq. (6.40) for f_{12}. Superimposed in the figure is the prediction of the Tsai-Hill theory. It is seen that the predictions of the two theories are almost indistinguishable for the tensile strength, but they deviate somewhat for the compressive strength. The off-axis strength is not very sensitive to the interaction coefficient f_{12}. Considering the usual scatter in measured strengths, the differences between the two theories are not significant in this case.

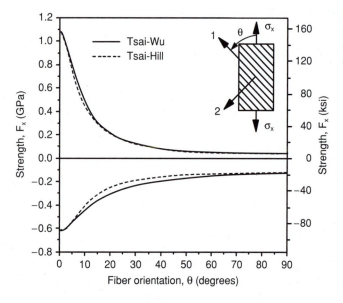

Fig. 6.9 Uniaxial strength of off-axis E-glass/epoxy unidirectional lamina as a function of fiber orientation (comparison of Tsai-Wu and Tsai-Hill failure criteria).

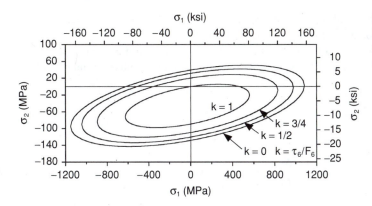

Fig. 6.10 Failure envelopes for unidirectional E-glass/epoxy lamina under biaxial loading with different levels of shear stress (Tsai-Wu criterion).

The reduced form of the Tsai-Wu criterion in two dimensions, Eq. (6.29), is a closed surface in the σ_1-σ_2-τ_6 space. Failure envelopes for constant values of shear stress $\tau_6 = kF_6$ have the form

$$f_1\sigma_1 + f_2\sigma_2 + f_{11}\sigma_1^2 + f_{22}\sigma_2^2 + 2f_{12}\sigma_1\sigma_2 = 1 - k^2 \qquad (6.48)$$

and are illustrated in Fig. 6.10 for the same E-glass/epoxy material (Table A.4).

The Tsai-Wu failure criterion has several desirable features:

1. It is operationally simple and readily amenable to computational procedures.
2. Like the Tsai-Hill theory, it is expressed in terms of a single criterion, instead of six subcriteria required in the maximum stress and maximum strain theories.
3. The stress interaction terms can be treated as independent material properties determined by appropriate experiments, unlike the Tsai-Hill theory where the interaction terms are fixed as functions of the other terms.
4. The theory, through its linear terms, accounts for the difference between tensile and compressive strengths.

The theory satisfies the invariant requirements of coordinate transformation, following normal tensor transformation laws. The strength tensors display similar symmetry properties as the stiffnesses and compliances.

6.7 FAILURE-MODE-BASED THEORIES (HASHIN-ROTEM)

In most composites reinforced with strong and stiff fibers, there is a clear distinction between fiber and interfiber (matrix, interface) failure modes. This allows for the treatment of failure in a simplified and more physically significant way than in the case of the maximum stress, maximum strain, and fully interactive theories.

For example, in the case of tensile off-axis loading, Ishai[11] noted that only the interfiber failure mode is dominant. He obtained experimental results for off-axis tensile strength of a unidirectional lamina for two types of composites, one with a brittle matrix and the other with a ductile one. He found that, in the case of the ductile matrix composite, the experimental failure envelope agreed with the von Mises prediction.

Hashin and Rotem[13] noted that failure of a lamina under a general in-plane loading can be characterized by two failure criteria, one for fiber failure and the other for interfiber failure, as follows:

$$\frac{|\sigma_1|}{F_1} = 1 \tag{6.49}$$

$$\left(\frac{\sigma_2}{F_2}\right)^2 + \left(\frac{\tau_6}{F_6}\right)^2 = 1 \tag{6.50}$$

These criteria can be extended for a general three-dimensional state of stress in terms of stresses acting on the three principal material planes and related to the expected failure modes on those planes.

$$\frac{|\sigma_1|}{F_1} = 1$$

$$\left(\frac{\sigma_2}{F_2}\right)^2 + \left(\frac{\tau_4}{F_4}\right)^2 + \left(\frac{\tau_6}{F_6}\right)^2 = 1 \tag{6.51}$$

$$\left(\frac{\sigma_3}{F_3}\right)^2 + \left(\frac{\tau_4}{F_4}\right)^2 + \left(\frac{\tau_5}{F_5}\right)^2 = 1$$

The above criteria correspond to fiber failure on the 2-3 plane and interfiber failures on the 1-3 and 1-2 planes. In the equations above the normal strengths F_i ($i = 1, 2, 3$) take the following values:

$$F_i = \begin{cases} F_{it} & \text{when} \quad \sigma_i > 0 \\ F_{ic} & \text{when} \quad \sigma_i < 0 \end{cases} \tag{6.52}$$

In the criteria above, the strength values are the ultimate values when the stress-strain behavior is linear to failure. In the case of nonlinear behavior, the strength values can be defined as the proportionality limits of the corresponding stress-strain curves.

In general, nine strength parameters are required. However, for a transversely isotropic material (with the 2-3 plane as the plane of isotropy),

$$F_{2t} = F_{3t}$$

$$F_{2c} = F_{3c} \tag{6.53}$$

$$F_5 = F_6$$

hence, six independent strength parameters suffice.

Other more elaborate failure theories have been proposed based on the concept of separating fiber and interfiber failure modes. Hashin[17] proposed a modification of the

Hashin-Rotem theory discussed before. He proposed more interactive criteria for tensile failure of the fiber and for combined transverse compression and shear. In the latter case he introduced the effect of transverse shear strength F_4 (F_{23}) in the criterion.

Puck and his coworkers have developed over the years a phenomenological failure theory based on the physical behavior of the composite. A thorough treatment of the theory was presented recently by Puck and Schürmann.[14] The theory accounts for nonlinear stress-strain relationships and, like the Hashin-Rotem theory, distinguishes between fiber and interfiber failures. They proposed interactive criteria for fiber failure, including effects of fiber properties and transverse composite stress. The prediction of interfiber failure is based on Mohr's hypothesis that fracture is produced only by the stresses acting on the fracture plane. They proposed three different failure criteria for interfiber failure, which include, in addition to macroscopic strength parameters, the inclination of the fracture plane.

The increased complexity of Hashin's and Puck's theories and the need for additional material parameters, without a proven advantage, make these theories less attractive than the simpler Hashin-Rotem theory.

6.8 FAILURE CRITERIA FOR TEXTILE COMPOSITES

The failure mechanisms of textile-reinforced composites depend on the weave style (plain, twill, satin) in addition to the fiber and matrix properties. One general characteristic of fabric composites is their nonlinear stress-strain behavior under normal stress. In the case of in-plane tensile loading along principal axes (warp and fill directions), the nonlinearity is due to matrix microcracking preceding ultimate failure.

Typical tensile stress-strain curves of glass and carbon fabric composites are shown in Figs. 6.11 and 6.12. Two characteristic strength parameters can be identified, the proportional limit and the ultimate strength. The woven glass/epoxy exhibits more pronounced nonlinearity than the woven carbon/epoxy. In a conservative approach for the glass fabric composite, the proportional limit associated with the initial tangent modulus can be defined as the strength parameter. In a less conservative approach, the ultimate value associated with the secant modulus can be used as the strength parameter. In the case of the carbon fabric composite, with a nearly linear stress-strain response to failure, the ultimate strength can be used in the failure criteria, as in the case of unidirectional composites.

In the general three-dimensional case, the material is characterized by nine strength parameters: F_{1t}, F_{1c}, F_{2t}, F_{2c}, F_{3t}, F_{3c}, F_4, F_5, F_6. In many fabric composites the properties in the warp direction are the same or nearly the same as those in the fill direction, that is, $F_{1t} \cong F_{2t}$, $F_{1c} \cong F_{2c}$, and $F_4 \cong F_5$. Then, the number of independent strength parameters is reduced to six.

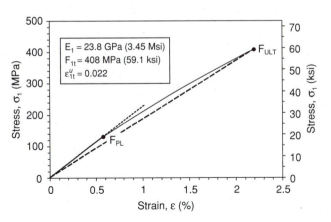

Fig. 6.11 Stress-strain curve of woven glass/epoxy composite under uniaxial tensile loading in the warp direction (eight-harness satin weave; M10E E-glass/epoxy).

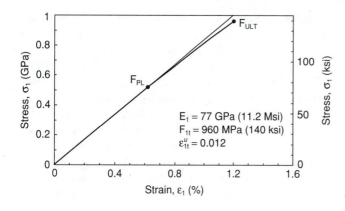

Fig. 6.12 Stress-strain curve of woven carbon/epoxy composite under uniaxial tensile loading in the warp direction (five-harness satin weave; AGP370-5H/3501-6S).

The failure theories discussed before can be applied to fabric composites as well, with certain modifications and approximations like the ones mentioned above. The Tsai-Hill interactive theory in three dimensions takes the form

$$\frac{\sigma_1^2 + \sigma_2^2}{F_1^2} + \left(\frac{\sigma_3}{F_3}\right)^2 + \frac{\tau_4^2 + \tau_5^2}{F_4^2} + \left(\frac{\tau_6}{F_6}\right)^2 - \frac{1}{F_1^2}(\sigma_1\sigma_2 + \sigma_2\sigma_3 + \sigma_3\sigma_1) = 1 \quad (6.54)$$

As discussed before, the Tsai-Hill formulation does not make an automatic distinction between tensile and compressive strengths. Appropriate values must be used for the strengths depending on the sign of the normal stresses, that is,

$$F_i = \begin{cases} F_{it} & \text{when} \quad \sigma_i > 0 \\ F_{ic} & \text{when} \quad \sigma_i < 0 \end{cases} \quad (6.52 \text{ bis})$$

The three-dimensional failure criterion of Eq. (6.54) could be further simplified by noting that, for some satin-weave fabric composites such as AGP370-5H/3501-6S carbon/epoxy, $F_1 \cong F_2 \cong F_{3c} \gg F_{3t}$ and $F_{3t} \cong F_4 \cong F_5 \cong F_6$. For in-plane loading, $\sigma_3 = \tau_4 = \tau_5 = 0$, and the above criterion is reduced to

$$\frac{1}{F_1^2}(\sigma_1^2 + \sigma_2^2 - \sigma_1\sigma_2) + \left(\frac{\tau_6}{F_6}\right)^2 = 1 \quad (6.55)$$

The in-plane off-axis tensile strength of the woven carbon/epoxy composite discussed before was calculated by the criterion of Eq. (6.55) and plotted as a function of warp filament orientation in Fig. 6.13. The superimposed experimental results are in excellent agreement with predictions.

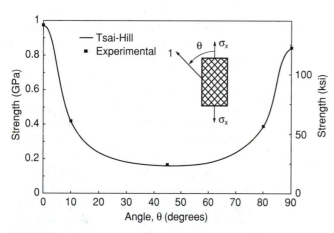

Fig. 6.13 Tensile strength of off-axis woven carbon/epoxy composite as a function of warp filament orientation (comparison of Tsai-Hill criterion and experimental results; AGP370-5H/3501-6S).

6.9 COMPUTATIONAL PROCEDURE FOR DETERMINATION OF LAMINA STRENGTH— TSAI-WU CRITERION (PLANE STRESS CONDITIONS)

As mentioned before, the Tsai-Wu criterion is operationally simple; therefore, it is a desirable one for computation. The goal of this computation is twofold: to determine the safety factor for a given loading and to determine the strength components of the lamina referred to any system of coordinates.

The safety factor S_f for a given two-dimensional state of stress $\sigma_i(\sigma_1, \sigma_2, \tau_6)$ is a multiplier that is applied to all stress components to produce a critical or failure state as defined by the selected failure criterion, say, the Tsai-Wu criterion. Thus, for a given state of stress $(\sigma_1, \sigma_2, \tau_6)$, the state of stress at failure is $(S_f\sigma_1, S_f\sigma_2, S_f\tau_6)$. Substitution of the critical stresses in the Tsai-Wu criterion in Eq. (6.29) yields

$$f_1 S_f \sigma_1 + f_2 S_f \sigma_2 + f_{11} S_f^2 \sigma_1^2 + f_{22} S_f^2 \sigma_2^2 + f_{66} S_f^2 \tau_6^2 + 2f_{12} S_f^2 \sigma_1 \sigma_2 = 1 \qquad (6.56)$$

or

$$a S_f^2 + b S_f - 1 = 0 \qquad (6.57)$$

where

$$a = f_{11}\sigma_1^2 + f_{22}\sigma_2^2 + f_{66}\tau_6^2 + 2f_{12}\sigma_1\sigma_2$$
$$b = f_1\sigma_1 + f_2\sigma_2 \qquad (6.58)$$

Thus, the problem of determining the safety factor is reduced to that of solving the quadratic Eq. (6.57). The roots of Eq. (6.57) are

$$S_{fa} = \frac{-b + \sqrt{b^2 + 4a}}{2a} \qquad \text{(actual state of stress)} \qquad (6.59)$$

$$S_{fr} = \left| \frac{-b - \sqrt{b^2 + 4a}}{2a} \right| \qquad \text{(reversed-in-sign state of stress)} \qquad (6.60)$$

A flowchart for computation of the safety factor as well as the strength components of a unidirectional lamina based on the Tsai-Wu criterion is shown in Fig. 6.14. The procedure for determination of safety factors consists of the following steps:

Step 1 Enter the given stress components σ_x, σ_y, and τ_s referred to the (x-y) coordinate system.

Step 2 Enter the fiber orientation θ, that is, the angle between the x-axis and the fiber direction (1-direction) measured positive counterclockwise.

Step 3 Calculate the stress components σ_1, σ_2, and τ_6 referred to the principal material directions using the stress transformation relations in Eq. (4.57).

Step 4 Enter the basic lamina strengths F_{1t}, F_{1c}, F_{2t}, F_{2c}, and F_6 for the material.

Step 5 Compute the Tsai-Wu coefficients $f_1, f_2, f_{11}, f_{22}, f_{66}$, and f_{12} using Eqs. (6.32), (6.33), (6.35), and (6.40).

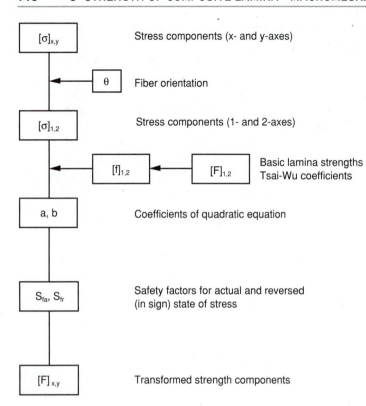

Fig. 6.14 Flowchart for computation of safety factors and transformed lamina strengths based on the Tsai-Wu failure criterion.

Step 6 Compute coefficients a and b of the quadratic Eq. (6.57) using Eq. (6.58).

Step 7 Obtain the roots of the quadratic equation S_{fa} and S_{fr} for the actual and reversed (in sign) states of stress, respectively.

The procedure above is modified slightly for computation of the transformed lamina strengths, by introducing separately a unit applied normal stress in the x- and y-directions and a unit applied shear stress.

To obtain the strength components F_{xt} and F_{xc} along the x-axis the following state of stress is entered.

$$\sigma_x = 1$$
$$\sigma_y = 0 \qquad (6.61)$$
$$\tau_s = 0$$

Then, the roots of the quadratic equation will give the strength components along the x-axis as

$$F_{xt} = S_{fa} \qquad \text{(tensile strength)}$$
$$F_{xc} = S_{fr} \qquad \text{(compressive strength)}$$

To obtain the strength components along the y-axis, the following stresses are entered:

$$\sigma_x = 0$$
$$\sigma_y = 1 \tag{6.62}$$
$$\tau_s = 0$$

Then, the strength components are

$$F_{yt} = S_{fa} \quad \text{(tensile strength)}$$
$$F_{yc} = S_{fr} \quad \text{(compressive strength)}$$

To obtain the shear strength F_s referred to the x- and y-axes, the following stresses are entered:

$$\sigma_x = 0$$
$$\sigma_y = 0 \tag{6.63}$$
$$\tau_s = 1$$

which yield the following shear strengths:

$$F_s^{(+)} = S_{fa} \quad \text{(positive shear strength)}$$
$$F_s^{(-)} = S_{fr} \quad \text{(negative shear strength)}$$

SAMPLE PROBLEM 6.1
Transformation of Shear Strength

Given the basic lamina strengths F_1, F_2, and F_6, it is required to determine the shear strength F_s at 45° to the fiber direction according to the Tsai-Hill criterion (Fig. 6.15). The transformed stresses along the principal material directions are

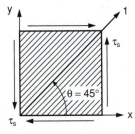

Fig. 6.15 Unidirectional lamina under pure shear loading at 45° to the fiber direction.

$$\sigma_1 = 2mn\tau_s = \tau_s$$
$$\sigma_2 = -2mn\tau_s = -\tau_s \tag{6.64}$$
$$\tau_6 = (m^2 - n^2)\tau_s = 0$$

Substitution into the Tsai-Hill criterion in Eq. (6.16) yields

$$\frac{\tau_s^2}{F_1^2} + \frac{\tau_s^2}{F_2^2} + \frac{\tau_s^2}{F_1^2} = 1$$

At failure, $\tau_s = F_s$ and

$$\frac{1}{F_s^2} = \frac{2}{F_1^2} + \frac{1}{F_2^2} \qquad (6.65)$$

which can be solved for F_s.

For a high-fiber-strength composite with $F_1 \gg F_2$,

$$F_s \cong F_2 \qquad (6.66)$$

or

$$F_s^{(+)} \cong F_{2c}$$
$$F_s^{(-)} \cong F_{2t} \qquad (6.67)$$

As discussed in Section 6.1, this result shows that the positive shear strength at 45° to the fiber direction is controlled by the transverse compressive strength F_{2c} of the lamina, whereas the negative shear strength is controlled by the transverse tensile strength F_{2t}. The same approximate result is obtained by using the Tsai-Wu criterion, the Hashin-Rotem criteria, or the maximum stress theory.

SAMPLE PROBLEM 6.2
Biaxial Strength

Consider a unidirectional lamina loaded under equal biaxial normal stress $\sigma_x = \sigma_y = \sigma_0$ at an angle θ with the fiber direction (Fig. 6.16). It is required to determine the biaxial strength F_0 according to the Tsai-Wu criterion.

The transformed stresses along the principal material directions are

$$\sigma_1 = \sigma_x m^2 + \sigma_y n^2 = \sigma_0$$
$$\sigma_2 = \sigma_x n^2 + \sigma_y m^2 = \sigma_0 \qquad (6.68)$$
$$\tau_6 = (\sigma_y - \sigma_x)mn = 0$$

that is, transformation does not change the hydrostatic nature of the state of stress. At failure,

$$\sigma_1 = \sigma_2 = \sigma_0 = F_0 \qquad (6.69)$$

Substitution into the Tsai-Wu criterion, Eq. (6.29), yields

$$F_0(f_1 + f_2) + F_0^2(f_{11} + f_{22} + 2f_{12}) = 1 \qquad (6.70)$$

which can be solved for F_0.

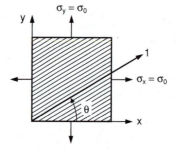

Fig. 6.16 Unidirectional lamina under equal biaxial normal loading.

For a high-strength composite with $F_1 \gg F_2$, it follows that

$$f_1 \ll f_2$$
$$f_{11} \ll f_{22} \tag{6.71}$$
$$f_{12} \ll f_{22}$$

and Eq. (6.70) yields

$$F_0 \cong \frac{-f_2 \pm \sqrt{f_2^2 + 4f_{22}}}{2f_{22}} \tag{6.72}$$

After substitution of f_2 and f_{22} from relations in Eq. (6.33), the two roots of Eq. (6.72) yield the tensile and compressive biaxial strengths

$$F_{0t} \cong F_{2t}$$
$$F_{0c} \cong F_{2c} \tag{6.73}$$

As expected, the strength under equal biaxial normal loading, which produces a hydrostatic state of stress, is an isotropic property.

6.10 EVALUATION AND APPLICABILITY OF LAMINA FAILURE THEORIES

As mentioned previously in Section 6.2 numerous failure theories have been proposed and are available to the composite structural designer. They are classified into three broad categories. The validity and applicability of a given theory depend on the convenience of application and agreement with experimental results. The plethora of theories is accompanied by a dearth of suitable and reliable experimental data, which makes the selection of one theory over another rather difficult. Considerable effort has been devoted recently to alleviate this difficulty.

C. T. Sun[5] reviewed six failure theories and showed comparisons of theoretical predictions with experimental results for six different composite material systems and various loading conditions. The latter included uniaxial (normal and shear) loading, off-axis loading, and biaxial (normal and shear) loading. It was found, as observed before, that most theories differed little from each other in the first quadrant (tension-tension). The biggest differences among theories occurred under combined transverse compression and shear. In this case, predictions of the Tsai-Wu interactive theory were in better agreement with experimental results than other theories.

A "Worldwide Failure Exercise" was organized and conducted by Hinton, Soden, and Kaddour over a twelve-year period for the purpose of assessing the predictive capabilities of current failure theories. The results of this exercise, encompassing nineteen theories and fourteen problems, appeared in three special issues of the journal *Composites Science and Technology*.[18–20] All results were collected and organized into a book.[7] In the first phase of this exercise (Part A), the developers or advocates of twelve different theories, given the basic lamina properties, presented failure predictions for four material systems, six laminate configurations, and various loading conditions.[18] One observation of this exercise was that, even for the unidirectional lamina, predictions of the various theories differed by up to 200–300% from each other.[21]

In the second phase of the exercise (Part B), the above theoretical predictions, including those of modified versions of the original theories, were compared with experimental results provided by the initiators of the exercise.[19] Comparisons covered a wide range of biaxial stress conditions from which failure envelopes and stress-strain curves to failure were obtained. The third phase of the exercise (Part C) introduced four additional theories and their comparisons with experiments.[20] Overall evaluation and conclusions were presented by the initiators of the exercise.[22,23] Regarding failure of the unidirectional lamina, based on eighteen test cases, best agreement with experimental results was shown by the fully interactive Tsai-Wu approach and the theories by Puck and Schürmann[14] and Cuntze and Freund.[24]

Of the failure theories reviewed and discussed in the literature, five representative and widely used theories were described in detail in this chapter. Comparison of these theories can be graphically illustrated by means of failure envelopes for different materials and different biaxial states of stress. Figures 6.17 and 6.18 show failure envelopes obtained by the five failure theories discussed before, that is, maximum stress, maximum strain, Tsai-Hill, Tsai-Wu, and Hashin-Rotem, for two biaxial states of stress (σ_1-σ_2 and σ_2-τ_6), for a unidirectional carbon/epoxy composite (AS4/3501-6).

The failure envelopes of the interactive theories have quasi-elliptical shapes, whereas those of the noninteractive theories are rectangular or parallelogram-shaped. In the case of biaxial normal loading (σ_1-σ_2), the highest predictions in the tension-tension quadrant ($\sigma_1 > 0$, $\sigma_2 > 0$) are given by the maximum strain theory; in the tension-compression quadrants ($\sigma_1 > 0$, $\sigma_2 < 0$ and $\sigma_1 < 0$, $\sigma_2 > 0$) by the maximum stress theory; and in the compression-compression quadrant ($\sigma_1 < 0$, $\sigma_2 < 0$) by the Tsai-Wu interaction theory.

Since failure modes depend greatly on material properties and type of loading, it would seem that the

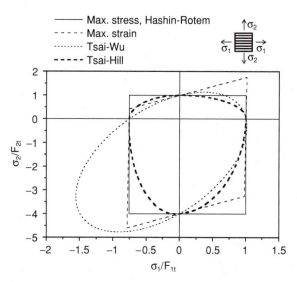

Fig. 6.17 Failure envelopes for unidirectional carbon/epoxy (AS4/3501-6) lamina under biaxial normal loading, obtained by various failure theories.

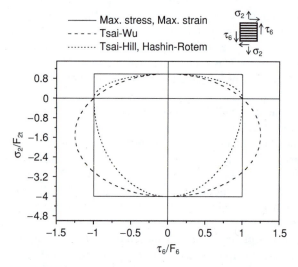

Fig. 6.18 Failure envelopes for unidirectional carbon/epoxy (AS4/3501-6) lamina under transverse normal and shear loading, obtained by various failure theories.

TABLE 6.1 Comparison of Failure Theories

Type	Theory	Physical Basis	Operational Convenience	Required Experimental Characterization
Limit or noninteractive	Maximum stress	Tensile behavior of brittle material $\sigma_1 > 0, \sigma_2 > 0$ No stress interaction	Inconvenient	Few parameters by simple testing
	Maximum strain	Tensile behavior of brittle material $\sigma_1 > 0, \sigma_2 > 0$ Some stress interaction	Inconvenient	Few parameters by simple testing
Interactive	Strain-energy-based (Tsai-Hill)	Ductile behavior of anisotropic materials $\sigma_{1,2} < 0, \tau_6 \neq 0$ "Curve fitting" for heterogeneous brittle composites	Can be programmed Different functions required for tensile and compressive strengths (for each quadrant)	Biaxial testing is needed in addition to uniaxial testing
	Interactive tensor polynomial (Tsai-Wu)	Mathematically consistent Reliable "curve fitting"	General and comprehensive; operationally simple	Numerous parameters Comprehensive experimental program needed
Mixed	Failure mode separation (Hashin-Rotem)	Distinct separation between fiber and interfiber failures	Somewhat inconvenient	Few parameters by simple testing

applicability of the various theories is also related to the type of material and state of stress. A comparison of the five selected failure theories is summarized in Table 6.1 from the points of view of physical basis, operational convenience, and required experimental input.

For example, the maximum stress and strain theories are more applicable when brittle behavior is predominant, typically in the first quadrant of the failure envelope (i.e., $\sigma_1 > 0, \sigma_2 > 0$). Of these two theories, only the maximum strain theory allows for a small degree of stress interaction through Poisson's ratio effect. These theories, although conceptually simple, are inconvenient for computational operations because they consist of several conditional subcriteria.

The interactive theories, such as the Tsai-Hill and Tsai-Wu theories, are more applicable when ductile behavior under shear or compression loading is predominant. The Tsai-Hill theory is based on Hill's theory for ductile anisotropic materials and adapted to the more brittle heterogeneous composites by a form of "curve fitting." Although suitable for computational operations, each quadrant of the failure envelope in the σ_1-σ_2 space requires a different input because of the inability of the theory to distinguish automatically between tensile and compressive strengths. Although the coefficient accounting for the σ_1-σ_2 interaction can be approximated, a more precise determination requires some form of biaxial testing.

The Tsai-Wu theory is mathematically consistent and operationally simple. The additional coefficients in the criterion allow for distinction between tensile and compressive strengths. It represents a more reliable form of "curve fitting." Although the interaction coefficient f_{12} can be approximated, a more precise determination requires biaxial testing. However, determination of f_{12} by means of a biaxial tensile test or an off-axis test as discussed before is not very accurate, because the Tsai-Wu criterion under these loading conditions is not very sensitive to the value of f_{12}. This interaction coefficient is more important when compression and/or shear are present. For this reason, it should be determined by means of a test that produces such a state of stress, for example, an off-axis compressive test. Because of the fully interactive nature of the Tsai-Wu criterion, some stress components may have a greater weight under certain loading conditions. This is the case, for example, with off-axis shear loading for fiber orientations in the range $\theta = 20\text{--}30°$, which maximizes the influence of the shear strength F_6.[5] For this reason, the property F_6 must be measured accurately.

The Hashin-Rotem theory uses both limit and interaction criteria. Sometimes it coincides with the maximum stress criterion, as in the case of biaxial normal loading (σ_1, σ_2) (Fig. 6.17), and sometimes it coincides with the Tsai-Hill criterion, as in the case of transverse normal and shear (σ_2, τ_6) biaxial loading (Fig. 6.18).

Composite materials that exhibit pronounced transitions between brittle and ductile behavior with type of loading would be best described by hybrid or combined failure criteria. Figure 6.19 shows such a hybrid failure envelope combining the maximum strain theory in the first quadrant ($\sigma_1 > 0$, $\sigma_2 > 0$) and an interactive theory in the remaining quadrants. When the material behavior and mode of failure are not known and when a conservative approach is desired, it is recommended to check a number of theories and select the most conservative envelope in each quadrant (Fig. 6.20).

The accuracy and predictive capabilities of the various theories can be assessed by comparing theoretical predictions with experimental results for various biaxial (or multiaxial) states of stress. Uniaxial off-axis tensile tests usually yield reliable, albeit limited in scope, biaxial data corresponding to the tension-tension quadrant of the failure envelope. Figure 6.21 shows the variation of the off-axis tensile strength with fiber orientation as predicted by various theories and compared with experimental results for a boron/epoxy composite (Sun[5] and Pipes and Cole[25]). It is seen that for the orientation range of $15° < \theta < 75°$, the interaction criteria of the Tsai-Wu, Tsai-Hill, and Hashin-Rotem theories agree well with experimental results. The limit criteria

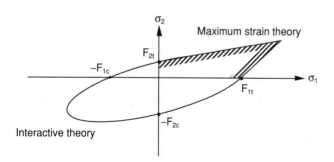

Fig. 6.19 Hybrid failure envelope incorporating two failure criteria.

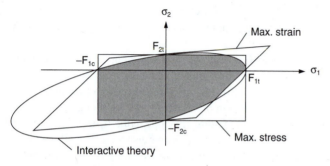

Fig. 6.20 Illustration of "conservative approach" in design by using different failure theories.

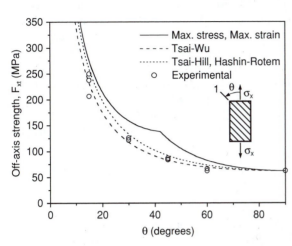

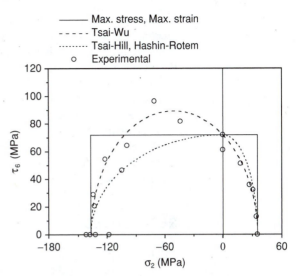

Fig. 6.21 Comparison of various failure theories and off-axis strength data for unidirectional boron/epoxy (AVCO 5505) lamina under off-axis tensile loading (Sun[5] and Pipes and Cole[25]).

Fig. 6.22 Failure envelopes for unidirectional glass/epoxy (E-glass/LY556) lamina under transverse normal and shear loading, obtained by various failure theories, compared with experimental results (Hinton et al.[22] and Hütter et al.[26]).

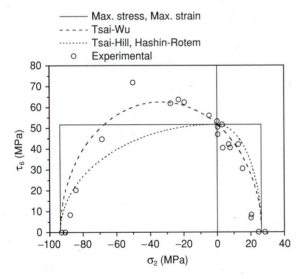

Fig. 6.23 Failure envelopes for unidirectional carbon/epoxy (AS4/55A) lamina under transverse normal and shear loading, obtained by various failure theories, compared with experimental results (Sun[5] and Swanson et al.[27])

(maximum stress, maximum strain) deviate from the experimental results, especially in the region of $\theta \cong 40°$, where there is a transition in the dominant failure mode from shear to transverse tension (see Fig. 6.6).

A more critical test is provided by the interaction of in-plane shear stress, τ_6, and transverse normal stress, σ_2. Comparisons between theoretical predictions and experimental results for two materials are shown in Figs. 6.22 and 6.23. As can be seen, in the region of transverse normal tension ($\sigma_2 > 0$) all the interactive criteria (Tsai-Hill, Tsai-Wu, and Hashin-Rotem) are in good agreement with experimental results. In the region of transverse normal compression ($\sigma_2 < 0$) only the Tsai-Wu criterion is in reasonable agreement with experimental results. The compressive normal stress (σ_2) allows for shear stresses to reach values in excess of the in-plane shear strength of the material, a fact not predicted by the Tsai-Hill and Hashin-Rotem theories. It appears that in the case above, the fully interactive Tsai-Wu theory is in better agreement with experiments although the failure mode is interfiber (matrix dominated) and the contribution of the longitudinal normal stress σ_1 can be neglected.

Another challenging test of the theories would be in the compression-compression quadrant ($\sigma_1 < 0$, $\sigma_2 < 0$) as seen in Fig. 6.17. However, no suitable experimental data are available for a unidirectional lamina. Another case of discrepancy between theories is in the regime of high tensile longitudinal stress (σ_1) and shear stress (τ_6), which occurs in off-axis testing at small angles ($\theta < 10°$). Hashin,[17] in his modified theory, proposed an interaction criterion for σ_1 and τ_6 that is relevant for this case.

In general, interactive theories based on curve fitting, like the Tsai-Wu theory, are better at predicting failure of a single lamina. All comparisons were made with very limited available experimental data in two-dimensional form. A comprehensive evaluation of the failure theories in three-dimensional form would require data under three-dimensional states of stress.

REFERENCES

1. R. S. Sandhu, "A Survey of Failure Theories of Isotropic and Anisotropic Materials," Air Force Flight Dynamics Laboratory, Technical Report AFFDL-TR-72-71, Wright Aeronautical Labs, Dayton, OH, 1972.
2. M. J. Owen and D. J. Rice, "Biaxial Strength Behavior of Glass Reinforced Polyester Resins," *Composite Materials: Testing and Design* (Sixth Conference), ASTM STP 787, I. M. Daniel, Ed., American Society for Testing and Materials, West Conshohocken, PA, 1982, pp. 124–144.
3. S. W. Tsai, "A Survey of Macroscopic Failure Criteria for Composite Materials," *J. Reinforced Plastics and Composites*, Vol. 3, 1984, pp. 40–62.
4. A. S. Chen and F. L. Matthews, "A Review of Multiaxial/Biaxial Loading Tests for Composite Materials," *Composites*, Vol. 24, 1993, pp. 395–406.
5. C. T. Sun, "Strength Analysis of Unidirectional Composites and Laminates," in *Comprehensive Composite Materials*, A. Kelly and C. Zweben, Ed., Ch. 1.20, Elsevier Science, Ltd., Oxford, 2000.
6. R. M. Christensen, "A Survey of and Evaluation Methodology for Fiber Composite Material Failure Theories," in *Mechanics for a New Millennium*, H. Aref and J. W. Phillips, Ed., Kluwer Academic Publishers, Dordrecht, 2001.
7. M. J. Hinton, P. D. Soden, and A. S. Kaddour, *Failure Criteria in Fiber-Reinforced-Polymer Composites*, Elsevier, Oxford, 2004.
8. V. D. Azzi and S. W. Tsai, "Anisotropic Strength of Composites," *Exp. Mech.*, Vol. 5, No. 9, 1965, pp. 283–288.
9. R. Hill, "A Theory of the Yielding and Plastic Flow of Anisotropic Metals," *Proceedings of the Royal Society*, Series A, Vol. 193, 1948.
10. A. Kelly, *Strong Solids*, Clarendon Press, Oxford, 1966.
11. O. Ishai, "Failure of Unidirectional Composites in Tension," *J. Eng. Mech. Div.*, Proc. of the ASCE, Vol. 97, No. EM2, 1971, pp. 205–221.
12. S. W. Tsai and E. M. Wu, "A General Theory of Strength for Anisotropic Materials," *J. Composite Materials*, Vol. 5, 1971, pp. 58–80.
13. Z. Hashin and A. Rotem, "A Fatigue Failure Criterion for Fiber Reinforced Materials," *J. Composite Materials*, Vol. 7, 1973, pp. 448–464.
14. A. Puck and H. Schürmann, "Failure Theories of FRP Laminates by Means of Physically Based Phenomenological Models," *Composites Science and Technology*, Vol. 58, 1998, pp. 1045–1067.
15. I. I. Gol'denblat and V. A. Kopnov, "Strength of Glass-Reinforced Plastics in Complex Stress State," *Mekhanika Polimerov*, Vol. 1, 1965, pp. 70–78 (in Russian); English translation: *Polymer Mechanics*, Vol. 1, Faraday Press, 1966, p. 54.

16. K.-S. Liu and S. W. Tsai, "A Progressive Quadratic Failure Criterion for a Laminate," *Composites Science and Technology*, Vol. 58, 1998, pp. 1023–1032.

17. Z. Hashin, "Failure Criteria for Unidirectional Fiber Composites," *J. Applied Mechanics*, Vol. 47, 1980, pp. 329–334.

18. *Composites Science and Technology*, Vol. 58, 1998, pp. 999–1254.

19. *Composites Science and Technology*, Vol. 62, 2002, pp. 1479–1797.

20. *Composites Science and Technology*, Vol. 64, 2004, pp. 319–605.

21. P. D. Soden, M. J. Hinton, and A. S. Kaddour, "A Comparison of the Predictive Capabilities of Current Failure Theories for Composite Laminates," *Composites Science and Technology*, Vol. 58, 1998, pp. 1225–1254.

22. M. J. Hinton, A. S. Kaddour, and P. D. Soden, "A Comparison of the Predictive Capabilities of Current Failure Theories for Composite Laminates, Judged Against Experimental Evidence," *Composites Science and Technology*, Vol. 62, 2002, pp. 1725–1798.

23. M. J. Hinton, A. S. Kaddour, and P. D. Soden, "A Further Assessment of the Predictive Capabilities of Current Failure Theories for Composite Laminates: Comparison with Experimental Evidence," *Composites Science and Technology*, Vol. 64, 2004, pp. 549–588.

24. R. G. Cuntze and A. Freund, "The Predictive Capability of Failure Mode Concept-Based Strength Criteria for Multidirectional Laminates," *Composites Science and Technology*, Vol. 64, 2004, pp. 343–377.

25. R. B. Pipes and B. W. Cole, "On the Off-Axis Strength Test for Anisotropic Materials," *J. Composite Materials*, Vol. 7, 1973, pp. 246–256.

26. U. Hütter, H. Schelling, and H. Krauss, "An Experimental Study to Determine Failure Envelope of Composite Materials with Tubular Specimen Under Combined Loads and Comparison with Several Classical Criteria," in *Failure Modes of Composite Materials with Organic Matrices and Their Consequences on Design*, NATO, AGARD, Conf. Proc. No. 163, Munich, Germany, 13–19 October 1974.

27. S. R. Swanson, M. J. Messick, and Z. Tian, "Failure of Carbon/Epoxy Lamina Under Combined Stress," *J. Composite Materials*, Vol. 21, 1987, pp. 619–630.

PROBLEMS

6.1 A unidirectional lamina is under biaxial normal loading $\sigma_x = -2\sigma_y = 2\sigma_0$ at 45° to the fiber direction as shown in Fig. P6.1. The basic strength properties of the material are $F_{1t} = F_{1c} = 3F_{2c} = 5F_6 = 12F_{2t} =$ 600 MPa (87 ksi). Determine the stress level σ_0'' at failure of the lamina according to the maximum stress theory. What is the failure mode?

6.2 For the off-axis lamina under positive and negative shear loading as shown in Fig. P6.2, express the

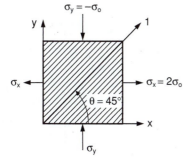

Fig. P6.1

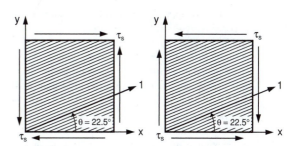

Fig. P6.2

positive and negative shear strengths $F_s^{(+)}$ and $F_s^{(-)}$ in terms of the basic lamina strengths $(F_{1t}, F_{1c}, \ldots)$ using the maximum stress theory. Obtain a relationship between $F_s^{(+)}$ and $F_s^{(-)}$ assuming $F_{1t} > F_{1c} \gg F_{2c} > F_6 > F_{2t}$ and $F_6 = 2F_{2t}$.

6.3 A thin-wall cylindrical pressure vessel, made of a unidirectional composite lamina with the fiber direction at 45° to the cylinder axis, is loaded under internal pressure p (Fig. P6.3). Using the maximum stress criterion, determine the pressure p at failure in terms of geometrical (D, h) and strength $(F_{1t}, F_{1c}, F_{2t}, F_{2c}, F_6)$ parameters, assuming $F_{1t} \gg F_6 > F_{2t}$.

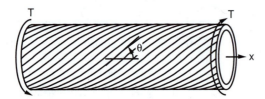

Fig. P6.3

6.4 A thin-wall tube made of a unidirectional composite lamina with a fiber direction θ with its axis is loaded in torsion as shown in Fig. P6.4. Using the maximum stress criterion determine the transition angle θ_{sc} at which the failure mode shifts from shear to transverse compression, for the following material properties: $E_1 = 5E_2$, $\nu_{12} = 0.3$, $F_{1t} = 1.5F_{1c} = 7.5F_{2c} = 10F_6 = 20F_{2t} = 1000$ MPa.

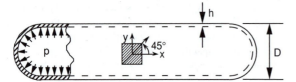

Fig. P6.4

6.5 A thin-wall pipe is made of a unidirectional glass/epoxy with the fiber direction at 30° to its axis and has an allowable pressure p_{1a} (Fig. P6.5). A cylindrical container (closed ends) made of the same material with the fiber direction at 45° to its axis has the same outer diameter but twice the wall thickness of the pipe $(d_2 = d_1, h_2 = 2h_1)$. Relate the allowable pressure p_{2a} of the container to that of the pipe p_{1a} assuming the same safety factor based on the maximum stress theory with the following relations: $F_{1t} \gg F_6 > F_{2t}$.

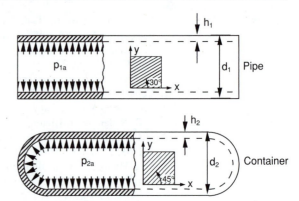

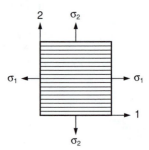

Fig. P6.5

6.6 A unidirectional lamina is under biaxial tensile normal loading as shown in Fig. P6.6. For the material properties of AS4/3501-6 carbon/epoxy given in Table A.4, determine the maximum value that σ_2 can reach at failure, based on the maximum strain theory.

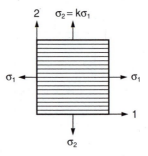

Fig. P6.6

6.7 A unidirectional lamina is under biaxial normal loading (σ_1, σ_2) such that $\sigma_2 = k\sigma_1$ (Fig. P6.7). Using the maximum strain criterion, determine the value

Fig. P6.7

of k (in terms of material parameters E_1, E_2, ν_{12}, F_{1t}, F_{1c}, F_{2t}, and so on) for which the maximum value of σ_1 is reached at failure.

6.8 A unidirectional S-glass/epoxy lamina is loaded in tension at an angle to the fiber direction (Fig. P6.8). Using the maximum strain criterion, determine the off-axis strength, F_{xt}, and the fiber orientation θ at which the predictions of in-plane shear and transverse tensile failure coincide.

$$F_{1t} = 1280 \text{ MPa (185 ksi)}$$
$$F_{2t} = 50 \text{ MPa (7.2 ksi)}$$
$$F_6 = 70 \text{ MPa (10 ksi)}$$
$$\nu_{12} = 0.27$$
$$\nu_{21} = 0.06$$

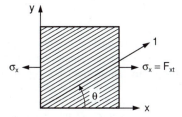

Fig. P6.8

6.9 A 45° off-axis lamina is loaded under uniaxial tension and shear of equal magnitude, $\sigma_x = \tau_s$ (Fig. P6.9). Using the maximum strain theory, determine the strength $\sigma_x^u = \tau_s^u = F_0$ in terms of the shear strength F_6 for a material with properties $F_{1t} = 10F_{2c} = 30F_6$, $\nu_{12} = 0.30$, and $\nu_{21} = 0.02$.

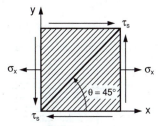

Fig. P6.9

6.10 For the off-axis lamina under positive and negative shear stress as shown in Fig. P6.10, and using the maximum strain theory, express the positive and negative shear strengths, $F_s^{(+)}$ and $F_s^{(-)}$, in terms of the basic lamina strengths (F_{1t}, F_{1c}, ...) and material Poisson's ratios. Assume $F_{1t} > F_{1c} \gg F_{2c} > F_{2t}$, $F_6 = F_{2t}$, and $F_{2c} = 3F_{2t}$.

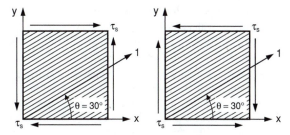

Fig. P6.10

6.11 An off-axis lamina is loaded under biaxial compression $\sigma_x = 2\sigma_y = -F_0$ as shown in Fig. P6.11. Calculate the ultimate value F_0 using the maximum strain theory for the AS4/3501-6 carbon/epoxy (see Table A.4).

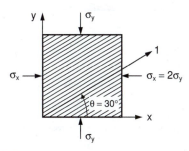

Fig. P6.11

6.12 A unidirectional carbon/epoxy lamina with fiber direction at 30° to the reference x-axis is loaded under pure shear (case 1) and equal biaxial normal tensile stress (case 2) as shown in Fig. P6.12. It was found for case 1 that the positive shear strength was four times the negative shear strength, $F_s^{(+)} = 4F_s^{(-)}$. Based on this finding, calculate the biaxial strength

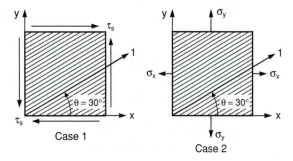

Fig. P6.12

$\sigma_x^u = \sigma_y^u = F_{(xy)}$ for case 2 using the maximum strain theory with the following material properties: $F_{1t} = F_{1c} = 10F_{2c} = 16F_6 = 1600$ MPa (230 ksi) $\gg F_{2t}$; $E_1 \gg E_2$; $v_{12} = 0.25$.

6.13 A thin-wall tube made of a unidirectional lamina with a fiber direction θ to its axis is loaded in torsion as shown in Fig. P6.13. Using the maximum strain theory, plot torsional strength, T_f, versus θ (for $0° \le \theta \le 90°$) for the following material properties: $E_1 = 10E_2$, $v_{12} = 0.25$, $F_{1t} = F_{1c} = 2F_{2c} = 4F_6 = 10F_{2t} = 1000$ MPa (145 ksi).

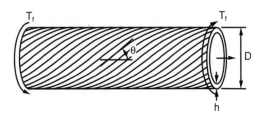

Fig. P6.13

6.14 For the data in Problem 6.13, calculate the transition fiber direction angles θ_1 and θ_2 at which the failure mode shifts from shear to transverse compression and from transverse compression to shear, respectively.

6.15 For the geometry and loading of Problem 6.13 and using the maximum strain criterion, determine the transition angle θ_{sc} at which the failure mode shifts from shear to transverse compression, for the following material properties: $E_1 = 5E_2$, $v_{12} = 0.3$, $F_{1t} = 1.5F_{1c} = 7.5F_{2c} = 10F_6 = 20F_{2t} = 1000$ MPa.

6.16 Express the Tsai-Hill failure criterion for pure shear loading of a lamina at an angle θ to the principal material axes and find an expression for the shear strength $\tau_s^u = F_s$ in terms of F_1, F_2, and F_6 (Fig. P6.16).

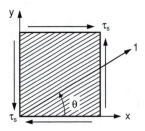

Fig. P6.16

6.17 A unidirectional lamina is loaded in pure shear τ_s at an angle $\theta = 30°$ to the fiber direction (Fig. P6.17). Determine the shear stress at failure $\tau_s^u = F_s$ using the Tsai-Hill failure criterion for AS4/3501-6 carbon/epoxy (see Table A.4).

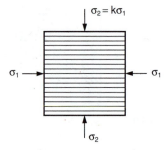

Fig. P6.17

6.18 The unidirectional carbon/epoxy lamina of Problem 6.17 is loaded under pure shear. It was found that the positive shear strength was four times the negative shear strength, $F_s^{(+)} = 4F_s^{(-)}$. Based on this finding, obtain an approximate value for the transverse tensile strength F_{2t} of the material using the Tsai-Hill theory with the following material properties: $F_{1t} = F_{1c} = 10F_{2c} = 16F_6 = 1600$ MPa (230 ksi) $\gg F_{2t}$; $E_1 \gg E_2$; $v_{12} = 0.25$.

6.19 For the lamina shown in Fig. P6.19 loaded under biaxial compression, $\sigma_2 = k\sigma_1$, determine the value of k (in terms of strength parameters F_{1t}, F_{1c}, and so on) for which the maximum value of $|\sigma_1|$ is reached at failure, using the Tsai-Hill criterion.

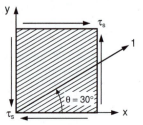

Fig. P6.19

6.20 An off-axis lamina is loaded as shown in Fig. P6.20. Determine $\sigma_x^u = 2\sigma_y^u = -F_0$ at failure using the Tsai-Hill and maximum stress criteria for AS4/3501-6 carbon/epoxy (Table A.4).

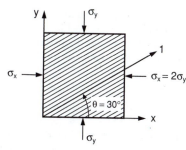

Fig. P6.20

6.21 Using the Tsai-Hill failure criterion, determine the strength of a lamina under uniaxial tension and shear of equal magnitude at 45° to the fiber direction, that is, determine $\sigma_x^u = \tau_s^u = F_0$ at failure in terms of F_1, F_2, and F_6 (Fig. P6.21).

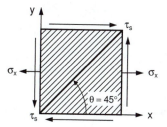

Fig. P6.21

6.22 Using the Tsai-Hill failure criterion, determine the strength of a lamina under equal biaxial tension and shear at 45° to the fiber direction $\sigma_x = \sigma_y = 2\tau_s$, that is, determine $\sigma_x^u = \sigma_y^u = 2\tau_s^u = F_0$ at failure in terms of F_1 and F_2 (Fig. P6.22).

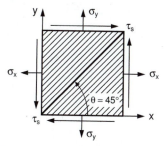

Fig. P6.22

6.23 A 45° off-axis lamina is loaded under the biaxial stresses $\sigma_x = -\sigma_y = 2\tau_s$ (Fig. P6.23). Using the Tsai-Hill failure criterion, determine the strength

$\sigma_x^u = -\sigma_y^u = 2\tau_s^u = F_0$ for an E-glass/epoxy material with properties listed in Table A.4.

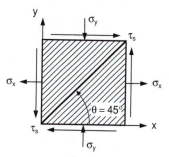

Fig. P6.23

6.24 A thin-wall tube is made of a unidirectional carbon/ epoxy with fiber direction at 45° to its axis (Fig. P6.24). The tube is loaded under combined internal pressure and torsion producing a normal hoop stress $\sigma_y = \sigma_0$ and a shear stress $\tau_s = 2\sigma_0$. Calculate the ultimate value of σ_0^u at failure of the tube according to the Tsai-Hill failure criterion for the following material properties: $F_{1t} = F_{1c} = 10F_{2c} = 20F_6 = 30F_{2t} = 1800$ MPa (260 ksi).

Fig. P6.24

6.25 The off-axis strength of a unidirectional lamina can be higher than F_1 at some angle between 0° and 90°. Using the Tsai-Hill failure criterion, Eq. (6.19), find a relationship among F_1, F_2, and F_6 such that $F_x > F_1$ for some angle $0° < \theta < 90°$.

6.26 For the same conditions above find a relationship among F_1, F_2, and F_6 such that $F_x < F_2$ for some angle $0° < \theta < 90°$.

6.27 Using the Tsai-Wu failure criterion for pure shear loading of a lamina at 45° to the fiber direction, express the shear stress at failure $\tau_s^u = F_s$ in terms of the Tsai-Wu coefficients (Fig. P6.27). Then, obtain an approximate expression when $F_{1t} > F_{1c} \gg F_{2c} > F_{2t}$.

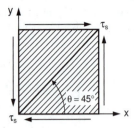

Fig. P6.27

6.28 Using the Tsai-Wu failure criterion for pure shear loading of a lamina at an angle θ to the fiber direction, express the shear stress at failure $\tau_s^u = F_s$ in terms of the Tsai-Wu coefficients (Fig. P6.28). Find approximate relation for composites with much higher longitudinal than transverse strengths.

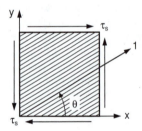

Fig. P6.28

6.29 For the off-axis lamina under positive and negative shear stress as shown in Fig. P6.29, express the positive and negative shear strengths $F_s^{(+)}$ and $F_s^{(-)}$, using the Tsai-Wu failure theory in terms of the polynomial coefficients (f_1, f_2, f_{11}, f_{22}, f_{12}). Then, obtain approximate values for $F_s^{(+)}$ and $F_s^{(-)}$ in terms of the basic lamina strengths (F_{1t}, F_{1c}, . . .) by assuming $f_1 \ll f_2$, $f_{11} \ll f_{22}$, and $f_{12} \ll f_{22}$.

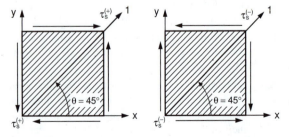

Fig. P6.29

6.30 For the off-axis lamina of Problem 6.29, obtain expressions for the coefficient f_{12} of the Tsai-Wu criterion in terms of the basic strength parameters and the positive or negative shear strength, $F_s^{(+)}$ or $F_s^{(-)}$. Compare the values of f_{12} based on $F_s^{(+)}$ and $F_s^{(-)}$ by assuming $f_1 \ll f_2$ and $f_{11} \ll f_{22}$.

6.31 Using the Tsai-Wu criterion, determine the strength of a lamina under the loading $-\sigma_x = \sigma_y = F_0$ (Fig. P6.31). Obtain first the exact solution for the ultimate value of F_0 in terms of the Tsai-Wu coefficients f_1, f_2, f_{11}, and so on. Then, obtain an approximate solution in terms of lamina strengths (F_{1t}, F_{1c}, F_{2t}, and so on) for high-strength composites, that is, when $f_1 \ll f_2$, $f_{11} \ll f_{22}$, and $f_{12} \ll f_{22}$.

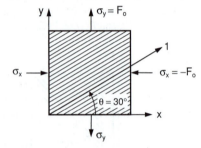

Fig. P6.31

6.32 An off-axis lamina is under biaxial loading $\sigma_x = 40$ MPa (5.8 ksi), $\sigma_y = -120$ MPa (-17.4 ksi) (Fig. P6.32). Determine the safety factor by using the Tsai-Wu criterion and the maximum stress theory for an S-glass/epoxy material (Table A.4).

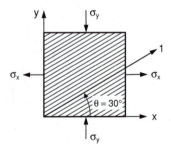

Fig. P6.32

6.33 An off-axis lamina is under biaxial loading $\sigma_x = -\sigma_y = -\sqrt{3}\,\tau_s = 50$ MPa (7.2 ksi) (Fig. P6.33). Determine the safety factor using the Tsai-Wu criterion for a carbon/epoxy material (AS4/3501-6, Table A.4). Show that the answer is the same for all other four theories discussed.

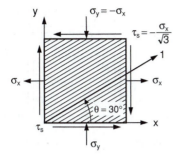

Fig. P6.33

6.34 An off-axis lamina is under biaxial loading $-\sigma_x = \sigma_y = \tau_s = 50$ MPa (7.2 ksi) (Fig. P6.34). Determine the safety factor using the Tsai-Wu and Tsai-Hill criteria for a carbon/epoxy composite (AS4/3501-6, Table A.4).

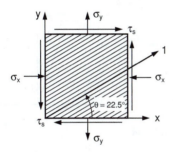

Fig. P6.34

6.35 Three failure envelopes are illustrated in Fig. P6.35 for a given unidirectional lamina based on three

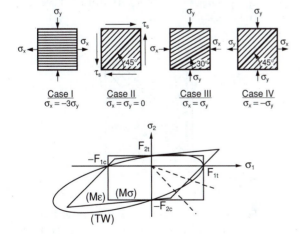

Fig. P6.35

failure theories: maximum stress ($M\sigma$), maximum strain ($M\varepsilon$), and Tsai-Wu (TW). Rank the three theories from the most conservative to the least conservative for each one of the loading cases shown in Fig. P6.35. (Hint: Plot σ_1 versus σ_2, the loading path, for each loading case and relate to the failure envelopes.)

6.36 For the off-axis lamina under positive and negative shear stress as shown in Fig. P6.36, and using the Hashin-Rotem criteria, express the positive and negative shear strengths, $F_s^{(+)}$ and $F_s^{(-)}$, in terms of the basic lamina strengths ($F_{1t}, F_{1c}, \dots$). Obtain first exact expressions, and then approximations for a high-strength composite ($F_1 \gg F_6, F_1 \gg F_2$).

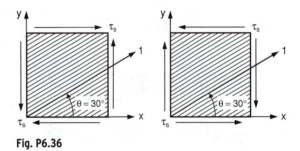

Fig. P6.36

6.37 For the lamina and loading of Problem 6.32, determine the safety factor using the Hashin-Rotem criteria.

6.38 A 30° off-axis lamina is loaded as shown in Fig. P6.38. Using the Hashin-Rotem criteria determine $\sigma_x^u = \tau_s^u = F_0$ at failure for the lamina properties

$$F_{1t} = 855 \text{ MPa (125 ksi)}$$
$$F_{1c} = 1030 \text{ MPa (150 ksi)}$$
$$F_{2t} = 48 \text{ MPa (7 ksi)}$$
$$F_{2c} = 165 \text{ MPa (24 ksi)}$$
$$F_6 = 52 \text{ MPa (7.5 ksi)}$$

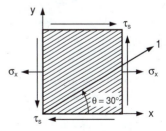

Fig. P6.38

6.39 For the same lamina and loading as in Problem 6.23, determine F_0 at failure using the Hashin-Rotem criteria.

6.40 For the same lamina and loading as in Problem 6.34, determine the safety factor using the Hashin-Rotem criteria.

6.41 The composite thin-wall pressure vessel shown in Fig. P6.41 is made of a unidirectional glass/epoxy material with the fiber direction at 60° to its axis and loaded by internal pressure p. Find the maximum pressure at failure by the Hashin-Rotem criteria for the data

$$D = 5 \text{ cm } (1.97 \text{ in})$$
$$h = 1 \text{ mm } (0.04 \text{ in})$$
$$F_{1t} = 1080 \text{ MPa } (155 \text{ ksi})$$
$$F_{2t} = 50 \text{ MPa } (7.2 \text{ ksi})$$
$$F_6 = 90 \text{ MPa } (13.0 \text{ ksi})$$

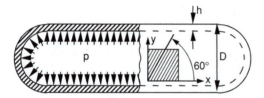

Fig. P6.41

6.42 Plot the shear strength F_s of a unidirectional E-glass/epoxy laminate (see Table A.4) as a function of fiber orientation θ, using the maximum stress, maximum strain, Tsai-Hill, and Tsai-Wu theories (Fig. P6.42).

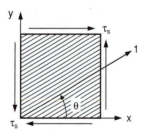

Fig. P6.42

6.43 Determine the shear strength F_s of an off-axis lamina with a fiber orientation $\theta = 35°$ using the maximum stress, Tsai-Wu, and Hashin-Rotem theories for a carbon/epoxy composite (AS4/3501-6, Table A.4) (Fig. P6.43). Compare results.

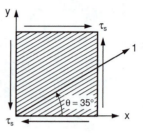

Fig. P6.43

6.44 Determine the shear strength F_s of an off-axis lamina with a fiber orientation $\theta = 110°$ using the maximum stress, Tsai-Wu, and Tsai-Hill theories for a carbon/epoxy composite (IM7/977-3, Table A.4) (Fig. P6.44). Compare results.

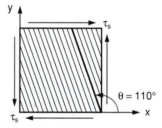

Fig. P6.44

6.45 Determine the shear strength F_s of an off-axis lamina with principal material axis at $\theta = 45°$ using the maximum stress, maximum strain, Tsai-Wu, Tsai-Hill, and Hashin-Rotem theories for a woven carbon/epoxy composite (AGP370-5H/3501-6S, Table A.5) (Fig. P6.45).

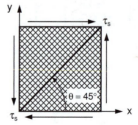

Fig. P6.45

6.46 For the same lamina and loading as in Problem 6.11, determine the ultimate value F_0 using the maximum stress, Tsai-Hill, Tsai-Wu, and Hashin-

Rotem theories. Compare all results with that of Problem 6.11.

6.47 For the 45° off-axis lamina under biaxial stress $\sigma_x = \frac{1}{2}\tau_s = F_0$ determine the ultimate value of F_0 using the maximum stress, Tsai-Wu, and Tsai-Hill theories for the AS4/3501-6 carbon/epoxy (Table A.4) (Fig. P6.47).

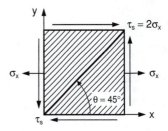

Fig. P6.47

7 Elastic Behavior of Multidirectional Laminates

It is apparent that the overall behavior of a multidirectional laminate is a function of the properties and stacking sequence of the individual layers. The so-called *classical lamination theory* predicts the behavior of the laminate within the framework of the following assumptions and restrictions:[1–3]

1. Each layer (lamina) of the laminate is quasi-homogeneous and orthotropic.
2. The laminate is thin with its lateral dimensions much larger than its thickness and is loaded in its plane only, that is, the laminate and its layers (except for their edges) are in a state of plane stress ($\sigma_z = \tau_{xz} = \tau_{yz} = 0$).
3. All displacements are small compared with the thickness of the laminate ($|u|$, $|v|$, $|w| \ll h$).
4. Displacements are continuous throughout the laminate.
5. In-plane displacements vary linearly through the thickness of the laminate, that is, u and v displacements in the x- and y-directions are linear functions of z.
6. Straight lines normal to the middle surface remain straight and normal to that surface after deformation. This implies that transverse shear strains γ_{xz} and γ_{yz} are zero.
7. Strain-displacement and stress-strain relations are linear.
8. Normal distances from the middle surface remain constant, that is, the transverse normal strain ε_z is zero. This implies that the transverse displacement w is independent of the thickness coordinate z.

7.2 STRAIN-DISPLACEMENT RELATIONS

Figure 7.1 shows a section of the laminate normal to the y-axis before and after deformation. The x-y plane is equidistant from the top and bottom surfaces of the laminate and is called the *midplane* or *reference plane*.

The reference plane displacements u_o and v_o in the x- and y-directions and the out-of-plane displacement w in the z-direction are functions of x and y only:

$$u_o = u_o(x, y)$$

$$v_o = v_o(x, y) \qquad (7.1)$$

$$w = f(x, y)$$

The rotations of the x- and y-axes are

$$\alpha_x = \frac{\partial w}{\partial x}$$

$$\qquad (7.2)$$

$$\alpha_y = \frac{\partial w}{\partial y}$$

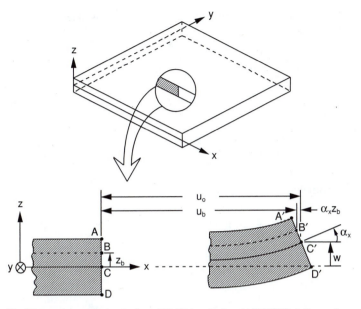

Fig. 7.1 Laminate section before (ABCD) and after (A′B′C′D′) deformation.

The in-plane displacement components of a point B of coordinate z_b (Fig. 7.1) are

$$u_b = u_o - \alpha_x z_b$$

$$\qquad (7.3)$$

$$v_b = v_o - \alpha_y z_b$$

and in general,

$$u = u_o - z\frac{\partial w}{\partial x}$$

$$\qquad (7.4)$$

$$v = v_o - z\frac{\partial w}{\partial y}$$

where z is the through-the-thickness coordinate of a general point of the cross section.

For small displacements, the classical strain-displacement relations of elasticity yield

$$\varepsilon_x = \frac{\partial u}{\partial x} = \frac{\partial u_o}{\partial x} - z\frac{\partial^2 w}{\partial x^2}$$

$$\varepsilon_y = \frac{\partial v}{\partial y} = \frac{\partial v_o}{\partial y} - z\frac{\partial^2 w}{\partial y^2} \qquad (7.5)$$

$$\gamma_{xy} = \gamma_s = \frac{\partial u}{\partial y} + \frac{\partial v}{\partial x} = \frac{\partial u_o}{\partial y} + \frac{\partial v_o}{\partial x} - 2z\frac{\partial^2 w}{\partial x \partial y}$$

Noting that the strain components on the reference plane are expressed as

$$\varepsilon_x^o = \frac{\partial u_o}{\partial x}$$

$$\varepsilon_y^o = \frac{\partial v_o}{\partial y}$$ (7.6)

$$\gamma_{xy}^o = \gamma_s^o = \frac{\partial u_o}{\partial y} + \frac{\partial v_o}{\partial x}$$

and the curvatures of the laminate as

$$\kappa_x = -\frac{\partial^2 w}{\partial x^2}$$

$$\kappa_y = -\frac{\partial^2 w}{\partial y^2}$$ (7.7)

$$\kappa_{xy} = \kappa_s = -\frac{2\partial^2 w}{\partial x \partial y}$$

we can relate the strains at any point in the laminate to the reference plane strains and the laminate curvatures as follows:

$$\begin{bmatrix} \varepsilon_x \\ \varepsilon_y \\ \gamma_s \end{bmatrix} = \begin{bmatrix} \varepsilon_x^o \\ \varepsilon_y^o \\ \gamma_s^o \end{bmatrix} + z \begin{bmatrix} \kappa_x \\ \kappa_y \\ \kappa_s \end{bmatrix}$$ (7.8)

7.3 STRESS-STRAIN RELATIONS OF A LAYER WITHIN A LAMINATE

Consider an individual layer k in a multidirectional laminate whose midplane is at a distance $\bar{z}_k$ from the laminate reference plane (Fig. 7.2). The stress-strain relations for this layer referred to its material axes are

$$\begin{bmatrix} \sigma_1 \\ \sigma_2 \\ \tau_6 \end{bmatrix}_k = \begin{bmatrix} Q_{11} & Q_{12} & 0 \\ Q_{21} & Q_{22} & 0 \\ 0 & 0 & Q_{66} \end{bmatrix}_k \begin{bmatrix} \varepsilon_1 \\ \varepsilon_2 \\ \gamma_6 \end{bmatrix}_k$$ (7.9)

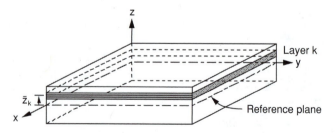

Fig. 7.2 Layer k within laminate.

and after transformation to the laminate coordinate system,

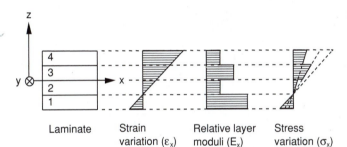

Fig. 7.3 Illustration of linear strain variation and discontinuous stress variation in multidirectional laminate.

Laminate — Strain variation (ε_x) — Relative layer moduli (E_x) — Stress variation (σ_x)

$$
\begin{bmatrix} \sigma_x \\ \sigma_y \\ \tau_s \end{bmatrix}_k = \begin{bmatrix} Q_{xx} & Q_{xy} & Q_{xs} \\ Q_{yx} & Q_{yy} & Q_{ys} \\ Q_{sx} & Q_{sy} & Q_{ss} \end{bmatrix}_k \begin{bmatrix} \varepsilon_x \\ \varepsilon_y \\ \gamma_s \end{bmatrix}_k \tag{7.10}
$$

Substituting the expressions for the strains from Eq. (7.8), we obtain

$$
\begin{bmatrix} \sigma_x \\ \sigma_y \\ \tau_s \end{bmatrix}_k = \begin{bmatrix} Q_{xx} & Q_{xy} & Q_{xs} \\ Q_{yx} & Q_{yy} & Q_{ys} \\ Q_{sx} & Q_{sy} & Q_{ss} \end{bmatrix}_k \begin{bmatrix} \varepsilon_x^o \\ \varepsilon_y^o \\ \gamma_s^o \end{bmatrix}_k + z \begin{bmatrix} Q_{xx} & Q_{xy} & Q_{xs} \\ Q_{yx} & Q_{yy} & Q_{ys} \\ Q_{sx} & Q_{sy} & Q_{ss} \end{bmatrix}_k \begin{bmatrix} \kappa_x \\ \kappa_y \\ \kappa_s \end{bmatrix} \tag{7.11}
$$

or, in brief,

$$
[\sigma]_{x,y}^k = [Q]_{x,y}^k [\varepsilon^o]_{x,y} + z[Q]_{x,y}^k [\kappa]_{x,y}
$$

From Eqs. (7.8) and (7.11) it is seen that, whereas the strains vary linearly through the thickness, the stresses do not. Because of the discontinuous variation of the transformed stiffness matrix $[Q]_{x,y}$ from layer to layer, the stresses may also vary discontinuously from layer to layer. This is illustrated by the hypothetical four-layer laminate in Fig. 7.3 under uniaxial stress in the x-direction. For a certain linear strain variation through the thickness, which can result from axial and flexural loading, the variation of the modulus E_x from layer to layer can cause the discontinuous stress variation illustrated. In many applications the stress gradient through the layer thickness is disregarded. The average stresses in each layer are determined by knowing the reference plane strains $[\varepsilon^o]_{x,y}$, the curvatures $[\kappa]_{x,y}$ of the laminate, the location of the layer midplane $\bar{z}_k$, and its transformed stiffness matrix $[Q]_{x,y}$.

7.4 FORCE AND MOMENT RESULTANTS

Because of the discontinuous variation of stresses from layer to layer, it is more convenient to deal with the integrated effect of these stresses on the laminate. Thus, we seek expressions relating forces and moments to laminate deformation. The stresses acting on a layer k of a laminate (Fig. 7.2) given by Eq. (7.11) can be replaced by resultant forces and moments as shown in Fig. 7.4, as

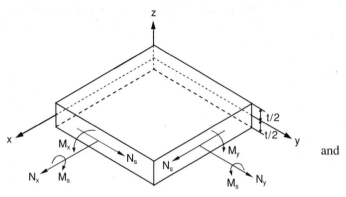

$$N_x^k = \int_{-t/2}^{t/2} \sigma_x dz$$

$$N_y^k = \int_{-t/2}^{t/2} \sigma_y dz \qquad (7.12)$$

$$N_s^k = \int_{-t/2}^{t/2} \tau_s dz$$

Fig. 7.4 Element of single layer with force and moment resultants.

and

$$M_x^k = \int_{-t/2}^{t/2} \sigma_x z dz$$

$$M_y^k = \int_{-t/2}^{t/2} \sigma_y z dz \qquad (7.13)$$

$$M_s^k = \int_{-t/2}^{t/2} \tau_s z dz$$

where

$z =$ through-the-thickness coordinate of a point in the cross section

$t =$ layer thickness

$N_x^k, N_y^k =$ normal forces per unit length

$N_s^k =$ shear force per unit length

$M_x^k, M_y^k =$ bending moments per unit length

$M_s^k =$ twisting moment per unit length

In the case of a multilayer laminate, the total force and moment resultants are obtained by summing the effects for all layers. Thus, for the n-ply laminate in Fig. 7.5, the force and moment resultants are obtained as

$$\begin{bmatrix} N_x \\ N_y \\ N_s \end{bmatrix} = \sum_{k=1}^{n} \int_{z_{k-1}}^{z_k} \begin{bmatrix} \sigma_x \\ \sigma_y \\ \tau_s \end{bmatrix}_k dz \qquad (7.14)$$

and

$$\begin{bmatrix} M_x \\ M_y \\ M_s \end{bmatrix} = \sum_{k=1}^{n} \int_{z_{k-1}}^{z_k} \begin{bmatrix} \sigma_x \\ \sigma_y \\ \tau_s \end{bmatrix}_k z dz \qquad (7.15)$$

where z_k and z_{k-1} are the z-coordinates of the upper and lower surfaces of layer k.

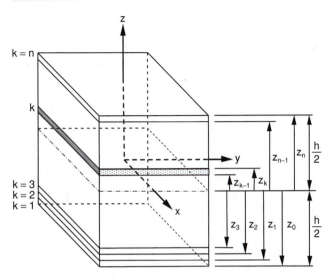

7.5 GENERAL LOAD-DEFORMATION RELATIONS: LAMINATE STIFFNESSES

Substituting Eq. (7.11) for the layer stresses in Eqs. (7.14) and (7.15), we obtain

$$
\begin{bmatrix} N_x \\ N_y \\ N_s \end{bmatrix} = \sum_{k=1}^{n} \left\{ \begin{bmatrix} Q_{xx} & Q_{xy} & Q_{xs} \\ Q_{yx} & Q_{yy} & Q_{ys} \\ Q_{sx} & Q_{sy} & Q_{ss} \end{bmatrix}_k \begin{bmatrix} \varepsilon_x^o \\ \varepsilon_y^o \\ \gamma_s^o \end{bmatrix} \int_{z_{k-1}}^{z_k} dz + \begin{bmatrix} Q_{xx} & Q_{xy} & Q_{xs} \\ Q_{yx} & Q_{yy} & Q_{ys} \\ Q_{sx} & Q_{sy} & Q_{ss} \end{bmatrix}_k \begin{bmatrix} \kappa_x \\ \kappa_y \\ \kappa_s \end{bmatrix} \int_{z_{k-1}}^{z_k} z\,dz \right\} \quad (7.16)
$$

and

$$
\begin{bmatrix} M_x \\ M_y \\ M_s \end{bmatrix} = \sum_{k=1}^{n} \left\{ \begin{bmatrix} Q_{xx} & Q_{xy} & Q_{xs} \\ Q_{yx} & Q_{yy} & Q_{ys} \\ Q_{sx} & Q_{sy} & Q_{ss} \end{bmatrix}_k \begin{bmatrix} \varepsilon_x^o \\ \varepsilon_y^o \\ \gamma_s^o \end{bmatrix} \int_{z_{k-1}}^{z_k} z\,dz + \begin{bmatrix} Q_{xx} & Q_{xy} & Q_{xs} \\ Q_{yx} & Q_{yy} & Q_{ys} \\ Q_{sx} & Q_{sy} & Q_{ss} \end{bmatrix}_k \begin{bmatrix} \kappa_x \\ \kappa_y \\ \kappa_s \end{bmatrix} \int_{z_{k-1}}^{z_k} z^2\,dz \right\} \quad (7.17)
$$

In the expressions above, the stiffnesses $[Q]_{x,y}^k$, reference plane strains $[\varepsilon^o]_{x,y}$, and curvatures $[\kappa]_{x,y}$ are taken outside the integration operation since they are not functions of z. Of these quantities only the stiffnesses are unique for each layer k, whereas the reference plane strains and curvatures refer to the entire laminate and are the same for all plies. Thus, $[\varepsilon^o]_{x,y}$ and $[\kappa]_{x,y}$ can be factored outside the summation sign as follows:

$$
[N]_{x,y} = \left[\sum_{k=1}^{n} [Q]_{x,y}^k \int_{z_{k-1}}^{z_k} dz \right] [\varepsilon^o]_{x,y} + \left[\sum_{k=1}^{n} [Q]_{x,y}^k \int_{z_{k-1}}^{z_k} z\,dz \right] [\kappa]_{x,y}
$$

$$
= \left[\sum_{k=1}^{n} [Q]_{x,y}^k (z_k - z_{k-1}) \right] [\varepsilon^o]_{x,y} + \left[\frac{1}{2} \sum_{k=1}^{n} [Q]_{x,y}^k (z_k^2 - z_{k-1}^2) \right] [\kappa]_{x,y}
$$

$$
= [A]_{x,y} [\varepsilon^o]_{x,y} + [B]_{x,y} [\kappa]_{x,y} \quad (7.18)
$$

and

$$[M]_{x,y} = \left[\frac{1}{2}\sum_{k=1}^{n}[Q]_{x,y}^{k}(z_k^2 - z_{k-1}^2)\right][\varepsilon^o]_{x,y} + \left[\frac{1}{3}\sum_{k=1}^{n}[Q]_{x,y}^{k}(z_k^3 - z_{k-1}^3)\right][\kappa]_{x,y}$$

$$= [B]_{x,y}[\varepsilon^o]_{x,y} + [D]_{x,y}[\kappa]_{x,y} \tag{7.19}$$

where

$$A_{ij} = \sum_{k=1}^{n}Q_{ij}^k(z_k - z_{k-1})$$

$$B_{ij} = \frac{1}{2}\sum_{k=1}^{n}Q_{ij}^k(z_k^2 - z_{k-1}^2) \tag{7.20}$$

$$D_{ij} = \frac{1}{3}\sum_{k=1}^{n}Q_{ij}^k(z_k^3 - z_{k-1}^3)$$

with $i, j = x, y, s$.

Thus, in full form the force-deformation relations are

$$\begin{bmatrix} N_x \\ N_y \\ N_s \end{bmatrix} = \begin{bmatrix} A_{xx} & A_{xy} & A_{xs} \\ A_{yx} & A_{yy} & A_{ys} \\ A_{sx} & A_{sy} & A_{ss} \end{bmatrix}\begin{bmatrix} \varepsilon_x^o \\ \varepsilon_y^o \\ \gamma_s^o \end{bmatrix} + \begin{bmatrix} B_{xx} & B_{xy} & B_{xs} \\ B_{yx} & B_{yy} & B_{ys} \\ B_{sx} & B_{sy} & B_{ss} \end{bmatrix}\begin{bmatrix} \kappa_x \\ \kappa_y \\ \kappa_s \end{bmatrix} \tag{7.21}$$

and the moment-deformation relations are

$$\begin{bmatrix} M_x \\ M_y \\ M_s \end{bmatrix} = \begin{bmatrix} B_{xx} & B_{xy} & B_{xs} \\ B_{yx} & B_{yy} & B_{ys} \\ B_{sx} & B_{sy} & B_{ss} \end{bmatrix}\begin{bmatrix} \varepsilon_x^o \\ \varepsilon_y^o \\ \gamma_s^o \end{bmatrix} + \begin{bmatrix} D_{xx} & D_{xy} & D_{xs} \\ D_{yx} & D_{yy} & D_{ys} \\ D_{sx} & D_{sy} & D_{ss} \end{bmatrix}\begin{bmatrix} \kappa_x \\ \kappa_y \\ \kappa_s \end{bmatrix} \tag{7.22}$$

The expressions above can be combined into one general expression relating in-plane forces and moments to reference plane strains and curvatures:

$$\begin{bmatrix} N_x \\ N_y \\ N_s \\ M_x \\ M_y \\ M_s \end{bmatrix} = \begin{bmatrix} A_{xx} & A_{xy} & A_{xs} & B_{xx} & B_{xy} & B_{xs} \\ A_{yx} & A_{yy} & A_{ys} & B_{yx} & B_{yy} & B_{ys} \\ A_{sx} & A_{sy} & A_{ss} & B_{sx} & B_{sy} & B_{ss} \\ B_{xx} & B_{xy} & B_{xs} & D_{xx} & D_{xy} & D_{xs} \\ B_{yx} & B_{yy} & B_{ys} & D_{yx} & D_{yy} & D_{ys} \\ B_{sx} & B_{sy} & B_{ss} & D_{sx} & D_{sy} & D_{ss} \end{bmatrix}\begin{bmatrix} \varepsilon_x^o \\ \varepsilon_y^o \\ \gamma_s^o \\ \kappa_x \\ \kappa_y \\ \kappa_s \end{bmatrix} \tag{7.23}$$

or, in brief,

$$\begin{bmatrix} N \\ M \end{bmatrix} = \begin{bmatrix} A & B \\ B & D \end{bmatrix}\begin{bmatrix} \varepsilon^o \\ \kappa \end{bmatrix} \tag{7.24}$$

It should be noted that all the above matrices are symmetric, that is,

$$A_{ij} = A_{ji}$$

$$B_{ij} = B_{ji} \qquad (i, j = x, y, s)$$

$$D_{ij} = D_{ji}$$

The relations above are expressed in terms of three laminate stiffness matrices, $[A]$, $[B]$, and $[D]$, which are functions of the geometry, material properties, and stacking sequence of the individual plies, as defined in Eq. (7.20). They are the average elastic parameters of the multidirectional laminate with the following significance:

- A_{ij} are extensional stiffnesses, or in-plane laminate moduli, relating in-plane loads to in-plane strains.
- B_{ij} are coupling stiffnesses, or in-plane/flexure coupling laminate moduli, relating in-plane loads to curvatures and moments to in-plane strains. Thus, if $B_{ij} \neq 0$, in-plane forces produce flexural and twisting deformation in addition to in-plane deformation; moments produce extensional and shear deformation of the middle surface in addition to flexural and twisting deformation.
- D_{ij} are bending or flexural laminate stiffnesses relating moments to curvatures.

7.6 INVERSION OF LOAD-DEFORMATION RELATIONS: LAMINATE COMPLIANCES

Since multidirectional laminates are characterized by stress discontinuities from ply to ply, it is preferable to work with strains, which are continuous through the thickness. For this reason it is necessary to invert the load-deformation relations, Eqs. (7.23), and express strains and curvatures as a function of applied loads and moments.

Equations (7.23) can be rewritten as follows by performing the appropriate matrix inversions:

$$
\begin{bmatrix} \varepsilon_x^o \\ \varepsilon_y^o \\ \gamma_s^o \\ \kappa_x \\ \kappa_y \\ \kappa_s \end{bmatrix}
=
\begin{bmatrix}
a_{xx} & a_{xy} & a_{xs} & b_{xx} & b_{xy} & b_{xs} \\
a_{yx} & a_{yy} & a_{ys} & b_{yx} & b_{yy} & b_{ys} \\
a_{sx} & a_{sy} & a_{ss} & b_{sx} & b_{sy} & b_{ss} \\
c_{xx} & c_{xy} & c_{xs} & d_{xx} & d_{xy} & d_{xs} \\
c_{yx} & c_{yy} & c_{ys} & d_{yx} & d_{yy} & d_{ys} \\
c_{sx} & c_{sy} & c_{ss} & d_{sx} & d_{sy} & d_{ss}
\end{bmatrix}
\begin{bmatrix} N_x \\ N_y \\ N_s \\ M_x \\ M_y \\ M_s \end{bmatrix}
\tag{7.25}
$$

or, in brief,

$$
\begin{bmatrix} \varepsilon^o \\ \kappa \end{bmatrix}
=
\begin{bmatrix} a & b \\ c & d \end{bmatrix}
\begin{bmatrix} N \\ M \end{bmatrix}
\tag{7.26}
$$

Here, matrices $[a]$, $[b]$, $[c]$, and $[d]$ are the laminate compliance matrices obtained from the stiffness matrices as follows:[3]

$$[a] = [A^{-1}] - \{[B^*][D^{*-1}]\}[C^*]$$

$$[b] = [B^*][D^{*-1}]$$

$$[c] = -[D^{*-1}][C^*]$$

$$[d] = [D^{*-1}]$$

(7.27)

where

$$[A^{-1}] = \text{inverse of matrix } [A]$$

$$[B^*] = -[A^{-1}][B]$$

$$[C^*] = [B][A^{-1}]$$

$$[D^*] = [D] - \{[B][A^{-1}]\}[B]$$

From the inverse relations of Eqs. (7.24) and (7.26), it follows that

$$\begin{bmatrix} a & b \\ c & d \end{bmatrix} = \begin{bmatrix} A & B \\ B & D \end{bmatrix}^{-1}$$

(7.28)

that is, the 6×6 compliance matrix is the inverse of the 6×6 stiffness matrix. Since the stiffness matrices $[A]$, $[B]$, and $[D]$ and the combined 6×6 stiffness matrix are symmetric, the 6×6 compliance matrix in Eq. (7.28) and the individual matrices $[a]$ and $[d]$ are also symmetric. However, $[b]$ and $[c]$ need not be symmetric nor equal to each other. In fact $[c] = [b]^T$, that is, matrix $[c]$ is the transpose of matrix $[b]$ obtained from it by interchanging columns and rows.

The observations above can be illustrated for a specific laminate. For example, the calculated stiffness matrices of a $[0/\pm30/90]$ carbon/epoxy laminate (AS4/3501-6, Table A.4 in Appendix A) are

$$[A] = \begin{bmatrix} 42.95 & 7.35 & 0 \\ 7.35 & 25.50 & 0 \\ 0 & 0 & 9.49 \end{bmatrix} \text{(MN/m)}$$

$$[B] = \begin{bmatrix} -3.32 & 0 & -0.70 \\ 0 & 3.32 & -0.26 \\ -0.70 & -0.26 & 0 \end{bmatrix} \text{(kN)}$$

$$[D] = \begin{bmatrix} 0.88 & 0.06 & 0 \\ 0.06 & 0.78 & 0 \\ 0 & 0 & 0.11 \end{bmatrix} \text{(N-m)}$$

As mentioned before, these matrices as well as the combined 6×6 matrix shown in Eqs. (7.23) and (7.24) are symmetric.

The calculated compliance matrices are

$$[a] = \begin{bmatrix} 4.43 & -2.00 & 1.45 \\ -2.00 & 10.36 & -1.61 \\ 1.45 & -1.61 & 12.03 \end{bmatrix} (10^{-8}\,\text{m/N})$$

$$[b] = \begin{bmatrix} 17.29 & 7.55 & 22.54 \\ -5.68 & -43.71 & 11.92 \\ 14.23 & 9.65 & 5.15 \end{bmatrix} (10^{-5}/\text{N})$$

$$[c] = \begin{bmatrix} 17.29 & -5.68 & 14.23 \\ 7.55 & -43.71 & 9.65 \\ 22.54 & 11.92 & 5.15 \end{bmatrix} (10^{-5}/\text{N})$$

$$[d] = \begin{bmatrix} 1.93 & 0.13 & 0.93 \\ 0.13 & 3.18 & -0.56 \\ 0.93 & -0.56 & 10.83 \end{bmatrix} (1/\text{N-m})$$

As mentioned before, it is seen in this example that matrices $[a]$ and $[d]$ are symmetric, whereas $[b]$ and $[c]$ are not. Matrix $[c]$ is the transpose of matrix $[b]$. However, the combined 6×6 compliance matrix in Eq. (7.25) is symmetric and is the inverse of the 6×6 stiffness matrix in Eq. (7.23), as indicated in Eq. (7.28).

7.7 SYMMETRIC LAMINATES

A laminate is called *symmetric* when for each layer on one side of a reference plane (middle surface) there is a corresponding layer at an equal distance from the reference plane on the other side with identical thickness, orientation, and properties. The laminate is symmetric in both geometry and material properties.

Consider the n-layer laminate in Fig. 7.6, where identical layers k and k' are symmetrically situated about the reference plane. Then,

$$t_k = t_{k'}$$
$$Q_{ij}^k = Q_{ij}^{k'} \qquad (i, j = x, y, s) \qquad (7.29)$$
$$\bar{z}_k = -\bar{z}_{k'} \qquad (z\text{-coordinates of lamina midplanes})$$

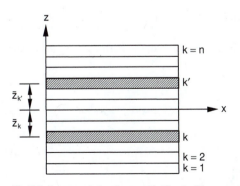

Fig. 7.6 Symmetric laminate with identical layers k and k'.

and according to the definition of Eq. (7.20), the coupling stiffnesses are

$$B_{ij} = \frac{1}{2} \sum_{k=1}^{n} Q_{ij}^k (z_k^2 - z_{k-1}^2) = \frac{1}{2} \sum_{k=1}^{n} Q_{ij}^k (z_k + z_{k-1})(z_k - z_{k-1}) = \sum_{k=1}^{n} Q_{ij}^k \bar{z}_k t_k$$

since

$$\bar{z}_k = \frac{1}{2}(z_k + z_{k-1})$$

and

$$t_k = z_k - z_{k-1}$$

For the conditions of symmetry stated before, the above sum will consist of pairs of terms of equal absolute value and opposite sign. Thus, for a symmetric laminate,

$$B_{ij} = 0 \qquad (i, j = x, y, s) \tag{7.30}$$

and the load-deformation relations are reduced to

$$\begin{bmatrix} N_x \\ N_y \\ N_s \end{bmatrix} = \begin{bmatrix} A_{xx} & A_{xy} & A_{xs} \\ A_{yx} & A_{yy} & A_{ys} \\ A_{sx} & A_{sy} & A_{ss} \end{bmatrix} \begin{bmatrix} \varepsilon_x^o \\ \varepsilon_y^o \\ \gamma_s^o \end{bmatrix} \tag{7.31}$$

and

$$\begin{bmatrix} M_x \\ M_y \\ M_s \end{bmatrix} = \begin{bmatrix} D_{xx} & D_{xy} & D_{xs} \\ D_{yx} & D_{yy} & D_{ys} \\ D_{sx} & D_{sy} & D_{ss} \end{bmatrix} \begin{bmatrix} \kappa_x \\ \kappa_y \\ \kappa_s \end{bmatrix} \tag{7.32}$$

In symmetric laminates no coupling exists between in-plane loading and out-of-plane deformation (curvatures) and between bending and twisting moments and in-plane deformation. These laminates exhibit no distortion or warpage after fabrication and are easier to analyze. Some special types of symmetric laminates are discussed below.

7.7.1 Symmetric Laminates with Isotropic Layers

If the layers are isotropic although not all of the same material, the stiffnesses of each pair of symmetrically situated layers k and k' are

$$Q_{xx}^k = Q_{yy}^k = Q_{xx}^{k'} = Q_{yy}^{k'} = \frac{E_k}{1 - v_k^2}$$

$$Q_{xs}^k = Q_{ys}^k = Q_{xs}^{k'} = Q_{ys}^{k'} = 0$$

$$Q_{xy}^k = Q_{xy}^{k'} = \frac{v_k E_k}{1 - v_k^2} \tag{7.33}$$

$$Q_{ss}^k = Q_{ss}^{k'} = \frac{E_k}{2(1 + v_k)}$$

The above relations lead to

$$A_{xx} = A_{yy}$$
$$A_{xs} = A_{ys} = 0$$
$$D_{xx} = D_{yy}$$
$$D_{xs} = D_{ys} = 0$$
(7.34)

and the load-deformation relations take the form

$$\begin{bmatrix} N_x \\ N_y \\ N_s \end{bmatrix} = \begin{bmatrix} A_{xx} & A_{xy} & 0 \\ A_{xy} & A_{xx} & 0 \\ 0 & 0 & A_{ss} \end{bmatrix} \begin{bmatrix} \varepsilon_x^o \\ \varepsilon_y^o \\ \gamma_s^o \end{bmatrix}$$
(7.35)

and

$$\begin{bmatrix} M_x \\ M_y \\ M_s \end{bmatrix} = \begin{bmatrix} D_{xx} & D_{xy} & 0 \\ D_{xy} & D_{xx} & 0 \\ 0 & 0 & D_{ss} \end{bmatrix} \begin{bmatrix} \kappa_x \\ \kappa_y \\ \kappa_s \end{bmatrix}$$
(7.36)

7.7.2 Symmetric Laminates with Specially Orthotropic Layers (Symmetric Crossply Laminates)

In a symmetric laminate with specially orthotropic layers, the principal material axes of each layer coincide with the laminate axes, for example, [0/90/0] and [0/90]$_{ns}$. Because of symmetry, the coupling stiffnesses $B_{ij} = 0$, that is, there is no coupling between in-plane loading and out-of-plane deformation.

Assuming that the kth layer is oriented with its principal 1-direction along the x-axis, we have

$$Q_{xx}^k = Q_{11}^k = \frac{E_1^k}{1 - \nu_{12}^k \nu_{21}^k}$$

$$Q_{xy}^k = Q_{12}^k = \frac{\nu_{21}^k E_1^k}{1 - \nu_{12}^k \nu_{21}^k}$$

$$Q_{yy}^k = Q_{22}^k = \frac{E_2^k}{1 - \nu_{12}^k \nu_{21}^k}$$
(7.37)

$$Q_{xs}^k = Q_{16}^k = 0$$

$$Q_{ys}^k = Q_{26}^k = 0$$

$$Q_{ss}^k = Q_{66}^k = G_{12}^k$$

From the above it follows that

$$A_{xs} = A_{ys} = 0$$
$$D_{xs} = D_{ys} = 0 \tag{7.38}$$

The load-deformation relations then are reduced to

$$
\begin{bmatrix} N_x \\ N_y \\ N_s \end{bmatrix} =
\begin{bmatrix} A_{xx} & A_{xy} & 0 \\ A_{yx} & A_{yy} & 0 \\ 0 & 0 & A_{ss} \end{bmatrix}
\begin{bmatrix} \varepsilon_x^o \\ \varepsilon_y^o \\ \gamma_s^o \end{bmatrix} \tag{7.39}
$$

and

$$
\begin{bmatrix} M_x \\ M_y \\ M_s \end{bmatrix} =
\begin{bmatrix} D_{xx} & D_{xy} & 0 \\ D_{yx} & D_{yy} & 0 \\ 0 & 0 & D_{ss} \end{bmatrix}
\begin{bmatrix} \kappa_x \\ \kappa_y \\ \kappa_s \end{bmatrix} \tag{7.40}
$$

7.7.3 Symmetric Angle-Ply Laminates

Laminates containing plies oriented at $+\theta$ and $-\theta$ directions are called *angle-ply laminates*. They can be symmetric or asymmetric. If such a laminate consists of an odd number of alternating $+\theta$ and $-\theta$ plies of equal thickness, then it is symmetric, for example, $[\theta/-\theta/\theta/-\theta/\theta] = [\pm\theta/\overline{\theta}]_s$.

In a laminate with an odd number of plies n, $(n-1)/2$ plies have orientation θ, $(n-1)/2$ plies have orientation $-\theta$, and one ply has orientation θ. For a ply thickness t and laminate thickness $h = nt$, the shear coupling terms A_{is} $(i = x, y)$ are

$$
A_{is} = \sum_{k=1}^{n} Q_{is}^k(z_k - z_{k-1}) = \sum_{k=1}^{n} Q_{is}^k t = \sum_{k=1}^{(n-1)/2} Q_{is}^k(\theta)t + \sum_{k=1}^{(n-1)/2} Q_{is}^k(-\theta)t + Q_{is}(\theta)t
$$

$$
= Q_{is}(\theta)t = Q_{is}(\theta)\frac{h}{n} \tag{7.41}
$$

since $Q_{is}(\theta) = -Q_{is}(-\theta)$ from Eq. (4.67).

Similarly, it can be shown that $D_{is} \neq 0$. The nonzero A_{is} and D_{is} terms above decrease in inverse proportion to the number of plies for the same overall laminate thickness.

For an angle-ply laminate with an even number of plies, for example, $[\pm\theta]_{ps}$, the in-plane shear coupling terms A_{is} are

$$
A_{is} = \sum_{k=1}^{n} Q_{is}^k t = \sum_{k=1}^{n/2} Q_{is}^k(\theta)t + \sum_{k=1}^{n/2} Q_{is}^k(-\theta)t = 0 \tag{7.42}
$$

However, the coupling terms D_{is} are not zero, but their magnitude decreases in inverse proportion to the number of plies as in the previous case.

7.8 BALANCED LAMINATES

A laminate is balanced if it consists of pairs of layers with identical thickness and elastic properties but having $+\theta$ and $-\theta$ orientations of their principal material axes with respect to the laminate principal axes.

For each balanced pair of layers k and k'

$$t_k = t_{k'}$$

$$\theta_k = -\theta_{k'}$$

Then, it follows from the transformation relations of Eq. (4.67) that

$$Q_{is}(\theta) = -Q_{is}(-\theta)$$

and

$$A_{is} = \sum_{k=1}^{n} Q_{is}^k t_k = 0 \qquad (i = x, y) \tag{7.43}$$

The fact that the in-plane shear coupling stiffnesses A_{is} are zero is a defining characteristic of a balanced laminate. Thus, a laminate consisting of pairs of identical layers of θ and $-\theta$ orientations plus any number of $0°$ or $90°$ plies is also balanced.

A balanced laminate can be symmetric, antisymmetric, or asymmetric. For example, a laminate consisting of pairs of θ_1 and $-\theta_1$ and θ_2 and $-\theta_2$ plies can be arranged in the following layups:

Symmetric: $[\pm\theta_1/\pm\theta_2]_s$

Antisymmetric: $[\theta_1/\theta_2/-\theta_2/-\theta_1]$

Asymmetric: $[\theta_1/\theta_2/-\theta_1/-\theta_2]$

In general, the bending/twisting coupling stiffnesses D_{is} are not zero unless the laminate is antisymmetric, as we will see below.

The general load-deformation relations for this class of laminates are

$$
\begin{bmatrix} N_x \\ N_y \\ N_s \\ \hline M_x \\ M_y \\ M_s \end{bmatrix} =
\begin{bmatrix}
A_{xx} & A_{xy} & 0 & B_{xx} & B_{xy} & B_{xs} \\
A_{yx} & A_{yy} & 0 & B_{yx} & B_{yy} & B_{ys} \\
0 & 0 & A_{ss} & B_{sx} & B_{sy} & B_{ss} \\
\hline
B_{xx} & B_{xy} & B_{xs} & D_{xx} & D_{xy} & D_{xs} \\
B_{yx} & B_{yy} & B_{ys} & D_{yx} & D_{yy} & D_{ys} \\
B_{sx} & B_{sy} & B_{ss} & D_{sx} & D_{sy} & D_{ss}
\end{bmatrix}
\begin{bmatrix} \varepsilon_x^o \\ \varepsilon_y^o \\ \gamma_s^o \\ \hline \kappa_x \\ \kappa_y \\ \kappa_s \end{bmatrix}
\tag{7.44}
$$

7.8.1 Antisymmetric Laminates

An antisymmetric laminate is a special case of a balanced laminate, having its balanced $+\theta$ and $-\theta$ pairs symmetrically situated about the middle surface.

In this case the bending/twisting coupling stiffnesses are

$$D_{is} = \frac{1}{3}\sum_{k=1}^{n} Q_{is}^k(z_k^3 - z_{k-1}^3) = 0 \tag{7.45}$$

since

$$(z_k^3 - z_{k-1}^3) = (z_{k'}^3 - z_{k'-1}^3)$$

and

$$Q_{is}^k = -Q_{is}^{k'}$$

for the symmetrically situated balanced pair of k and k' (or θ and $-\theta$) layers.

The coupling stiffnesses B_{ij} for antisymmetric laminates are in general nonzero, and they vary according to the specific layup. The overall load-deformation relations for this class of laminates are

$$\begin{bmatrix} N_x \\ N_y \\ N_s \\ \hline M_x \\ M_y \\ M_s \end{bmatrix} = \begin{bmatrix} A_{xx} & A_{xy} & 0 & B_{xx} & B_{xy} & B_{xs} \\ A_{yx} & A_{yy} & 0 & B_{yx} & B_{yy} & B_{ys} \\ 0 & 0 & A_{ss} & B_{sx} & B_{sy} & B_{ss} \\ \hline B_{xx} & B_{xy} & B_{xs} & D_{xx} & D_{xy} & 0 \\ B_{yx} & B_{yy} & B_{ys} & D_{yx} & D_{yy} & 0 \\ B_{sx} & B_{sy} & B_{ss} & 0 & 0 & D_{ss} \end{bmatrix} \begin{bmatrix} \varepsilon_x^o \\ \varepsilon_y^o \\ \gamma_{ys}^o \\ \kappa_x \\ \kappa_y \\ \kappa_s \end{bmatrix} \tag{7.46}$$

7.8.2 Antisymmetric Crossply Laminates

Antisymmetric crossply laminates consist of 0° and 90° plies arranged in such a way that for every 0° ply at a distance z from the midplane there is a 90° ply of the same material and thickness at a distance $-z$ from the midplane. By definition then, this laminate has an even number of plies. Antisymmetric laminates with +45° and −45° plies fall in this category, since the reference axes of the laminate can be rotated by 45°.

For every pair k and k' of 0° and 90° plies we have

$$\bar{z}_k = -\bar{z}_{k'}$$

$$t_k = t_{k'}$$

$$Q_{xx}^k = Q_{yy}^{k'}$$

$$Q_{yy}^k = Q_{xx}^{k'}$$

$$Q_{xy}^k = Q_{xy}^{k'}$$

$$Q_{xs}^k = Q_{ys}^k = Q_{xs}^{k'} = Q_{ys}^{k'} = 0$$

$$\tag{7.47}$$

Then, it follows from the definitions of laminate stiffnesses that

$$A_{xx} = A_{yy}$$
$$A_{xs} = A_{ys} = 0$$
$$B_{xx} = -B_{yy}$$
$$B_{xy} = B_{xs} = B_{ys} = B_{ss} = 0 \qquad (7.48)$$
$$D_{xx} = D_{yy}$$
$$D_{xs} = D_{ys} = 0$$

The overall load-deformation relations are

$$
\begin{bmatrix} N_x \\ N_y \\ N_s \\ M_x \\ M_y \\ M_s \end{bmatrix}
=
\begin{bmatrix}
A_{xx} & A_{xy} & 0 & B_{xx} & 0 & 0 \\
A_{yx} & A_{xx} & 0 & 0 & -B_{xx} & 0 \\
0 & 0 & A_{ss} & 0 & 0 & 0 \\
B_{xx} & 0 & 0 & D_{xx} & D_{xy} & 0 \\
0 & -B_{xx} & 0 & D_{yx} & D_{xx} & 0 \\
0 & 0 & 0 & 0 & 0 & D_{ss}
\end{bmatrix}
\begin{bmatrix} \varepsilon_x^o \\ \varepsilon_y^o \\ \gamma_s^o \\ \kappa_x \\ \kappa_y \\ \kappa_s \end{bmatrix}
\qquad (7.49)
$$

For crossply laminates with an even number of alternating 0° and 90° plies, the coupling stiffness B_{xx} decreases in inverse proportion to the number of plies for the same laminate thickness.

SAMPLE PROBLEM 7.1
Stiffnesses of Antisymmetric Crossply Laminate

It is required to derive approximate expressions for stiffnesses A_{xx}, A_{xy}, A_{ss}, B_{xx}, D_{xx}, D_{xy}, and D_{ss} of a [0/90] antisymmetric crossply laminate in terms of the basic lamina properties (E_1, E_2, G_{12}, and ν_{12}) and the lamina thickness t. It is assumed that the composite material contains high-stiffness fibers such that $E_1 \gg E_2$ and $\nu_{21} \ll 1$.

From the definitions of laminate stiffnesses we obtain

$$A_{xx} = \sum_{k=1}^{n} Q_{xx}^k(z_k - z_{k-1}) = (Q_{11} + Q_{22})t = \frac{t}{1 - \nu_{12}\nu_{21}}(E_1 + E_2) \cong (E_1 + E_2)t \cong E_1 t \quad (7.50)$$

$$A_{xy} = \sum_{k=1}^{n} Q_{xy}^k(z_k - z_{k-1}) = 2Q_{12}t = \frac{2\nu_{12}E_2 t}{1 - \nu_{12}\nu_{21}} \cong 2\nu_{12}E_2 t \qquad (7.51)$$

$$A_{ss} = \sum_{k=1}^{n} Q_{ss}^k(z_k - z_{k-1}) = 2G_{12}t \qquad (7.52)$$

$$B_{xx} = \frac{1}{2}\sum_{k=1}^{n} Q_{xx}^k(z_k^2 - z_{k-1}^2) \cong \frac{t^2}{2}(E_1 - E_2) \cong \frac{t^2}{2}E_1 \qquad (7.53)$$

$$D_{xx} = \frac{1}{3}\sum_{k=1}^{n} Q_{xx}^k(z_k^3 - z_{k-1}^3) \cong \frac{t^3}{3}(E_1 + E_2) \cong \frac{t^3}{3}E_1 \tag{7.54}$$

$$D_{xy} = \frac{1}{3}\sum_{k=1}^{n} Q_{xy}^k(z_k^3 - z_{k-1}^3) \cong \frac{2}{3}t^3\nu_{12}E_2 \tag{7.55}$$

$$D_{ss} = \frac{1}{3}\sum_{k=1}^{n} Q_{ss}^k(z_k^3 - z_{k-1}^3) \cong \frac{2}{3}t^3 G_{12} \tag{7.56}$$

7.8.3 Antisymmetric Angle-Ply Laminates

Antisymmetric angle-ply laminates consist of pairs of plies of $+\theta_i$ and $-\theta_i$ orientations $(0 < \theta_i < 90)$, symmetrically situated about the middle plane and having the same thickness and elastic properties. Because of antisymmetry

$$A_{is} = D_{is} = 0 \qquad (i = x, y)$$

For every balanced pair of k and k' plies with orientations θ and $-\theta$ we have

$$\begin{aligned}
\bar{z}_k &= -\bar{z}_{k'} \\
\theta_k &= -\theta_{k'} \\
t_k &= t_{k'} \\
Q_{xx}^k &= Q_{xx}^{k'} \\
Q_{yy}^k &= Q_{yy}^{k'} \\
Q_{xy}^k &= Q_{xy}^{k'} \\
Q_{xs}^k &= -Q_{xs}^{k'} \\
Q_{ys}^k &= -Q_{ys}^{k'} \\
Q_{ss}^k &= Q_{ss}^{k'}
\end{aligned} \tag{7.57}$$

Then, from the definition of B_{ij} it follows that

$$B_{xx} = B_{yy} = B_{xy} = B_{ss} = 0$$

and the overall load-deformation relations take the form

$$\begin{bmatrix} N_x \\ N_y \\ N_s \\ M_x \\ M_y \\ M_s \end{bmatrix} = \begin{bmatrix} A_{xx} & A_{xy} & 0 & 0 & 0 & B_{xs} \\ A_{yx} & A_{yy} & 0 & 0 & 0 & B_{ys} \\ 0 & 0 & A_{ss} & B_{sx} & B_{sy} & 0 \\ 0 & 0 & B_{xs} & D_{xx} & D_{xy} & 0 \\ 0 & 0 & B_{ys} & D_{yx} & D_{yy} & 0 \\ B_{sx} & B_{sy} & 0 & 0 & 0 & D_{ss} \end{bmatrix} \begin{bmatrix} \varepsilon_x^o \\ \varepsilon_y^o \\ \gamma_s^o \\ \kappa_x \\ \kappa_y \\ \kappa_s \end{bmatrix} \tag{7.58}$$

A more special case of this class of laminates is the antisymmetric regular angle-ply laminate, consisting of an even number of plies alternating between θ and $-\theta$ in orientation, that is, $[\theta/-\theta/\theta/ \ldots /\theta/-\theta]$ or $[\pm\theta]_n$.

The nonzero coupling stiffnesses B_{xs} and B_{ys} decrease in inverse proportion to the number of plies for the same overall laminate thickness.

7.9 ORTHOTROPIC LAMINATES: TRANSFORMATION OF LAMINATE STIFFNESSES AND COMPLIANCES

A symmetric and balanced laminate has its ply orientations parallel to or balanced about two perpendicular axes $\bar{x}$ and $\bar{y}$, referred to as *principal laminate axes* (Fig. 7.7). On a macroscopic scale, this laminate can be treated as a homogeneous orthotropic material with the $\bar{x}$ and $\bar{y}$ axes as the principal material axes. This type of laminate is called an *orthotropic laminate*.

By definition, the in-plane/flexure coupling stiffnesses and the in-plane shear coupling stiffnesses are zero, that is,

$$B_{ij} = 0 \qquad (i, j = \bar{x}, \bar{y}, \bar{s})$$

and

$$A_{i\bar{s}} = 0 \qquad (i = \bar{x}, \bar{y})$$

Thus, the force-deformation relations referred to the $\bar{x}$-$\bar{y}$ system of coordinates are

$$\begin{bmatrix} N_{\bar{x}} \\ N_{\bar{y}} \\ N_{\bar{s}} \end{bmatrix} = \begin{bmatrix} A_{\bar{x}\bar{x}} & A_{\bar{x}\bar{y}} & 0 \\ A_{\bar{y}\bar{x}} & A_{\bar{y}\bar{y}} & 0 \\ 0 & 0 & A_{\bar{s}\bar{s}} \end{bmatrix} \begin{bmatrix} \varepsilon_{\bar{x}}^o \\ \varepsilon_{\bar{y}}^o \\ \gamma_{\bar{s}}^o \end{bmatrix} \qquad (7.59)$$

These relations, when referred to an arbitrary x-y system (Fig. 7.7), take the form

$$\begin{bmatrix} N_x \\ N_y \\ N_s \end{bmatrix} = \begin{bmatrix} A_{xx} & A_{xy} & A_{xs} \\ A_{yx} & A_{yy} & A_{ys} \\ A_{sx} & A_{sy} & A_{ss} \end{bmatrix} \begin{bmatrix} \varepsilon_x^o \\ \varepsilon_y^o \\ \gamma_x^o \end{bmatrix} \qquad (7.60)$$

or, in brief,

$$[N]_{x,y} = [A]_{x,y}[\varepsilon^o]_{x,y}$$

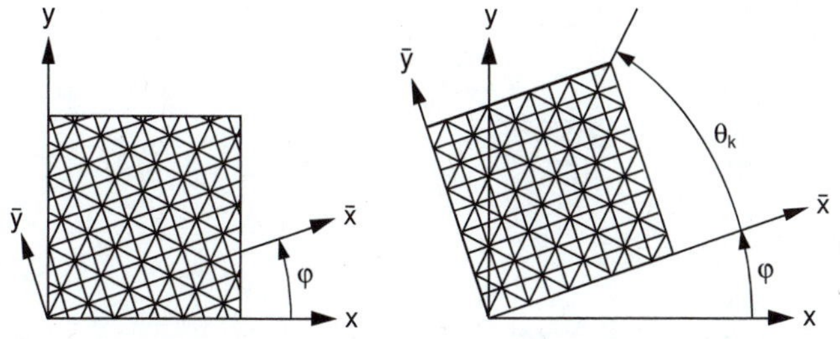

Equations (7.59) and (7.60) are entirely analogous to Eqs. (4.31) and (4.63) for a single lamina, repeated here:

$$\begin{bmatrix} \sigma_1 \\ \sigma_2 \\ \tau_6 \end{bmatrix} = \begin{bmatrix} Q_{11} & Q_{12} & 0 \\ Q_{21} & Q_{22} & 0 \\ 0 & 0 & Q_{66} \end{bmatrix} \begin{bmatrix} \varepsilon_1 \\ \varepsilon_2 \\ \gamma_6 \end{bmatrix} \qquad (4.31 \text{ bis})$$

Fig. 7.7 Notation for coordinate transformation in orthotropic laminate.

TABLE 7.1 Relations for Stiffness and Compliance Transformation of Orthotropic Laminates

	$a_{\bar{x}\bar{x}}(A_{\bar{x}\bar{x}})$	$a_{\bar{y}\bar{y}}(A_{\bar{y}\bar{y}})$	$a_{\bar{x}\bar{y}}(A_{\bar{x}\bar{y}})$	$a_{\bar{s}\bar{s}}(4A_{\bar{s}\bar{s}})$
$a_{xx}(A_{xx})$	m^4	n^4	$2m^2n^2$	m^2n^2
$a_{yy}(A_{yy})$	n^4	m^4	$2m^2n^2$	m^2n^2
$a_{xy}(A_{xy})$	m^2n^2	m^2n^2	$(m^4 + n^4)$	$-m^2n^2$
$a_{ss}(4A_{ss})$	$4m^2n^2$	$4m^2n^2$	$-8m^2n^2$	$(m^2 - n^2)^2$
$a_{xs}(2A_{xs})$	$2m^3n$	$-2mn^3$	$-2mn(m^2 - n^2)$	$-mn(m^2 - n^2)$
$a_{ys}(2A_{ys})$	$2mn^3$	$-2m^3n$	$2mn(m^2 - n^2)$	$mn(m^2 - n^2)$

$m = \cos\varphi, n = \sin\varphi$

and

$$\begin{bmatrix} \sigma_x \\ \sigma_y \\ \tau_s \end{bmatrix} = \begin{bmatrix} Q_{xx} & Q_{xy} & Q_{xs} \\ Q_{yx} & Q_{yy} & Q_{ys} \\ Q_{sx} & Q_{sy} & Q_{ss} \end{bmatrix} \begin{bmatrix} \varepsilon_x \\ \varepsilon_y \\ \gamma_s \end{bmatrix}$$

(4.63 bis)

Then, the transformation relations for the laminate stiffnesses and compliances (A_{ij} and a_{ij} with $i, j = x, y, s$) are identical to those for the lamina stiffnesses and compliances (Q_{ij} and S_{ij} with $i, j = x, y, s$). These relations, presented in Table 7.1, are identical to those of Table 4.2.

The transformation relations of Table 7.1 could be extended to three dimensions by analogy with the three-dimensional transformations for the elastic parameters of a composite lamina (see Table 4.3).

There is a special class of orthotropic laminates for which the elastic properties along its principal directions $\bar{x}$ and $\bar{y}$ are equal. Referring to Fig. 7.7, this condition is expressed as

$$A_{\bar{x}\bar{x}} = A_{\bar{y}\bar{y}}$$

(7.61)

This type of laminate has properties analogous to those of a tetragonal crystal and has *tetragonal symmetry*. Then, from the transformation relations of Table 7.1 we obtain, for any orthogonal system (x, y),

$$A_{xx} = A_{yy}$$

(7.62)

Equation (7.61) can be written as

$$\sum_{k=1}^{n} (Q_{\bar{x}\bar{x}})_k t_k = \sum_{k=1}^{n} (Q_{\bar{y}\bar{y}}) t_k$$

or, assuming equal lamina thicknesses,

$$\sum_{k=1}^{n} (Q_{\bar{x}\bar{x}})_k = \sum_{k=1}^{n} (Q_{\bar{y}\bar{y}})_k$$

Using the transformation relations of Eqs. (4.67), we obtain

$$\sum_{k=1}^{n} (m^4 Q_{11} + n^4 Q_{22} + 2m^2 n^2 Q_{12} + 4m^2 n^2 Q_{66})_k$$

$$= \sum_{k=1}^{n} (n^4 Q_{11} + m^4 Q_{22} + 2m^2 n^2 Q_{12} + 4m^2 n^2 Q_{66})_k$$

or

$$\sum_{k=1}^{n} (m^4 - n^4)_k (Q_{11} - Q_{22}) = 0$$

or

$$\sum_{k=1}^{n} (m^2 - n^2)_k (Q_{11} - Q_{22}) = 0$$

and, since $(Q_{11} - Q_{22})$ is a constant,

$$\sum_{k=1}^{n} (m^2 - n^2)_k = 0 \tag{7.63}$$

or

$$\sum_{k=1}^{n} \cos 2\theta_k = 0 \tag{7.64}$$

where θ_k is the orientation of lamina k. The lamina orientations must satisfy the above relations for condition of Eq. (7.61) to hold, that is, Eq. (7.63) or (7.64) is a necessary and sufficient condition for a laminate to have tetragonal symmetry.

7.10 QUASI-ISOTROPIC LAMINATES

There is a special class of orthotropic laminates for which the elastic properties are independent of orientation, that is, the in-plane stiffnesses and compliances and all engineering elastic constants are identical in all directions. Referring to Fig. 7.8, this condition is expressed as

$$[A]_{\bar{x},\bar{y}} = [A]_{x,y} = \text{constant}$$

$$[a]_{\bar{x},\bar{y}} = [a]_{x,y} = \text{constant} \tag{7.65}$$

$$A_{\bar{x}\bar{s}} = A_{\bar{y}\bar{s}} = A_{xs} = A_{ys} = 0$$

or, in terms of engineering constants,

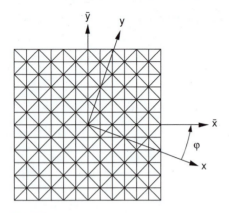

Fig. 7.8 Quasi-isotropic laminate.

$$\bar{E}_{\bar{x}} = \bar{E}_x = \text{constant}$$

$$\bar{G}_{\bar{x}\bar{y}} = \bar{G}_{xy} = \text{constant}$$

$$\bar{\nu}_{\bar{x}\bar{y}} = \bar{\nu}_{xy} = \text{constant} \tag{7.66}$$

$$\bar{\eta}_{\bar{x}\bar{s}} = \bar{\eta}_{\bar{y}\bar{s}} = \bar{\eta}_{xs} = \bar{\eta}_{ys} = 0$$

where symbols with an overbar denote effective laminate properties. All of the above properties are invariant with respect to orientation φ.

The simplest type of quasi-isotropic laminate is one of $[0/60/-60]_s$ layup. Another type is the so-called *π/4 quasi-isotropic laminate*, or $[0/\pm 45/90]_s$. In general any laminate of

$$\left[0/\frac{\pi}{n}/\frac{2\pi}{n}/\ldots/\frac{n-1}{n}\pi \right]_s$$

or

$$\left[\frac{\pi}{n}/\frac{2\pi}{n}/\ldots/\pi \right]_s$$

layup is quasi-isotropic for any integer *n* greater than 2.

SAMPLE PROBLEM 7.2
Quasi-isotropic $[0/\pm 45/90]_s$ Laminate

It is required to prove that the $[0/\pm 45/90]_s$ laminate is quasi-isotropic. Referring to Fig. 7.8 and the specific laminate layup, we observe that

$$A_{\bar{x}\bar{x}} = A_{\bar{y}\bar{y}} = (A_{xx})_{\varphi=45°}$$

$$A_{\bar{s}\bar{s}} = (A_{ss})_{\varphi=45°} \tag{7.67}$$

From the stiffness transformation relations (Table 7.1) and Eq. (7.67) we obtain for φ = 45°

$$(A_{ss})_{\varphi=45°} = A_{\bar{s}\bar{s}} = \frac{1}{2}(A_{\bar{x}\bar{x}} - A_{\bar{x}\bar{y}}) \tag{7.68}$$

The stiffness A_{xx} along any arbitrary direction φ is

$$A_{xx} = (m^4 + n^4)A_{\bar{x}\bar{x}} + 2m^2n^2A_{\bar{x}\bar{y}} + 4m^2n^2A_{\bar{s}\bar{s}}$$

and, in view of Eq. (7.68),

$$A_{xx} = (m^4 + n^4)A_{\bar{x}\bar{x}} + 2m^2n^2(A_{\bar{x}\bar{y}} + A_{\bar{s}\bar{s}} - A_{\bar{x}\bar{y}})$$
$$= (m^4 + n^4 + 2m^2n^2)A_{\bar{x}\bar{x}} = (m^2 + n^2)^2 A_{\bar{x}\bar{x}} = A_{\bar{x}\bar{x}}$$

(7.69)

Similarly we can prove that

$$A_{xy} = A_{\bar{x}\bar{y}}$$

(7.70)

$$A_{ss} = A_{\bar{s}\bar{s}}$$

(7.71)

$$A_{xs} = A_{ys} = 0$$

(7.72)

7.11 DESIGN CONSIDERATIONS

A summary of the characteristic properties of the various types of laminates is given in Table 7.2. It is important to keep these properties in mind in the selection of the appropriate layup and stacking sequence of a laminate. The coupling stiffnesses B_{ij}, A_{is}, and D_{is} complicate the analysis and design of multilayer and multidirectional laminates.

The extension/bending coupling stiffnesses B_{ij} $(i, j = x, y, s)$, coupling in-plane loading with out-of-plane deformation, are responsible for laminate warpage due to cooldown after curing and to hygrothermal environment variations. It is therefore desirable in general to eliminate this type of coupling by selecting a symmetric layup. It is conceivable that in some special designs such coupling might be used to advantage, such as laminate shells subjected to aerodynamic loading (*smart composites*). In other cases of nonsymmetric laminates, it is possible to minimize bending coupling by selecting an appropriate stacking sequence.

TABLE 7.2 Summary of Laminate Characteristics

Symmetric ($B_{ij} = 0$)	Balanced ($A_{xs} = A_{ys} = 0$)	Symmetric/Balanced ($B_{ij} = 0$; $A_{xs} = A_{ys} = 0$)	General ($A_{ij}, B_{ij}, D_{ij} \neq 0$)
Angle-ply, n = odd	Antisymmetric	Crossply symmetric	Isotropic layers
$[\theta/-\theta/\theta/\ldots/\theta]$	$D_{xs} = D_{ys} = 0$	$D_{xs} = D_{ys} = 0$	$A_{xs} = A_{ys} = 0$
$A_{xs}, A_{ys} \neq 0$	$B_{ij} \neq 0$		$B_{xs} = B_{ys} = 0$
$D_{xs}, D_{ys} \neq 0$		Angle-ply	$D_{xs} = D_{ys} = 0$
	Crossply antisymmetric	$[\pm\theta]_{ps}$	$A_{xx} = A_{yy}$
$A_{xs}, A_{ys}, D_{xs}, D_{ys} \propto \dfrac{1}{n}$	$D_{xs} = D_{ys} = 0$	$D_{xs}, D_{ys} \neq 0$	$B_{xx} = B_{yy}$
	$B_{xx} = -B_{yy}$		$D_{xx} = D_{yy}$
	All other $B_{ij} = 0$	$D_{xs}, D_{ys} \propto \dfrac{1}{n}$	
			Specially orthotropic layers
	Angle-ply antisymmetric	Tetragonal	$A_{xs} = A_{ys} = 0$
	$[\pm\theta]_p$	$A_{xx} = A_{yy}$	$B_{xs} = B_{ys} = 0$
	$D_{xs} = D_{ys} = 0$	Quasi-isotropic	$D_{xs} = D_{ys} = 0$
	$B_{xs}, B_{ys} \neq 0$	$A_{ij}, a_{ij}, E_{i,j}$	
	All other $B_{ij} = 0$	Independent of reference axes	

Shear coupling stiffnesses A_{xs} and A_{ys} cause in-plane shear deformations under in-plane normal loading, and normal in-plane deformation under in-plane shear loading. These stiffnesses become zero for a balanced or a crossply layup. For unbalanced laminates it is advisable to select a stacking sequence that minimizes these shear coupling stiffnesses.

Torsion coupling stiffnesses D_{xs} and D_{ys} are responsible for twisting deformation under cylindrical flexure and may produce interlaminar stresses under bending. These stiffnesses are zero only for antisymmetric (special case of balanced) or crossply layups. In other cases it is possible to minimize the torsion coupling stiffnesses by a proper choice of stacking sequence.

The only layup for which all three types of coupling stiffnesses B_{ij}, A_{is}, and D_{is} are zero is the crossply symmetric layup, for example, $[0/90]_s$. One can design symmetric and balanced laminates where $B_{ij} = 0$ and $A_{is} = 0$, but in general $D_{is} \neq 0$. However, by proper selection of the stacking sequence, for example, by increasing the number of layers for the same overall laminate thickness, D_{is} can be minimized.

Consider, for example, a laminate consisting of ten 0° plies, four 45° plies, and four −45° plies arranged in the following three stacking sequences:

- Balanced/asymmetric: $[0_5/45_4/{-}45_4/0_5]$

This stacking sequence is definitely not recommended because of its asymmetry and coarse ply distribution. The shear coupling stiffnesses A_{is} and torsion coupling stiffnesses D_{is} are zero, but the extension/bending coupling stiffnesses B_{ij} are nonzero. The coarseness of the ply distribution would increase the interlaminar edge stresses. It has been shown analytically that interlaminar edge stresses are a function of the stacking sequence and decrease as the thickness of the various layers decreases.[4]

- Balanced/symmetric: $[0_5/45_2/{-}45_2]_s$

This stacking sequence is balanced and symmetric; therefore, $B_{ij} = 0$ and $A_{is} = 0$. However, the torsion coupling stiffnesses D_{is} are nonzero and are relatively high due to the coarseness of the ply distribution. It represents an adequate but not the best design.

- Balanced/symmetric: $[0_2/45_2/0_2/{-}45_2/0]_s$

The bending coupling stiffnesses B_{ij} and shear coupling stiffnesses A_{is} are zero again. The nonzero torsion coupling stiffnesses D_{is} are relatively low because of the finer ply distribution. This design is recommended. The minimum for D_{is} and corresponding interlaminar stresses is obtained by selecting a finer ply distribution, for example, $[0/45/0/-45/0/45/0/-45/0]_s$. Although this design represents an optimum mechanically, it may make the fabrication process more complicated and expensive.

Whenever possible it is recommended to select a symmetric and balanced layup with fine ply interdispersion in order to eliminate extension/bending coupling and shear coupling and to minimize torsion coupling. Thus, warpage and unexpected distortions will be avoided, interlaminar stresses will be reduced, and the analysis will become considerably simpler.

7.12 LAMINATE ENGINEERING PROPERTIES

7.12.1 Symmetric Balanced Laminates

Simple relations can be derived for engineering properties as a function of laminate stiffnesses for the special case of a symmetric balanced laminate. Consider an element of such a laminate under uniaxial loading N_x as shown in Fig. 7.9. Then, by definition, the Young's modulus $\bar{E}_x$ and Poisson's ratio $\bar{v}_{xy}$ of the laminate are given by

$$\bar{E}_x = \frac{N_x}{h\varepsilon_x^o}$$
$$\bar{v}_{xy} = -\frac{\varepsilon_y^o}{\varepsilon_x^o} \tag{7.73}$$

where ε_x^o and ε_y^o are the normal strains in the x- and y-directions, respectively, and h is the laminate thickness. Symbols with an overbar denote effective laminate properties, and the x- and y-axes here are the principal laminate axes (denoted before as $\bar{x}$ and $\bar{y}$).

The force-deformation relations are

$$\begin{bmatrix} N_x \\ 0 \\ 0 \end{bmatrix} = \begin{bmatrix} A_{xx} & A_{xy} & 0 \\ A_{yx} & A_{yy} & 0 \\ 0 & 0 & A_{ss} \end{bmatrix} \begin{bmatrix} \varepsilon_x^o \\ \varepsilon_y^o \\ 0 \end{bmatrix} \tag{7.74}$$

There is no shear strain γ_s^o under the applied loading because the laminate is balanced.

Equations (7.74) in expanded form ($A_{xy} = A_{yx}$) are

$$N_x = A_{xx}\varepsilon_x^o + A_{xy}\varepsilon_y^o$$
$$0 = A_{xy}\varepsilon_x^o + A_{yy}\varepsilon_y^o \tag{7.75}$$

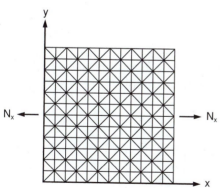

Fig. 7.9 Symmetric balanced laminate under uniaxial loading.

From Eqs. (7.75) and (7.73) we obtain

$$\bar{E}_x = \frac{1}{h}\left[A_{xx} - \frac{A_{xy}^2}{A_{yy}} \right]$$

$$\bar{v}_{xy} = \frac{A_{xy}}{A_{yy}}$$

(7.76)

Similarly, by considering a uniaxial loading N_y in the y-direction, we obtain

$$\bar{E}_y = \frac{1}{h}\left[A_{yy} - \frac{A_{xy}^2}{A_{xx}} \right]$$

$$\bar{v}_{yx} = \frac{A_{xy}}{A_{xx}}$$

(7.77)

For pure shear loading N_s we obtain

$$\bar{G}_{xy} = \frac{A_{ss}}{h}$$

(7.78)

The shear coupling coefficients for this balanced laminate are zero:

$$\bar{\eta}_{xs} = \bar{\eta}_{ys} = \bar{\eta}_{sx} = \bar{\eta}_{sy} = 0$$

(7.79)

It follows from the above that

$$\frac{\bar{E}_x}{\bar{E}_y} = \frac{A_{xx}}{A_{yy}} = \frac{\bar{v}_{xy}}{\bar{v}_{yx}}$$

7.12.2 Symmetric Laminates

Expressions for engineering properties in terms of laminate stiffnesses are more complicated for more general types of laminates, such as symmetric but not balanced laminates. In that case it is preferable to develop relations in terms of laminate compliances. For symmetric laminates the extension/bending coupling stiffnesses B_{ij} and compliances b_{ij} (with $i, j = x, y, s$) are zero. Thus, the reference plane strains are related only to in-plane forces as follows:

$$\begin{bmatrix} \varepsilon_x^o \\ \varepsilon_y^o \\ \gamma_s^o \end{bmatrix} = \begin{bmatrix} a_{xx} & a_{xy} & a_{xs} \\ a_{yx} & a_{yy} & a_{ys} \\ a_{sx} & a_{sy} & a_{ss} \end{bmatrix} \begin{bmatrix} N_x \\ N_y \\ N_s \end{bmatrix}$$

(7.80)

where $[a]$ is the extensional laminate compliance matrix, which in this case is the inverse of the corresponding stiffness matrix

$$[a] = [A^{-1}] \tag{7.81}$$

A symmetric laminate may be treated on a macroscopic scale as a homogeneous anisotropic material. Its elastic behavior is analogous to that of a unidirectional lamina, and thus similar expressions can be used between average stresses and strains and effective laminate constants. Thus, Eq. (7.80) can be written in terms of engineering constants by replacing the lamina constants in Eq. (4.77) with corresponding laminate moduli and noting that the average laminate stresses are

$$
\begin{bmatrix} \bar{\sigma}_x \\ \bar{\sigma}_y \\ \bar{\tau}_s \end{bmatrix} = \begin{bmatrix} N_x \\ N_y \\ N_s \end{bmatrix} \frac{1}{h} \tag{7.82}
$$

By analogy with the single lamina, the strain-force relations for the laminate are written in terms of engineering constants as follows:

$$
\begin{bmatrix} \varepsilon_x^o \\ \varepsilon_y^o \\ \gamma_s^o \end{bmatrix} = \begin{bmatrix} \dfrac{1}{\bar{E}_x} & -\dfrac{\bar{\nu}_{yx}}{\bar{E}_y} & \dfrac{\bar{\eta}_{sx}}{\bar{G}_{xy}} \\[2mm] -\dfrac{\bar{\nu}_{xy}}{\bar{E}_x} & \dfrac{1}{\bar{E}_y} & \dfrac{\bar{\eta}_{sy}}{\bar{G}_{xy}} \\[2mm] \dfrac{\bar{\eta}_{xs}}{\bar{E}_x} & \dfrac{\bar{\eta}_{ys}}{\bar{E}_y} & \dfrac{1}{\bar{G}_{xy}} \end{bmatrix} \begin{bmatrix} N_x \\ N_y \\ N_s \end{bmatrix} \frac{1}{h} \tag{7.83}
$$

where

$\bar{E}_x, \bar{E}_y =$ effective laminate Young's moduli in the x- and y-directions, respectively

$\bar{G}_{xy} =$ effective laminate shear modulus

$\bar{\nu}_{xy}, \bar{\nu}_{yx} =$ effective laminate Poisson's ratios

$\bar{\eta}_{xs}, \bar{\eta}_{ys}, \bar{\eta}_{sx}, \bar{\eta}_{sy} =$ effective laminate shear coupling coefficients

By equating corresponding terms in the compliance matrices of Eqs. (7.80) and (7.83), we obtain the following relations, analogous to Eqs. (4.80):

$$
\bar{E}_x = \frac{1}{ha_{xx}} \qquad \bar{E}_y = \frac{1}{ha_{yy}} \qquad \bar{G}_{xy} = \frac{1}{ha_{ss}}
$$

$$
\bar{\nu}_{xy} = -\frac{a_{yx}}{a_{xx}} \qquad \bar{\nu}_{yx} = -\frac{a_{xy}}{a_{yy}} \qquad \bar{\eta}_{sx} = \frac{a_{xs}}{a_{ss}} \tag{7.84}
$$

$$
\bar{\eta}_{xs} = \frac{a_{sx}}{a_{xx}} \qquad \bar{\eta}_{ys} = \frac{a_{sy}}{a_{yy}} \qquad \bar{\eta}_{sy} = \frac{a_{ys}}{a_{ss}}
$$

The symmetry of the compliance matrix implies the following relations among the laminate engineering constants, as in the case of the single unidirectional lamina, Eqs. (4.78):

$$\frac{\bar{v}_{xy}}{\bar{E}_x} = \frac{\bar{v}_{yx}}{\bar{E}_y}$$

$$\frac{\bar{\eta}_{xs}}{\bar{E}_x} = \frac{\bar{\eta}_{sx}}{\bar{G}_{xy}} \qquad (7.85)$$

$$\frac{\bar{\eta}_{ys}}{\bar{E}_y} = \frac{\bar{\eta}_{sy}}{\bar{G}_{xy}}$$

In order to calculate the engineering properties of a symmetric laminate using Eqs. (7.84) one needs to calculate the extensional stiffness matrix $[A]$ and then invert it to obtain the compliance matrix $[a]$.

7.12.3 General Laminates

Expressions for engineering constants of general asymmetric laminates can be obtained from the general strain-load relations, Eq. (7.25), that are reduced to Eq. (7.80) for in-plane loading. Thus, using the normal definitions of engineering properties in terms of average in-plane stresses and strains, one obtains that relations in Eqs. (7.84) are equally valid for general laminates. For example, comparison of engineering constants of a [0/90] asymmetric and a [0/90]$_s$ symmetric laminate leads to the correct conclusion that the Young's moduli of the two laminates are different. This is because in Eqs. (7.84) the compliance matrices $[a_{ij}]$ of the two laminates are different due to coupling effects.

SAMPLE PROBLEM 7.3
Axial Modulus of Angle-Ply Laminate

It is required to determine Young's modulus $\bar{E}_x$ of a $[\pm 45]_{ns}$ laminate in terms of the basic lamina properties (Fig. 7.10). For this symmetric balanced laminate we can apply Eq. (7.76),

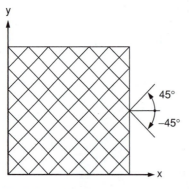

Fig. 7.10 $[\pm 45]_{ns}$ angle-ply laminate.

$$\bar{E}_x = \frac{1}{h}\left[A_{xx} - \frac{A_{xy}^2}{A_{yy}} \right] \qquad (7.76 \text{ bis})$$

where

$$A_{xx} = \sum_{k=1}^{4n} (Q_{xx})^k t_k = hQ_{xx}$$

$$A_{xy} = \sum_{k=1}^{4n} (Q_{xy})^k t_k = hQ_{xy} \qquad (7.86)$$

$$A_{yy} = \sum_{k=1}^{4n} (Q_{yy})^k t_k = hQ_{yy}$$

where t_k is the ply thickness, h the laminate thickness, and Q_{xx}, Q_{yy}, and Q_{xy} are the transformed 45° lamina stiffnesses.

From the stiffness transformation relations in Eq. (4.67) we obtain

$$(Q_{xx})_{\theta=\pm45°} = (Q_{yy})_{\theta=\pm45°} = \frac{1}{4}(Q_{11} + Q_{22} + 2Q_{12} + 4Q_{66}) \qquad (7.87)$$

$$(Q_{xy})_{\theta=\pm45°} = \frac{1}{4}(Q_{11} + Q_{22} + 2Q_{12} - 4Q_{66}) \qquad (7.88)$$

Then,

$$A_{xx} = A_{yy} = \frac{h}{4}(Q_{11} + Q_{22} + 2Q_{12} + 4Q_{66}) \qquad (7.89)$$

$$A_{xy} = \frac{h}{4}(Q_{11} + Q_{22} + 2Q_{12} - 4Q_{66}) \qquad (7.90)$$

From Eqs. (7.76) and (7.86) it follows that

$$\bar{E}_x = \left(Q_{xx} - \frac{Q_{xy}^2}{Q_{yy}}\right)_{\theta=\pm45°} = \left(Q_{xx} - \frac{Q_{xy}^2}{Q_{xx}}\right)_{\theta=\pm45°}$$

$$= \frac{1}{Q_{xx}}(Q_{xx} + Q_{xy})(Q_{xx} - Q_{xy}) \qquad (\text{for } \theta = \pm45) \qquad (7.91)$$

and from Eqs. (7.87) and (7.88)

$$\bar{E}_x = \frac{4(Q_{11} + Q_{22} + 2Q_{12})Q_{66}}{Q_{11} + Q_{22} + 2Q_{12} + 4Q_{66}} \qquad (7.92)$$

For composites with high-stiffness fibers

$$Q_{11} \gg Q_{22}$$
$$Q_{11} \gg Q_{12} \qquad (7.93)$$
$$Q_{11} \gg Q_{66}$$

and then Eq. (7.92) is reduced to

$$\bar{E}_x \cong \frac{4Q_{11}Q_{66}}{Q_{11}} = 4Q_{66} = 4G_{12} \qquad (7.94)$$

This result shows that the axial Young's modulus of a $[\pm45]_{ns}$ laminate is a matrix-dominated property since it depends primarily on the in-plane shear modulus G_{12} of the lamina. This was also true for the case of the [45] off-axis lamina (see Eq. (4.83)).

SAMPLE PROBLEM 7.4
Shear Modulus of Angle-Ply Laminate

It is required to determine the shear modulus $\bar{G}_{xy}$ of a $[\pm 45]_{ns}$ laminate as a function of the basic lamina properties (Fig. 7.10). This is obtained from Eq. (7.78),

$$\bar{G}_{xy} = \frac{A_{ss}}{h} \qquad (7.78 \text{ bis})$$

where

$$A_{ss} = \sum_{k=1}^{4n} (Q_{ss})^k t_k = h(Q_{ss})_{\theta=\pm 45°} \qquad (7.95)$$

From the stiffness transformation relations in Eq. (4.67) we obtain

$$(Q_{ss})_{\theta=\pm 45°} = \frac{1}{4}(Q_{11} + Q_{22} - 2Q_{12}) \qquad (7.96)$$

and then, using Eqs. (7.78) and (7.95),

$$\bar{G}_{xy} = (Q_{ss})_{\theta=\pm 45°} = \frac{1}{4}(Q_{11} + Q_{22} - 2Q_{12}) \qquad (7.97)$$

For composites with high-stiffness fibers,

$$\bar{G}_{xy} \cong \frac{Q_{11}}{4} \cong \frac{E_1}{4} \qquad (7.98)$$

Thus, the shear modulus of a $[\pm 45]_{ns}$ laminate is a fiber-dominated property since it depends primarily on the longitudinal lamina modulus E_1.

SAMPLE PROBLEM 7.5
Poisson's Ratio of Angle-Ply Laminate

It is required to determine Poisson's ratio $\bar{v}_{xy}$ of a $[\pm 45]_{ns}$ laminate as a function of the basic lamina properties (Fig. 7.10). This is obtained from Eq. (7.76),

$$\bar{v}_{xy} = \frac{A_{xy}}{A_{yy}} \qquad (7.76 \text{ bis})$$

Substituting Eqs. (7.89) and (7.90) into Eq. (7.76) yields

$$\bar{\nu}_{xy} = \frac{Q_{11} + Q_{22} + 2Q_{12} - 4Q_{66}}{Q_{11} + Q_{22} + 2Q_{12} + 4Q_{66}} \tag{7.99}$$

For composites with high-stiffness fibers,

$$\bar{\nu}_{xy} \cong \frac{Q_{11} - 4Q_{66}}{Q_{11} + 4Q_{66}} \cong \frac{E_1 - 4G_{12}}{E_1 + 4G_{12}} \tag{7.100}$$

SAMPLE PROBLEM 7.6
Engineering Constants of [0/±45/90]$_s$ Quasi-isotropic Laminate

For the [0/±45/90]$_s$ laminate discussed before,

$$A_{ij} = \sum_{k=1}^{n} Q_{ij}^k t_k = t \sum_{k=1}^{n} Q_{ij}^k = \frac{h}{n} \sum_{k=1}^{n} Q_{ij}^k \tag{7.101}$$

where

$$i, j = x, y, s$$

$$t = \text{lamina thickness}$$

$$h = \text{laminate thickness}$$

The transformed lamina stiffnesses Q_{xx} for the four different ply orientations are

$$Q_{xx(\theta=0°)} = Q_{11}$$

$$Q_{xx(\theta=90°)} = Q_{22}$$

$$Q_{xx(\theta=45°)} = Q_{xx(\theta=-45°)} = \frac{1}{4}(Q_{11} + Q_{22} + 2Q_{12} + 4Q_{66})$$

Thus, from Eq. (7.101) we obtain

$$A_{xx} = A_{yy} = \frac{h}{8}(3Q_{11} + 3Q_{22} + 2Q_{12} + 4Q_{66}) \tag{7.102}$$

Similarly, from the relations

$$Q_{xy(\theta=0°)} = Q_{xy(\theta=90°)} = Q_{12}$$

$$Q_{xy(\theta=45°)} = Q_{xy(\theta=-45°)} = \frac{1}{4}(Q_{11} + Q_{22} + 2Q_{12} - 4Q_{66})$$

we obtain

$$A_{xy} = \frac{h}{8}(Q_{11} + Q_{22} + 6Q_{12} - 4Q_{66}) \tag{7.103}$$

and from the relations

$$Q_{ss(\theta=0°)} = Q_{ss(\theta=90°)} = Q_{66}$$

$$Q_{ss(\theta=45°)} = Q_{ss(\theta=-45°)} = \frac{1}{4}(Q_{11} + Q_{22} - 2Q_{12})$$

we obtain

$$A_{ss} = \frac{h}{8}(Q_{11} + Q_{22} - 2Q_{12} + 4Q_{66}) \tag{7.104}$$

Introducing Eqs. (7.102) and (7.103) into (7.76), we obtain

$$
\begin{aligned}
\bar{E}_x &= \frac{1}{hA_{xx}}(A_{xx} + A_{xy})(A_{xx} - A_{xy}) \\
&= \frac{(Q_{11} + Q_{22} + 2Q_{12})(Q_{11} + Q_{22} - 2Q_{12} + 4Q_{66})}{3Q_{11} + 3Q_{22} + 2Q_{12} + 4Q_{66}}
\end{aligned}
\tag{7.105}
$$

For a high-stiffness composite

$$E_1 \gg E_2 \qquad \nu_{21} \ll 1 \qquad Q_{11} \gg Q_{22} \qquad Q_{11} \gg Q_{12} \qquad Q_{11} \cong E_1$$

and Eq. (7.105) becomes

$$\bar{E}_x = \bar{E}_y \cong \frac{E_1(E_1 + 4G_{12})}{3E_1 + 4G_{12}} \tag{7.106}$$

Similarly, from the same Eqs. (7.102), (7.103), and (7.76) we obtain

$$\bar{\nu}_{xy} = \frac{A_{xy}}{A_{yy}} = \frac{Q_{11} + Q_{22} + 6Q_{12} - 4Q_{66}}{3Q_{11} + 3Q_{22} + 2Q_{12} + 4Q_{66}} \tag{7.107}$$

which, for a high-stiffness composite, becomes

$$\bar{\nu}_{xy} \cong \frac{E_1 - 4G_{12}}{3E_1 + 4G_{12}} \tag{7.108}$$

Finally, from Eqs. (7.78) and (7.104) we obtain

$$\bar{G}_{xy} = \frac{A_{ss}}{h} = \frac{1}{8}(Q_{11} + Q_{22} - 2Q_{12} + 4Q_{66}) \tag{7.109}$$

which, for a high-stiffness composite, becomes

$$\bar{G}_{xy} \cong \frac{1}{8}(E_1 + 4G_{12}) \tag{7.110}$$

7.13 COMPUTATIONAL PROCEDURE FOR DETERMINATION OF ENGINEERING ELASTIC PROPERTIES

A flowchart for the determination of engineering properties of multidirectional laminates is given in Fig. 7.11. It consists of the following steps:

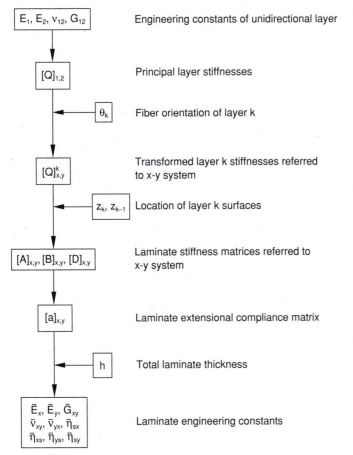

$E_1, E_2, \nu_{12}, G_{12}$ — Engineering constants of unidirectional layer

$[Q]_{1,2}$ — Principal layer stiffnesses

θ_k — Fiber orientation of layer k

$[Q]^k_{x,y}$ — Transformed layer k stiffnesses referred to x-y system

z_k, z_{k-1} — Location of layer k surfaces

$[A]_{x,y}, [B]_{x,y}, [D]_{x,y}$ — Laminate stiffness matrices referred to x-y system

$[a]_{x,y}$ — Laminate extensional compliance matrix

h — Total laminate thickness

$\bar{E}_x, \bar{E}_y, \bar{G}_{xy}$
$\bar{\nu}_{xy}, \bar{\nu}_{yx}, \bar{\eta}_{sx}$
$\bar{\eta}_{xs}, \bar{\eta}_{ys}, \bar{\eta}_{sy}$ — Laminate engineering constants

Fig. 7.11 Flowchart for computation of engineering elastic properties of multidirectional laminates.

Step 1 Enter the engineering properties of the unidirectional layer, E_1, E_2, ν_{12}, and G_{12}.

Step 2 Calculate the layer stiffnesses Q_{11}, Q_{22}, Q_{12}, and Q_{66} referred to the principal material axes using Eq. (4.56).

Step 3 Enter the fiber orientation or principal material axis orientation, θ_k, of layer k.

Step 4 Calculate the transformed stiffnesses $[Q]_{x,y}$ of layer k referred to the laminate coordinate system (x, y), using Eq. (4.67).

Step 5 Enter the through-the-thickness coordinates of layer k surfaces z_k and z_{k-1}.

Step 6 Calculate the laminate stiffness matrices $[A]$, $[B]$, and $[D]$ using Eqs. (7.20).

Step 7 Calculate the laminate compliance matrix $[a]$ using relations in Eq. (7.27), or by inversion of the 6×6 stiffness matrix of Eq. (7.23).

Step 8 Enter the total laminate thickness, h.

Step 9 Calculate the laminate engineering properties referred to the x- and y-axes using Eq. (7.84).

7.14 COMPARISON OF ELASTIC PARAMETERS OF UNIDIRECTIONAL AND ANGLE-PLY LAMINATES

Exact and approximate expressions for the elastic properties of a $[\pm45]_{ns}$ angle-ply laminate were obtained before [see Eqs. (7.92), (7.94), and (7.97) through (7.100)]. In addition to these results it should be mentioned that the laminate shear coupling coefficients $\bar{\eta}_{xs}$, $\bar{\eta}_{sx}$, $\bar{\eta}_{ys}$, and $\bar{\eta}_{sy}$ are zero since the laminate is balanced. It is of interest to compare the properties of the angle-ply laminate with those of the 45° off-axis lamina obtained before [see Eqs. (4.82) through (4.90)].

Numerical results for all properties above, exact and approximate, were obtained for a typical carbon/epoxy material (AS4/3501-6) and compared in Table 7.3. It is seen that values obtained by the approximation formulas are close to the exact values. The differences between the unidirectional lamina and angle-ply laminate are illustrated in Figs. 7.12–7.15. The Young's modulus, shear modulus, Poisson's ratio, and shear coupling coefficient are plotted as a function of fiber orientation for the AS4/3501-6 carbon/epoxy material. Results for a quasi-isotropic laminate are also plotted for reference.

TABLE 7.3 Comparison of Engineering Constants of [45] Unidirectional and [±45]$_s$ Angle-Ply Carbon/Epoxy Laminates (AS4/3501-6)

Property	[45]			[±45]$_s$		
	Approximation Formula	Approximate Value	Exact Value	Approximation Formula	Approximate Value	Exact Value
$\bar{E}_x$, GPa (Msi)	$\dfrac{4G_{12}E_2}{G_{12}+E_2}$	16.7 (2.42)	16.6 (2.40)	$4G_{12}$	28.0 (4.06)	23.9 (3.47)
$\bar{G}_{xy}$, GPa (Msi)	E_2	10.4 (1.50)	9.3 (1.35)	$E_1/4$	36.7 (5.32)	38.2 (5.53)
$\bar{\nu}_{xy}$	$\dfrac{E_2-G_{12}}{E_2+G_{12}}$	0.19	0.18	$\dfrac{E_1-4G_{12}}{E_1+4G_{12}}$	0.68	0.71
$\bar{\eta}_{xs}$	$-\dfrac{2G_{12}}{E_2+G_{12}}$	−0.81	−0.74	0	0	0
$\bar{\eta}_{sx}$	−0.50	−0.50	−0.42	0	0	0

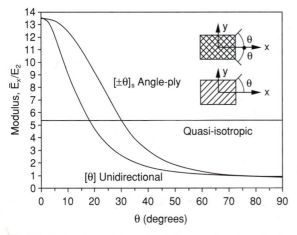

Fig. 7.12 Young's modulus of unidirectional and angle-ply laminates as a function of fiber orientation compared with that of a quasi-isotropic laminate (AS4/3501-6 carbon/epoxy).

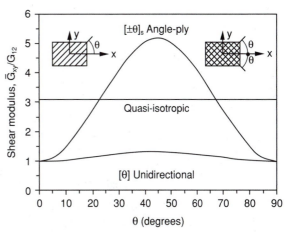

Fig. 7.13 Shear modulus of unidirectional and angle-ply laminates as a function of fiber orientation compared with that of a quasi-isotropic laminate (AS4/3501-6 carbon/epoxy).

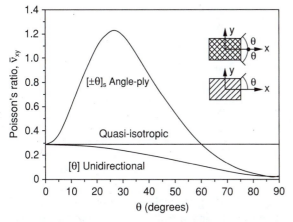

Fig. 7.14 Poisson's ratio of unidirectional and angle-ply laminates as a function of fiber orientation compared with that of a quasi-isotropic laminate (AS4/3501-6 carbon/epoxy).

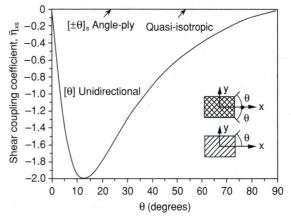

Fig. 7.15 Shear coupling coefficient of unidirectional and angle-ply laminates as a function of fiber orientation compared with that of a quasi-isotropic laminate (AS4/3501-6 carbon/epoxy).

7.15 CARPET PLOTS FOR MULTIDIRECTIONAL LAMINATES

Many design applications involve layups consisting of various numbers of 0°, 90°, and ±45° plies. These layups are balanced since the +45° plies are balanced by an equal number of −45° plies. A designation for such a general layup is $[0_m/90_n/(\pm45)_p]_s$ where m, n, and p denote the number of 0°, 90°, and ±45° plies, respectively. The in-plane engineering constants of a symmetric laminate depend only on the proportion of the various plies in the entire laminate and not on the exact stacking sequence. Thus, in-plane engineering constants are a function of the fractional values α, β, and γ, where

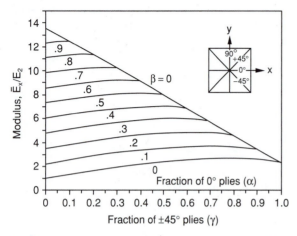

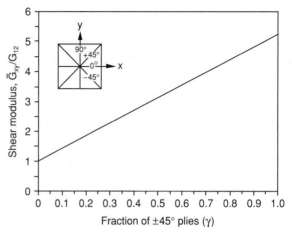

Fig. 7.16 Carpet plot for Young's modulus of $[0_m/90_n/(\pm 45)_p]_s$ carbon/epoxy laminates (AS4/3501-6) (α, β, and γ are fractions of 0°, 90°, and ±45° plies, respectively).

Fig. 7.17 Carpet plot for shear modulus of $[0_m/90_n/(\pm 45)_p]_s$ carbon/epoxy laminates (AS4/3501-6) (α, β, and γ are fractions of 0°, 90°, and ±45° plies, respectively).

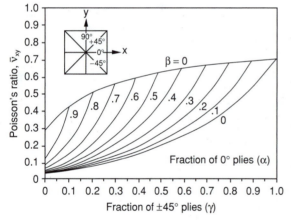

Fig. 7.18 Carpet plot for Poisson's ratio of $[0_m/90_n/(\pm 45)_p]_s$ carbon/epoxy laminates (AS4/3501-6) (α, β, and γ are fractions of 0°, 90°, and ±45° plies, respectively).

$$\alpha = \frac{2m}{N}$$

$$\beta = \frac{2n}{N} \tag{7.111}$$

$$\gamma = \frac{4p}{N}$$

$$N = 2(m + n + 2p) = \text{total number of plies}$$

For a given material with known basic lamina properties, in-plane properties for the general $[0_m/90_n/(\pm 45)_p]_s$ laminate can be obtained as a function of α, β, and γ. Although such computations can be performed readily using available computer programs, it is sometimes useful and practical for the designer to have so-called *carpet plots*. A carpet plot is a parametric family of curves with one of the fractions α, β, or γ as a variable and the other two as parameters, keeping in mind that $\alpha + \beta + \gamma = 1$. Such plots for Young's modulus, shear modulus, and Poisson's ratio are shown in Figs. 7.16–7.18 for a carbon/epoxy material (AS4/3501-6).

7.16 TEXTILE COMPOSITE LAMINATES

Many analytical models exist for the calculation of average stiffness matrices of textile laminates.[5–10] They are mostly based on micromechanics analyses of a representative unit cell, which is different for every weave pattern. Many of these models do not consider the distribution and effect of the crimp (or undulation) regions in the fabric and the resulting nonzero coupling terms in the fabric layer stiffness matrices. A modified lamination theory was proposed recently, where the single fabric layer is treated as a fundamental unit (sublaminate) with its own average stiffness properties determined experimentally.[11,12]

Textile composites have distinct mechanical properties different from those of unidirectional composites. Some common weave patterns, such as five-harness or eight-harness satin weaves (Figs. 2.10 and 2.11), lack certain symmetry properties. A single layer of these fabrics is neither symmetric nor balanced. It is similar but not exactly equivalent to an antisymmetric [0/90] laminate with unidirectional plies. These problems are minimized or resolved by appropriate stacking of fabric layers. For example, a pair of five-harness satin layers stacked with their warp or fill tows in contact (flipped pair) is not quite symmetric but is balanced with respect to the crimp regions.[11,12] A stacked pair of these layers rotated by 90° with respect to each other is symmetric but not quite balanced. A stack of the two pairs of layers above would produce in theory the thinnest symmetric and balanced laminate of five-harness satin fabric composite.

In general, multilayer textile composites can be treated in an approximate way by the classical lamination theory discussed before, by considering each single fabric layer as a unidirectional (orthotropic) one, with the warp and fill directions as the principal material axes 1 and 2, respectively. This is more justified as the number of layers in the laminate increases.

7.17 MODIFIED LAMINATION THEORY—EFFECTS OF TRANSVERSE SHEAR

In the classical lamination theory discussed before, it was assumed that the laminate is thin compared to its lateral dimensions and that straight lines normal to the middle surface remain straight and normal to that surface after deformation. As a result, the transverse shear stresses (τ_{xz}, τ_{yz}) and shear strains (γ_{xz}, γ_{yz}) are zero. These assumptions are not valid in the case of thicker laminates and laminates with low-stiffness central plies undergoing significant transverse shear deformations. In the theory discussed below, referred to as *first-order shear deformation laminated plate theory*, the assumption of normality of straight lines is removed, that is, straight lines normal to the middle surface remain straight but not normal to that surface after deformation.[13]

Figure 7.19 shows a section of a laminate normal to the y-axis before and after deformation, including the effects of transverse shear. The result of the latter is to rotate the cross section ABCD by an angle α_x to a location A′B′C′D′, which is not normal to the deformed middle surface.

The displacements of a generic point B can be expressed as[13]

$$u(x, y, z) = u_o(x, y) - z\alpha_x(x, y)$$
$$v(x, y, z) = v_o(x, y) - z\alpha_y(x, y) \tag{7.112}$$
$$w(x, y, z) = w(x, y)$$

where u_o, v_o, and w_o are the reference plane displacements and α_x and α_y are the rotations of the cross sections normal to the x- and y-axes.

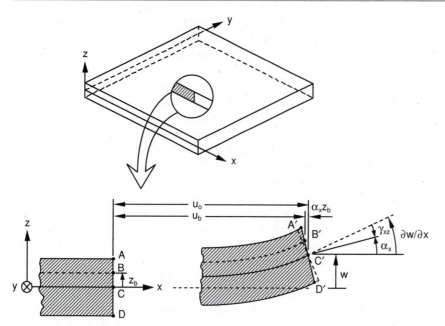

Fig. 7.19 Laminate section before (ABCD) and after (A'B'C'D') deformation with noticeable effects of transverse shear.

The classical strain-displacement relations yield the strains

$$\varepsilon_x = \frac{\partial u}{\partial x} = \frac{\partial u_o}{\partial x} - z\frac{\partial \alpha_x}{\partial x} = \varepsilon_x^o - z\frac{\partial \alpha_x}{\partial x}$$

$$\varepsilon_y = \frac{\partial v}{\partial y} = \frac{\partial v_o}{\partial y} - z\frac{\partial \alpha_y}{\partial y} = \varepsilon_y^o - z\frac{\partial \alpha_y}{\partial y}$$

$$\varepsilon_z = \frac{\partial w}{\partial z} = 0$$

$$\gamma_{yz} = \gamma_q = \frac{\partial v}{\partial z} + \frac{\partial w}{\partial y} = -\alpha_y + \frac{\partial w}{\partial y}$$

$$\gamma_{xz} = \gamma_r = \frac{\partial u}{\partial z} + \frac{\partial w}{\partial x} = -\alpha_x + \frac{\partial w}{\partial x}$$

$$\gamma_{xy} = \gamma_s = \frac{\partial u}{\partial y} + \frac{\partial v}{\partial x} = \frac{\partial u_o}{\partial y} + \frac{\partial v_o}{\partial x} - z\left(\frac{\partial \alpha_x}{\partial y} + \frac{\partial \alpha_y}{\partial x}\right) = \gamma_s^o - z\left(\frac{\partial \alpha_x}{\partial y} + \frac{\partial \alpha_y}{\partial x}\right)$$

(7.113)

From the above we can observe that the transverse shear strains γ_{xz} and γ_{yz} are equal to the rotations of the cross sections relative to the normals to the middle surface after deformation.

The in-plane stress-strain relations of a layer k within the laminate are decoupled from the transverse shear terms (if the z-direction is a principal material direction for the layer) and are the same as in Eq. (7.10):

$$\begin{bmatrix} \sigma_x \\ \sigma_y \\ \tau_s \end{bmatrix}_k = \begin{bmatrix} Q_{xx} & Q_{xy} & Q_{xs} \\ Q_{yx} & Q_{yy} & Q_{ys} \\ Q_{sx} & Q_{sy} & Q_{ss} \end{bmatrix}_k \begin{bmatrix} \varepsilon_x \\ \varepsilon_y \\ \gamma_s \end{bmatrix}_k \tag{7.10 bis}$$

The transverse shear stress-strain relations, which are decoupled from the in-plane stress and strain terms are

$$\begin{bmatrix} \tau_{yz} \\ \tau_{xz} \end{bmatrix}_k = \begin{bmatrix} \tau_q \\ \tau_r \end{bmatrix}_k = \begin{bmatrix} C_{qq} & C_{qr} \\ C_{rq} & C_{rr} \end{bmatrix}_k \begin{bmatrix} \gamma_q \\ \gamma_r \end{bmatrix}_k \tag{7.114}$$

The in-plane force and moment resultants are obtained by the same relations as in Eqs. (7.14) and (7.15), and the transverse shear force resultants are obtained from the following:

$$\begin{bmatrix} V_{yz} \\ V_{xz} \end{bmatrix} = \begin{bmatrix} V_q \\ V_r \end{bmatrix} = K \sum_{k=1}^{n} \int_{z_{k-1}}^{z_k} \begin{bmatrix} \tau_q \\ \tau_r \end{bmatrix}_k dz \tag{7.115}$$

where K is the so-called *shear correction factor* introduced to account for the nonuniform distribution of transverse shear stresses through the thickness of the layer. In many cases, depending on the laminate and material, it may be sufficiently accurate to assume $K = 1$.

Substituting the stress-strain relations in Eqs. (7.10) and (7.114) in Eqs. (7.14), (7.15), and (7.115), and using the strain expressions in Eqs. (7.113), we obtain

$$\begin{bmatrix} N_x \\ N_y \\ N_s \end{bmatrix} = \begin{bmatrix} A_{xx} & A_{xy} & A_{xs} \\ A_{yx} & A_{yy} & A_{ys} \\ A_{sx} & A_{sy} & A_{ss} \end{bmatrix} \begin{bmatrix} \varepsilon_x^o \\ \varepsilon_y^o \\ \gamma_s^o \end{bmatrix} - \begin{bmatrix} B_{xx} & B_{xy} & B_{xs} \\ B_{yx} & B_{yy} & B_{ys} \\ B_{sx} & B_{sy} & B_{ss} \end{bmatrix} \begin{bmatrix} \dfrac{\partial \alpha_x}{\partial x} \\[2mm] \dfrac{\partial \alpha_y}{\partial y} \\[2mm] \dfrac{\partial \alpha_x}{\partial y} + \dfrac{\partial \alpha_y}{\partial x} \end{bmatrix} \tag{7.116}$$

$$\begin{bmatrix} M_x \\ M_y \\ M_s \end{bmatrix} = \begin{bmatrix} B_{xx} & B_{xy} & B_{xs} \\ B_{yx} & B_{yy} & B_{ys} \\ B_{sx} & B_{sy} & B_{ss} \end{bmatrix} \begin{bmatrix} \varepsilon_x^o \\ \varepsilon_y^o \\ \gamma_s^o \end{bmatrix} - \begin{bmatrix} D_{xx} & D_{xy} & D_{xs} \\ D_{yx} & D_{yy} & D_{ys} \\ D_{sx} & D_{sy} & D_{ss} \end{bmatrix} \begin{bmatrix} \dfrac{\partial \alpha_x}{\partial x} \\[2mm] \dfrac{\partial \alpha_y}{\partial y} \\[2mm] \dfrac{\partial \alpha_x}{\partial y} + \dfrac{\partial \alpha_y}{\partial x} \end{bmatrix} \tag{7.117}$$

$$\begin{bmatrix} V_q \\ V_r \end{bmatrix} = \begin{bmatrix} A_{qq} & A_{qr} \\ A_{rq} & A_{rr} \end{bmatrix} \begin{bmatrix} \dfrac{\partial w}{\partial y} - \alpha_y \\[2mm] \dfrac{\partial w}{\partial x} - \alpha_x \end{bmatrix} \tag{7.118}$$

where the laminate stiffnesses A_{ij}, B_{ij}, and D_{ij} ($i, j = x, y, s$) are defined as before in Eqs. (7.20). The transverse shear laminate stiffnesses A_{ij} ($i, j = q, r$) are defined as

$$A_{ij} = \sum_{k=1}^{n} C_{ij}^k t_k \qquad (i, j = q, r) \tag{7.119}$$

The stiffnesses C_{qq}, C_{qr}, and C_{rr} are related to stiffnesses C_{44} and C_{55} or the engineering shear moduli G_{23} and G_{13}, as shown in Eqs. (4.103) and Table 4.3.

In Eqs. (7.116) and (7.117) the last deformation terms represent the gradients of the total rotations of the sections that were initially normal to the reference surface. In Eq. (7.118) the deformation terms represent the relative rotations of the sections initially normal to the middle surface with respect to the deformed middle surface.

7.18 SANDWICH PLATES

Sandwich construction is of particular interest and widely used, because the concept is very suitable for lightweight structures with high in-plane and flexural stiffness. A sandwich panel is a special type of laminate consisting typically of two thin facesheets (facings or skins) and a lightweight, thicker, and low-stiffness core. Commonly used materials for facings are composite laminates and metals, while cores are made of metallic or non-metallic honeycombs, cellular foams, balsa wood, and trusses. Properties of typical core materials are given in Table A.9 in Appendix A. The facings carry almost all of the bending and in-plane loads, and the core helps to stabilize the facings against buckling and defines the flexural stiffness and out-of-plane shear and compressive behavior.

Depending on the geometry (aspect ratio, core thickness), constituent material properties, and type of loading, a sandwich plate can be analyzed by the classical lamination theory or the modified lamination theory including transverse shear effects. Figure 7.20 shows a section of a sandwich plate under the action of forces and moments. This sandwich plate can be treated as a three-ply laminate consisting of the core and the two facesheets.[14]

If the two facesheets are identical in thickness and properties, the laminate is symmetric, that is,

$$B_{ij} = 0$$

The in-plane extensional laminate (sandwich) stiffnesses are determined as

$$A_{ij} = \sum_{k=1}^{3} Q_{ij}^k t_k = 2(Q_{ij})_f h_f + (Q_{ij})_c h_c \tag{7.120}$$

where

$$i, j = x, y, s$$

h_f, h_c = facesheet and core thickness, respectively

$(Q_{ij})_f, (Q_{ij})_c$ = facesheet and core stiffnesses, respectively

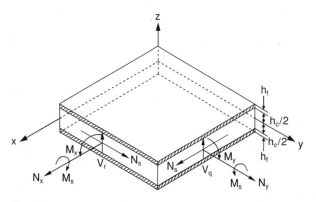

Fig. 7.20 Element of sandwich plate with force and moment resultants.

The flexural laminate stiffnesses are

$$D_{ij} = \frac{1}{3}\sum_{k=1}^{3} Q_{ij}^k(z_k^3 - z_{k-1}^3) = \frac{1}{2}(Q_{ij})_f\, h_f\, h_c^2\left(1 + 2\frac{h_f}{h_c} + \frac{4}{3}\frac{h_f^2}{h_c^2}\right) + \frac{1}{12}(Q_{ij})_c\, h_c^3 \quad (7.121)$$

The transverse shear laminate stiffnesses are, from Eq. (7.119),

$$A_{ij} = \sum_{k=1}^{3} C_{ij}^k t_k \qquad (i, j = q, r)$$

or

$$\begin{bmatrix} A_{qq} & A_{qr} \\ A_{rq} & A_{rr} \end{bmatrix} = \sum_{k=1}^{3} \begin{bmatrix} C_{qq} & C_{qr} \\ C_{rq} & C_{rr} \end{bmatrix}_k t_k \qquad (7.122)$$

or

$$A_{qq} = 2(C_{qq})_f h_f + (C_{qq})_c h_c$$
$$A_{qr} = 2(C_{qr})_f h_f + (C_{qr})_c h_c$$
$$A_{rr} = 2(C_{rr})_f h_f + (C_{rr})_c h_c$$

where C_{qq}, C_{qr}, and C_{rr} are the transformed transverse shear stiffnesses of a layer, corresponding to the x-y-z coordinate system. These are obtained by transformation of the transverse shear stiffnesses C_{44}, C_{45}, and C_{55} using Eqs. (4.103).

If the facesheets are much thinner and stiffer than the core, that is,

$$h_f \ll h_c$$
$$(Q_{ij})_f \gg (Q_{ij})_c$$

and if

$$2(Q_{ij})_f h_f \gg (Q_{ij})_c h_c \qquad (7.123)$$

the expressions for the in-plane extensional and the flexural laminate stiffnesses in Eqs. (7.120) and (7.121) can be approximated as

$$A_{ij} \cong 2(Q_{ij})_f h_f \qquad (i, j = x, y, s) \qquad (7.124)$$

$$D_{ij} \cong \frac{1}{2}(Q_{ij})_f h_f h_c^2 \cong \frac{1}{4}A_{ij}h_c^2 \qquad (i, j = x, y, s) \qquad (7.125)$$

If, in addition to the previous assumptions, Eq. (7.123), the facesheets are unidirectional laminae with their principal directions 1 and 2 along the x- and y-axes, then

$$A_{ij} \cong \begin{bmatrix} Q_{11} & Q_{12} & 0 \\ Q_{21} & Q_{22} & 0 \\ 0 & 0 & Q_{66} \end{bmatrix}_f 2h_f \qquad (i, j = 1, 2, 6) \qquad (7.126)$$

and

$$D_{ij} \cong \frac{1}{2} \begin{bmatrix} Q_{11} & Q_{12} & 0 \\ Q_{21} & Q_{22} & 0 \\ 0 & 0 & Q_{66} \end{bmatrix}_f h_f h_c^2 \quad (i, j = 1, 2, 6) \tag{7.127}$$

where

$$Q_{11} = \frac{E_1}{1 - v_{12}v_{21}}$$

$$Q_{12} = \frac{v_{21}E_1}{1 - v_{12}v_{21}} = \frac{v_{12}E_2}{1 - v_{12}v_{21}}$$

$$Q_{22} = \frac{E_2}{1 - v_{12}v_{21}}$$

$$Q_{66} = G_{12}$$

$$\tag{4.56 bis}$$

The transverse shear stiffnesses for an isotropic core are expressed as

$$A_{44} = 2(G_{23})_f h_f + G_c h_c$$
$$A_{55} = 2(G_{13})_f h_f + G_c h_c$$

$$\tag{7.128}$$

For isotropic facesheets

$$(Q_{11})_f = (Q_{22})_f = \frac{E_f}{1 - v_f^2}$$

$$(Q_{12})_f = \frac{v_f E_f}{1 - v_f^2}$$

$$(Q_{66})_f = G_f = \frac{E_f}{2(1 + v_f)}$$

The load-deformation relations for a symmetric sandwich plate are the same as those for a symmetric laminate, Eqs. (7.31) and (7.32), that is,

$$[N] = [A][\varepsilon^o] \tag{7.31 bis}$$

$$[M] = [D][\kappa] \tag{7.32 bis}$$

Strains and curvatures, obtained by inverting these relations, are

$$[\varepsilon^o] = [A^{-1}][N] = [a][N] \tag{7.129}$$

$$[\kappa] = [D^{-1}][M] = [d][M] \tag{7.130}$$

SAMPLE PROBLEM 7.7
Load-Deformation Relations for Sandwich Beam

A composite sandwich beam, consisting of identical unidirectional lamina facesheets and an isotropic core, is subjected to a general type of loading as shown in Fig. 7.21. It is required to determine the load-deformation relations in terms of material and geometric properties.

Assuming that the principal directions 1 and 2 of the unidirectional facesheets coincide with x- and y-directions and that $h_f \ll h_c$, we obtain from Eqs. (7.31), (7.32), (7.126), (7.127), and (7.128),

$$N_x = A_{11}\varepsilon_1^o + A_{12}\varepsilon_2^o \cong 2h_f(Q_{11}\varepsilon_1^o + Q_{12}\varepsilon_2^o) \tag{7.131}$$

$$M_x = D_{11}\kappa_1 + D_{12}\kappa_2 \cong \frac{1}{2}h_f h_c^2(Q_{11}\kappa_1 + Q_{12}\kappa_2) \tag{7.132}$$

$$V_r = V_{xz} = V_{13} \cong A_{55}\gamma_{13} = [2(G_{13})_f h_f + G_c h_c]\gamma_{13} \tag{7.133}$$

For a high-stiffness composite facesheet, that is, $E_1 \gg E_2$, Eqs. (7.131) and (7.132) can be approximated as

$$N_x \cong 2h_f Q_{11}\varepsilon_1^o \tag{7.134}$$

and

$$M_x \cong \frac{1}{2}h_f h_c^2 Q_{11}\kappa_1 \cong \frac{1}{2}h_f h_c Q_{11}(\varepsilon_1^t - \varepsilon_1^b) \tag{7.135}$$

where

$$\varepsilon_1^t, \varepsilon_1^b = \text{strains at top and bottom facesheets}$$

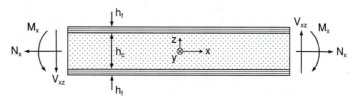

Fig. 7.21 Composite sandwich beam under general type of loading.

REFERENCES

1. K. S. Pister and S. B. Dong, "Elastic Bending of Layered Plates," *J. Eng. Mech. Division, ASCE*, 1959, pp. 1–10.
2. E. Reissner and Y. Stavsky, "Bending and Stretching of Certain Types of Heterogeneous Aeolotropic Elastic Plates," *J. Appl. Mech.*, 1961, pp. 402–408.
3. R. M. Jones, *Mechanics of Composite Materials*, Second Edition, Taylor and Francis, Philadelphia, 1998.
4. N. J. Pagano and R. B. Pipes, "The Influence of Stacking Sequence on Laminate Strength," *J. Comp. Materials*, Vol. 5, 1971, pp. 50–57.
5. T. Ishikawa and T. W. Chou, "Elastic Behavior of Woven Hybrid Composites," *J. Comp. Materials*, Vol. 16, 1982, pp. 2–19.
6. T. W. Chou, *Microstructural Design of Fiber Composites*, Cambridge University Press, New York, 1992.
7. N. K. Naik and V. K. Gamesh, "Prediction of On-Axes Elastic Properties of Plain Weave Composites," *Composites Science and Technology*, Vol. 45, 1992, pp. 135–152.
8. I. S. Raju and J. T. Wang, "Classical Laminate Theory Models for Woven Fabric Composites," *J. of Comp. Technology and Research*, Vol. 16, No. 4, 1994, pp. 289–303.
9. T. J. Walsh and O. O. Ochoa, "Analytical and Experimental Mechanics of Woven Fabric Composites," *Mech. of Comp. Materials and Structures*, Vol. 3, 1996, pp. 133–152.
10. P. Tang, L. Tong, and G. P. Steven, "Modeling for Predicting the Mechanical Properties of Textile Composites—A Review," *Composites, Part A*, Vol. 28, No. 11, 1997, pp. 903–922.
11. A. J. Jacobsen, J. J. Luo, and I. M. Daniel, "Characterization of Constitutive Behavior of Satin-Weave Fabric Composite," *Journal of Comp. Materials*, Vol. 38, No. 7, 2004, pp. 555–565.
12. J. J. Luo and I. M. Daniel, "Sublaminate-based Lamination Theory and Symmetry Properties of Textile Composite Laminates," *Composites, Part B*, Vol. 35, No. 6–8, 2004, pp. 483–496.
13. J. N. Reddy, *Mechanics of Laminated Composite Plates*, Second Edition, CRC Press, New York, 2004.
14. J. R. Vinson, *The Behavior of Sandwich Structures of Isotropic and Composite Materials*, Technomic Pub. Co., Lancaster, 1999.

PROBLEMS

7.1 Show that the laminate stiffnesses can be written as

$$A_{ij} = \sum_{k=1}^{n} Q_{ij}^k t_k$$

$$B_{ij} = \sum_{k=1}^{n} Q_{ij}^k t_k \bar{z}_k$$

$$D_{ij} = \sum_{k=1}^{n} Q_{ij}^k \left(t_k \bar{z}_k^2 + \frac{t_k^3}{12}\right)$$

where

t_k = thickness of layer k
$\bar{z}_k$ = distance from the midplane to the centroid of layer k
$i, j = x, y, s$

7.2 Identify each of the following laminates by name (i.e., symmetric, balanced, and so on) and indicate

which terms of the $[A]$, $[B]$, and $[D]$ matrices are zero for each laminate.

(a) [0/±45/∓45/0]
(b) [0/±45]
(c) [0/90/0/90/0]
(d) [0/45/90/−45]
(e) [0/±30/∓30/0]
(f) [0/90/0/90]
(g) [30/−45/−30/45]
(h) [−45/30/−30/45]
(i) [±45/∓45]
(j) [0/±45/90/∓45/0]

7.3 Name two types of laminates, with specific examples, for which

$$B_{ij} = 0$$
$$A_{xs} = A_{ys} = D_{xs} = D_{ys} = 0$$

7.4   Which terms of the $[A]$, $[B]$, and $[D]$ matrices are zero for the following laminates?

(a) $[\pm\theta/\theta/\mp\theta]$

(b) $[\pm\theta]_s$

(c) $[\pm\theta]_2$

(d) $[\theta_1/-\theta_2/\theta_2/-\theta_1]$

7.5   Determine and compare the coupling terms B_{xs} of the $[\pm\theta]$ and $[\pm\theta/\pm\theta]$ laminates having the same total thickness, h, but different ply thickness.

7.6   For a $[\pm\theta]_p$ laminate, determine B_{xs} in terms of the total laminate thickness, h, the number of plies, $n = 2p$, the basic lamina stiffnesses (Q_{11}, Q_{12}, Q_{22}, Q_{66}), and the ply orientation θ.

7.7   A laminate consists of two equal-thickness isotropic layers of properties as shown in Fig. P7.7. Determine all terms of the $[A]$, $[B]$, and $[D]$ matrices in terms of E_i, ν_i, and t.

Fig. P7.7

7.8   Compute all terms of the $[A]$, $[B]$, and $[D]$ matrices for a $[0/90]$ laminate with the lamina properties

$E_1 = 145$ GPa (21 Msi)

$E_2 = 10.5$ GPa (1.5 Msi)

$G_{12} = 7.0$ GPa (1.0 Msi)

$\nu_{12} = 0.28$

$t = 0.25$ mm (0.01 in)   (lamina thickness)

7.9   Compute all terms of the $[A]$ and $[B]$ matrices for a $[+45/-45]$ laminate with the lamina properties of Problem 7.8.

7.10   Evaluate the $[B]$ matrix for a $[+45/-45]$ laminate in terms of the lamina engineering properties E_1, E_2, ν_{12}, G_{12}, and the lamina thickness t.

7.11   Compute all terms of the $[A]$ matrix for a $[0/\pm45]$ laminate with the lamina properties of Problem 7.8.

7.12   Determine and compare the terms of the $[B]$ matrix for laminates $[0_2/90_2]$ and $[0/90]_2$ in terms of lamina properties Q_{ij} and thickness t.

7.13   An antisymmetric $[\pm45]$ laminate consists of two carbon/epoxy layers of thickness t. The elastic properties of the lamina are

$$E_1 = 15E_o$$
$$E_2 = E_o$$
$$G_{12} = 0.6E_o$$
$$\nu_{12} = 0.3$$

Determine the laminate stiffnesses $[A]$, $[B]$, and $[D]$ in terms of E_o and t. Obtain approximate expressions by assuming $E_1 \gg E_2 > G_{12}$.

7.14   Show that the $[0/\pm60]_s$ laminate is quasi-isotropic, that is, $A_{xx} = A_{\bar{x}\bar{x}}$ for any angle φ (Fig. P7.14).

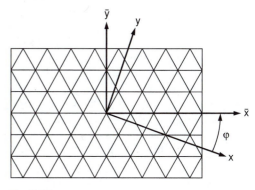

Fig. P7.14

7.15   Prove Eqs. (7.70) through (7.72) in Sample Problem 7.2.

7.16   Prove that a laminate of

$$\left[\frac{\pi}{n}/\frac{2}{n}\pi/\ldots\frac{k\pi}{n}/\ldots/\frac{n-1}{n}\pi/\pi\right]_s$$

layup is quasi-isotropic, that is, that all extensional stiffnesses A_{ij} are independent of orientation of reference axes.

7.17   Prove that an orthotropic laminate is quasi-isotropic if $A_{\bar{x}\bar{x}} = A_{\bar{y}\bar{y}}$ and $2A_{\bar{s}\bar{s}} = A_{\bar{x}\bar{x}} - A_{\bar{x}\bar{y}}$, where $\bar{x}$ and $\bar{y}$ are its principal directions (Fig. P7.17).

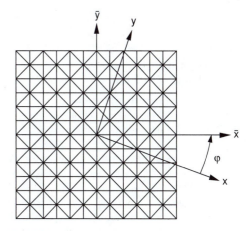

Fig. P7.17

7.18   Given a symmetric crossply laminate $[0/90]_s$, determine

(a) the stiffnesses ($A_{\bar{x}\bar{x}}$, $A_{\bar{y}\bar{y}}$, $A_{\bar{x}\bar{y}}$, $A_{\bar{x}\bar{s}}$) and moduli ($\bar{E}_x$, $\bar{E}_y$, $\bar{G}_{xy}$) referred to the $\bar{x}$-$\bar{y}$ system of coordinates in terms of the reduced lamina stiffnesses (Q_{11}, Q_{12}, Q_{22}, Q_{66}) and the lamina thickness t

(b) the approximate stiffnesses (A_{xx}, A_{yy}, A_{xy}, A_{xs}) and moduli ($\bar{E}_x$, $\bar{E}_y$, $\bar{G}_{xy}$) referred to a system of coordinates rotated by an angle θ with respect to the $\bar{x}$-$\bar{y}$ system, in terms of the lamina properties (Q_{11}, Q_{12}, Q_{22}, Q_{66}, t) for a high-stiffness composite ($E_1 \gg E_2$)

7.19   A symmetric crossply laminate is designed such that $\bar{E}_x = 5\bar{E}_y$. The relative lamina moduli are $E_1 = 14E_2$. Determine the approximate ratio r of the number of $0°$ plies to that of the $90°$ plies.

7.20   Two samples were cut from a $[0/90]_s$ laminate, one along the $0°$ direction and the other at $45°$. The measured moduli from these samples were $(\bar{E}_x)_{\theta=0°} = 80$ GPa (11.6 Msi) and $(\bar{E}_x)_{\theta=45°} = 24$ GPa (3.5 Msi). Determine the lamina moduli E_1 and G_{12} assuming $E_1 \gg E_2$.

7.21   Two samples were cut from a $[0/90]_s$ laminate, one along the $0°$ direction and the other at $45°$. The measured properties from the first sample were $\bar{E}_x = 80$ GPa (11.6 Msi) and $\bar{\nu}_{\bar{x}\bar{y}} \ll 1$. The second sample gave strains $\varepsilon_x = -\varepsilon_y = 5 \times 10^{-3}$ under a uniaxial loading of 120 MPa (17.4 ksi). Determine the lamina moduli E_1 and G_{12} assuming $E_1 \gg E_2$.

7.22   Prove that for a $[\pm45]_s$ angle-ply laminate, the in-plane lamina shear modulus G_{12} is related to the laminate modulus and Poisson's ratio as follows:

$$G_{12} = \frac{\bar{E}_x}{2(1 + \bar{\nu}_{xy})}$$

(Hint: See Sample Problems 7.3 and 7.5.)

7.23   Determine $\bar{E}_x$, $\bar{G}_{xy}$, and $\bar{\nu}_{xy}$ for a $[\pm45]_s$ angle-ply laminate with the lamina properties of Problem 7.8.

7.24   (a) Give general expressions for Poisson's ratio of a [45] lamina and a $[\pm45]_s$ laminate in terms of lamina engineering properties (E_1, E_2, G_{12}, ν_{12}, ν_{21}).

(b) Express approximate formulas for Poisson's ratios of the same lamina and laminate for a high-stiffness composite ($E_1 \gg E_2$).

(c) Compare exact and approximate values of Poisson's ratio for each case, [45] and $[\pm45]_s$, for a material with the properties

$E_1 = 191$ GPa (27.7 Msi)
$E_2 = 10$ GPa (1.45 Msi)
$G_{12} = 7.8$ GPa (1.13 Msi)
$\nu_{12} = 0.35$

7.25   Determine Poisson's ratio $\bar{\nu}_{xy}$ for a $[\pm30]_s$ angle-ply laminate with the lamina properties of Problem 7.8.

7.26   Determine the shear modulus $\bar{G}_{xy}$ of a $[0/\pm45]_s$ laminate in terms of lamina engineering properties (E_1, E_2, G_{12}, ν_{12}) both exactly and also using approximations for a high-stiffness composite, that is, $E_1 \gg E_2$, $\nu_{21} \ll 1$.

7.27   Determine Poisson's ratio $\bar{\nu}_{xy}$ for a $[0/\pm45]_s$ laminate with the lamina properties of Problem 7.8.

7.28   Show that for a quasi-isotropic laminate the following relation holds among axial modulus, shear modulus and Poisson's ratio:

$$\bar{G}_{xy} = \frac{\bar{E}_x}{2(1 + \bar{\nu}_{xy})}$$

7.29   Based on the results of Problem 7.20, compute the approximate properties $\bar{E}_{\bar{x}}$, $\bar{G}_{\bar{x}\bar{y}}$, and $\bar{\nu}_{\bar{x}\bar{y}}$ of a $[0/\pm45/90]_s$ quasi-isotropic laminate of the same material. Verify the relationship of Problem 7.28 above.

7.30   Prove that, for an orthotropic laminate, if $\bar{E}_{\bar{x}} = \bar{E}_{\bar{y}}$ referred to the principal system of coordinates ($\bar{x}$, $\bar{y}$) then $\bar{E}_x = \bar{E}_y$ referred to any other orthogonal system (x, y).

7.31   Determine the fiber orientation θ in terms of lamina properties (Q_{11}, Q_{12}, Q_{22}, Q_{66}) so that the moduli of lamina [θ] and laminate [$\pm\theta$]$_s$ are related as $\bar{E}_x = 2E_x$.

7.32   (a) Show that the force N_x for a $[\theta/-\theta]$ laminate takes the form

$$N_x = A_{xx}\varepsilon_x^o + A_{xy}\varepsilon_y^o + B_{xs}\kappa_s$$

(b) Express A_{xx}, A_{xy}, and B_{xx} in terms of transformed lamina stiffnesses (Q_{xx}, Q_{xy}, Q_{yy}, Q_{xs}) of the θ-ply and the ply thickness t.

(c) Obtain an approximate relation between N_x and ε_x^o, ε_y^o, and κ_s in terms of E_1 and t for $E_1 \gg E_2$ and $\theta = 45°$.

7.33   A laminate of $[0/\pm45/90]_s$ layup is loaded under in-plane biaxial loading

$N_x = N_o$
$N_y = 2N_o$
$N_s = 0$

and the resulting strains are

$$\varepsilon_x = 1.3 \times 10^{-3}$$
$$\varepsilon_y = 5.1 \times 10^{-3}$$

Determine Poisson's ratio $\bar{v}_{xy}$.

7.34 An asymmetric [0/90] laminate is subjected to biaxial loading $N_x = N_y = N_o$, and the following strains and curvatures are obtained: $\varepsilon_x^o = \varepsilon_y^o = \varepsilon_o$, $\kappa_x = -\kappa_y = \kappa_o$. Assuming $\bar{v}_{xy} \ll 1$, obtain an expression for the laminate modulus $\bar{E}_x$ as a function of its relevant stiffness parameters A_{ij}, B_{ij}, and D_{ij}.

7.35 For the laminate of Problem 7.34 made of AS4/3501-6 carbon/epoxy (see Table A.4) determine the load N_o corresponding to a curvature $\kappa_x = 2 \times 10^{-3}$ mm^{-1} (0.051 in^{-1}). The layer thickness is $t = 0.1$ mm (0.004 in).

7.36 An antisymmetric [0/90] laminate is loaded under axial load N_x. Determine curvatures κ_x and κ_y in terms of the load N_x and stiffnesses A_{ij}, B_{ij}, and D_{ij} ($i, j = x, y, s$).

7.37 A symmetric and unbalanced laminate for which $\bar{E}_x = \bar{E}_y$ and $\bar{\eta}_{xs} = \bar{\eta}_{ys}$ is loaded in pure shear $N_s = N_o$ and the normal strain ε_x is measured.
   (a) Determine the shear strain γ_o produced under biaxial compressive loading $N_x = N_y = -N_o$ of the same laminate as a function of ε_x.
   (b) Determine Poisson's ratio $\bar{v}_{xy}$ of this laminate if the normal strain ε_x under equal normal biaxial loading is zero.

7.38 Determine the [A] and [D] stiffness matrices of a sandwich panel with the following properties:

Facesheets
   E-glass fabric/epoxy (M10E/3783, Table A.6), 1.52 mm (0.060 in) thick

Core
   Balsa wood CK57 (Table A.9), $v_{12c} = 0.4$, 25 mm (1 in) thick

Compare the exact results with those obtained by neglecting the contribution of the core.

7.39 Determine the [A] and [D] stiffness matrices of a sandwich panel with the following properties:

Facesheets
   Carbon fabric/epoxy (AGP370-5H/3501-6S, Table A.6), 1.0 mm (0.040 in) thick

Core
   PVC foam H250 (Table A.9), $v_{12c} = 0.4$, 25 mm (1 in) thick

Determine stiffnesses exactly and also by neglecting the contribution of the core.

7.40 Consider a sandwich beam cut from the panel of Problem 7.38 with the 1-direction (warp) of the composite facesheet coinciding with the x-axis. Obtain a relationship between the axial load N_x (per unit width of the beam) and the average strain ε_x^o, with and without consideration of the core contribution.

7.41 Consider a sandwich beam cut from the panel of Problem 7.39 with the 1-direction (warp) of the composite facesheet coinciding with the x-axis. Obtain a relationship between the bending moment M_x (per unit width of the beam) and the maximum tensile strain $(\varepsilon_1)_{max}$, with and without consideration of the core contribution.

8 Hygrothermal Effects

8.1 INTRODUCTION

The fabrication process of composite materials introduces reversible and irreversible effects due to the processing thermal cycle and chemical changes, and due to the mismatch in thermal properties of the constituents. The most common manifestation of these effects are residual stresses and warpage.

After fabrication, composite structures operate in a variety of thermal and moisture environments that may have a pronounced impact on their performance. These hygrothermal effects are a result of the temperature and moisture content variations and are related to the difference in thermal and hygric properties of the constituents.

Processing and environmental effects are similar in nature. They can be viewed and analyzed from the microscopic point of view, on the scale of the fiber diameter, or from the macroscopic point of view, by considering the overall effects on the lamina, which is treated as a homogeneous material.

Analysis of the processing and hygrothermal effects is an important component of the overall structural design and analysis. The performance of a composite structure is a function of its environmental history, temperature and moisture distributions, processing and hygrothermal stresses, and property variations with temperature and moisture.

In general, it is assumed that the composite material is exposed to an environment of known temperature and moisture histories, $T(t)$ and $H(t)$, respectively. For the given hygrothermal conditions, the hygrothermal analysis seeks the following:

1. temperature inside the material as a function of location and time
2. moisture concentration inside the material as a function of location and time
3. changes in material properties as a function of time (e.g., strength, stiffness, glass transition temperature, thermal conductivity, thermal expansion)
4. hygrothermal stresses in the material as a function of location and time
5. dimensional changes and deformation of material (e.g., warpage) as a function of location and time

Although temperature and moisture concentration and their effects vary with location and time, only uniform and steady-state conditions will be discussed in this chapter. Hygrothermal effects can be categorized as follows.

8.1.1 Physical and Chemical Effects

Numerous studies have been reported on the diffusion of water into composites.[1-3] Moisture absorption and desorption processes in polymeric composites depend on the current hygrothermal state and on the environment. The glass transition temperature of the polymeric matrix varies with moisture content.[4] Polymerization processes are a function of the hygrothermal properties of the polymer matrix and the current hygrothermal state. Material degradation and corrosion can be related to hygrothermal factors.

8.1.2 Effects on Mechanical Properties

Elastic and viscoelastic (time-dependent) properties may vary with temperature and moisture concentration. Strength and failure characteristics, especially interfacial and matrix-dominated ones, may vary with temperature and moisture content.

8.1.3 Hygrothermoelastic (HTE) Effects

The composite material undergoes reversible deformations related to thermal expansion (α) and moisture expansion (β) coefficients. Intralaminar and interlaminar stresses are developed as a result of the thermoelastic and hygroelastic inhomogeneity and anisotropy of the material.

8.2 HYGROTHERMAL EFFECTS ON MECHANICAL BEHAVIOR

The hygrothermal state affects the stress-strain behavior of composite materials in two different ways: the properties of the constituents may vary with temperature and moisture concentration, and fabrication residual stresses may be altered by the hygrothermal state. Since the fibers are usually the least sensitive to environment, hygrothermal effects are most noticeable in matrix-dominated properties, for example, transverse tensile, transverse compressive, in-plane shear, and longitudinal compressive properties.

Effects of temperature on stress-strain behavior of typical composites are illustrated in Figs. 8.1–8.4. Figure 8.1 shows the transverse stress-strain behavior of a carbon/epoxy (AS4/3501-6) composite at various temperatures.[5] It is seen that the transverse modulus decreases steadily with increasing temperature, although the strength and ultimate strain are not affected much. Figure 8.2 shows similar stress-strain curves for in-plane shear loading. The in-plane shear modulus and strength decrease with increasing temperature, but the ultimate shear strain remains nearly constant. Figure 8.3 shows the effect of temperature on the transverse

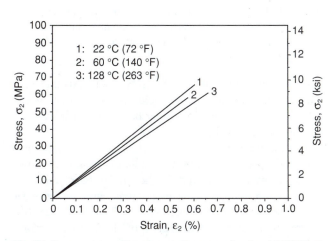

Fig. 8.1 Transverse tensile stress-strain curves for dry AS4/3501-6 carbon/epoxy composite at various temperatures.[5]

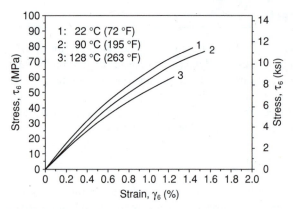

Fig. 8.2 In-plane shear stress-strain curves for unidirectional AS4/3501-6 carbon/epoxy composite at various temperatures.[5]

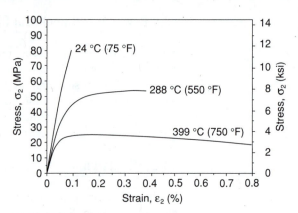

Fig. 8.3 Transverse tensile stress-strain curves for unidirectional silicon carbide/aluminum (SCS-2/6061-Al) composite at various temperatures.

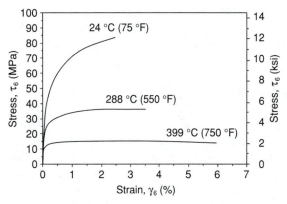

Fig. 8.4 In-plane shear stress-strain curves for unidirectional silicon carbide/aluminum (SCS-2/6061-Al) composite at various temperatures.

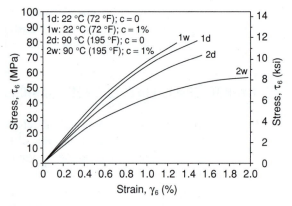

Fig. 8.5 In-plane shear stress-strain curves for unidirectional AS4/3501-6 carbon/epoxy composite illustrating effects of temperature and moisture concentration.

stress-strain behavior of a silicon carbide/aluminum (SiC/Al; SCS-2/6061-Al) unidirectional composite. The initial modulus is not affected, but the yield stress decreases and the ultimate strain increases with increasing temperature. A similar behavior is displayed by the same material under in-plane shear loading (Fig. 8.4).

The influence of moisture concentration is similar to that of temperature on polymer-matrix composites, and it is more pronounced at elevated temperatures. Figure 8.5 shows that a 1% moisture concentration produces a very small difference in the shear stress versus shear strain behavior of a carbon/epoxy composite at room temperature; however, it has a noticeable effect at the elevated temperature of 90 °C (195 °F). The most deleterious effects on stiffness and strength are produced by a combination of elevated temperature and high moisture concentration.[6,7] This is further illustrated in Fig. 8.6, where it is shown

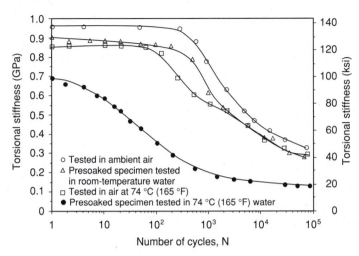

Fig. 8.6 Torsional stiffness degradation of carbon/epoxy composite under cyclic loading under various hygrothermal conditions.

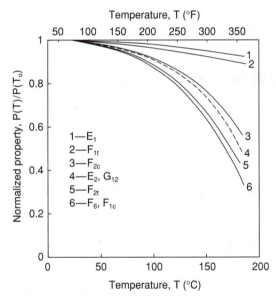

Fig. 8.7 Temperature effects on elastic and strength properties of a typical carbon/epoxy composite[8] (T_o, reference temperature; T, current temperature).

that the torsional stiffness under cyclic loading degrades most at high temperature and moisture concentration.

Figure 8.7 illustrates the degradation of stiffness and strength properties with temperature of an intermediate-strength carbon/epoxy.[8] It is seen that the fiber-dominated properties of longitudinal modulus and tensile strength decrease by approximately 5% and 10% at 121 °C (250 °F) and 177 °C (350 °F), respectively. Matrix-dominated properties, such as transverse and in-plane shear moduli and transverse tensile and compressive strengths, in-plane shear strength and longitudinal compressive strength degrade by 40 to 60% at 177 °C (350 °F). The strengths degrade somewhat more than the corresponding moduli. Property degradation with temperature is less pronounced in composites having a matrix with a higher glass transition temperature, T_g. Table 8.1 shows properties of a carbon/epoxy (IM7/977-3) measured at three different temperatures, 24 °C (75 °F), 93 °C (200 °F) and 149 °C (300 °F).

Numerous studies have been published on the effects of hygrothermal exposure and aging on the deformation, damage development, strength, and durability of composites.[9] This is particularly important due to the continuously increasing applications of composites to offshore structures, submersibles, and civil infrastructure.[10]

Glass fiber composites with rubber-modified epoxy matrix exhibited high strength degradation when exposed to hot and humid ambient conditions, compared with composites having a standard (unmodified) "brittle" epoxy matrix, which showed a slight improvement in strength.[7,11] Exposure to boiling water degraded mechanical properties of glass/vinylester composites due to ingress of moisture through the fiber/matrix interface.[12] Fatigue damage in unidirectional composites, characterized by the accumulation of fiber breaks, increased with environmental exposure.[13] Polyimide-matrix composites have shown degradation in microcracking fracture toughness when exposed to water at high temperature.[14] A master plot for hygrothermal aging can be generated from a limited number of experiments, leading to accelerated test methods for characterization of hygrothermal stability.[14]

TABLE 8.1 Properties of Unidirectional Carbon/Epoxy Composite at Different Temperatures (IM7/977-3)

Property	24 °C (75 °F)	93 °C (200 °F)	149 °C (300 °F)
Longitudinal modulus, E_1, GPa (Msi)	191 (27.7)	186 (27.0)	179 (26.0)
Transverse modulus, E_2, GPa (Msi)	9.89 (1.44)	9.48 (1.38)	8.93 (1.30)
In-plane shear modulus, G_{12}, GPa (Msi)	7.79 (1.13)	7.45 (1.08)	6.53 (0.95)
Major Poisson's ratio, ν_{12}	0.35	0.35	0.34
Minor Poisson's ratio, ν_{21}	0.02	0.02	0.02
Longitudinal tensile strength, F_{1t}, MPa (ksi)	3250 (470)	3167 (460)	2961 (430)
Ultimate longitudinal tensile strain, ε_{1t}^u	0.016	0.015	0.014
Transverse tensile strength, F_{2t}, MPa (ksi)	62 (8.9)	55 (7.9)	51 (7.3)
Ultimate transverse tensile strain, ε_{2t}^u	0.006	0.006	0.006
In-plane shear strength, F_6, MPa (ksi)	75 (10.9)	61 (8.9)	50 (7.2)
Ultimate in-plane shear strain, γ_6^u	0.014	0.013	0.012
Longitudinal thermal expansion coefficient, α_1, $10^{-6}/°C$ ($10^{-6}/°F$)	−0.9 (−0.5)	−0.9 (−0.5)	−0.9 (−0.5)
Transverse thermal expansion coefficient, α_2, $10^{-6}/°C$ ($10^{-6}/°F$)	22.3 (12.4)	23.2 (12.9)	25.9 (14.4)

8.3 COEFFICIENTS OF THERMAL AND MOISTURE EXPANSION OF A UNIDIRECTIONAL LAMINA

The hygrothermal behavior of a unidirectional lamina is fully characterized in terms of two principal coefficients of thermal expansion (CTEs), α_1 and α_2, and two principal coefficients of moisture expansion (CMEs), β_1 and β_2 (Fig. 8.8). These coefficients can be related to the geometric and material properties of the constituents.

Approximate micromechanical relations for the coefficients of thermal expansion were given by Schapery[15] for isotropic constituents. The longitudinal coefficient of a continuous-fiber composite is given by the relation

$$\alpha_1 = \frac{E_f \alpha_f V_f + E_m \alpha_m V_m}{E_f V_f + E_m V_m} = \frac{(E\alpha)_1}{E_1} \qquad (8.1)$$

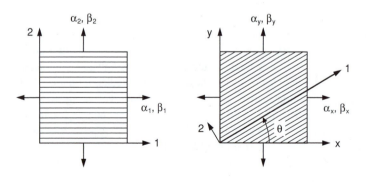

Fig. 8.8 Coefficients of thermal and moisture expansion of a unidirectional lamina.

where

E_f, E_m = fiber and matrix moduli, respectively

α_f, α_m = fiber and matrix coefficients of thermal expansion, respectively

V_f, V_m = fiber and matrix volume ratios, respectively

$(E\alpha)_1 = E_f \alpha_f V_f + E_m \alpha_m V_m$

E_1 = longitudinal composite modulus (by rule of mixtures)

This relation is similar to the rule of mixtures for longitudinal modulus and gives fairly accurate results. It is identical to that obtained by the self-consistent scheme discussed before.

The relation for the transverse coefficient of thermal expansion based on energy principles is[15]

$$\alpha_2 = \alpha_f V_f(1 + \nu_f) + \alpha_m V_m(1 + \nu_m) - \nu_{12}\alpha_1 \tag{8.2}$$

where

ν_f, ν_m = Poisson's ratios of fiber and matrix, respectively

$\nu_{12} = \nu_f V_f + \nu_m V_m$ = major Poisson's ratio of composite lamina as obtained by the rule of mixtures, Eq. (3.24)

α_1 = longitudinal CTE of lamina as obtained by Eq. (8.1)

In many cases, such as carbon and aramid composites, the fibers are orthotropic, that is, they have different properties in the axial and transverse directions. Properties for composites with orthotropic constituents were obtained by Hashin.[16] The relation for the transverse CTE for orthotropic fibers and isotropic matrix is

$$\alpha_2 = \alpha_{2f} V_f\left(1 + \nu_{12f}\frac{\alpha_{1f}}{\alpha_{2f}}\right) + \alpha_m V_m(1 + \nu_m) - (\nu_{12f} V_f + \nu_m V_m)\frac{(E\alpha)_1}{E_1} \tag{8.3}$$

where

α_{1f}, α_{2f} = axial and transverse CTEs of fiber, respectively

α_m = CTE of matrix

ν_{12f}, ν_m = Poisson's ratios of fiber and matrix, respectively

$$(E\alpha)_1 = E_{1f}\alpha_{1f}V_f + E_m\alpha_m V_m \tag{8.4}$$

$$E_1 = E_{1f}V_f + E_m V_m \tag{8.5}$$

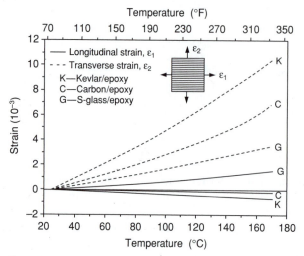

Fig. 8.9 Thermal strains as a function of temperature for three representative unidirectional composites.

In the thickness direction (3-direction) it can be assumed that for a unidirectional lamina $\alpha_3 = \alpha_2$. This is not true in the case of a fabric composite layer.

The coefficients of thermal expansion are obtained by measuring strains as a function of temperature and determining the slopes of the thermal strain versus temperature curves. Thermal strains versus temperature curves are shown in Fig. 8.9 for three material systems, S-glass/epoxy, carbon/epoxy, and Kevlar/epoxy.[17] It is seen that the CTEs in general are not constant but vary with temperature. The S-glass/epoxy system is the least anisotropic material thermally with both principal coefficients positive. The carbon/epoxy and Kevlar/epoxy systems display negative longitudinal coefficients of thermal expansion, with the Kevlar/epoxy being the most thermally anisotropic material. Coefficients of thermal expansion of some typical composite materials are listed in Table 8.2. Similar coefficients for many constituent and composite materials are included in Tables A.2–A.7 in Appendix A.

Once the principal coefficients α_1 and α_2 are known ($\alpha_6 = 0$), the coefficients referred to any system of coordinates (x, y) can be obtained by the following transformation relations, which are the same as those for strain transformation, Eqs. (4.61) (Fig. 8.8):

$$\alpha_x = \alpha_1 m^2 + \alpha_2 n^2$$

$$\alpha_y = \alpha_1 n^2 + \alpha_2 m^2 \tag{8.6}$$

$$\alpha_{xy} = \alpha_s = 2(\alpha_1 - \alpha_2)mn$$

TABLE 8.2 Coefficients of Thermal Expansion of Typical Unidirectional Composite Materials[17]

Material	Longitudinal Coefficient of Thermal Expansion, α_1 $10^{-6}/°C$ ($10^{-6}/°F$)		Transverse Coefficient of Thermal Expansion, α_2 $10^{-6}/°C$ ($10^{-6}/°F$)	
	24 °C (75 °F)	177 °C (350 °F)	24 °C (75 °F)	177 °C (350 °F)
Boron/epoxy (boron/AVCO 5505)	6.1 (3.4)	6.1 (3.4)	30.3 (16.9)	37.8 (21.0)
Boron/polyimide (boron/WRD 9371)	4.9 (2.7)	4.9 (2.7)	28.4 (15.8)	28.4 (15.8)
Carbon/epoxy (AS4/3501-6)	−0.9 (−0.5)	−0.9 (−0.5)	27.0 (15.0)	34.2 (19.0)
Carbon/epoxy (IM7/977-3)	−0.9 (−0.5)	−0.9 (−0.5)	22.3 (12.4)	27.0 (15.0)
Carbon/polyimide (modmor I/WRD 9371)	−0.4 (−0.2)	−0.4 (−0.2)	25.3 (14.1)	25.3 (14.1)
S-glass/epoxy (Scotchply 1009-26-5901)	3.8 (2.1)	3.8 (2.1)	16.7 (9.3)	54.9 (30.5)
S-glass/epoxy (S-glass/ERLA 4617)	6.6 (3.7)	14.1 (7.9)	19.7 (10.9)	26.5 (14.7)
Kevlar/epoxy (Kevlar 49/ERLA 4617)	−4.0 (−2.2)	−5.7 (−3.2)	57.6 (32.0)	82.8 (46.0)

where

$$m = \cos \theta$$

$$n = \sin \theta$$

Micromechanical relations for the coefficients of moisture expansion are entirely analogous. However, some simplification results by taking into consideration the fact that in most cases the fiber (graphite, boron, glass) does not absorb moisture, that is, $\beta_f = \beta_{1f} = \beta_{2f} = 0$. Then the relations for the longitudinal and transverse coefficients of moisture expansion for isotropic constituents take the form

$$\beta_1 = \beta_m \frac{E_m V_m}{E_f V_f + E_m V_m} = \beta_m \frac{E_m V_m}{E_1} \tag{8.7}$$

and

$$\beta_2 = \beta_m \frac{V_m}{E_1} [E_1 + V_f (\nu_m E_f - \nu_f E_m)] \tag{8.8}$$

For a composite with isotropic matrix but orthotropic fibers, β_1 is given by Eq. (8.7), and β_2 takes the form

$$\beta_2 = \beta_m \frac{V_m}{E_1} [E_1 + V_f (\nu_m E_{1f} - \nu_{12f} E_m)] \tag{8.9}$$

The transformation relations for β_1 and β_2 are entirely analogous to Eq. (8.6) for the CTEs.

Methods for determination of hygrothermal properties are discussed in Sections 10.3.4 and 10.3.5 in Chapter 10. Results for a unidirectional and a fabric carbon/epoxy composite are shown in Table 8.3.

TABLE 8.3 Hygrothermal Properties of Typical Unidirectional and Fabric Carbon/Epoxy Composites

Property	Unidirectional (AS4/3501-6)	Woven Fabric (AGP370-5H/3501-6S)
Longitudinal thermal expansion coefficient, α_1, $10^{-6}/°C$ ($10^{-6}/°F$)	−0.9 (−0.5)	3.4 (1.9)
Transverse thermal expansion coefficient, α_2, $10^{-6}/°C$ ($10^{-6}/°F$)	27 (15)	3.7 (2.1)
Out-of-plane thermal expansion coefficient, α_3, $10^{-6}/°C$ ($10^{-6}/°F$)	27 (15)	52 (29)
Longitudinal moisture expansion coefficient, β_1	0.01	0.05
Transverse moisture expansion coefficient, β_2	0.20	0.05
Out-of-plane moisture expansion coefficient, β_3	0.20	0.27

8.4 HYGROTHERMAL STRAINS IN A UNIDIRECTIONAL LAMINA

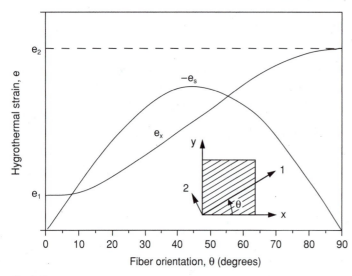

Fig. 8.10 Variation of hygrothermal strains with fiber orientation in unidirectional lamina.

A lamina undergoes hygrothermal deformation when subjected to a uniform change in temperature $\Delta T = T - T_o$ and change in moisture concentration $\Delta c = c - c_o$, where (T_o, c_o) is a reference hygrothermal state. Assuming the thermal and moisture deformations to be uncoupled and the thermal and moisture expansion coefficients to be constant (which is a good approximation for most composites under normal service conditions), the hygrothermal strains referred to the principal material axes of the lamina are

$$e_1 = \alpha_1 \Delta T + \beta_1 \Delta c$$
$$e_2 = \alpha_2 \Delta T + \beta_2 \Delta c \qquad (8.10)$$
$$e_6 = 0$$

The transformed hygrothermal strains referred to the x-y coordinate system of Fig. 8.8 are

$$e_x = e_1 m^2 + e_2 n^2$$
$$e_y = e_1 n^2 + e_2 m^2 \qquad (8.11)$$
$$e_{xy} = e_s = 2(e_1 - e_2)mn$$

where

$$m = \cos \theta$$
$$n = \sin \theta$$

The variation of the normal and shear hygrothermal strains e_x and e_s with fiber orientation θ is shown in Fig. 8.10.

Substituting e_1 and e_2 from Eq. (8.10) into Eq. (8.11), and in view of relations in Eq. (8.6), we obtain

$$e_x = \alpha_x \Delta T + \beta_x \Delta c$$
$$e_y = \alpha_y \Delta T + \beta_y \Delta c \qquad (8.12)$$
$$e_s = \alpha_s \Delta T + \beta_s \Delta c$$

where α_x, α_y, α_s and β_x, β_y, β_s are the transformed lamina coefficients of thermal and moisture expansion.

8.5 HYGROTHERMOELASTIC LOAD-DEFORMATION RELATIONS

When a multidirectional laminate is subjected to mechanical ($[N]$, $[M]$) and hygrothermal (ΔT, Δc) loading, a lamina k within the laminate is under a state of stress $[\sigma]_{x,y}^k$ and deformation $[\varepsilon]_{x,y}^k$. The hygrothermoelastic superposition principle states that the strains $[\varepsilon]_{x,y}^k$ in lamina k within the laminate are equal to the sum of the strains produced by the stresses $[\sigma]_{x,y}^k$ acting in the lamina and the free (unrestrained) hygrothermal strains of the lamina, that is,

$$\begin{bmatrix} \varepsilon_x \\ \varepsilon_y \\ \gamma_s \end{bmatrix}_k = \begin{bmatrix} S_{xx} & S_{xy} & S_{xs} \\ S_{yx} & S_{yy} & S_{ys} \\ S_{sx} & S_{sy} & S_{ss} \end{bmatrix}_k \begin{bmatrix} \sigma_x \\ \sigma_y \\ \tau_s \end{bmatrix}_k + \begin{bmatrix} e_x \\ e_y \\ e_s \end{bmatrix}_k \tag{8.13}$$

or, in brief,

$$[\varepsilon]_{x,y}^k = [S]_{x,y}^k[\sigma]_{x,y}^k + [e]_{x,y}^k$$

Equations (8.13) can be inverted to give the stresses in lamina k as follows:

$$\begin{bmatrix} \sigma_x \\ \sigma_y \\ \tau_s \end{bmatrix}_k = \begin{bmatrix} Q_{xx} & Q_{xy} & Q_{xs} \\ Q_{yx} & Q_{yy} & Q_{ys} \\ Q_{sx} & Q_{sy} & Q_{ss} \end{bmatrix}_k \begin{bmatrix} \varepsilon_x - e_x \\ \varepsilon_y - e_y \\ \gamma_s - e_s \end{bmatrix}_k \tag{8.14}$$

or, in view of Eq. (7.8),

$$\begin{bmatrix} \sigma_x \\ \sigma_y \\ \tau_s \end{bmatrix}_k = \begin{bmatrix} Q_{xx} & Q_{xy} & Q_{xs} \\ Q_{yx} & Q_{yy} & Q_{ys} \\ Q_{sx} & Q_{sy} & Q_{ss} \end{bmatrix}_k \begin{bmatrix} \varepsilon_x^o + z\kappa_x - e_x \\ \varepsilon_y^o + z\kappa_y - e_y \\ \gamma_x^o + z\kappa_s - e_s \end{bmatrix}_k \tag{8.15}$$

or, in brief,

$$[\sigma]_{x,y}^k = [Q]_{x,y}^k[\varepsilon^o]_{x,y} + [Q]_{x,y}^k[\kappa]_{x,y}z - [Q]_{x,y}^k[e]_{x,y}^k \tag{8.16}$$

As mentioned in Chapter 7, stresses in the laminate may vary discontinuously from lamina to lamina. Therefore, a more convenient form of expressing the stress-strain relations in Eq. (8.15) is in terms of force and moment resultants for the entire laminate.

Integration of the stresses from Eq. (8.15) across the thickness of each lamina k and summation for all laminae in the laminate gives the force resultants:

$$[N]_{x,y} = \sum_{k=1}^{n} \int_{z_{k-1}}^{z_k} [\sigma]_{x,y}^k dz = \sum_{k=1}^{n} \int_{z_{k-1}}^{z_k} [Q]_{x,y}^k \{[\varepsilon^o]_{x,y} + z[\kappa]_{x,y} - [e]_{x,y}^k\} dz$$

or

$$\begin{bmatrix} N_x \\ N_y \\ N_s \end{bmatrix} = \sum_{k=1}^{n} \int_{z_{k-1}}^{z_k} \begin{bmatrix} Q_{xx} & Q_{xy} & Q_{xs} \\ Q_{yx} & Q_{yy} & Q_{ys} \\ Q_{sx} & Q_{sy} & Q_{ss} \end{bmatrix}_k \left\{ \begin{bmatrix} \varepsilon_x^o \\ \varepsilon_y^o \\ \gamma_s^o \end{bmatrix} + z \begin{bmatrix} \kappa_x \\ \kappa_y \\ \kappa_s \end{bmatrix} - \begin{bmatrix} e_x \\ e_y \\ e_s \end{bmatrix}_k \right\} dz$$

or, in view of Eqs. (7.18) through (7.22),

$$
\begin{bmatrix} N_x \\ N_y \\ N_s \end{bmatrix} = \begin{bmatrix} A_{xx} & A_{xy} & A_{xs} \\ A_{yx} & A_{yy} & A_{ys} \\ A_{sx} & A_{sy} & A_{ss} \end{bmatrix} \begin{bmatrix} \varepsilon_x^o \\ \varepsilon_y^o \\ \gamma_s^o \end{bmatrix} + \begin{bmatrix} B_{xx} & B_{xy} & B_{xs} \\ B_{yx} & B_{yy} & B_{ys} \\ B_{sx} & B_{sy} & B_{ss} \end{bmatrix} \begin{bmatrix} \kappa_x \\ \kappa_y \\ \kappa_s \end{bmatrix} - \begin{bmatrix} N_x^{HT} \\ N_y^{HT} \\ N_s^{HT} \end{bmatrix} \qquad (8.17)
$$

where the laminate stiffness matrices $[A]$ and $[B]$ are defined as in Eqs. (7.20), and $[N^{HT}]_{x,y}$ are the hygrothermal force resultants defined as

$$
\begin{bmatrix} N_x^{HT} \\ N_y^{HT} \\ N_s^{HT} \end{bmatrix} = \sum_{k=1}^{n} \begin{bmatrix} Q_{xx} & Q_{xy} & Q_{xs} \\ Q_{yx} & Q_{yy} & Q_{ys} \\ Q_{sx} & Q_{sy} & Q_{ss} \end{bmatrix}_k \begin{bmatrix} e_x \\ e_y \\ e_s \end{bmatrix}_k t_k \qquad (8.18)
$$

and $t_k = z_k - z_{k-1}$ is the thickness of lamina k.

In a similar fashion, integration of the stresses from Eq. (8.15) multiplied by the z-coordinate across the thickness of the lamina k and summation for all laminae in the laminate gives the moment resultants:

$$
[M]_{x,y} = \sum_{k=1}^{n} \int_{z_{k-1}}^{z_k} [\sigma]_{x,y}^k z \, dz = \sum_{k=1}^{n} \int_{z_{k-1}}^{z_k} [Q]_{x,y}^k \{ [\varepsilon^o]_{x,y} + z[\kappa]_{x,y} - [e]_{x,y}^k \} z \, dz
$$

or, in view of Eqs. (7.18) through (7.22),

$$
\begin{bmatrix} M_x \\ M_y \\ M_s \end{bmatrix} = \begin{bmatrix} B_{xx} & B_{xy} & B_{xs} \\ B_{yx} & B_{yy} & B_{ys} \\ B_{sx} & B_{sy} & B_{ss} \end{bmatrix} \begin{bmatrix} \varepsilon_x^o \\ \varepsilon_y^o \\ \gamma_s^o \end{bmatrix} + \begin{bmatrix} D_{xx} & D_{xy} & D_{xs} \\ D_{yx} & D_{yy} & D_{ys} \\ D_{sx} & D_{sy} & D_{ss} \end{bmatrix} \begin{bmatrix} \kappa_x \\ \kappa_y \\ \kappa_s \end{bmatrix} - \begin{bmatrix} M_x^{HT} \\ M_y^{HT} \\ M_s^{HT} \end{bmatrix} \qquad (8.19)
$$

where the laminate stiffness matrices $[B]$ and $[D]$ are defined as before in Eqs. (7.20), and $[M^{HT}]_{x,y}$ are the hygrothermal moment resultants defined as

$$
\begin{bmatrix} M_x^{HT} \\ M_y^{HT} \\ M_s^{HT} \end{bmatrix} = \sum_{k=1}^{n} \begin{bmatrix} Q_{xx} & Q_{xy} & Q_{xs} \\ Q_{yx} & Q_{yy} & Q_{ys} \\ Q_{sx} & Q_{sy} & Q_{ss} \end{bmatrix}_k \begin{bmatrix} e_x \\ e_y \\ e_s \end{bmatrix}_k \bar{z}_k t_k \qquad (8.20)
$$

where t_k is the lamina k thickness, and $\bar{z}_k = (z_k + z_{k-1})/2$ is the z-coordinate of the midplane of lamina k. Equations (8.17) and (8.19) can be rewritten in the form

$$
\begin{bmatrix} N_x \\ N_y \\ N_s \end{bmatrix} + \begin{bmatrix} N_x^{HT} \\ N_y^{HT} \\ N_s^{HT} \end{bmatrix} = \begin{bmatrix} \bar{N}_x \\ \bar{N}_y \\ \bar{N}_s \end{bmatrix} = \begin{bmatrix} A_{xx} & A_{xy} & A_{xs} \\ A_{yx} & A_{yy} & A_{ys} \\ A_{sx} & A_{sy} & A_{ss} \end{bmatrix} \begin{bmatrix} \varepsilon_x^o \\ \varepsilon_y^o \\ \gamma_s^o \end{bmatrix} + \begin{bmatrix} B_{xx} & B_{xy} & B_{xs} \\ B_{yx} & B_{yy} & B_{ys} \\ B_{sx} & B_{sy} & B_{ss} \end{bmatrix} \begin{bmatrix} \kappa_x \\ \kappa_y \\ \kappa_s \end{bmatrix} \qquad (8.21)
$$

$$
\begin{bmatrix} M_x \\ M_y \\ M_s \end{bmatrix} + \begin{bmatrix} M_x^{HT} \\ M_y^{HT} \\ M_s^{HT} \end{bmatrix} = \begin{bmatrix} \bar{M}_x \\ \bar{M}_y \\ \bar{M}_s \end{bmatrix} = \begin{bmatrix} B_{xx} & B_{xy} & B_{xs} \\ B_{yx} & B_{yy} & B_{ys} \\ B_{sx} & B_{sy} & B_{ss} \end{bmatrix} \begin{bmatrix} \varepsilon_x^o \\ \varepsilon_y^o \\ \gamma_s^o \end{bmatrix} + \begin{bmatrix} D_{xx} & D_{xy} & D_{xs} \\ D_{yx} & D_{yy} & D_{ys} \\ D_{sx} & D_{sy} & D_{ss} \end{bmatrix} \begin{bmatrix} \kappa_x \\ \kappa_y \\ \kappa_s \end{bmatrix} \qquad (8.22)
$$

or, in brief,

$$[\bar{N}]_{x,y} = [N]_{x,y} + [N^{HT}]_{x,y} = [A][\varepsilon^o]_{x,y} + [B][\kappa]_{x,y} \tag{8.23}$$

$$[\bar{M}]_{x,y} = [M]_{x,y} + [M^{HT}]_{x,y} = [B][\varepsilon^o]_{x,y} + [D][\kappa]_{x,y} \tag{8.24}$$

where $[\bar{N}]$ and $[\bar{M}]$ are total force and moment resultants equal to the respective sums of their mechanical and hygrothermal components.

The above relations can also be presented in a combined concise form as

$$
\begin{bmatrix} \bar{N} \\ \bar{M} \end{bmatrix} = \begin{bmatrix} A & B \\ B & D \end{bmatrix} \begin{bmatrix} \varepsilon^o \\ \kappa \end{bmatrix} \tag{8.25}
$$

Thus, the force-deformation and moment-deformation relations are identical to those derived before for mechanical loading only Eqs. (7.21) to (7.24), except for the fact that here the hygrothermal forces and moments are added to the mechanically applied forces and moments.

8.6 HYGROTHERMOELASTIC DEFORMATION-LOAD RELATIONS

As mentioned in Chapter 7, it is preferable to work with strains because they are continuous through the laminate thickness and because the applied loads are independent variables. Inversion of the load-deformation and moment-deformation relations in Eqs. (8.21) and (8.22) yields the following relations, which are identical to Eq. (7.25) except for the substitution of total forces and moments for the mechanical forces and moments.

$$
\begin{bmatrix} \varepsilon_x^o \\ \varepsilon_y^o \\ \gamma_s^o \\ \kappa_x \\ \kappa_y \\ \kappa_s \end{bmatrix} = \begin{bmatrix} a_{xx} & a_{xy} & a_{xs} & b_{xx} & b_{xy} & b_{xs} \\ a_{yx} & a_{yy} & a_{ys} & b_{yx} & b_{yy} & b_{ys} \\ a_{sx} & a_{sy} & a_{ss} & b_{sx} & b_{sy} & b_{ss} \\ c_{xx} & c_{xy} & c_{xs} & d_{xx} & d_{xy} & d_{xs} \\ c_{yx} & c_{yy} & c_{ys} & d_{yx} & d_{yy} & d_{ys} \\ c_{sx} & c_{sy} & c_{ss} & d_{sx} & d_{sy} & d_{ss} \end{bmatrix} \begin{bmatrix} \bar{N}_x \\ \bar{N}_y \\ \bar{N}_s \\ \bar{M}_x \\ \bar{M}_y \\ \bar{M}_s \end{bmatrix} \tag{8.26}
$$

or, in brief,

$$
\begin{bmatrix} \varepsilon^o \\ \kappa \end{bmatrix} = \begin{bmatrix} a & b \\ c & d \end{bmatrix} \begin{bmatrix} \bar{N} \\ \bar{M} \end{bmatrix} \tag{8.27}
$$

where $[a]$, $[b]$, $[c]$, and $[d]$ are the laminate compliance matrices related to the stiffness matrices $[A]$, $[B]$, and $[D]$ by Eq. (7.27) or obtained by direct inversion of the 6×6 stiffness matrix in Eq. (7.28).

8.7 HYGROTHERMAL LOAD-DEFORMATION RELATIONS

In the absence of externally applied mechanical forces and moments, that is, when $[N] = 0$ and $[M] = 0$, the hygrothermoelastic relations in Eqs. (8.21) and (8.22) are reduced to

$$\begin{bmatrix} N_x^{HT} \\ N_y^{HT} \\ N_s^{HT} \end{bmatrix} = \begin{bmatrix} A_{xx} & A_{xy} & A_{xs} \\ A_{yx} & A_{yy} & A_{ys} \\ A_{sx} & A_{sy} & A_{ss} \end{bmatrix} \begin{bmatrix} \varepsilon_x^o \\ \varepsilon_y^o \\ \gamma_s^o \end{bmatrix} + \begin{bmatrix} B_{xx} & B_{xy} & B_{xs} \\ B_{yx} & B_{yy} & B_{ys} \\ B_{sx} & B_{sy} & B_{ss} \end{bmatrix} \begin{bmatrix} \kappa_x \\ \kappa_y \\ \kappa_s \end{bmatrix} \tag{8.28}$$

$$\begin{bmatrix} M_x^{HT} \\ M_y^{HT} \\ M_s^{HT} \end{bmatrix} = \begin{bmatrix} B_{xx} & B_{xy} & B_{xs} \\ B_{yx} & B_{yy} & B_{ys} \\ B_{sx} & B_{sy} & B_{ss} \end{bmatrix} \begin{bmatrix} \varepsilon_x^o \\ \varepsilon_y^o \\ \gamma_s^o \end{bmatrix} + \begin{bmatrix} D_{xx} & D_{xy} & D_{xs} \\ D_{yx} & D_{yy} & D_{ys} \\ D_{sx} & D_{sy} & D_{ss} \end{bmatrix} \begin{bmatrix} \kappa_x \\ \kappa_y \\ \kappa_s \end{bmatrix} \tag{8.29}$$

The hygrothermal forces $[N^{HT}]$ and moments $[M^{HT}]$ are defined as in Eqs. (8.18) and (8.20).

Inversion of the relations above yields the reference plane strains $[\varepsilon^o]$ and curvatures $[\kappa]$ produced by purely hygrothermal changes:

$$\begin{bmatrix} \varepsilon_x^o \\ \varepsilon_y^o \\ \gamma_s^o \end{bmatrix} = \begin{bmatrix} a_{xx} & a_{xy} & a_{xs} \\ a_{yx} & a_{yy} & a_{ys} \\ a_{sx} & a_{sy} & a_{ss} \end{bmatrix} \begin{bmatrix} N_x^{HT} \\ N_y^{HT} \\ N_s^{HT} \end{bmatrix} + \begin{bmatrix} b_{xx} & b_{xy} & b_{xs} \\ b_{yx} & b_{yy} & b_{ys} \\ b_{sx} & b_{sy} & b_{ss} \end{bmatrix} \begin{bmatrix} M_x^{HT} \\ M_y^{HT} \\ M_s^{HT} \end{bmatrix} \tag{8.30}$$

$$\begin{bmatrix} \kappa_x \\ \kappa_y \\ \kappa_s \end{bmatrix} = \begin{bmatrix} c_{xx} & c_{xy} & c_{xs} \\ c_{yx} & c_{yy} & c_{ys} \\ c_{sx} & c_{sy} & c_{ss} \end{bmatrix} \begin{bmatrix} N_x^{HT} \\ N_y^{HT} \\ N_s^{HT} \end{bmatrix} + \begin{bmatrix} d_{xx} & d_{xy} & d_{xs} \\ d_{yx} & d_{yy} & d_{ys} \\ d_{sx} & d_{sy} & d_{ss} \end{bmatrix} \begin{bmatrix} M_x^{HT} \\ M_y^{HT} \\ M_s^{HT} \end{bmatrix} \tag{8.31}$$

8.8 COEFFICIENTS OF THERMAL AND MOISTURE EXPANSION OF MULTIDIRECTIONAL LAMINATES

For purely hygrothermal loading, that is, $[N] = 0$ and $[M] = 0$, the reference plane strains can also be related to the effective, or laminate, coefficients of thermal and moisture expansion as

$$\begin{bmatrix} \varepsilon_x^o \\ \varepsilon_y^o \\ \gamma_s^o \end{bmatrix} = \begin{bmatrix} \bar{\alpha}_x \\ \bar{\alpha}_y \\ \bar{\alpha}_s \end{bmatrix} \Delta T + \begin{bmatrix} \bar{\beta}_x \\ \bar{\beta}_y \\ \bar{\beta}_s \end{bmatrix} \Delta c \tag{8.32}$$

where $[\bar{\alpha}]_{x,y}$ and $[\bar{\beta}]_{x,y}$ are the coefficients of thermal and moisture (hygric) expansion of the laminate, respectively. These can be determined by comparing Eqs. (8.30) and (8.32) and equating the right-hand sides of those equations.

The coefficients of thermal expansion are obtained by setting $\Delta T = 1$ and $\Delta c = 0$ in the above relations. Thus,

$$
\begin{bmatrix} \bar{\alpha}_x \\ \bar{\alpha}_y \\ \bar{\alpha}_s \end{bmatrix} = \begin{bmatrix} a_{xx} & a_{xy} & a_{xs} \\ a_{yx} & a_{yy} & a_{ys} \\ a_{sx} & a_{sy} & a_{ss} \end{bmatrix} \begin{bmatrix} N_x^T \\ N_y^T \\ N_s^T \end{bmatrix} + \begin{bmatrix} b_{xx} & b_{xy} & b_{xs} \\ b_{yx} & b_{yy} & b_{ys} \\ b_{sx} & b_{sy} & b_{ss} \end{bmatrix} \begin{bmatrix} M_x^T \\ M_y^T \\ M_s^T \end{bmatrix} \tag{8.33}
$$

where $[N^T]_{x,y}$ and $[M^T]_{x,y}$ are the resultant thermal forces and moments as defined in Eqs. (8.18) and (8.20) in the absence of moisture concentration change ($\Delta c = 0$).

The coefficients of moisture expansion are obtained by setting in Eq. (8.32) $\Delta T = 0$ and $\Delta c = 1$. Then

$$
\begin{bmatrix} \bar{\beta}_x \\ \bar{\beta}_y \\ \bar{\beta}_s \end{bmatrix} = \begin{bmatrix} a_{xx} & a_{xy} & a_{xs} \\ a_{yx} & a_{yy} & a_{ys} \\ a_{sx} & a_{sy} & a_{ss} \end{bmatrix} \begin{bmatrix} N_x^H \\ N_y^H \\ N_s^H \end{bmatrix} + \begin{bmatrix} b_{xx} & b_{xy} & b_{xs} \\ b_{yx} & b_{yy} & b_{ys} \\ b_{sx} & b_{sy} & b_{ss} \end{bmatrix} \begin{bmatrix} M_x^H \\ M_y^H \\ M_s^H \end{bmatrix} \tag{8.34}
$$

where $[N^H]_{x,y}$ and $[M^H]_{x,y}$ are the resultant hygric forces and moments as defined in Eqs. (8.18) and (8.20) in the absence of temperature change ($\Delta T = 0$).

In the case of symmetric laminates Eqs. (8.33) and (8.34) are simplified by noting that the coupling compliance matrices $[b]$ and $[c]$ are zero.

8.9 COEFFICIENTS OF THERMAL AND MOISTURE EXPANSION OF BALANCED/SYMMETRIC LAMINATES

In the case of symmetric and balanced laminates a more direct determination can be made of the coefficients of thermal and moisture expansion in terms of the laminate stiffnesses. The hygrothermal load-deformation relations in Eq. (8.28) referred to the principal laminate axes $\bar{x}$ and $\bar{y}$ are reduced to

$$
\begin{bmatrix} N_{\bar{x}}^{HT} \\ N_{\bar{y}}^{HT} \\ N_{\bar{s}}^{HT} \end{bmatrix} = \begin{bmatrix} A_{\bar{x}\bar{x}} & A_{\bar{x}\bar{y}} & 0 \\ A_{\bar{y}\bar{x}} & A_{\bar{y}\bar{y}} & 0 \\ 0 & 0 & A_{\bar{s}\bar{s}} \end{bmatrix} \begin{bmatrix} \varepsilon_{\bar{x}}^o \\ \varepsilon_{\bar{y}}^o \\ \gamma_{\bar{s}}^o \end{bmatrix} = \begin{bmatrix} A_{\bar{x}\bar{x}} & A_{\bar{x}\bar{y}} & 0 \\ A_{\bar{y}\bar{x}} & A_{\bar{y}\bar{y}} & 0 \\ 0 & 0 & A_{\bar{s}\bar{s}} \end{bmatrix} \begin{bmatrix} \bar{\alpha}_{\bar{x}}\Delta T + \bar{\beta}_{\bar{x}}\Delta c \\ \bar{\alpha}_{\bar{y}}\Delta T + \bar{\beta}_{\bar{y}}\Delta c \\ \bar{\alpha}_{\bar{s}}\Delta T + \bar{\beta}_{\bar{s}}\Delta c \end{bmatrix} \tag{8.35}
$$

By setting $\Delta T = 1$ and $\Delta c = 0$, the relations above can be written as follows in an expanded form ($A_{\bar{x}\bar{y}} = A_{\bar{y}\bar{x}}$):

$$
A_{\bar{x}\bar{x}}\bar{\alpha}_{\bar{x}} + A_{\bar{x}\bar{y}}\bar{\alpha}_{\bar{y}} = N_{\bar{x}}^T
$$
$$
A_{\bar{x}\bar{y}}\bar{\alpha}_{\bar{x}} + A_{\bar{y}\bar{y}}\bar{\alpha}_{\bar{y}} = N_{\bar{y}}^T \tag{8.36}
$$
$$
A_{\bar{s}\bar{s}}\bar{\alpha}_{\bar{s}} = N_{\bar{s}}^T
$$

From the definition of thermal force resultants in Eq. (8.18), by setting $\Delta T = 1$ and $\Delta c = 0$, we obtain for the balanced laminate

$$
N_s^T = \sum_{k=1}^{n} [Q_{\bar{s}\bar{x}}\alpha_{\bar{x}} + Q_{\bar{s}\bar{y}}\alpha_{\bar{y}} + Q_{\bar{s}\bar{s}}\alpha_{\bar{s}}]_k t_k = 0
$$

since for each pair of plies of orientations θ and $-\theta$

$$Q_{\bar{x}\bar{x}}(\theta) = -Q_{\bar{x}\bar{x}}(-\theta) \qquad \text{and} \qquad \alpha_{\bar{x}}(\theta) = \alpha_{\bar{x}}(-\theta)$$

$$Q_{\bar{s}\bar{y}}(\theta) = -Q_{\bar{s}\bar{y}}(-\theta) \qquad \text{and} \qquad \alpha_{\bar{y}}(\theta) = \alpha_{\bar{y}}(-\theta)$$

$$Q_{\bar{s}\bar{s}}(\theta) = Q_{\bar{s}\bar{s}}(-\theta) \qquad \text{and} \qquad \alpha_{\bar{s}}(\theta) = -\alpha_{\bar{s}}(-\theta)$$

and for plies of 0° and 90° orientation

$$Q_{\bar{x}\bar{s}} = Q_{\bar{s}\bar{y}} = 0 \qquad \text{and} \qquad \alpha_{\bar{s}} = 0$$

The coefficients of thermal expansion along the principal laminate axes are obtained from the system of Eq. (8.36) as follows:

$$\bar{\alpha}_{\bar{x}} = \frac{A_{\bar{y}\bar{y}}N_{\bar{x}}^{T} - A_{\bar{x}\bar{y}}N_{\bar{y}}^{T}}{A_{\bar{x}\bar{x}}A_{\bar{y}\bar{y}} - A_{\bar{x}\bar{y}}^{2}}$$

$$\bar{\alpha}_{\bar{y}} = \frac{A_{\bar{x}\bar{x}}N_{\bar{y}}^{T} - A_{\bar{x}\bar{y}}N_{\bar{x}}^{T}}{A_{\bar{x}\bar{x}}A_{\bar{y}\bar{y}} - A_{\bar{x}\bar{y}}^{2}} \qquad (8.37)$$

$$\bar{\alpha}_{\bar{s}} = 0$$

These coefficients can be transformed to an arbitrary system of coordinates (x, y) making an angle φ with the principal laminate axes $\bar{x}$ and $\bar{y}$ (Fig. 7.7) as follows:

$$\bar{\alpha}_{x} = m^{2}\bar{\alpha}_{\bar{x}} + n^{2}\bar{\alpha}_{\bar{y}}$$

$$\bar{\alpha}_{y} = n^{2}\bar{\alpha}_{\bar{x}} + m^{2}\bar{\alpha}_{\bar{y}} \qquad (8.38)$$

$$\bar{\alpha}_{s} = 2mn(\bar{\alpha}_{\bar{x}} - \bar{\alpha}_{\bar{y}})$$

where

$$m = \cos \varphi$$

$$n = \sin \varphi$$

The principal coefficients of moisture expansion for the laminate are obtained in an entirely analogous manner by setting $\Delta T = 0$ and $\Delta c = 1$:

$$\bar{\beta}_{\bar{x}} = \frac{A_{\bar{y}\bar{y}}N_{\bar{x}}^{H} - A_{\bar{x}\bar{y}}N_{\bar{y}}^{H}}{A_{\bar{x}\bar{x}}A_{\bar{y}\bar{y}} - A_{\bar{x}\bar{y}}^{2}}$$

$$\bar{\beta}_{\bar{y}} = \frac{A_{\bar{x}\bar{x}}N_{\bar{y}}^{H} - A_{\bar{x}\bar{y}}N_{\bar{x}}^{H}}{A_{\bar{x}\bar{x}}A_{\bar{y}\bar{y}} - A_{\bar{x}\bar{y}}^{2}} \qquad (8.39)$$

$$\bar{\beta}_{\bar{s}} = 0$$

The transformed coefficients of moisture expansion are

$$\bar{\beta}_x = m^2\bar{\beta}_{\bar{x}} + n^2\bar{\beta}_{\bar{y}}$$
$$\bar{\beta}_y = n^2\bar{\beta}_{\bar{x}} + m^2\bar{\beta}_{\bar{y}}$$
$$\bar{\beta}_s = 2mn(\bar{\beta}_{\bar{x}} - \bar{\beta}_{\bar{y}})$$

(8.40)

In some cases, such as those of thick composites or laminates with curvature, it is important to know the coefficients of thermal (and moisture) expansion in the thickness direction (z-direction). Such coefficients can be calculated exactly in terms of lamina properties and the laminate configuration.[18]

8.10 PHYSICAL SIGNIFICANCE OF HYGROTHERMAL FORCES AND MOMENTS

The hygrothermal forces and moments (N_i^{HT}, M_i^{HT}), as defined in Eqs. (8.18) and (8.20), are not just a mathematical abstraction, but can be understood in physical terms. They may be defined as the equivalent mechanical forces and moments (N_i, M_i), which produce laminate deformations (ε_i, κ_i) equal to the hygrothermal deformations (ε_i^{HT}, κ_i^{HT}) resulting from a purely hygrothermal loading (ΔT, Δc) ($i = x, y, s$).

For purely hygrothermal loading (ΔT, Δc) and in the absence of mechanical loading ($N_i = M_i = 0$), the hygrothermal force-deformation relations given by Eqs. (8.28) and (8.29) can be written as

$$\begin{bmatrix} N^{HT} \\ M^{HT} \end{bmatrix} = \begin{bmatrix} A & B \\ B & D \end{bmatrix} \begin{bmatrix} \varepsilon^o \\ \kappa \end{bmatrix}^{HT}$$

(8.41)

Similarly, for purely mechanical loading ($\Delta T = \Delta c = 0$), the force-deformation relations are given by Eqs. (7.24)

$$\begin{bmatrix} N \\ M \end{bmatrix} = \begin{bmatrix} A & B \\ B & D \end{bmatrix} \begin{bmatrix} \varepsilon^o \\ \kappa \end{bmatrix}$$

(7.24 bis)

Hence, if

$$\varepsilon_i^o = \varepsilon_i^{HT}$$
$$\kappa_i = \kappa_i^{HT} \qquad (i = x, y, s)$$

(8.42)

then

$$N_i = N_i^{HT}$$
$$M_i = M_i^{HT} \qquad (i = x, y, s)$$

(8.43)

This equivalence can be demonstrated for the case of an orthotropic (balanced/symmetric) laminate under thermal loading ΔT (Fig. 8.11). The reference plane strains

Thermal loading

$$\Delta T = \Delta T_o$$

$$N_{\bar{x}} = N_{\bar{y}} = 0$$

$$N_{\bar{x}}^T = (A_{\bar{x}\bar{x}}\bar{\alpha}_{\bar{x}} + A_{\bar{x}\bar{y}}\bar{\alpha}_{\bar{y}})\Delta T_o$$

$$N_{\bar{y}}^T = (A_{\bar{y}\bar{x}}\bar{\alpha}_{\bar{x}} + A_{\bar{y}\bar{y}}\bar{\alpha}_{\bar{y}})\Delta T_o$$

$$\varepsilon_{\bar{x}}^o = \bar{\alpha}_{\bar{x}}\Delta T_o$$

$$\varepsilon_{\bar{y}}^o = \bar{\alpha}_{\bar{y}}\Delta T_o$$

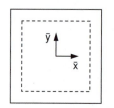

Mechanical loading

$$\Delta T = 0$$

$$N_{\bar{x}} = N_{\bar{x}}^T$$

$$N_y = N_{\bar{y}}^T$$

$$\varepsilon_{\bar{x}}^o = \bar{\alpha}_{\bar{x}}\Delta T_o$$

$$\varepsilon_{\bar{y}}^o = \bar{\alpha}_{\bar{y}}\Delta T_o$$

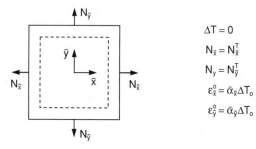

Fig. 8.11 Illustration of the physical significance of thermal forces.

in Eq. (8.41) can be written in terms of the laminate coefficients of thermal expansion (see Eq. (8.32)) along the principal laminate axes $(\bar{x}, \bar{y})$ as

$$\begin{bmatrix} \varepsilon_{\bar{x}}^o \\ \varepsilon_{\bar{y}}^o \\ 0 \end{bmatrix}^T = \begin{bmatrix} \bar{\alpha}_{\bar{x}} \\ \bar{\alpha}_{\bar{y}} \\ 0 \end{bmatrix} \Delta T \qquad (8.44)$$

and from Eqs. (8.41) we obtain for the orthotropic laminate

$$\begin{bmatrix} N_{\bar{x}}^T \\ N_{\bar{y}}^T \\ 0 \end{bmatrix} = \begin{bmatrix} A_{\bar{x}\bar{x}} & A_{\bar{x}\bar{y}} & 0 \\ A_{\bar{y}\bar{x}} & A_{\bar{y}\bar{y}} & 0 \\ 0 & 0 & A_{\bar{s}\bar{s}} \end{bmatrix} \begin{bmatrix} \bar{\alpha}_{\bar{x}} \\ \bar{\alpha}_{\bar{y}} \\ 0 \end{bmatrix} \Delta T \qquad (8.45)$$

The force-deformation relations for mechanical loading, obtained from Eqs. (7.24), are

$$\begin{bmatrix} N_{\bar{x}} \\ N_{\bar{y}} \\ 0 \end{bmatrix} = \begin{bmatrix} A_{\bar{x}\bar{x}} & A_{\bar{x}\bar{y}} & 0 \\ A_{\bar{y}\bar{x}} & A_{\bar{y}\bar{y}} & 0 \\ 0 & 0 & A_{\bar{s}\bar{s}} \end{bmatrix} \begin{bmatrix} \varepsilon_{\bar{x}}^o \\ \varepsilon_{\bar{y}}^o \\ 0 \end{bmatrix} \qquad (8.46)$$

If the strains in Eqs. (8.45) and (8.46) are equal, the corresponding forces are also equal. Thus, the thermal forces $N_{\bar{x}}^T$ and $N_{\bar{y}}^T$ are equal to the mechanical forces $N_{\bar{x}}$ and $N_{\bar{y}}$ that produce strains equal to the purely thermal strains.

$$\varepsilon_{\bar{x}}^o = \bar{\alpha}_{\bar{x}}\Delta T$$

$$\varepsilon_{\bar{y}}^o = \bar{\alpha}_{\bar{y}}\Delta T$$

8.11 HYGROTHERMAL ISOTROPY AND STABILITY

An orthotropic laminate with *tetragonal symmetry*, that is, with identical stiffnesses along its *two principal axes* $(\bar{x}, \bar{y})$, has identical hygrothermal properties in *all directions* and is called *hygrothermal isotropic*.

Consider, for example, the orthotropic laminate in Fig. 7.7 with $\bar{x}$ and $\bar{y}$ the principal laminate axes. For the case when $\Delta T = 1$, $\Delta c = 0$, and $t_k = 1$, the thermal force resultants referred to the principal laminate axes are (see Eq. (8.18)).

$$\begin{bmatrix} N_{\bar{x}}^T \\ N_{\bar{y}}^T \\ N_{\bar{s}}^T \end{bmatrix} = \sum_{k=1}^{n} \begin{bmatrix} Q_{\bar{x}\bar{x}} & Q_{\bar{x}\bar{y}} & Q_{\bar{x}\bar{s}} \\ Q_{\bar{y}\bar{x}} & Q_{\bar{y}\bar{y}} & Q_{\bar{y}\bar{s}} \\ Q_{\bar{s}\bar{x}} & Q_{\bar{s}\bar{y}} & Q_{\bar{s}\bar{s}} \end{bmatrix}_k \begin{bmatrix} \alpha_{\bar{x}} \\ \alpha_{\bar{y}} \\ \alpha_{\bar{s}} \end{bmatrix}_k \qquad (8.47)$$

or

$$N_{\bar{x}}^T = \sum_{k=1}^{n} (Q_{\bar{x}\bar{x}}\alpha_{\bar{x}} + Q_{\bar{x}\bar{y}}\alpha_{\bar{y}} + Q_{\bar{x}\bar{s}}\alpha_{\bar{s}})_k$$

$$N_{\bar{y}}^T = \sum_{k=1}^{n} (Q_{\bar{y}\bar{x}}\alpha_{\bar{x}} + Q_{\bar{y}\bar{y}}\alpha_{\bar{y}} + Q_{\bar{y}\bar{s}}\alpha_{\bar{s}})_k \qquad (8.48)$$

$$N_{\bar{s}}^T = \sum_{k=1}^{n} (Q_{\bar{s}\bar{x}}\alpha_{\bar{x}} + Q_{\bar{s}\bar{y}}\alpha_{\bar{y}} + Q_{\bar{s}\bar{s}}\alpha_{\bar{s}})_k$$

It was shown in the discussion following Eq. (8.36) that $N_s^T = 0$ because of the balanced nature of the laminate. Using the transformation relations in Eqs. (4.67) and Eq. (8.6) for the stiffnesses and coefficients of thermal expansion of the lamina, we obtain

$$\begin{aligned}
N_{\bar{x}}^T - N_{\bar{y}}^T = \sum_{k=1}^{n} [&(m^4 Q_{11} + n^4 Q_{22} + 2m^2 n^2 Q_{12} + 4m^2 n^2 Q_{66})\,(\alpha_1 m^2 + \alpha_2 n^2) \\
&+ (m^2 n^2 Q_{11} + m^2 n^2 Q_{22} + (m^4 + n^4)Q_{12} - 4m^2 n^2 Q_{66})(\alpha_1 n^2 + \alpha_2 m^2) \\
&+ (m^3 n Q_{11} - mn^3 Q_{22} - mn(m^2 - n^2)Q_{12} \\
&- 2mn(m^2 - n^2)Q_{66})2mn(\alpha_1 - \alpha_2) - (m^2 n^2 Q_{11} + m^2 n^2 Q_{22} \\
&+ (m^4 + n^4)Q_{12} - 4m^2 n^2 Q_{66})(\alpha_1 m^2 + \alpha_2 n^2) - (n^4 Q_{11} + m^4 Q_{22} \\
&+ 2m^2 n^2 Q_{12} + 4m^2 n^2 Q_{66})(\alpha_1 n^2 + \alpha_2 m^2) - (mn^3 Q_{11} - m^3 n Q_{22} \\
&+ mn(m^2 - n^2)Q_{12} + 2mn(m^2 - n^2)Q_{66})2mn(\alpha_1 - \alpha_2)]_k
\end{aligned}$$

which reduces to

$$N_{\bar{x}}^T - N_{\bar{y}}^T = [(Q_{11} - Q_{12})\alpha_1 - (Q_{22} - Q_{12})\alpha_2] \sum_{k=1}^{n} (m^2 - n^2)_k \qquad (8.49)$$

If the laminate has tetragonal symmetry, then, it satisfies the condition

$$\sum_{k=1}^{n} (m^2 - n^2)_k = \sum_{k=1}^{n} \cos 2\theta_k = 0 \qquad (7.64 \text{ bis})$$

For the above condition,

$$A_{\bar{x}\bar{x}} = A_{\bar{y}\bar{y}} \qquad (7.61 \text{ bis})$$

and, from Eq. (8.49),

$$N_{\bar{x}}^T = N_{\bar{y}}^T \qquad (8.50)$$

From Eqs. (8.37) it follows that

$$\bar{\alpha}_{\bar{x}} = \bar{\alpha}_{\bar{y}} = \bar{\alpha}$$
$$\bar{\alpha}_{\bar{s}} = 0 \qquad (8.51)$$

Then, from the transformation relations in Eq. (8.38), we obtain for any arbitrary direction x at an angle φ with the $\bar{x}$-axis (Fig. 7.7)

$$\bar{\alpha}_x = m^2\bar{\alpha}_{\bar{x}} + n^2\bar{\alpha}_{\bar{y}} = \bar{\alpha}(m^2 + n^2) = \bar{\alpha}$$

$$\bar{\alpha}_s = 2mn(\bar{\alpha}_{\bar{x}} - \bar{\alpha}_{\bar{y}}) = 0$$

$$(8.52)$$

where

$$m = \cos\varphi$$

$$n = \sin\varphi$$

$$\bar{\alpha} = \text{invariant coefficient of thermal expansion}$$

Thus, a laminate with tetragonal symmetry has a constant coefficient of thermal expansion in all directions. It can be shown in an entirely similar way that the same laminate has a constant coefficient of hygric expansion in all directions. Therefore, all *laminates with tetragonal symmetry are hygrothermal isotropic*. Examples of such laminates (with tetragonal symmetry) are those of $[\pm 45]_{ns}$, $[0/90]_{ns}$, and $[0/\pm 30/90_2]_{ns}$ layup.

Composite materials such as carbon/epoxy and Kevlar/epoxy with $E_1 \gg E_2$ and $|\alpha_1| \ll \alpha_2$ have the additional characteristic that

$$\bar{\alpha}_{\bar{x}} = \bar{\alpha}_{\bar{y}} = \bar{\alpha} \ll \alpha_2$$

Such materials are defined as *thermal isotropic and stable*, meaning that they have a very low coefficient of thermal expansion in all directions, and therefore they are nearly dimensionally stable. Similarly, for hygric characteristics, materials having the property

$$\bar{\beta}_{\bar{x}} = \bar{\beta}_{\bar{y}} = \bar{\beta} \ll \beta_2$$

are defined as *hygric isotropic and stable*.

SAMPLE PROBLEM 8.1
Coefficient of Thermal Expansion of $[\pm 45]_s$ Laminate

It is required to calculate the coefficient of thermal expansion $\bar{\alpha}_x$ of a carbon/epoxy $[\pm 45]_s$ laminate. For this material the following observations can be made regarding the lamina properties:

$$Q_{11} \cong E_1 \qquad Q_{22} \cong E_2 \ll E_1$$

$$Q_{12} \cong \nu_{12}E_2 = \nu_{21}E_1 < E_2 \ll E_1$$

$$Q_{66} = G_{12} < E_2 \ll E_1 \qquad (8.53)$$

$$\alpha_1 \cong 0 \qquad \alpha_2 = \alpha$$

The transformed coefficients of thermal expansion of the 45° layer are

$$
\begin{bmatrix} \alpha_x \\ \alpha_y \\ \frac{1}{2}\alpha_s \end{bmatrix} = \begin{bmatrix} m^2 & n^2 & -2mn \\ n^2 & m^2 & 2mn \\ mn & -mn & (m^2 - n^2) \end{bmatrix} \begin{bmatrix} \alpha_1 \\ \alpha_2 \\ 0 \end{bmatrix} = \begin{bmatrix} \frac{1}{2} & \frac{1}{2} & -1 \\ \frac{1}{2} & \frac{1}{2} & 1 \\ \frac{1}{2} & -\frac{1}{2} & 0 \end{bmatrix} \begin{bmatrix} 0 \\ \alpha \\ 0 \end{bmatrix}
\tag{8.54}
$$

Thus,

$$
(\alpha_x)_{45°} = (\alpha_y)_{45°} = (\alpha_x)_{-45°} = (\alpha_y)_{-45°} = \frac{1}{2}\alpha
$$

$$
(\alpha_s)_{45°} = -(\alpha_s)_{-45°} = -\alpha
\tag{8.55}
$$

The transformed lamina stiffnesses are

$$
(Q_{xx})_{45°} = (Q_{yy})_{45°} = (Q_{xx})_{-45°} = (Q_{yy})_{-45°} = \frac{1}{4}(Q_{11} + Q_{22} + 2Q_{12} + 4Q_{66})
$$

$$
(Q_{xy})_{45°} = (Q_{xy})_{-45°} = \frac{1}{4}(Q_{11} + Q_{22} + 2Q_{12} - 4Q_{66})
$$

$$
(Q_{ss})_{45°} = (Q_{ss})_{-45°} = \frac{1}{4}(Q_{11} + Q_{22} - 2Q_{12})
\tag{8.56}
$$

$$
(Q_{xs})_{45°} = (Q_{ys})_{45°} = -(Q_{xs})_{-45°} = -(Q_{ys})_{-45°} = \frac{1}{4}(Q_{11} - Q_{22})
$$

From the definition of thermal forces we have

$$
\begin{bmatrix} N_x^T \\ N_y^T \\ N_s^T \end{bmatrix} = \begin{bmatrix} Q_{xx} & Q_{xy} & Q_{xs} \\ Q_{yx} & Q_{yy} & Q_{ys} \\ Q_{sx} & Q_{sy} & Q_{ss} \end{bmatrix}_{45°} \begin{bmatrix} \alpha_x \\ \alpha_y \\ \alpha_s \end{bmatrix}_{45°} 2t\Delta T + \begin{bmatrix} Q_{xx} & Q_{xy} & Q_{xs} \\ Q_{yx} & Q_{yy} & Q_{ys} \\ Q_{sx} & Q_{sy} & Q_{ss} \end{bmatrix}_{-45°} \begin{bmatrix} \alpha_x \\ \alpha_y \\ \alpha_s \end{bmatrix}_{-45°} 2t\Delta T
\tag{8.57}
$$

where t is the ply thickness.

Substituting the relations in Eqs. (8.55) and (8.56) in the above we obtain

$$
N_x^T = N_y^T = (Q_{xx}\alpha_x + Q_{xy}\alpha_y + Q_{xs}\alpha_s)_{45°} h\Delta T
$$

$$
N_s^T = 0
\tag{8.58}
$$

where $h = 4t$ is the laminate thickness.

From Eq. (8.37) (in which $\Delta T = 1$), we obtain

$$
\bar{\alpha}_x = \bar{\alpha}_y = \frac{N_x^T}{A_{xx} + A_{xy}}
$$

$$
\bar{\alpha}_s = 0
\tag{8.59}
$$

Noting that in this case

$$(A_{xx} + A_{xy}) = h(Q_{xx} + Q_{xy})_{45°}$$

and substituting Eq. (8.58) into Eq. (8.59) with $\Delta T = 1$, we obtain

$$\bar{\alpha}_x = \bar{\alpha}_y = \left[\frac{Q_{xx}\alpha_x + Q_{xy}\alpha_y + Q_{xs}\alpha_s}{Q_{xx} + Q_{xy}} \right]_{45°} \tag{8.60}$$

Taking Eqs. (8.55) and (8.56) into consideration, we obtain

$$\bar{\alpha}_x = \bar{\alpha}_y = \frac{\dfrac{\alpha}{2}(Q_{xx} + Q_{xy} - 2Q_{xs})}{Q_{xx} + Q_{xy}} = \frac{\alpha}{2} \frac{(Q_{11} + Q_{22} + 2Q_{12} - Q_{11} + Q_{22})}{Q_{11} + Q_{22} + 2Q_{12}} = \alpha \frac{(Q_{22} + Q_{12})}{Q_{11} + Q_{22} + 2Q_{12}} \tag{8.61}$$

from which it follows, for high-fiber-stiffness composites, that

$$\bar{\alpha}_x = \bar{\alpha}_y = \frac{E_2(1 + \nu_{12})\alpha_2}{E_2(1 + \nu_{12}) + E_1(1 + \nu_{21})} \cong \frac{E_2(1 + \nu_{12})\alpha_2}{E_1} \tag{8.62}$$

and

$$\bar{\alpha}_x = \bar{\alpha}_y \ll \alpha_2$$

Thus, the $[\pm 45]_s$ laminate is thermoelastic isotropic, and in the case of carbon/epoxy, its coefficient of thermal expansion in any direction is much less than the transverse CTE α_2. For carbon/epoxy (AS4/3501-6) in particular $\bar{\alpha}_x \cong 0.093\alpha_2 \cong \alpha_2/11$, that is, the CTE of a $[\pm 45]_s$ carbon/epoxy laminate is fiber dominated.

8.12 COEFFICIENTS OF THERMAL EXPANSION OF UNIDIRECTIONAL AND MULTIDIRECTIONAL CARBON/EPOXY LAMINATES

A typical carbon/epoxy composite (AS4/3501-6), has the following principal coefficients of thermal expansion:

$$\alpha_1 = -0.9 \times 10^{-6}/°C \ (-0.5 \times 10^{-6}/°F)$$

$$\alpha_2 = 27 \times 10^{-6}/°C \ (15 \times 10^{-6}/°F)$$

The variation of the transformed CTE, α_x, for a unidirectional lamina with fiber orientation θ is illustrated in Fig. 8.12, where it is compared with the $\bar{\alpha}_x$ of the angle-ply laminate $[\pm\theta]_s$ and the thermally isotropic laminate. It is interesting to note that the variation of $\bar{\alpha}_x$ with angle θ for the angle-ply laminate is not monotonic and has a local minimum at $\theta \cong 28°$.

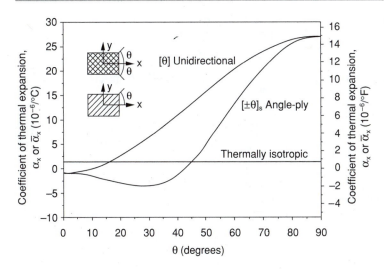

Fig. 8.12 Coefficients of thermal expansion of unidirectional and angle-ply laminates as a function of fiber orientation compared with that of a thermally isotropic laminate (e.g., $[0/90]_s$) (AS4/3501-6 carbon/epoxy).

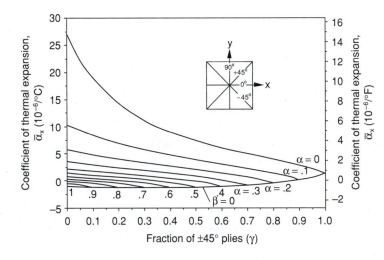

Fig. 8.13 Carpet plot for principal coefficient of thermal expansion of $[0_m/90_n/(\pm 45)_p]_s$ carbon/epoxy laminates (AS4/3501-6) (α, β, and γ are fractions of $0°$, $90°$, and $\pm 45°$ plies, respectively).

A carpet plot of the CTE for the general class of $[0_m/90_n/(\pm 45)_p]_s$ laminates is shown in Fig. 8.13 as a function of the percentages of the $0°$, $\pm 45°$, and $90°$ plies. It can be seen that there are many laminates with exactly zero CTE, (in the x-direction), a fact of great importance in the design of dimensionally stable structures, such as space telescopes, antennae, mirrors, and other aerospace components and structures.

8.13 HYGROTHERMOELASTIC STRESS ANALYSIS OF MULTIDIRECTIONAL LAMINATES

Given a hygrothermomechanical loading $[\bar{N}]$ and $[\bar{M}]$, as defined in Eqs. (8.21) and (8.22), the reference plane strains and curvatures of the laminate are obtained from Eq. (8.27) repeated here:

$$\begin{bmatrix} \varepsilon^o \\ \hline \kappa \end{bmatrix} = \begin{bmatrix} a & \vdots & b \\ \hline c & \vdots & d \end{bmatrix} \begin{bmatrix} \bar{N} \\ \hline \bar{M} \end{bmatrix} \tag{8.27 bis}$$

The total (net) strains in layer k at a distance z from the reference plane are

$$\begin{bmatrix} \varepsilon_x \\ \varepsilon_y \\ \gamma_s \end{bmatrix}_k = \begin{bmatrix} \varepsilon_x^o \\ \varepsilon_y^o \\ \gamma_s^o \end{bmatrix} + z \begin{bmatrix} \kappa_x \\ \kappa_y \\ \kappa_s \end{bmatrix} \tag{7.8 bis}$$

Based on the hygrothermoelastic principle stated in Eq. (8.13), the stress-induced (elastic) strains in layer k are obtained as

$$\begin{bmatrix} \varepsilon_{xe} \\ \varepsilon_{ye} \\ \gamma_{se} \end{bmatrix}_k = \begin{bmatrix} \varepsilon_x \\ \varepsilon_y \\ \gamma_s \end{bmatrix}_k - \begin{bmatrix} e_x \\ e_y \\ e_s \end{bmatrix}_k \tag{8.63}$$

The stresses in layer k referred to the laminate coordinate axes (x, y) can be obtained by

$$\begin{bmatrix} \sigma_x \\ \sigma_y \\ \tau_s \end{bmatrix}_k = \begin{bmatrix} Q_{xx} & Q_{xy} & Q_{xs} \\ Q_{yx} & Q_{yy} & Q_{ys} \\ Q_{sx} & Q_{sy} & Q_{ss} \end{bmatrix}_k \begin{bmatrix} \varepsilon_{xe} \\ \varepsilon_{ye} \\ \gamma_{se} \end{bmatrix}_k \tag{8.64}$$

These stresses can then be transformed to the lamina axes $(1, 2)$ by the transformation relation

$$\begin{bmatrix} \sigma_1 \\ \sigma_2 \\ \tau_6 \end{bmatrix} = [T] \begin{bmatrix} \sigma_x \\ \sigma_y \\ \tau_s \end{bmatrix} \tag{4.57 bis}$$

where

$$[T] = \begin{bmatrix} m^2 & n^2 & 2mn \\ n^2 & m^2 & -2mn \\ -mn & mn & m^2 - n^2 \end{bmatrix}$$

$$m = \cos \theta$$

$$n = \sin \theta$$

Alternatively, these stresses can be obtained by first transforming the stress-induced (elastic) strains from Eq. (8.63) to the lamina principal axes and then using the stress-strain relations for the lamina referred to its principal axes as follows:

$$
\begin{bmatrix} \varepsilon_{1e} \\ \varepsilon_{2e} \\ \frac{1}{2}\gamma_{6e} \end{bmatrix}_k = \begin{bmatrix} \varepsilon_1 - e_1 \\ \varepsilon_2 - e_2 \\ \frac{1}{2}\gamma_6 \end{bmatrix}_k = [T] \begin{bmatrix} \varepsilon_{xe} \\ \varepsilon_{ye} \\ \frac{1}{2}\gamma_{se} \end{bmatrix}_k \tag{8.65}
$$

and

$$
\begin{bmatrix} \sigma_1 \\ \sigma_2 \\ \tau_6 \end{bmatrix}_k = \begin{bmatrix} Q_{11} & Q_{12} & 0 \\ Q_{21} & Q_{22} & 0 \\ 0 & 0 & Q_{66} \end{bmatrix}_k \begin{bmatrix} \varepsilon_{1e} \\ \varepsilon_{2e} \\ \gamma_{6e} \end{bmatrix}_k \tag{8.66}
$$

In the absence of mechanical loading, the total loading $[\bar{N}]$ and $[\bar{M}]$ in Eq. (8.27) is replaced by the hygrothermal forces and moments $[N^{HT}]$ and $[M^{HT}]$. Then, Eqs. (8.64) and (8.66) above give the stresses in each layer due to purely hygrothermal loading.

8.14 RESIDUAL STRESSES

Residual stresses are introduced in composite laminates during fabrication. On a micromechanical scale, residual stresses are introduced in unidirectional layers in and around individual fibers due to the mismatch in thermal properties of the constituents. These stresses are accounted for in the analysis and design of laminates by using lamina properties determined macroscopically by standard testing. In addition, there exist residual stresses on a macroscopic level, the so-called *lamination residual stresses*, due to the thermal anisotropy of the layers and the heterogeneity of the laminate. These stresses are similar in nature to the hygrothermal stresses discussed before.

During processing at elevated temperatures, there is a certain temperature level at which the composite material is assumed to be stress free. This temperature level may be taken as the glass transition temperature of the polymer matrix, or the melting temperature of the metal matrix, or the sintering temperature of the ceramic matrix. Residual stresses develop in the initially stress-free laminate if the thermally anisotropic plies are oriented along different directions. Residual stresses are a function of many parameters, such as ply orientation and stacking sequence, curing process, fiber volume ratio, and other material and processing variables.

Lamination residual stresses can be analyzed by using lamination theory and lamina properties.[19–23] Experimentally, they can be determined using embedded electrical-resistance strain gages and fiber-optic strain sensors, or measuring out-of-plane deflections.[17,22–25] The effect of curing cycle on the development of residual strains and stresses was studied experimentally.[24] It has been shown by viscoelastic analysis that residual stresses are also a function of the cooldown path in the curing cycle.[26,27] For a given temperature drop over a specified length of time, there is an optimum cooldown path that minimizes curing residual stresses. A more precise analysis of residual stresses should take into consideration the viscoelastic properties and the irreversible polymerization (or chemical) shrinkage of the matrix.[26–30]

The procedure for elastic analysis of thermal residual stresses consists of the following steps:

Step 1   Determine the free (unrestrained) thermal strains in each layer ($e_i = \alpha_i \Delta T$) by introducing the difference ΔT between ambient and stress-free temperature in Eqs. (8.12) ($i = x, y, s$).

Step 2   Calculate the thermal forces and moments N_i^T and M_i^T using the definitions in Eqs. (8.18) and (8.20).

Step 3   Determine the midplane strains ε_i^o and laminate curvatures κ_i from the thermal strain-stress relations, Eqs. (8.30) and (8.31).

$$[\varepsilon^o]_{x,y} = [a][N^T]_{x,y} + [b][M^T]_{x,y} \tag{8.30 bis}$$

$$[\kappa]_{x,y} = [c][N^T]_{x,y} + [d][M^T]_{x,y} \tag{8.31 bis}$$

Step 4   Determine the stress-induced (elastic) strains in each layer k using Eq. (8.63).

$$[\varepsilon_e]_{x,y}^k = [\varepsilon]_{x,y}^k - [e]_{x,y}^k \tag{8.63 bis}$$

Step 5   Transform these strains to the lamina principal axes as shown by Eq. (8.65).

$$[\varepsilon_e]_{1,2}^k = [T][\varepsilon_e]_{x,y}^k \tag{8.65 bis}$$

Step 6   Calculate the residual (thermal) stresses using the stress-strain relations in Eq. (8.66).

$$[\sigma]_{1,2}^k = [Q]_{1,2}^k [\varepsilon_e]_{1,2}^k \tag{8.66 bis}$$

As an example, residual stresses in a ply of a $[\pm\theta]_s$ angle-ply carbon/epoxy laminate were calculated as a function of the fiber orientation θ and plotted in Fig. 8.14. The properties of carbon/epoxy (AS4/3501-6) in Table A.4 in Appendix A with a temperature difference of $\Delta T = -110$ °C (-200 °F) were used. The normal lamina stresses σ_1 and σ_2 are equal and opposite in sign and reach maximum absolute values for $\theta = 45°$. In the case of a $[\pm45]_s$ laminate, which is the same as a $[0/90]_s$ laminate rotated by 45°, the maximum transverse residual stress σ_2 is nearly equal to half the transverse lamina strength, F_{2t}.

A clear manifestation of lamination residual stresses is shown for hybrid laminates in Fig. 8.15.[17] Before failure the laminate was flat under the self-equilibrated residual stresses. In the $[\pm45^C/0^G/0^C]_s$ laminate (where superscripts C and G denote carbon and glass, respectively) failure of the 0° carbon plies caused delamination of the outer three-ply ($\pm45^C/0^G$) sublaminate.

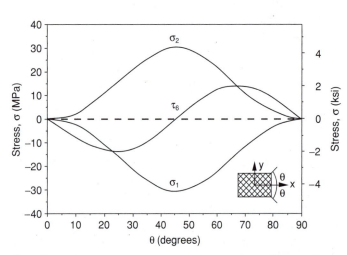

Fig. 8.14   Residual stresses at room temperature in layer of $[\pm\theta]_s$ carbon/epoxy laminate (AS4/3501-6; a temperature difference $\Delta T = -110$ °C (-200 °F) was assumed).

$[\pm 45^{C}/0^{G}/0^{C}]_{s}$

Fig. 8.15 Characteristic failure patterns in two carbon/S-glass/epoxy specimens under uniaxial tensile loading, illustrating presence of residual stresses.[17]

$[0^{G}/\pm 45^{C}/0^{G}]_{s}$

The compressive residual stresses in the $\pm 45^\circ$ carbon plies and tensile residual stress in the 0° glass ply of this sublaminate caused the warpage shown in the asymmetric sublaminate. In the $[0^{G}/\pm 45^{C}/0^{G}]_{s}$ laminates, the outer three-ply ($0^{G}/\pm 45^{C}$) sublaminate delaminated and warped as shown due to the tensile residual stress in the 0° glass ply and compressive residual stress in the $\pm 45^\circ$ carbon plies of the sublaminate.

SAMPLE PROBLEM 8.2
Residual Stresses in Crossply Symmetric Laminate

It is required to calculate the residual stresses in a $[0/90]_{s}$ crossply symmetric laminate in terms of the temperature difference ΔT and the lamina mechanical and thermal properties (Fig. 8.16).[21]

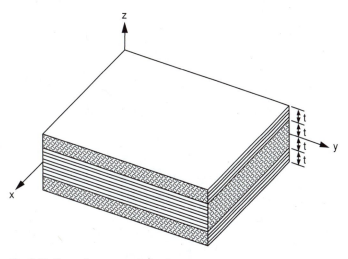

Fig. 8.16 Crossply symmetric laminate.

The general procedure is to determine the thermal forces $[N^T]$, the reference plane strains from Eq. (8.27), and the thermal strains and stresses using Eqs. (8.63) and (8.64) or Eqs. (8.65) and (8.66). In the present case a more direct approach may be taken by making some observations.

The laminate layup implies that

$$\varepsilon_x^o = \varepsilon_y^o \qquad \gamma_s^o = 0$$
$$\kappa_x = \kappa_y = \kappa_s = 0 \tag{8.67}$$

Also, from the laminate symmetry and layup, it follows that

$$A_{xx} = A_{yy} \qquad A_{xs} = A_{ys} = 0$$
$$[B] = 0$$
$$N_x^T = N_y^T \qquad N_s^T = 0 \tag{8.68}$$
$$M_x^T = M_y^T = M_s^T = 0$$

The hygrothermal load-deformation relations in Eq. (8.28) are reduced to

$$\begin{bmatrix} N_x^T \\ N_y^T \\ 0 \end{bmatrix} = \begin{bmatrix} A_{xx} & A_{xy} & 0 \\ A_{yx} & A_{yy} & 0 \\ 0 & 0 & A_{ss} \end{bmatrix} \begin{bmatrix} \varepsilon_x^o \\ \varepsilon_y^o \\ 0 \end{bmatrix} \tag{8.69}$$

which, in view of Eqs. (8.67) and (8.68) are reduced to

$$N_x^T = N_y^T = (A_{xx} + A_{xy})\varepsilon_x^o$$

or

$$\varepsilon_x^o = \varepsilon_y^o = \frac{N_x^T}{A_{xx} + A_{xy}} \tag{8.70}$$

The thermal force N_x^T is obtained from its defining relation, Eq. (8.18), as

$$N_x^T = 2t(Q_{11}\alpha_1 + Q_{12}\alpha_2 + Q_{12}\alpha_1 + Q_{22}\alpha_2)\Delta T \tag{8.71}$$

where t is the lamina thickness.

Substituting the above into Eq. (8.70) and evaluating A_{xx} and A_{xy} for the laminate, we obtain

$$\varepsilon_x^o = \varepsilon_y^o = \frac{(Q_{11} + Q_{12})\alpha_1 + (Q_{22} + Q_{12})\alpha_2}{Q_{11} + Q_{22} + 2Q_{12}}\Delta T \tag{8.72}$$

The stress-induced residual (thermal) strains in the $0°$ ply, for example, are then

$$\varepsilon_{1e} = \varepsilon_x^o - e_1 = \varepsilon_x^o - \alpha_1 \Delta T$$
$$\varepsilon_{2e} = \varepsilon_y^o - e_2 = \varepsilon_y^o - \alpha_2 \Delta T$$

(8.73)

and, substituting from Eq. (8.72),

$$\varepsilon_{1e} = \Delta T(\alpha_2 - \alpha_1) \frac{Q_{22} + Q_{12}}{Q_{11} + Q_{22} + 2Q_{12}}$$

$$\varepsilon_{2e} = \Delta T(\alpha_1 - \alpha_2) \frac{Q_{11} + Q_{12}}{Q_{11} + Q_{22} + 2Q_{12}}$$

(8.74)

Finally, the residual stresses in the $0°$ ply are obtained as follows:

$$\sigma_1 = Q_{11}\varepsilon_{1e} + Q_{12}\varepsilon_{2e} = \Delta T(\alpha_2 - \alpha_1) \frac{Q_{11}Q_{22} - Q_{12}^2}{Q_{11} + Q_{22} + 2Q_{12}}$$

$$\sigma_2 = Q_{12}\varepsilon_{1e} + Q_{22}\varepsilon_{2e} = \Delta T(\alpha_1 - \alpha_2) \frac{Q_{11}Q_{22} - Q_{12}^2}{Q_{11} + Q_{22} + 2Q_{12}}$$

(8.75)

For a high-fiber-stiffness composite the stresses above can be approximated as

$$\sigma_1 = \Delta T(\alpha_2 - \alpha_1) \frac{Q_{11}Q_{22}(1 - v_{12}v_{21})}{Q_{11}(1 + v_{21}) + Q_{22}(1 + v_{12})}$$

$$\cong \Delta T(\alpha_2 - \alpha_1) \frac{E_2}{1 + \dfrac{E_2}{E_1}(1 + v_{12})} \cong \Delta T(\alpha_2 - \alpha_1)E_2$$

(8.76)

$$\sigma_2 \cong \Delta T(\alpha_1 - \alpha_2) \frac{E_2}{1 + \dfrac{E_2}{E_1}(1 + v_{12})} \cong \Delta T(\alpha_1 - \alpha_2)E_2$$

For carbon fiber composites where $|\alpha_1| \ll \alpha_2$,

$$\sigma_1 = -\sigma_2 \cong \Delta T \alpha_2 E_2$$

It is seen that the residual stresses in the fiber and transverse to the fiber directions are of equal magnitude and opposite sign, which satisfies equilibrium conditions. For a temperature drop during cooldown, this means that there is a transverse tensile stress and longitudinal compressive stress, since $\Delta T < 0$ and $\alpha_2 > \alpha_1$. The stresses in the $90°$ ply referred to its principal axes are identical to those of the $0°$ ply. The results above obtained for a $[0/90]_s$ layup are identical to those for a $[\pm 45]_s$ laminate.

8.15 WARPAGE

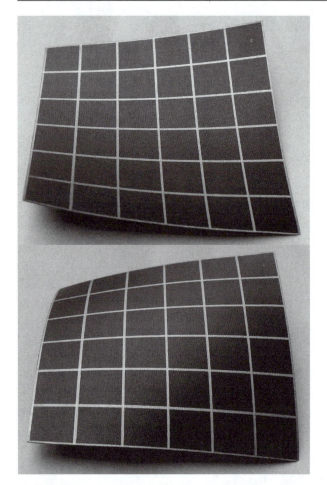

Fig. 8.17 Warpage of $[20_2/-20_2]$ carbon/epoxy laminate (AS4/3501-6).

Warpage or geometrical distortions occur in composite laminates after curing or exposure to hygrothermal environments. They result from a combination of the induced hygrothermal stresses and laminate asymmetry or nonuniformity (Fig. 8.17). Factors contributing to warpage include resin shrinkage, which is partly irreversible (chemical) and partly reversible (thermal), difference between in-plane and out-of-plane chemical and hygrothermal properties of the laminate, nonuniform fiber volume ratio, nonuniform gelation and curing, and mold constraints combined with viscoelastic effects. Bending asymmetries in laminates may result from an intentionally designed asymmetric layup, or material nonuniformities as well as thermal and moisture gradients through the thickness.

Small warpage of asymmetric laminates, when the out-of-plane deformation is relatively small compared to the laminate thickness and lateral dimensions, can be calculated by classical lamination theory or numerical models, using lamina properties.[21,29,31–34]

Integration of the curvature-deflection relations

$$\kappa_x = -\frac{\partial^2 w}{\partial x^2}$$

$$\kappa_y = -\frac{\partial^2 w}{\partial y^2} \qquad \text{(7.7 bis)}$$

$$\kappa_s = -\frac{2\partial^2 w}{\partial x \partial y}$$

yields the out-of-plane deflection w:

$$w = -\frac{1}{2}(\kappa_x x^2 + \kappa_y y^2 + \kappa_s xy) + \text{(rigid body motion)} \qquad (8.77)$$

Warpage is described by the quadratic part of this expression.

The curvatures are also related to the hygrothermal forces and moments as shown in Eqs. (8.27):

$$[\kappa]_{x,y} = [c][N^{HT}]_{x,y} + [d][M^{HT}]_{x,y} \qquad (8.78)$$

which, substituted into Eq. (8.77), yields the specific equation for w describing the warpage.

The analysis above is linear. However, when the panel size is large compared to its thickness, or when the hygrothermal loading (ΔT, Δc) is high, large deformations and rotations induce geometrical nonlinearity.[35] Hyer developed a nonlinear theory based on a polynomial approximation of the displacement, and extended the classical lamination theory to include geometric nonlinearities.[32] In the case of antisymmetric angle-ply laminates (Fig. 8.17), the linear solution obtained by Eq. (8.77) predicts an anticlastic (saddle-like) deformed shape. The nonlinear analysis predicts two cylindrical shapes, one obtained from the other by a simple snap-through action (Fig. 8.17).

The original nonlinear theory was expressed in terms of lamina properties (Q_{ij}) and was not easily extendable to general laminates. The results are usually expressed in numerical format. A new closed-form general nonlinear theory was developed recently, which is expressed in terms of laminate stiffnesses (A_{ij}, B_{ij}, D_{ij}) and explains and quantifies the transition from linear to nonlinear behavior.[36]

Composite laminates with changing orientation, such as angle or channel sections, and curved laminates undergo a special kind of warpage during curing or hygrothermal loading, the so-called *spring-in effect*.[37] This effect is due to a change in curvature related to the difference in hygrothermal strains (e.g., thermal expansion coefficients) between the in-plane and out-of-plane directions.

SAMPLE PROBLEM 8.3
Warpage of Crossply Antisymmetric Laminate

It is required to calculate the warpage of a [0/90] crossply antisymmetric laminate in terms of the uniform temperature change ΔT and the lamina mechanical and thermal properties (Fig. 8.18)[21,33]

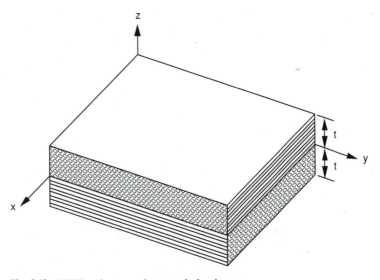

Fig. 8.18 [0/90] antisymmetric crossply laminate.

The thermal force and moment resultants are

$$
\begin{bmatrix} N_x^T \\ N_y^T \\ 0 \end{bmatrix} = \frac{h\Delta T}{2} \begin{bmatrix} Q_{11} & Q_{12} & 0 \\ Q_{21} & Q_{22} & 0 \\ 0 & 0 & Q_{66} \end{bmatrix} \begin{bmatrix} \alpha_1 \\ \alpha_2 \\ 0 \end{bmatrix} + \frac{h\Delta T}{2} \begin{bmatrix} Q_{22} & Q_{21} & 0 \\ Q_{12} & Q_{11} & 0 \\ 0 & 0 & Q_{66} \end{bmatrix} \begin{bmatrix} \alpha_2 \\ \alpha_1 \\ 0 \end{bmatrix}
\tag{8.79}
$$

and

$$
\begin{bmatrix} M_x^T \\ M_y^T \\ 0 \end{bmatrix} = \frac{h^2\Delta T}{8} \begin{bmatrix} Q_{11} & Q_{12} & 0 \\ Q_{21} & Q_{22} & 0 \\ 0 & 0 & Q_{66} \end{bmatrix} \begin{bmatrix} \alpha_1 \\ \alpha_2 \\ 0 \end{bmatrix} - \frac{h^2\Delta T}{8} \begin{bmatrix} Q_{22} & Q_{21} & 0 \\ Q_{12} & Q_{11} & 0 \\ 0 & 0 & Q_{66} \end{bmatrix} \begin{bmatrix} \alpha_2 \\ \alpha_1 \\ 0 \end{bmatrix}
\tag{8.80}
$$

The following observations can be made for the specific layup considered:

$$
\begin{aligned}
\varepsilon_x^o &= \varepsilon_y^o & \gamma_s^o &= 0 \\
\kappa_x &= -\kappa_y & \kappa_s &= 0 \\
N_x^T &= N_y^T & N_s^T &= 0 \\
M_x^T &= -M_y^T & M_s^T &= 0
\end{aligned}
\tag{8.81}
$$

Furthermore,

$$
\begin{aligned}
B_{xx} &= -B_{yy} \\
B_{xy} &= B_{xs} = B_{ss} = 0
\end{aligned}
\tag{8.82}
$$

The thermal force-deformation relations in Eqs. (8.28) and (8.29) are reduced to

$$
\begin{aligned}
N_x^T &= (A_{xx} + A_{xy})\varepsilon_x^o + B_{xx}\kappa_x \\
M_x^T &= B_{xx}\varepsilon_x^o + (D_{xx} - D_{xy})\kappa_x
\end{aligned}
\tag{8.83}
$$

The system of Eqs. (8.83) can be solved for the two unknowns ε_x^o and κ_x:

$$
\varepsilon_x^o = \varepsilon_y^o = \frac{(D_{xx} - D_{xy})N_x^T - B_{xx}M_x^T}{(A_{xx} + A_{xy})(D_{xx} - D_{xy}) - B_{xx}^2}
\tag{8.84}
$$

and

$$
\kappa_x = -\kappa_y = \frac{(A_{xx} + A_{xy})M_x^T - B_{xx}N_x^T}{(A_{xx} + A_{xy})(D_{xx} - D_{xy}) - B_{xx}^2}
\tag{8.85}
$$

Substitution of Eq. (8.85) into Eq. (8.77) yields the equation of the warped surface as

$$
w = -\frac{1}{2}\frac{(A_{xx} + A_{xy})M_x^T - B_{xx}N_x^T}{(A_{xx} + A_{xy})(D_{xx} - D_{xy}) - B_{xx}^2}(x^2 - y^2)
\tag{8.86}
$$

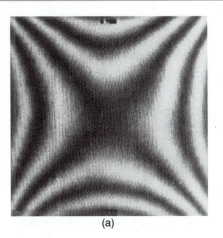 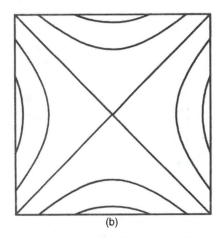

(a)   (b)

Fig. 8.19 Contours of warped surface obtained by shadow moiré method and predicted by analysis for $[0_6/90_6]$ woven glass/epoxy laminate: (a) experimental and (b) predicted. (Experimental and predicted fringes are loci of points of 0.60 mm [0.023 in] constant out-of-plane displacement with respect to points on the neighboring fringe.[33])

which represents a hyperbolic paraboloid or saddle-shaped surface. Predicted contours of this surface for a woven glass/epoxy antisymmetric laminate are shown in Fig. 8.19. They are in good agreement with the moiré fringe pattern, which depicts the same contours.[33]

8.16 COMPUTATIONAL PROCEDURE FOR HYGROTHERMOELASTIC ANALYSIS OF MULTIDIRECTIONAL LAMINATES

A flowchart for hygrothermoelastic analysis of multidirectional laminates is shown in Fig. 8.20. It consists of two main branches. The left-hand branch is similar to the flowchart in Fig. 7.11 used for computation of laminate stiffnesses and compliances and has been discussed before. The right-hand branch consists of the following steps:

Step 1   Enter the lamina coefficients of thermal and moisture expansion α_1, α_2, β_1, and β_2.

Step 2   Enter the temperature and moisture concentration differences, ΔT and Δc.

Step 3   Calculate the free lamina hygrothermal strains e_1 and e_2 referred to their principal material axes (1, 2) using Eq. (8.10).

Step 4   Enter the ply (or principal axis) orientation, θ_k, of lamina k.

Step 5   Calculate the transformed free lamina hygrothermal strains $(e_x, e_y, e_s)_k$ referred to laminate reference axes (x, y) using Eq. (8.11).

Step 6   Enter the transformed lamina stiffnesses $[Q]_{x,y}^k$ of lamina k from the left-hand branch.

Step 7   Enter the through-the-thickness coordinates z_k and z_{k-1} of lamina k surfaces and total number of plies n.

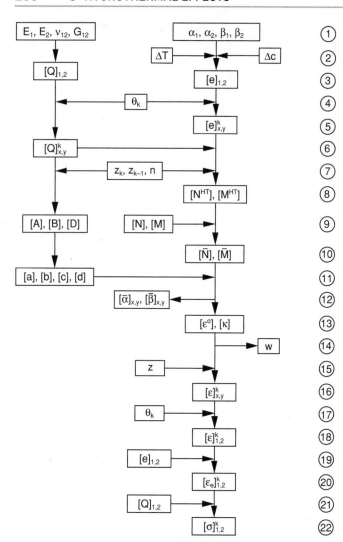

(2)
(3)
(4)
(5)
(6)
(7)
(8)
(9)
(10)
(11)
(12)
(13)
(14)
(15)
(16)
(17)
(18)
(19)
(20)
(21)
(22)

Fig. 8.20 Flowchart for hygrothermoelastic analysis of multidirectional laminates.

Step 8 Calculate the hygrothermal forces and moments, N^{HT} and M^{HT}, using Eqs. (8.18) and (8.20).

Step 9 Enter the mechanical loading $[N]$ and $[M]$.

Step 10 Combine the hygrothermal and mechanical loading, $[\bar{N}]$ and $[\bar{M}]$.

Step 11 Enter the laminate compliances $[a]$, $[b]$, $[c]$, and $[d]$ from the left-hand branch.

Step 12 Calculate the laminate coefficients of thermal and moisture expansion, $\bar{\alpha}_x$, $\bar{\alpha}_y$, $\bar{\beta}_x$, and $\bar{\beta}_y$, using Eqs. (8.33) and (8.34).

Step 13 Calculate the reference plane strains $[\varepsilon^o]$ and curvatures $[\kappa]$ using Eq. (8.26).

Step 14 Calculate the warpage w, using Eq. (8.77).

Step 15 Enter the through-the-thickness coordinate, z, of the point of interest in lamina k. For a laminate consisting of many thin (compared with the laminate thickness) layers or for any symmetric laminate under in-plane loading, the coordinate of the lamina midplane $z = \bar{z}_k$, is used. Then, the computed strains and stresses are the average through-the-thickness strains and stresses in the layer. However, when the laminate consists of few and thick (relative to the laminate thickness) layers and is asymmetric or subjected to bending and twisting, $z = z_k$ and $z = z_{k-1}$ are entered in order to determine the extreme values of the strains and stresses at the top and bottom surfaces of the layer.

Step 16   Calculate the total (net) strains $(\varepsilon_x, \varepsilon_y, \gamma_s)_k$ in lamina k using Eq. (7.8).

Step 17   Enter the orientation θ_k of lamina k.

Step 18   Transform the total lamina strains $[\varepsilon]_{x,y}^k$ to lamina coordinate axes (1, 2).

Step 19   Enter free lamina hygrothermal strains e_1 and e_2 referred to lamina axes (1, 2) from step 3.

Step 20   Calculate the stress-induced (elastic) strains $[\varepsilon_e]_{1,2}^k$ referred to lamina axes (1, 2) using Eq. (8.65).

Step 21   Enter the lamina stiffnesses referred to lamina axes (1, 2).

Step 22   Calculate the stresses $[\sigma]_{1,2}^k$ of lamina k referred to lamina axes (1, 2) using Eq. (8.66).

The procedure above reduces to purely hygrothermal stress analysis when the mechanical loading [N] and [M] is zero and to purely mechanical stress analysis when the hygrothermal loading ΔT and Δc is zero.

REFERENCES

1. J. R. Vinson, Ed., *Advanced Composite Materials—Environmental Effects*, ASTM STP 658, American Society for Testing and Materials, West Conshohocken, PA, 1978.

2. G. S. Springer, Ed., *Environmental Effects on Composite Materials*, Technomic, Lancaster, PA, Vol. 1, 1981, Vol. 2, 1984.

3. Y. J. Weitsman, "Effects of Fluids in Polymeric Composites—A Review," in *Comprehensive Composite Materials*, Vol. 2, *Polymer Matrix Composites*, R. Talreja and J.-A. E. Monson, Eds., Chapter 2.11, Elsevier Science, Ltd., Oxford, 2000, pp. 369–401.

4. A. Chateauminois, B. Chabert, J. P. Soulier, and L. Vincent, "Dynamic Mechanical Analysis of Epoxy Composites Plasticized by Water: Artifacts and Reality," *Polymer Composites*, Vol. 16, No. 4, 1995, pp. 288–296.

5. I. M. Daniel, G. Yaniv, and G. Peimanidis, "Hygrothermal and Strain Rate Effects on Properties of Graphite/Epoxy Composites," *J. Eng. Materials Technol.*, Vol. 110, 1988, pp. 169–173.

6. O. Ishai and A. Mazor, "The Effect of Environmental Loading History on Longitudinal Strength of Glass-Fiber Reinforced Plastics," *Rheol. Acta*, Vol. 13, 1974, pp. 381–394.

7. O. Ishai and U. Arnon, "The Effect of Hygrothermal History on Residual Strength of Glass-Fiber Reinforced Plastic Laminates," *J. Testing Evaluation*, Vol. 5, No. 4, American Society for Testing and Materials, West Conshohocken, PA, 1977, pp. 320–326.

8. T. T. Matoi, J. A. Rohlen, C. H. Hamilton, "Material Properties," Ch. 1.2 in *Advanced Composites Design Guide*, Third Edition, Vol. 1, *Design* (prepared by the Los Angeles Division of the Rockwell International Corp.), Air Force Materials Laboratory, Wright-Patterson Air Force Base, OH, 1973.

9. Y. J. Weitsman and M. Elahi, "Effects of Fluids on the Deformation, Strength and Durability of Polymeric Composites—An Overview," *Mechanics of Time-Dependent Materials*, Vol. 4, No. 2, 2000, pp. 107–126.

10. K. Liao, C. R. Schultheisz, D. L. Hunston, and L. C. Brinson, "Long-term Durability of Fiber-Reinforced Polymer-Matrix Composite Materials for Infrastructure Applications: A Review," SAMPE, *Journal of Advanced Materials*, Vol. 30, No. 4, 1998, pp. 2–40.

11. O. Ishai and U. Arnon, "Instantaneous Effect of Internal Moisture Conditions on Strength of Glass-Fiber Reinforced Plastics," *Advanced Composite Materials-Environmental Effects*, ASTM STP 658, J. R. Vinson, Ed., American Society for Testing and Materials, West Conshohocken, PA, 1978, pp. 267–276.

12. R. Roy, B. K. Sarkar, and N. R. Bose, "Effects of Moisture on the Mechanical Properties of Glass Fibre Reinforced Vinylester Resin Properties," *Bulletin of Material Science*, Vol. 24, No. 1, 2001, pp. 87–94.

13. E. Vauthier, J. C. Abry, T. Baillez, and A. Chateauminois, "Interactions Between Hygrothermal Aging and Fatigue Damage in Unidirectional Glass/Epoxy Composites," *Composites Science and Technology*, Vol. 58, No. 5, 1998, pp. 687–692.

14. M.-H. Han and J. A. Nairn, "Hygrothermal Aging of Polyimide Matrix Composite Laminates," *Composites, Part A*, Vol. 34, No. 10, 2003, pp. 979–986.

15. R. A. Schapery, "Thermal Expansion Coefficients of Composite Materials Based on Energy Principles," *J. Composite Materials*, Vol. 2, 1968, pp. 380–404.

16. Z. Hashin, "Analysis of Properties of Fiber Composites with Anisotropic Constituents," *J. Appl. Mechanics*, Vol. 46, 1979, pp. 543–550.

17. I. M. Daniel, "Thermal Deformations and Stresses in Composite Materials," *Thermal Stresses in Severe Environments*, D. P. H. Hasselman and R. A. Heller, Eds., Plenum Press, New York, 1980, pp. 607–628.

18. N. J. Pagano, "Thickness Expansion Coefficients of Composite Laminates," *Journal of Composite Materials*, Vol. 8, 1974, pp. 310–312.

19. C. C. Chamis, "Lamination Residual Stresses in Crossplied Fiber Composites," in *Proc. of 26th Annual Conf. of SPI, Reinforced Plastics/Composites Div.*, The Society of the Plastics Industry, New York, 1971, Sect. 17-D.

20. H. T. Hahn, "Residual Stresses in Polymer Matrix Composite Laminates," *J. Composite Materials*, Vol. 10, 1976, pp. 266–278.

21. I. G. Zewi, I. M. Daniel, and J. T. Gotro, "Residual Stresses and Warpage in Woven-Glass/Epoxy Laminates," *Exp. Mech.*, Vol. 27, 1987, pp. 44–50.

22. I. M. Daniel and T. Liber, "Effect of Laminate Construction on Residual Stresses in Graphite/Polyimide Composites," *Exp. Mech.*, Vol. 17, 1977, pp. 21–25.

23. I. M. Daniel and T. Liber, "Lamination Residual Strains and Stresses in Hybrid Laminates," in *Composite Materials: Testing and Design* (Fourth Conf.), ASTM STP 617, American Society for Testing and Materials, West Conshohocken, PA, 1977, pp. 331–343.

24. Y. K. Kim and I. M. Daniel, "Cure Cycle Effect on Composite Structures Manufactured by Resin Transfer Molding," *Journal of Composite Materials*, Vol. 36, No. 14, 2002, pp. 1725–1743.

25. G. Yaniv and O. Ishai, "Residual Thermal Stresses in Bonded Metal and FRP Systems," *Comp. Technology Review*, Vol. 1, No. 4, 1981, pp. 122–137.

26. Y. J. Weitsman, "Residual Thermal Stresses Due to Cool-Down of Epoxy-Resin Composites," *J. Appl. Mech.*, Vol. 46, 1979, pp. 563–567.

27. Y. J. Weitsman and B. D. Harper, "Optimal Cooling of Crossply Composite Laminates and Adhesive Joints," *J. Appl. Mech.*, Vol. 49, 1982, pp. 735–739.

28. I. M. Daniel, T. M. Wang, D. Karalekas, and J. T. Gotro, "Determination of Chemical Cure Shrinkage in Composite Laminates," *J. Composite Technol. Res.*, Vol. 12, 1990, pp. 172–176.

29. T. M. Wang and I. M. Daniel, "Thermoviscoelastic Analysis of Residual Stresses and Warpage in Composite Laminates," *J. Composite Materials*, Vol. 26, 1992, pp. 883–899.

30. S. R. White and Y. K. Kim, "Process-Induced Residual Stress Analysis of AS4/3501-6 Composite Material," *Mechanics of Composite Materials and Structures*, Vol. 4, 1997, pp. 361–387.

31. C. C. Chamis, "A Theory for Predicting Composite Laminate Warpage Resulting from Fabrication," in *Proc. 30th Tech. Conf., Reinforced Plastics/Composites Inst.*, Society of Plastics Industry, New York, 1975, Sect. 18-C.

32. M. W. Hyer, "Calculations of the Room-Temperature Shapes of Unsymmetric Laminates," *J. Composite Materials*, Vol. 15, 1981, pp. 296–309.

33. I. M. Daniel, T. M. Wang, and J. T. Gotro, "Thermomechanical Behavior of Multilayer Structures in Microelectronics," *J. Electronic Packaging, Trans. ASME*, Vol. 112, 1990, pp. 11–15.

34. A. Johnston, R. Vaziri, and A. Poursartip, "A Plane Strain Model for Process-Induced Deformation of Laminated Composite Structures," *Journal of Composite Materials*, Vol. 35, No. 16, 2001, pp. 1435–1469.

35. M. W. Hyer, "Some Observations on the Cured Shapes of Thin Unsymmetric Laminates," *Journal of Composite Materials*, Vol. 15, 1981, pp. 175–194.

36. J. J. Luo and I. M. Daniel, "Thermally Induced Deformation of Asymmetric Composite Laminates," *Proc. of the American Society for Composites*, 18th Technical Conference, ASC18, Gainesville, FL, 2003.

37. C. Albert and G. Fernlund, "Spring-In and Warpage of Angled Composite Laminates," *Composites Science and Technology*, Vol. 62, No. 14, 2002, pp. 1895–1912.

PROBLEMS

8.1 Determine the coefficients of thermal expansion α_1 and α_2 of a unidirectional glass/epoxy lamina of the properties

$$E_f = 69 \text{ GPa (10 Msi)}$$
$$E_m = 3.80 \text{ GPa (0.55 Msi)}$$
$$\nu_f = 0.22$$
$$\nu_m = 0.36$$
$$\alpha_f = 4.5 \times 10^{-6}/°C \ (2.5 \times 10^{-6}/°F)$$
$$\alpha_m = 90 \times 10^{-6}/°C \ (50 \times 10^{-6}/°F)$$
$$V_f = 0.55$$

8.2 Determine the coefficients of thermal expansion α_1 and α_2 of a unidirectional silicon carbide/aluminum (SCS-2/6061-A1) lamina of the properties

$$E_f = 410 \text{ GPa (60 Msi)}$$
$$E_m = 69 \text{ GPa (10 Msi)}$$
$$\nu_f = 0.20$$
$$\nu_m = 0.33$$
$$\alpha_f = 1.5 \times 10^{-6}/°C \ (0.83 \times 10^{-6}/°F)$$
$$\alpha_m = 23.4 \times 10^{-6}/°C \ (13 \times 10^{-6}/°F)$$
$$V_f = 0.44$$

8.3 Determine the coefficients of thermal expansion α_1 and α_2 of a unidirectional silicon carbide/glass-ceramic composite (SiC/CAS) of the properties

$$E_f = 170 \text{ GPa (25 Msi)}$$
$$E_m = 98 \text{ GPa (14.2 Msi)}$$
$$\nu_f = 0.20$$
$$\nu_m = 0.20$$
$$\alpha_f = 3.2 \times 10^{-6}/°C \ (1.8 \times 10^{-6}/°F)$$
$$\alpha_m = 5.0 \times 10^{-6}/°C \ (2.8 \times 10^{-6}/°F)$$
$$V_f = 0.39$$

8.4 The measured coefficients of thermal expansion of a unidirectional carbon/epoxy composite of fiber volume ratio $V_f = 0.65$ are

$$\alpha_1 = -0.9 \times 10^{-6}/°C \ (-0.5 \times 10^{-6}/°F)$$
$$\alpha_2 = 27 \times 10^{-6}/°C \ (15 \times 10^{-6}/°F)$$

Determine the coefficients α_{1f} and α_{2f} of the fiber from the above and the constituent properties

$$E_{1f} = 235 \text{ GPa } (34 \times 10^6 \text{ psi})$$
$$E_m = 4.1 \text{ GPa } (0.6 \times 10^6 \text{ psi})$$
$$\nu_f = 0.20$$
$$\nu_m = 0.34$$
$$\alpha_m = 41 \times 10^{-6}/°C \ (23 \times 10^{-6}/°F)$$

8.5 Determine the thermal forces N_x^T, N_y^T, and N_s^T for a $[\pm 45]_s$ carbon/epoxy laminate (AS4/3501-6, Table A.4) for a temperature difference $\Delta T = 56 \ °C \ (100 \ °F)$ in terms of the lamina thickness t.

8.6 Determine the thermal forces N_x^T, N_y^T, and N_s^T for a $[\pm 30]_s$ carbon/epoxy laminate (AS4/3501-6, Table A.4) for a temperature difference $\Delta T = 56$ °C (100 °F) and lamina thickness $t = 0.127$ mm (0.005 in).

8.7 Following a procedure similar to that described under Sample Problem 8.1, show that the coefficient of thermal expansion $\bar{\alpha}_x$ of a $[0/90]_s$ carbon/epoxy laminate is equal to $[E_2(1 + \nu_{12})\alpha_2]/E_1$, that is, it is equal to that of the $[\pm 45]_s$ laminate.

8.8 Determine the coefficient of thermal expansion $\bar{\alpha}_x$ of a $[\pm 30]_s$ carbon/epoxy laminate (AS4/3501-6, Table A.4).

8.9 Determine the coefficients of thermal expansion of a $[0/90]_s$ laminate as a function of the lamina engineering constants E_1, E_2, G_{12}, and ν_{12} and the lamina coefficients of thermal expansion α_1 and α_2. Find the relationship among the lamina properties such that the CTE of the laminate is zero in all directions.

8.10 For a $[0_m/90_p]_s$ crossply laminate obtain a general relation between $r = m/p$ and the lamina properties $(Q_{11}, Q_{12}, Q_{22}, Q_{66}, \alpha_1, \alpha_2)$ such that the coefficient of thermal expansion $\bar{\alpha}_x$ along the 0° direction is zero. Determine a specific value of r for a carbon/epoxy material (AS4/3501-6, Table A.4).

8.11 Calculate the coefficients of thermal expansion $\bar{\alpha}_x$ and $\bar{\alpha}_y$ of a $[0/\pm 45]_s$ S-glass/epoxy laminate with properties listed in Table A.4.

8.12 Calculate the coefficients of thermal expansion $\bar{\alpha}_x$ and $\bar{\alpha}_y$ of a $[0/\pm 45]_s$ carbon/epoxy (AS4/3601-6, Table A.4).

8.13 Prove that the quasi-isotropic laminate $[0/\pm 60]_s$ is thermally isotropic, that is, prove that $\bar{\alpha}_x$ is independent of rotation. (Hint: Use transformation relations and the observation that $\bar{\alpha}_{0°} = \bar{\alpha}_{60°} = \bar{\alpha}_{-60°}$.)

8.14 Show that the $[0/\pm 30/90_2]_s$ laminate, with tetragonal in-plane symmetry, is hygrothermally isotropic.

8.15 Independently of Sample Problem 8.2, show that the lamina stresses in a $[\pm 45]_s$ carbon/epoxy laminate subjected to a temperature change ΔT are

$$\sigma_1 \cong \alpha_2 E_2 \Delta T$$
$$\sigma_2 \cong -\alpha_2 E_2 \Delta T$$

assuming that $\alpha_1 \cong 0$ and $E_1 \gg E_2$.

8.16 A $[\pm 30]_s$ carbon/epoxy laminate was cured at 180 °C (356 °F) and cooled down to 23 °C (73 °F) and dry condition. Then, it was allowed to absorb moisture. Determine the moisture concentration Δc at which

the net hygrothermal stresses will be zero for the lamina properties

$$\alpha_1 = \beta_1 \cong 0$$
$$\alpha_2 = 30 \times 10^{-6}/°C \ (16.7 \times 10^{-6}/°F)$$
$$\beta_2 = 0.55$$

8.17 A $[0/90]$ crossply antisymmetric carbon/epoxy laminate is cooled down during curing from 180 °C (356 °F) to 30 °C (86 °F) (Fig. 8.18). Compute the curvatures κ_x and κ_y for the lamina properties

$$E_1 = 140 \text{ GPa (20.3 Msi)}$$
$$E_2 = 10 \text{ GPa (1.45 Msi)}$$
$$G_{12} = 6 \text{ GPa (0.87 Msi)}$$
$$\nu_{12} = 0.34$$
$$\alpha_1 = \beta_1 \cong 0$$
$$\alpha_2 = 30 \times 10^{-6}/°C \ (16.7 \times 10^{-6}/°F)$$
$$\beta_2 = 0.55$$
$$t = 1 \text{ mm (0.040 in)} \qquad \text{(layer thickness)}$$

8.18 Determine the maximum (in absolute terms) lamina stresses σ_1 and σ_2 for the laminate and conditions of Problem 8.17.

8.19 After cooldown, the dry laminate of Problem 8.17 was exposed to moisture absorption. Determine the moisture concentration at which the laminate will become flat ($\kappa_x = -\kappa_y = 0$).

8.20 A $[\pm 45]$ antisymmetric carbon/epoxy laminate was cooled down during curing from 150 °C (302 °F) to 50 °C (122 °F) and the following deformations were measured:

$$\varepsilon_x^o = \varepsilon_y^o = -7.54 \times 10^{-4}$$
$$\gamma_s^o = 0$$
$$\kappa_x = \kappa_y = 0$$
$$\kappa_s = 2.47 \text{ m}^{-1}$$

Compute the maximum lamina residual stresses σ_1, σ_2, and τ_6 for the lamina properties of Problem 8.17.

8.21 A $[\pm 30]_s$ laminate of AS4/3501-6 carbon/epoxy (Table A.4) was cured at 180 °C (356 °F) and cooled down to 30 °C (86 °F) where it absorbed 0.5% moisture. Determine the lamina stresses σ_1, σ_2, and τ_6 for the following measured hygrothermal properties of the laminate:

$$\bar{\alpha}_x = -3.63 \times 10^{-6}/°C \ (-2.02 \times 10^{-6}/°F)$$
$$\bar{\alpha}_y = 16.6 \times 10^{-6}/°C \ (7.50 \times 10^{-6}/°F)$$
$$\bar{\beta}_x = -0.0086$$
$$\bar{\beta}_y = 0.108$$

8.22 An off-axis lamina undergoes a temperature rise of ΔT (Fig. P8.22) Determine the uniaxial normal stress σ_x (tensile or compressive) that must be

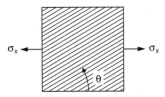

Fig. P8.22

applied in order to prevent shear distortion in the specimen. Find a general expression for σ_x in terms of material properties, ΔT and angle θ.

8.23 In Problem 8.22, determine the numerical value of σ_x for a carbon/epoxy lamina (AS4/3501-6, Table A.4) with $\theta = 45°$ and $\Delta T = 56\ °C$ (100 °F).

8.24 A 45 ° off-axis lamina is loaded under pure shear τ_s and subjected to a temperature rise ΔT (Fig. P8.24) Find a relationship between τ_s, ΔT, α_1, and α_2 and the engineering properties of the lamina such that there is no shear deformation.

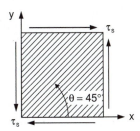

Fig. P8.24

8.25 A [0/90]$_s$ laminate is clamped on all sides to a rigid frame and undergoes a temperature change ΔT (Fig. P8.25) Obtain an expression for the force N_x developed in terms of ΔT, the engineering properties of the lamina (E_1, E_2, v_{12}, G_{12}) and its coefficients of thermal expansion, α_1 and α_2.

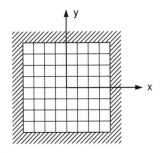

Fig. P8.25

8.26 In Problem 8.25, determine the force N_x for E-glass/epoxy (Table A.4), $\Delta T = -50\ °C$ (−90 °F) and lamina thickness $t = 0.165$ mm (0.0065 in).

8.27 In Problem 8.25, determine the force N_x for a silicon carbide/aluminum composite (Table A.4) with $\Delta T = -278\ °C$ (−500 °F) and $t = 0.178$ mm (0.007 in).

8.28 A [±45]$_s$ square laminate is clamped on all sides to a rigid frame and undergoes a temperature change ΔT (Fig. P8.28).

(a) Obtain an expression for the forces N_x and N_y developed in terms of ΔT, the laminate properties $\bar{E}_x$, $\bar{v}_{xy}$, and $\bar{\alpha}_x$, and lamina thickness t.

(b) Obtain a numerical value of N_x for carbon/epoxy (AS4/3501-6, Table A.4), $\Delta T = -50\ °C$ (−90 °F) and $t = 0.127$ mm (0.005 in). (Hint: Use approximate expressions for $\bar{E}_x$, $\bar{v}_{xy}$, and $\bar{\alpha}_x$ already derived in the text.)

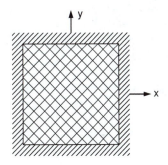

Fig. P8.28

8.29 In Problem 8.28, prove that the reaction forces N_x and N_y are equal to and opposite of the respective thermal forces N_x^T and N_y^T for the laminate.

8.30 The off-axis lamina shown in Fig. P8.30 is constrained without friction between two walls and is exposed to a temperature change ΔT. Determine the

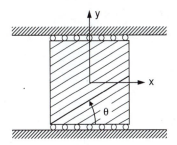

Fig. P8.30

stress σ_y developed as a function of engineering lamina properties $(E_1, E_2, \nu_{12}, G_{12}, \alpha_1, \alpha_2)$, the fiber orientation θ, and ΔT. Subsequently, obtain an approximate expression for σ_y for $E_1 \gg E_2$, $\theta = 45°$, and $\alpha_1 \cong 0$.

8.31 An asymmetric [0/90] square laminate is cured while clamped on all sides to a rigid frame so that it remains flat (Fig. P8.31). The temperature change between the stress-free state and room temperature is $\Delta T = -125$ °C (-225 °F).

(a) Obtain an expression for the reaction forces N_x and N_y and moments M_x and M_y developed in terms of ΔT, α_1, α_2, the lamina stiffnesses Q_{ij}, and the laminate thickness h.

(b) Obtain numerical values of forces and moments for a carbon/epoxy (AS4/3501-6, Table A.4),

$\Delta T = -125$ °C (-225 °F) and $t = 0.127$ mm (0.005 in).

8.32 The $[\pm45]_s$ laminate shown in Fig. P8.32 is constrained without friction between two walls and is exposed to a temperature change ΔT. Determine the stresses $(\sigma_1, \sigma_2, \tau_6)$ developed in each lamina (+45 or −45) as a function of lamina properties $(E_1, E_2, G_{12}, \nu_{12}, \alpha_2)$ and ΔT, assuming $E_1 \gg E_2$ and $\alpha_1 \cong 0$. (Hint: Utilize any properties or results already described in the text for the $[\pm45]_s$ laminate.)

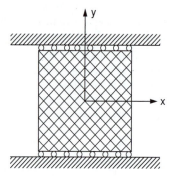

Fig. P8.32

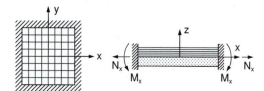

Fig. P8.31

9 Stress and Failure Analysis of Multidirectional Laminates

9.1 INTRODUCTION

In the classical lamination theory discussed in Chapter 7, stress-strain or load-deformation relations were developed for multidirectional laminates. It was shown how the laminate deformation can be fully described in terms of the reference plane strains and the curvatures, from which the strains can be obtained at any through-the-thickness location of the laminate. It was pointed out that, whereas strains are continuous (linear) through the thickness, stresses can be discontinuous from layer to layer, depending on the material properties and orientation of the layers (laminae).

Failure analysis of a laminate is much more complex than that of a single lamina. The stresses in the individual laminae are fundamental and control failure initiation and progression in the laminate. Failure of a lamina does not necessarily imply total failure of the laminate, but is only the beginning of an interactive failure process. Laminate strength theories, like lamina strength theories, are macroscopic and are expressed in terms of the basic lamina strength parameters discussed in Chapter 6. The strength of each individual lamina is assessed separately by referring its stresses to its principal axes (1, 2), which vary from lamina to lamina, and by applying a selected failure criterion.

The strength of a multidirectional laminate is a function of many factors, in addition to the fundamental lamina strengths. The varying lamina orientations, stiffnesses, strengths, and coefficients of thermal and moisture expansion affect the directional characteristics of laminate strength. The exact stacking sequence affects the bending and coupling stiffnesses (D and B) and hence the stresses and strength of the laminate. Finally, the fabrication process affects the residual stresses, which influence the overall strength. Failure or strength analysis can have one of two objectives:

1. analysis of an existing laminate and determination of ultimate loads or safety factors
2. design of a laminate for a given loading condition

9.2 TYPES OF FAILURE

Failure in a laminate may be caused by failure of individual laminae or plies within the laminate (*intralaminar* failure) or by separation of contiguous laminae or layers (*interlaminar* failure). Failure of a laminate may be defined as the initial failure or the ultimate failure, depending on the degree of conservatism applied. These definitions are analogous to the yield and ultimate stress criteria used in elastic/plastic analysis of metal structures.

In the first definition, called *first ply failure* (FPF), a laminate is considered failed when the first layer (or group of layers) fails. This is determined by conducting a stress analysis of the laminate under the given loading conditions, determining the state of stress in each individual layer, and assessing the strength of each layer by applying a selected failure criterion. This assumes that a layer, or lamina, within the laminate has the same properties and behaves in the same manner as an isolated unidirectional lamina. This is questionable, however, because the in situ properties of an embedded layer may be different from those of an isolated layer. Furthermore, a layer within the laminate is under a state of fabrication residual stresses, and its failure takes the form of dispersed damage (microcracking) rather than one major localized flaw or crack. Nevertheless, it is acceptable to use isolated lamina properties in existing lamina theories, such as those discussed in Chapter 6, for predicting first ply failure (FPF). The FPF approach is conservative, but it can be used with low safety factors. A general practice in design of primary structures is to keep working loads below levels producing first ply failure. For example, a general practice in the aircraft industry is to limit maximum transverse strains in carbon/epoxy plies below 0.4%.

In the case of ultimate laminate failure (ULF), there is no generally accepted definition of what constitutes such failure. It is generally accepted that a laminate is considered failed when the maximum load level is reached. Other definitions of laminate failure include a prescribed stiffness degradation, failure of the principal load-carrying plies (0° plies), failure of all plies, or a prescribed level of strain. Prediction of laminate failure entails, in addition to a lamina failure theory, a progressive damage scheme. Following each ply failure, the influence and contribution of the damaged ply on the remaining plies must be evaluated until final laminate failure according to the adopted progressive damage scheme. The ultimate laminate failure approach is more advanced and requires more precise knowledge of loading conditions and stress distributions, and thus it is used with higher safety factors.

Interlaminar failure is a special type of failure consisting of separation of contiguous layers, even when the layers themselves remain intact. This is a common form of failure at free edges or in regions of geometric or loading discontinuities. Prediction of this type of failure requires a three-dimensional stress and failure analysis including interlaminar strength and toughness properties of the laminate.

9.3 STRESS ANALYSIS AND SAFETY FACTORS FOR FIRST PLY FAILURE OF SYMMETRIC LAMINATES (IN-PLANE LOADING)

Given a symmetric laminate under general in-plane loading, the average laminate stresses (Fig. 9.1) are

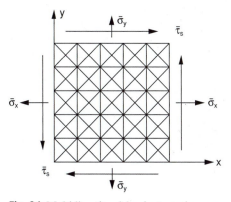

Fig. 9.1 Multidirectional laminate under general in-plane loading.

$$\begin{bmatrix} \bar{\sigma}_x \\ \bar{\sigma}_y \\ \bar{\tau}_s \end{bmatrix} = \frac{1}{h} \begin{bmatrix} N_x \\ N_y \\ N_s \end{bmatrix} \tag{7.82 bis}$$

where h is the laminate thickness.

The laminate strains, which are uniform through the thickness, are equal to the reference plane strains and to the strains of any layer k and are related to the applied forces by the force-deformation relations in Eq. (7.80):

$$\begin{bmatrix} \bar{\varepsilon}_x \\ \bar{\varepsilon}_y \\ \bar{\gamma}_s \end{bmatrix} = \begin{bmatrix} \varepsilon^o_x \\ \varepsilon^o_y \\ \gamma^o_s \end{bmatrix} = \begin{bmatrix} \varepsilon_x \\ \varepsilon_y \\ \gamma_s \end{bmatrix}_k = \begin{bmatrix} a_{xx} & a_{xy} & a_{xs} \\ a_{yx} & a_{yy} & a_{ys} \\ a_{sx} & a_{sy} & a_{ss} \end{bmatrix} \begin{bmatrix} N_x \\ N_y \\ N_s \end{bmatrix} \tag{9.1}$$

The strains in layer k referred to its principal material axes are obtained by transformation as

$$\begin{bmatrix} \varepsilon_1 \\ \varepsilon_2 \\ \frac{1}{2}\gamma_6 \end{bmatrix}_k = \begin{bmatrix} m^2 & n^2 & 2mn \\ n^2 & m^2 & -2mn \\ -mn & mn & (m^2 - n^2) \end{bmatrix}_k \begin{bmatrix} \varepsilon^o_x \\ \varepsilon^o_y \\ \frac{1}{2}\gamma^o_s \end{bmatrix} \tag{9.2}$$

and the corresponding stresses are

$$\begin{bmatrix} \sigma_1 \\ \sigma_2 \\ \tau_6 \end{bmatrix}_k = \begin{bmatrix} Q_{11} & Q_{12} & 0 \\ Q_{21} & Q_{22} & 0 \\ 0 & 0 & Q_{66} \end{bmatrix} \begin{bmatrix} \varepsilon_1 \\ \varepsilon_2 \\ \gamma_6 \end{bmatrix}_k \tag{9.3}$$

For the FPF approach, the selected failure criterion is applied to the state of stress in each layer separately. Thus, for a state of stress $(\sigma_1, \sigma_2, \tau_6)_k$ in layer k, the state of stress at failure of the same layer is $S_{fk}(\sigma_1, \sigma_2, \tau_6)_k$ where S_{fk} is the safety factor for layer k. Substitution of the critical (failure) state of stress in the Tsai-Wu failure criterion in Eq. (6.29) yields

$$f_1 S_{fk}\sigma_{1k} + f_2 S_{fk}\sigma_{2k} + f_{11}S^2_{fk}\sigma^2_{1k} + f_{22}S^2_{fk}\sigma^2_{2k} + f_{66}S^2_{fk}\tau^2_{6k} + 2f_{12}S^2_{fk}\sigma_{1k}\sigma_{2k} = 1 \tag{9.4}$$

or

$$aS^2_{fk} + bS_{fk} - 1 = 0 \tag{9.5}$$

where

$$a = f_{11}\sigma^2_{1k} + f_{22}\sigma^2_{2k} + f_{66}\tau^2_{6k} + 2f_{12}\sigma_{1k}\sigma_{2k}$$
$$b = f_1\sigma_{1k} + f_2\sigma_{2k} \tag{9.6}$$

The solutions of the quadratic Eq. (9.5) are

$$S_{fk} = \frac{-b \pm \sqrt{b^2 + 4a}}{2a}$$

or

$$S_{fka} = \frac{-b + \sqrt{b^2 + 4a}}{2a} \tag{9.7}$$

and

$$S_{fkr} = \left| \frac{-b - \sqrt{b^2 + 4a}}{2a} \right| \tag{9.8}$$

where S_{fka} is the safety factor of layer k for the actual state of stress $(\sigma_1, \sigma_2, \tau_6)_k$ and S_{fkr} is the safety factor of the same layer k for a state of stress with reversed sign, that is, $(-\sigma_1, -\sigma_2, -\tau_6)_k$.

The procedure above is carried out repeatedly for all layers of the laminate to find the minimum values of S_{fka} and S_{fkr}. These minimum values are the safety factors of the laminate based on the FPF approach, for the actual and reversed loadings. Thus,

$$\bar{S}_{fa} = (S_{fka})_{\min}$$
$$\bar{S}_{fr} = (S_{fkr})_{\min} \tag{9.9}$$

9.4 STRENGTH COMPONENTS FOR FIRST PLY FAILURE OF SYMMETRIC LAMINATES

Given a multidirectional laminate, it is required, for example, to determine its axial strength along the x-axis based on FPF. The laminate in Fig. 9.1 is assumed to be loaded under unit average stress in the x-direction, that is,

$$\bar{\sigma}_x = 1$$
$$\bar{\sigma}_y = 0$$
$$\bar{\tau}_s = 0$$

Then, the reference plane strains and layer k strains are obtained by Eqs. (7.80) and (9.1) as

$$\begin{bmatrix} \varepsilon_x^o \\ \varepsilon_y^o \\ \gamma_s^o \end{bmatrix} = \begin{bmatrix} \varepsilon_x \\ \varepsilon_y \\ \gamma_s \end{bmatrix}_k = \begin{bmatrix} a_{xx} & a_{xy} & a_{xs} \\ a_{yx} & a_{yy} & a_{ys} \\ a_{sx} & a_{sy} & a_{ss} \end{bmatrix} \begin{bmatrix} h \\ 0 \\ 0 \end{bmatrix}$$

or

$$\begin{bmatrix} \varepsilon_x^o \\ \varepsilon_y^o \\ \gamma_s^o \end{bmatrix} = \begin{bmatrix} \varepsilon_x \\ \varepsilon_y \\ \gamma_s \end{bmatrix}_k = \begin{bmatrix} a_{xx} \\ a_{yx} \\ a_{sx} \end{bmatrix} h \qquad (9.10)$$

The strains in layer k along its principal material axes are obtained by transformation as in Eq. (9.2), and the corresponding stresses are obtained as in Eq. (9.3). The critical (failure) state of stress for layer k is $(S_{fk}, \sigma_1, \sigma_2, T_6)_k$, which, substituted in the Tsai-Wu failure criterion Eq. (9.4), gives two safety factors, S_{fka} corresponding to $\bar{\sigma}_x = 1$, and S_{fkr} corresponding to $\bar{\sigma}_x = -1$.

The axial tensile and compressive strengths of the laminate are obtained as

$$\begin{aligned} \bar{F}_{xt} &= (S_{fka})_{\min} \\ \bar{F}_{xc} &= (S_{fkr})_{\min} \end{aligned} \qquad (9.11)$$

The axial strengths along the y-axis are obtained in a similar manner by assuming the following state of stress in the laminate:

$$\bar{\sigma}_x = 0$$

$$\bar{\sigma}_y = 1$$

$$\bar{\tau}_s = 0$$

The in-plane shear strength is obtained by assuming the following state of stress:

$$\bar{\sigma}_x = 0$$

$$\bar{\sigma}_y = 0$$

$$\bar{\tau}_s = 1$$

In the case of quasi-isotropic laminates discussed before (Section 7.10), the elastic properties are, by definition, independent of orientation. Thus, a uniaxial loading $\bar{\sigma}_x$ produces the same strains ε_x and ε_y in each layer regardless of the load direction x. Depending on the orientation of the principal layer directions with respect to the loading direction, a different state of stress exists in each layer. Thus, failure in each layer and the occurrence of first failure in any layer (FPF) depend on load orientation. This means that quasi-isotropic laminates are not isotropic as far as FPF is concerned. For example, the $[0/\pm60]_s$ quasi-isotropic laminate would be strongest along a fiber direction ($0°$, $60°$, or $-60°$) and weakest along a bisector between fiber directions of different plies ($30°$, $-30°$, or $90°$).

SAMPLE PROBLEM 9.1
Uniaxial Strength of Angle-Ply Laminate

It is required to determine the axial tensile strength $\bar{F}_{xt}$ of a $[\pm 45]_{ns}$ laminate using the Tsai-Hill criterion for FPF. Consider the laminate in Fig. 9.2 subjected to a uniaxial stress

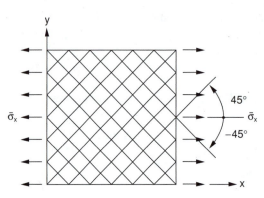

$$\bar{\sigma}_x = \frac{N_x}{h}$$

$$\bar{\sigma}_y = \bar{\tau}_s = 0$$

The strains, which are uniform through the thickness, can be related to the applied stress and the laminate engineering properties as

Fig. 9.2 $[\pm 45]_s$ angle-ply laminate under uniaxial loading.

$$\varepsilon_x^o = \frac{\bar{\sigma}_x}{\bar{E}_x}$$

$$\varepsilon_y^o = -\bar{\nu}_{xy}\frac{\bar{\sigma}_x}{\bar{E}_x} \qquad (9.12)$$

$$\gamma_s^o = 0$$

The strains in the $-45°$ lamina referred to its principal material axes are obtained from the strain transformation relations in Eq. (4.58) as

$$\varepsilon_1 = \varepsilon_2 = \tfrac{1}{2}(\varepsilon_x^o + \varepsilon_y^o)$$

$$\gamma_6 = \varepsilon_x^o - \varepsilon_y^o$$

and in view of Eq. (9.12)

$$\varepsilon_1 = \varepsilon_2 = \frac{\bar{\sigma}_x}{2\bar{E}_x}(1 - \bar{\nu}_{xy})$$

$$\gamma_6 = \frac{\bar{\sigma}_x}{\bar{E}_x}(1 + \bar{\nu}_{xy}) \qquad (9.13)$$

Referring to relations in Eqs. (7.92) and (7.99), derived for Sample Problems 7.3 and 7.5, we obtain further

$$\varepsilon_1 = \varepsilon_2 = \frac{\bar{\sigma}_x}{Q_{11} + Q_{22} + 2Q_{12}}$$

$$\gamma_6 = \frac{\bar{\sigma}_x}{2Q_{66}} \qquad (9.14)$$

The lamina stresses, obtained from Eq. (4.31), are

$$\sigma_1 = \varepsilon_1(Q_{11} + Q_{12}) = \bar{\sigma}_x \frac{Q_{11} + Q_{12}}{Q_{11} + Q_{22} + 2Q_{12}}$$

$$\sigma_2 = \varepsilon_1(Q_{12} + Q_{22}) = \bar{\sigma}_x \frac{Q_{12} + Q_{22}}{Q_{11} + Q_{22} + 2Q_{12}} \tag{9.15}$$

$$\tau_6 = \gamma_6 Q_{66} = \frac{\bar{\sigma}_x}{2}$$

The above stress components can be introduced into the Tsai-Hill criterion to yield an exact value for $\bar{\sigma}_x$ at failure, that is, $\bar{F}_x$. An approximate value for the strength can be obtained more easily for the case of a high-stiffness composite, that is, when $E_1 \gg E_2$. Then the lamina stress components in Eq. (9.15) can be approximated as

$$\sigma_1 \cong \bar{\sigma}_x$$

$$\sigma_2 \cong (1 + \nu_{12}) \frac{E_2}{E_1} \bar{\sigma}_x \tag{9.16}$$

$$\tau_6 = \frac{\bar{\sigma}_x}{2}$$

Hence $\sigma_2 \ll \sigma_1$ and $\sigma_2 < \tau_6$.

Substituting the above into the Tsai-Hill criterion,

$$\left(\frac{\sigma_1^u}{F_1}\right)^2 + \left(\frac{\sigma_2^u}{F_2}\right)^2 + \left(\frac{\tau_6^u}{F_6}\right)^2 - \frac{\sigma_1^u \sigma_2^u}{F_1^2} = 1$$

we obtain for $\bar{\sigma}_x^u = \bar{F}_{xt}$

$$\bar{F}_{xt}^2 \left[\frac{1}{F_1^2} + \frac{(1 + \nu_{12})^2}{F_2^2} \left(\frac{E_2}{E_1}\right)^2 + \frac{1}{4F_6^2} - \frac{(1 + \nu_{12})}{F_1^2} \left(\frac{E_2}{E_1}\right) \right] \cong 1 \tag{9.17}$$

For a high-strength composite with $F_1 \gg F_2$ and $F_1 \gg F_6$, the relation above is further approximated as

$$\bar{F}_{xt}^2 \left[\frac{(1 + \nu_{12})^2}{F_2^2} \left(\frac{E_2}{E_1}\right)^2 + \frac{1}{4F_6^2} \right] \cong 1 \tag{9.18}$$

and, because $E_2 \ll E_1$ and $F_2 \cong F_6$,

$$\bar{F}_{xt} \cong 2F_6 \tag{9.19}$$

Thus, the axial tensile strength of a $[\pm 45]_{ns}$ angle-ply laminate, for a high-stiffness, high-strength composite, is controlled primarily by the in-plane shear strength of the unidirectional lamina; hence it is a *matrix-dominated* property. A similar approximation and conclusion holds true for the axial compressive strength $\bar{F}_{xc}$. Similar results are obtained by using other failure theories.

SAMPLE PROBLEM 9.2
In-Plane Shear Strength of Angle-Ply Laminate

It is required to determine the in-plane shear strength $\bar{F}_s$ of a $[\pm 45]_{ns}$ laminate using the Tsai-Hill criterion for FPF. Consider the laminate in Fig. 9.2 under in-plane shear loading

$$\bar{\tau}_s = \frac{N_s}{h}$$

$$\bar{\sigma}_x = \bar{\sigma}_y = 0$$

The strains, which are uniform through the thickness, are

$$\varepsilon_x^o = \varepsilon_y^o = 0$$

$$\gamma_s^o = \frac{\bar{\tau}_s}{\bar{G}_{xy}} \tag{9.20}$$

The strains in the $+45°$ ply, obtained by transformation, are

$$\varepsilon_1 = mn\gamma_s^o = \frac{\bar{\tau}_s}{2\bar{G}_{xy}}$$

$$\varepsilon_2 = -mn\gamma_s^o = -\frac{\bar{\tau}_s}{2\bar{G}_{xy}} \tag{9.21}$$

$$\gamma_6 = 0$$

and the corresponding stresses are

$$\sigma_1 = \varepsilon_1 Q_{11} + \varepsilon_2 Q_{12} = (Q_{11} - Q_{12})\frac{\bar{\tau}_s}{2\bar{G}_{xy}}$$

$$\sigma_2 = \varepsilon_1 Q_{12} + \varepsilon_2 Q_{22} = (Q_{12} - Q_{22})\frac{\bar{\tau}_s}{2\bar{G}_{xy}} \tag{9.22}$$

$$\tau_6 = 0$$

or, using the relation in Eq. (7.97) in Sample Problem 7.4, we obtain

$$\sigma_1 = \frac{2(Q_{11} - Q_{12})\bar{\tau}_s}{Q_{11} + Q_{22} - 2Q_{12}}$$

$$\sigma_2 = \frac{2(Q_{12} - Q_{22})\bar{\tau}_s}{Q_{11} + Q_{22} - 2Q_{12}} \tag{9.23}$$

$$\tau_6 = 0$$

For a high-stiffness composite, with $E_1 \gg E_2$, the above stresses can be approximated as

$$\sigma_1 \cong 2\bar{\tau}_s$$

$$\sigma_2 \cong -2(1 - \nu_{12})\frac{E_2}{E_1}\bar{\tau}_s \qquad (9.24)$$

$$\tau_6 = 0$$

Thus, the $+45°$ ply is under longitudinal tension and transverse compression. The stresses in the $-45°$ ply are of the same magnitude but opposite sign as those of Eq. (9.24). This means that the $-45°$ ply is under longitudinal compression and transverse tension for the same positive shear loading. Since transverse tension is the most severe loading for a typical unidirectional lamina, FPF will occur in the $-45°$ ply.

Substituting the $-45°$ ply stresses in the Tsai-Hill criterion with appropriate consideration for tensile and compressive strengths, we obtain for the ultimate value $\bar{\tau}_s^u = \bar{F}_s$

$$\left(\frac{\sigma_1^u}{F_{1c}}\right)^2 + \left(\frac{\sigma_2^u}{F_{2t}}\right)^2 - \frac{\sigma_1^u \sigma_2^u}{F_{1c}^2} = 1$$

or

$$\frac{4\bar{F}_s^2}{F_{1c}^2}\left[1 + (1 - \nu_{12})^2\left(\frac{E_2}{E_1}\right)^2\left(\frac{F_{1c}}{F_{2t}}\right)^2 + (1 - \nu_{12})\left(\frac{E_2}{E_1}\right)\right] \cong 1$$

or, because $E_1 \gg E_2$ and assuming $F_{1c} \gg F_{2t}$,

$$\frac{4\bar{F}_s^2}{F_{1c}^2}\left[1 + (1 - \nu_{12})^2\left(\frac{E_2}{E_1}\right)^2\left(\frac{F_{1c}}{F_{2t}}\right)^2\right] \cong 1$$

from which we obtain

$$\bar{F}_s \cong \frac{F_{1c}}{2}\left[1 + (1 - \nu_{12})^2\left(\frac{E_2}{E_1}\right)^2\left(\frac{F_{1c}}{F_{2t}}\right)^2\right]^{-1/2} \qquad (9.25)$$

For a typical carbon/epoxy composite the quantity inside the brackets above is on the order of 3, which means that the shear strength $\bar{F}_s$ is roughly $F_{1c}/2\sqrt{3}$, or approximately 30% of the longitudinal compressive strength of the lamina. It may be concluded that, under shear loading, this angle-ply laminate is utilized more efficiently since at FPF, which is a transverse failure, the longitudinal stress in the ply reaches approximately 60% of its ultimate value ($\sigma_1 \cong 2\bar{F}_s$). Thus, shear strength of a $[\pm45]_s$ laminate for a high-stiffness, high-strength composite is primarily a fiber-dominated property.

9.5 COMPUTATIONAL PROCEDURE FOR STRESS AND FAILURE ANALYSIS OF GENERAL MULTIDIRECTIONAL LAMINATES (FIRST PLY FAILURE)

A flowchart for computation of safety factors and strength components of a general multi-directional laminate is shown in Fig. 9.3. It is based on the Tsai-Wu failure criterion and on the FPF approach. The procedure for determination of safety factors consists of the following steps:

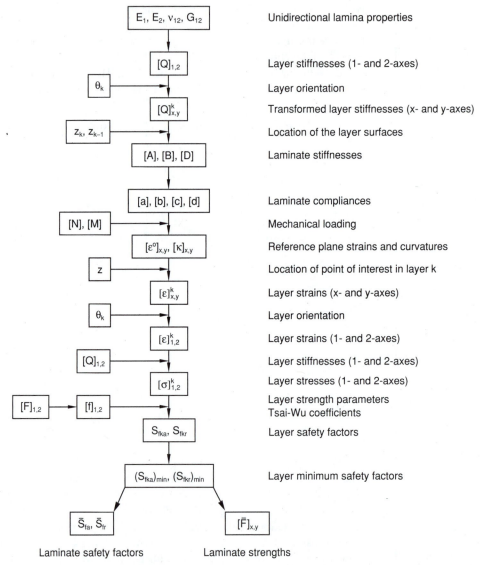

$E_1, E_2, \nu_{12}, G_{12}$	Unidirectional lamina properties
$[Q]_{1,2}$	Layer stiffnesses (1- and 2-axes)
θ_k	Layer orientation
$[Q]^k_{x,y}$	Transformed layer stiffnesses (x- and y-axes)
z_k, z_{k-1}	Location of the layer surfaces
$[A], [B], [D]$	Laminate stiffnesses
$[a], [b], [c], [d]$	Laminate compliances
$[N], [M]$	Mechanical loading
$[\varepsilon^o]_{x,y}, [\kappa]_{x,y}$	Reference plane strains and curvatures
z	Location of point of interest in layer k
$[\varepsilon]^k_{x,y}$	Layer strains (x- and y-axes)
θ_k	Layer orientation
$[\varepsilon]^k_{1,2}$	Layer strains (1- and 2-axes)
$[Q]_{1,2}$	Layer stiffnesses (1- and 2-axes)
$[\sigma]^k_{1,2}$	Layer stresses (1- and 2-axes)
$[F]_{1,2}, [f]_{1,2}$	Layer strength parameters Tsai-Wu coefficients
S_{fka}, S_{fkr}	Layer safety factors
$(S_{fka})_{min}, (S_{fkr})_{min}$	Layer minimum safety factors
$\bar{S}_{fa}, \bar{S}_{fr}$	Laminate safety factors
$[\bar{F}]_{x,y}$	Laminate strengths

Fig. 9.3 Flowchart for stress and failure analysis of multidirectional laminates (FPF; Tsai-Wu criterion).

Step 1   Enter the basic lamina properties (E_1, E_2, G_{12}, ν_{12}).

Step 2   Compute the ply stiffnesses $[Q]_{1,2}$ referred to their principal material axes, using relations in Eq. (4.56).

Step 3   Enter the orientation of the principal material axis, θ_k, of layer k.

Step 4   Calculate the transformed layer stiffnesses $[Q]_{x,y}^k$ of layer k referred to the laminate coordinate system (x, y), using Eq. (4.67).

Step 5   Enter the through-the-thickness coordinates z_k and z_{k-1} of layer k surfaces.

Step 6   Calculate the laminate stiffness matrices $[A]$, $[B]$, and $[D]$ using Eq. (7.20).

Step 7   Calculate the laminate compliance matrices $[a]$, $[b]$, $[c]$, and $[d]$ using Eq. (7.27).

Step 8   Enter the mechanical loading, that is, forces $[N]_{x,y}$ and moments $[M]_{x,y}$.

Step 9   Calculate the reference plane strains $[\varepsilon^o]_{x,y}$ and curvatures $[\kappa]_{x,y}$ using Eq. (7.25).

Step 10   Enter the through-the-thickness coordinate, z, of the point of interest in layer k. For a laminate consisting of many thin (compared with the laminate thickness) layers or for any symmetric laminate under in-plane loading, the coordinate of the lamina midplane, $z = \bar{z}_k$, is used. Then, the computed strains and stresses are the average through-the-thickness strains and stresses in the layer. However, when the laminate consists of few and thick (relative to the laminate thickness) layers and is asymmetric or subjected to bending and twisting, $z = z_k$ and $z = z_{k-1}$ are entered in order to determine the extreme values of the strains and stresses at the top and bottom surfaces of the layer.

Step 11   Calculate the layer strains $[\varepsilon]_{x,y}^k$ referred to laminate reference axes (x, y), using Eq. (7.8).

Step 12   Calculate the layer strains $[\varepsilon]_{1,2}^k$ referred to layer principal axes $(1, 2)$, using Eq. (4.58).

Step 13   Calculate the layer stresses $[\sigma]_{1,2}^k$ referred to the layer principal axes $(1, 2)$, using Eq. (4.31).

Step 14   Enter the lamina strengths $[F]_{1,2}$ and calculate Tsai-Wu coefficients f_i and f_{ij} using Eqs. (6.32), (6.33), (6.35), and (6.40).

Step 15   Calculate the layer safety factors S_{fka} and S_{fkr} using Eqs. (9.7) and (9.8).

Step 16a Determine the laminate safety factors $\bar{S}_{fa}$ and $\bar{S}_{fr}$ using Eq. (9.9).

Step 16b Determine the laminate strength components $[\bar{F}]_{x,y}$, by applying unit stress in the respective direction of each component and using Eq. (9.11).

9.6 COMPARISON OF STRENGTHS OF UNIDIRECTIONAL AND ANGLE-PLY LAMINATES (FIRST PLY FAILURE)

The effect of fiber orientation on the strength of unidirectional and angle-ply laminates is illustrated in Figs. 9.4–9.6 for a typical carbon/epoxy composite (AS4/3501-6). Figure 9.4 shows that the uniaxial tensile strength of the angle-ply laminate is much higher than that of the off-axis unidirectional material for fiber orientations between 5° and 20°.

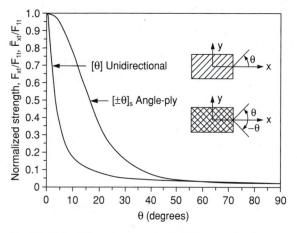

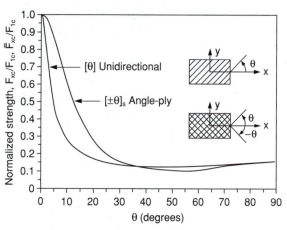

Fig. 9.4 Uniaxial tensile strength of unidirectional and angle-ply laminates as a function of fiber orientation (AS4/3501-6 carbon/epoxy; Tsai-Wu criterion; FPF).

Fig. 9.5 Uniaxial compressive strength of unidirectional and angle-ply laminates as a function of fiber orientation (AS4/3501-6 carbon/epoxy; Tsai-Wu criterion).

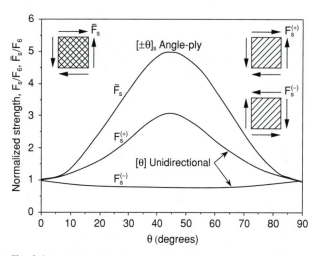

Fig. 9.6 In-plane shear strength of unidirectional and angle-ply laminates as a function of fiber orientation (AS4/3501-6 carbon/epoxy; Tsai-Wu criterion).

The maximum ratio of the two strengths occurs for a fiber orientation between 10° and 15°. Figure 9.5 shows the variation of uniaxial compressive strengths. The differences between the unidirectional and angle-ply laminates are less pronounced than in the case of tensile strength.

Figure 9.6 shows the variation of in-plane shear strength for unidirectional and angle-ply laminates. In the case of the unidirectional lamina there is a large difference between positive and negative shear strength for reasons discussed before (see Sample Problem 6.1). The shear strength of the angle-ply laminate is the same for both positive and negative shear loading and is much higher than either positive or negative shear strength of the unidirectional lamina.

9.7 CARPET PLOTS FOR STRENGTH OF MULTIDIRECTIONAL LAMINATES (FIRST PLY FAILURE)

Carpet plots, as in the case of elastic properties (Section 7.15), can be prepared for strength properties of general multidirectional laminates. A general laminate of $[0_m/90_n/(\pm45)_p]_s$ layup can be described in terms of the fractions α, β, and γ of the 0°, 90°, and ±45° plies, defined as before:

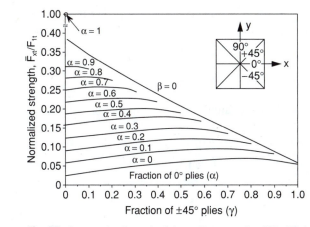

Fig. 9.7 Carpet plot for uniaxial tensile strength of $[0_m/90_n/(\pm45)_p]_s$ laminates (AS4/3501-6 carbon/epoxy; α, β, and γ are fractions of $0°$, $90°$, and $\pm45°$ plies, respectively; FPF).

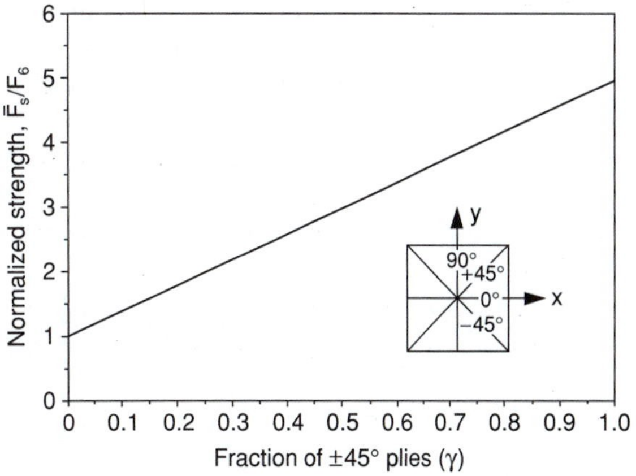

Fig. 9.8 Carpet plot for in-plane shear strength of $[0_m/90_n/(\pm45)_p]_s$ laminates (AS4/3501-6 carbon/epoxy; α, β, and γ are fractions of $0°$, $90°$, and $\pm45°$ plies, respectively; FPF).

$$\alpha = \frac{2m}{N}$$

$$\beta = \frac{2n}{N}$$

$$\gamma = \frac{4p}{N}$$

(7.111 bis)

$$N = 2(m + n + 2p) = \text{total number of plies}$$

Carpet plots for the uniaxial strength and in-plane shear strength are shown in Figs. 9.7 and 9.8 for carbon/epoxy laminates (AS4/3501-6). The gaps and discontinuities in the plot of Fig. 9.7 are due to the nature of the FPF approach, which causes finite jumps in the calculated strength near the $\beta = 0$ and $\gamma = 0$ values for infinitesimally small increments in β or γ.

9.8 EFFECT OF HYGROTHERMAL HISTORY ON STRENGTH OF MULTIDIRECTIONAL LAMINATES (FIRST PLY FAILURE; TSAI-WU CRITERION)

For a given laminate layup under mechanical loading $[N]$ and $[M]$, and hygrothermal loading ΔT and Δc, application of the failure criterion can yield the following results:

1. safety factor under given loading
2. laminate strength components
3. allowable laminate thickness (of the same basic layup) for a given allowable safety factor

The critical stresses at failure of layer k, $(\sigma_{if})_k$, are obtained by multiplying the stresses induced by mechanical loading (σ_i) by the safety factor S_{fk} and combining them with the hygrothermal stresses (σ_{ie}), where $i = 1, 2, 6$, that is,

$$(\sigma_{if})_k = S_{fk}(\sigma_i)_k + (\sigma_{ie})_k \qquad (i = 1, 2, 6)$$

Substitution of this state of stress in the Tsai-Wu failure criterion, Eq. (6.29), yields

$$f_1(S_f\sigma_1 + \sigma_{1e})_k + f_2(S_f\sigma_2 + \sigma_{2e})_k + f_{11}(S_f\sigma_1 + \sigma_{1e})_k^2 + f_{22}(S_f\sigma_2 + \sigma_{2e})_k^2$$
$$+ f_{66}(S_f\tau_6 + \tau_{6e})_k^2 + 2f_{12}(S_f\sigma_1 + \sigma_{1e})_k(S_f\sigma_2 + \sigma_{2e})_k - 1 = 0$$

or

$$aS_{fk}^2 + bS_{fk} + c = 0 \qquad (9.26)$$

where

$$a = f_{11}(\sigma_1^2)_k + f_{22}(\sigma_2^2)_k + f_{66}(\tau_6^2)_k + 2f_{12}(\sigma_1\sigma_2)_k$$
$$b = f_1(\sigma_1)_k + f_2(\sigma_2)_k + 2f_{11}(\sigma_1\sigma_{1e})_k + 2f_{22}(\sigma_2\sigma_{2e})_k + 2f_{66}(\tau_6\tau_{6e})_k + 2f_{12}(\sigma_1\sigma_{2e} + \sigma_2\sigma_{1e})_k$$
$$c = f_1(\sigma_{1e})_k + f_2(\sigma_{2e})_k + f_{11}(\sigma_{1e}^2)_k + f_{22}(\sigma_{2e}^2)_k + f_{66}(\tau_{6e}^2)_k + 2f_{12}(\sigma_{1e}\sigma_{2e})_k - 1$$

The solutions of the quadratic Eq. (9.26) yield the safety factors for the actual and reversed mechanical loading, S_{fka} and S_{fkr}, as

$$S_{fka} = \frac{-b + \sqrt{b^2 - 4ac}}{2a} \qquad (9.27)$$

$$S_{fkr} = \left| \frac{-b - \sqrt{b^2 - 4ac}}{2a} \right| \qquad (9.28)$$

The minimum values of these factors are the laminate safety factors as given in Eq. (9.9).

The procedure for determining the strength components of the laminate in the presence of hygrothermal stresses is the same as that discussed in Section 9.4. The laminate is assumed loaded by a unit average mechanical stress in the direction of the desired strength component. The laminate safety factors obtained for this unit stress loading combined with the hygrothermal loading yield the desired strength components as given in Eq. (9.11).

In design applications the objective is to determine a specific laminate configuration, the number and orientation of plies for the given mechanical and hygrothermal loading, and a given allowable safety factor. The complete design optimization involves many degrees of freedom, both number and orientation of plies. However, if a basic laminate layup is selected, the design objective is reduced to finding the allowable thickness of the laminate, consisting of a multiple of the basic layup. The approach in that case is to assume a unit laminate thickness $h = 1$. Then the average mechanical stresses applied to the laminate are

$$
\begin{bmatrix} \bar{\sigma}_x \\ \bar{\sigma}_y \\ \bar{\tau}_s \end{bmatrix} = \begin{bmatrix} N_x \\ N_y \\ N_s \end{bmatrix}
\tag{9.29}
$$

The above mechanical loading combined with the given hygrothermal loading gives in each lamina k the state of stress, $(\sigma_i + \sigma_{ie})_k$ for $h = 1$. Failure of the lamina occurs when the mechanical loads or stresses are multiplied by the safety factor, S_{fk}. Following the identical procedure described before, we determine the minimum safety factor, which is the laminate safety factor, $\bar{S}_f$ ($\bar{S}_{fa}$ or $\bar{S}_{fr}$). Then, the allowable laminate thickness is obtained as

$$
h_a = \frac{S_{all}}{\bar{S}_f}
\tag{9.30}
$$

where

$$h_a = \text{allowable laminate thickness}$$

$$S_{all} = \text{allowable safety factor}$$

This approach is limited to symmetric laminates ($[B] = 0$) under in-plane loading only ($[M] = 0$).

The results of the analysis discussed here, that is, safety factors, laminate strength, and laminate sizing are summarized in Table 9.1.

TABLE 9.1 Failure Analysis of Multidirectional Laminates Under Combined Mechanical and Hygrothermal Loading (First Ply Failure)

Input	Output	Remarks
Average laminate stresses: $\bar{\sigma}_x, \bar{\sigma}_y, \bar{\tau}_s$	Laminate safety factor: $\bar{S}_f = (S_{fk})_{min}$	Inputs: $N_i, M_i, \Delta T, \Delta c$
Unit laminate stress in given direction: $\bar{\sigma}_i = 1$	Laminate strength components: $\bar{F}_{it} = (S_{fka})_{min}$ $\bar{F}_{ic} = (S_{fkr})_{min}$	Inputs: $N_i = h$ $M_i = 0$ $\Delta T, \Delta c$
Average laminate stresses for unit thickness: $\bar{\sigma}_i = N_i$ Allowable safety factor: S_{all}	Allowable laminate thickness: $h_a = \dfrac{S_{all}}{(S_{fk})_{min}}$	Inputs: $N_i, \Delta T, \Delta c$ Limited to symmetric laminates under in-plane loading: $(M = 0, B_{ij} = 0)$

$i = x, y, s$

9.9 COMPUTATIONAL PROCEDURE FOR STRESS AND FAILURE ANALYSIS OF MULTIDIRECTIONAL LAMINATES UNDER COMBINED MECHANICAL AND HYGROTHERMAL LOADING (FIRST PLY FAILURE; TSAI-WU CRITERION)

A flowchart for stress and failure analysis under general combined mechanical and hygrothermal loading is shown in Fig. 9.9. For the most part the procedure is the same as that described in Section 9.5 for mechanical loading only. The first seven steps in the flow chart of Fig. 9.3 are omitted here, and the result, laminate compliances, is used as an input. The procedure illustrated in Fig. 9.9 consists of the following steps:

Step 1   Enter the mechanical loading, that is, forces $[N]_{x,y}$ and moments $[M]_{x,y}$.

Step 2   Enter the laminate compliances $[a]$, $[b]$, $[c]$, and $[d]$ (from step 7 of Fig. 9.3).

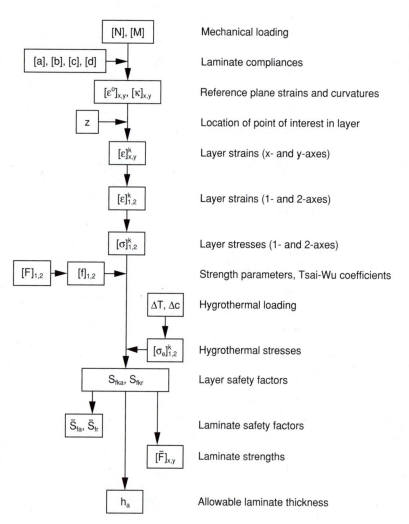

$[N], [M]$	Mechanical loading
$[a], [b], [c], [d]$	Laminate compliances
$[\varepsilon^o]_{x,y}, [\kappa]_{x,y}$	Reference plane strains and curvatures
z	Location of point of interest in layer
$[\varepsilon]^k_{x,y}$	Layer strains (x- and y-axes)
$[\varepsilon]^k_{1,2}$	Layer strains (1- and 2-axes)
$[\sigma]^k_{1,2}$	Layer stresses (1- and 2-axes)
$[F]_{1,2}$ $[f]_{1,2}$	Strength parameters, Tsai-Wu coefficients
$\Delta T, \Delta c$	Hygrothermal loading
$[\sigma_e]^k_{1,2}$	Hygrothermal stresses
S_{fka}, S_{fkr}	Layer safety factors
$\bar{S}_{fa}, \bar{S}_{fr}$	Laminate safety factors
$[\bar{F}]_{x,y}$	Laminate strengths
h_a	Allowable laminate thickness

Fig. 9.9 Flowchart for stress and failure analysis of general multidirectional laminates under combined mechanical and hygrothermal loading (FPF; Tsai-Wu criterion).

Step 3 Calculate the reference plane strains $[\varepsilon^o]_{x,y}$ and curvatures $[\kappa]_{x,y}$, using Eq. (7.25).

Step 4 Enter the through-the-thickness coordinate, z, of the point of interest in layer k. For a laminate consisting of many thin (compared with the laminate thickness) layers or for any symmetric laminate under in-plane loading, the coordinate of the lamina midplane, $z = \bar{z}_k$, is used. Then, the computed strains and stresses are the average through-the-thickness strains and stresses in the layer. However, when the laminate consists of few and thick (relative to the laminate thickness) layers and is asymmetric or subjected to bending and twisting, $z = z_k$ and $z = z_{k-1}$ are entered in order to determine the extreme values of the strains and stresses at the top and bottom surfaces of the layer.

Step 5 Calculate the layer strains referred to laminate axes (x, y), using Eq. (7.8).

Step 6 Transform the layer strains to the layer principal axes $[\varepsilon]^k_{1,2}$, using Eq. (4.58).

Step 7 Calculate the layer stresses $[\sigma]^k_{1,2}$, referred to the layer principal axes $(1, 2)$, using Eq. (4.31).

Step 8 Enter the lamina strengths $[F]_{1,2}$ and calculate Tsai-Wu coefficients f_i and f_{ij}, using relations in Eqs. (6.32), (6.33), (6.35), and (6.40).

Step 9 Enter the hygrothermal loading, ΔT and Δc.

Step 10 Calculate the hygrothermal stresses following the procedure outlined in the flowchart of Fig. 8.20.

Step 11 Introduce the combined mechanical and hygrothermal stresses in the Tsai-Wu failure criterion and determine the layer safety factors, using Eqs. (9.26) through (9.28).

Step 12a Determine the laminate safety factors $\bar{S}_{fa}$ and $\bar{S}_{fr}$, using Eq. (9.9).

Step 12b Determine the laminate strength components $[\bar{F}]_{x,y}$, using unit in-plane loading in the respective direction of each component and Eq. (9.11).

Step 12c Determine the allowable laminate thickness for the selected basic layup, using Eq. (9.30) for symmetric laminates under in-plane loading ($[B] = 0$, $[M] = 0$).

For a given laminate, the layer that fails first depends on its state of stress. As the laminate loading varies continuously, FPF moves from one layer to another. Failure envelopes for the laminate based on FPF consist of portions of failure envelopes of different individual layers. FPF envelopes were calculated for a $[0/\pm45/90]_s$ quasi-isotropic carbon/epoxy laminate (AS4/3501-6) under biaxial normal $(\bar{\sigma}_x, \bar{\sigma}_y)$ and biaxial shear/normal $(\bar{\tau}_s, \bar{\sigma}_x)$ loading. The curing residual stresses were accounted for by assuming a temperature difference $\Delta T = -100$ °C (-180 °F) between the stress-free and current temperatures. Results are shown in Figs. 9.10 and 9.11. In the first case ($\bar{\sigma}_y$ versus $\bar{\sigma}_x$), the theories are farthest apart in the compression-compression region ($\bar{\sigma}_x, \bar{\sigma}_y < 0$), with the Tsai-Wu theory predicting approximately twice as high strengths as the other theories (Fig. 9.10). In the case of shear/normal loading, all five theoretical FPF envelopes differ from each other, with the maximum stress theory predicting the highest strengths in the shear-compression region (Fig. 9.11). No experimental data are available for comparison because it is difficult to detect FPF accurately.

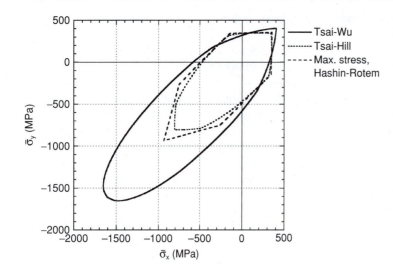

Fig. 9.10 First-ply-failure envelopes obtained by different theories for [0/±45/90]$_s$ carbon/epoxy laminate under biaxial normal loading (AS4/3501-6; stress-free temperature $T_o = 124$ °C, 255 °F).

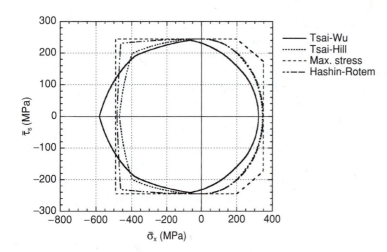

Fig. 9.11 First-ply-failure envelopes obtained by different theories for [0/±45/90]$_s$ carbon/epoxy laminate under biaxial shear/normal loading (AS4/3501-6; stress-free temperature $T_o = 124$ °C, 255 °F).

9.10 MICROMECHANICS OF PROGRESSIVE FAILURE

In most cases failure in a multidirectional laminate is initiated in the layer (or layers) with the highest stress normal to the fibers. Failure initiation takes the form of distributed microcracks, which coalesce into macrocracks as illustrated in the case of a unidirectional lamina under transverse tension (see Fig. 5.21). In the case of a single unsupported lamina, the first such macrocrack extending across its cross section corresponds to ultimate failure of the lamina. In the case of a multidirectional laminate, multiple macrocracking extending across the thickness of the layer develops and constitutes FPF as discussed before.

Although in most cases the formation of transverse cracks by itself does not cause catastrophic failure, their presence can influence the overall mechanical behavior of the

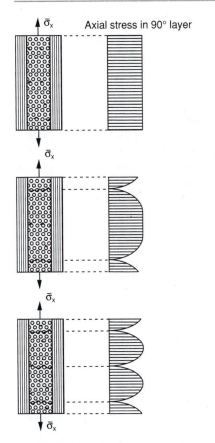

Fig. 9.12 Progressive failure and stress distributions in transverse ply of crossply laminate under uniaxial tension.

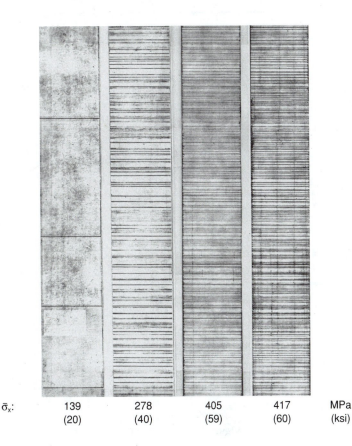

Fig. 9.13 X-radiographs of a $[0/90_4]_s$ carbon/epoxy laminate under uniaxial tensile loading at various applied stress levels.

laminate, such as stress redistribution, reduction of laminate stiffness, or initiation of other more severe damage modes. The failure process is best illustrated for the case of a crossply laminate under uniaxial tension (Fig. 9.12). The axial stress in the 90° layer is uniform initially, and when it reaches a critical value, it produces random microcracks as shown. The microcracks coalesce into roughly evenly spaced macrocracks, resulting in a redistribution of the average axial stress in the 90° ply with zero value at the crack faces and maximum value at the center between cracks. As the applied stress $\bar{\sigma}_x$ is increased, the maximum axial stress in the 90° ply increases up to a value equal to the transverse tensile strength of the lamina, F_{2t}, and further macrocracking takes place, resulting in further stress redistribution. This process continues up to a minimum crack spacing, at which point any further increase in applied stress cannot increase the axial stress in the 90° layer to the failure level F_{2t}. This state is called the *characteristic damage state* (CDS) of the laminate.[1] In most cases the minimum crack spacing is on the order of the layer thickness. The progression of cracking in the 90° layer of a $[0/90_2]_s$ carbon/epoxy laminate is illustrated by the X-radiographs in Fig. 9.13. The stress-strain behavior of a $[0/90_4]_s$ carbon/epoxy laminate in

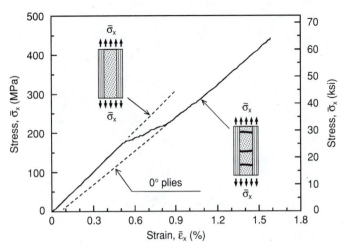

Fig. 9.14 Axial stress-strain curve of $[0/90_4]_s$ carbon/epoxy laminate (IM7/977-3).

Fig. 9.14 shows clearly the region of transverse matrix cracking and stiffness reduction.

Many investigations have been reported on the transverse cracking problem.[1-17] Analytical models have been proposed based on lamina strength[4,14-16] or fracture mechanics.[3,9-11,16] A shear lag analysis has been developed and used to obtain closed-form solutions for stress distributions, transverse matrix crack density, and reduced stiffnesses of the damaged plies and the entire laminate as a function of applied stress, properties of the constituent plies, and residual stresses.[14]

The axial (x-axis) stress distributions in the 0° and 90° layers of a crossply laminate segment containing two cracks (Fig. 9.15) are given as

$$\sigma_{1x} = \frac{E_1}{\bar{E}_x}\left[1 + \frac{E_2 h_2}{E_1 h_1}\frac{\cosh\alpha(l/2 - x)}{\cosh(\alpha l/2)}\right]\bar{\sigma}_x + \left[1 - \frac{\cosh\alpha(l/2 - x)}{\cosh(\alpha l/2)}\right]\sigma_{1r} \qquad (9.31)$$

$$\sigma_{2x} = \left[1 - \frac{\cosh\alpha(l/2 - x)}{\cosh(\alpha l/2)}\right]\left(\frac{E_2\bar{\sigma}_x}{\bar{E}_x} + \sigma_{2r}\right) \qquad (9.32)$$

where

E_1, E_2 = longitudinal and transverse lamina moduli

$\bar{E}_x = \dfrac{h_1 E_1 + h_2 E_2}{h_1 + h_2}$, laminate modulus (before damage)

$\sigma_{1x}, \sigma_{2x}, \bar{\sigma}_x$ = axial stresses in 0° layer, 90° layer, and laminate, respectively

σ_{1r}, σ_{2r} = residual stresses in 0° and 90° layers, respectively

l = crack spacing

h_1, h_2 = thicknesses of 0° and 90° layers, respectively

G_{12}, G_{23} = in-plane and out-of-plane lamina shear moduli

and

$$\alpha^2 = \frac{3(h_1 + h_2)\bar{E}_x}{h_1 h_2 E_1 E_2}\frac{G_{12}G_{23}}{h_1 G_{23} + h_2 G_{12}} \qquad (9.33)$$

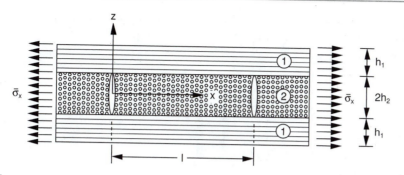

Fig. 9.15 Element of crossply laminate with transverse cracks in 90° layer.

As transverse cracking develops, the stiffness of the damaged layer as well as that of the entire laminate is reduced. The stiffness reduction caused by the matrix cracking was calculated based on the stress distributions above.[14] The modulus reduction ratio is expressed as a function of material and geometric parameters and the crack spacing as

$$\rho_E = \frac{\bar{E}_x'}{\bar{E}_x} = \left[1 + \frac{2}{al} \frac{E_2 h_2}{E_1 h_1} \tanh\frac{al}{2} + \frac{\sigma_{1r}}{\bar{\sigma}_x} \frac{\bar{E}_x}{E_1}\left(1 - \frac{2}{al} \tanh\frac{al}{2} \right) \right]^{-1} \qquad (9.34)$$

where

$$\bar{E}_x, \bar{E}_x' = \text{initial and reduced moduli of the laminate, respectively}$$

The reduced effective (in situ) modulus of the cracked transverse layer, E_2', is related to the modulus reduction ratio of the laminate as follows:

$$r_2 = \frac{E_2'}{E_2} = \rho_E - \frac{E_1 h_1}{E_2 h_2}(1 - \rho_E) \qquad (9.35)$$

The normalized reduced modulus of a $[0/90_2]_s$ carbon/epoxy laminate computed by Eq. (9.34) is plotted versus normalized transverse crack density in Fig. 9.16 and compared with experimental results. The normalized in situ modulus of the cracked 90° layer obtained from Eq. (9.35) is plotted versus normalized crack density in Fig. 9.17. It is seen that, at the limiting crack density, the laminate modulus is reduced to approximately 90% of its original value, and the in situ modulus of the 90° layer is reduced to approximately 25% of its original value.

Crack density can be predicted in a deterministic fashion by assuming uniform crack spacing and equating the value of σ_{2x} in Eq. (9.32) at $x = l/2$ to the transverse lamina strength F_{2t}:

$$\sigma_{2x(x=l/2)} = F_{2t} \qquad (9.36)$$

The average crack density, λ_m, is obtained as the mean of the crack densities for values of σ_{2x} slightly greater than and less than F_{2t}:

$$\lambda_m = \frac{3\alpha}{4\cosh^{-1}\xi} \qquad (9.37)$$

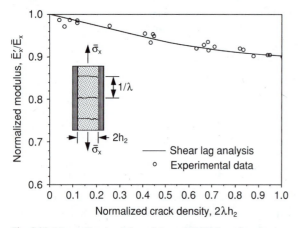

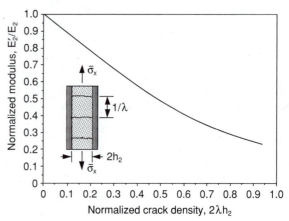

Fig. 9.16 Normalized axial modulus of $[0/90_2]_s$ carbon/epoxy (AS4/3501-6) laminate as a function of normalized transverse crack density.[14]

Fig. 9.17 Effective in situ normalized transverse modulus of cracked 90° layer in $[0/90_2]_s$ carbon/epoxy (AS4/3501-6) laminate as a function of normalized transverse crack density.[14]

where

$$\xi = \frac{E_2 \bar{\sigma}_x / \bar{E}_x + \sigma_{2r}}{E_2 \bar{\sigma}_x / \bar{E}_x + \sigma_{2r} - F_{2t}} = \frac{\bar{\sigma}_{2x}}{\bar{\sigma}_{2x} - F_{2t}} \tag{9.38}$$

with

$$\bar{\sigma}_{2x} = \frac{E_2}{\bar{E}_x} \bar{\sigma}_x + \sigma_{2r} = \text{average stress in 90° layer between cracks}$$

From Eqs. (9.37) and (9.38) we obtain

$$\bar{\bar{\sigma}}_{2x} = \frac{\bar{\sigma}_{2x}}{F_{2t}} = \frac{\cosh \dfrac{3}{4\bar{\lambda}}}{\cosh \dfrac{3}{4\bar{\lambda}} - 1} \tag{9.39}$$

where

$$\bar{\lambda} = \frac{\lambda_m}{\alpha}$$

Equation (9.39) is a generalized relation between normalized applied stress and normalized average crack density and is independent of crossply layup and transverse lamina strength (temperature). An experimental confirmation of this relationship is shown in Fig. 9.18.

Progressive failure under a general in-plane biaxial loading has also been analyzed.[18,19] Besides Young's modulus, the reduction of the in-plane shear modulus has been calculated

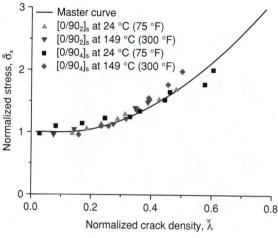

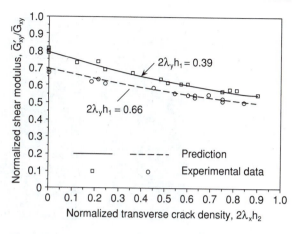

Fig. 9.18 Master curve for transverse cracking compared with experimental results.

Fig. 9.19 Shear modulus reduction as a function of normalized crack densities of $[0_3/90_3]_s$ crossply carbon/epoxy laminate.[20]

and correlated with the state of damage (cracking) in the laminate.[20] The variation of shear modulus for a $[0_3/90_3]_s$ carbon/epoxy laminate as a function of crack density is shown in Fig. 9.19.

9.11 PROGRESSIVE AND ULTIMATE LAMINATE FAILURE—LAMINATE EFFICIENCY

As discussed before, progressive failure of a lamina within the laminate consists of cracking of the lamina up to a characteristic limiting crack density. Following this FPF, the failure process continues up to ULF. The latter can occur at a much higher load than FPF.

Figure 9.20 illustrates the progression of damage in a $[0/\pm45/90]_s$ glass/epoxy laminate under uniaxial loading. The first failure mechanism consists of transverse cracking in the 90° layer. These cracks increase in density up to a limiting value or the characteristic damage state. Thereafter, as the load increases, cracking starts in the ±45 plies, precipitating ultimate failure.

Consider for example the stress-strain response of a multidirectional laminate under uniaxial tensile loading (Fig. 9.21). Initially, the laminate behaves linearly, with a stress-strain slope equal to the original laminate modulus $\bar{E}_x$ up to point 1, where the first ply fails. After this ply reaches its maximum crack density (characteristic damage state), its effective transverse modulus drops to E_2' and the laminate modulus drops to a value $\bar{E}_x^{(1)}$. If the material behaves in a brittle manner, the modulus drop will be sudden. It will be manifested by a horizontal or vertical shift in the stress-strain curve, depending on whether the test is conducted under load or strain control, respectively.

Under increasing load the specimen will respond linearly with a stress-strain slope equal to the reduced modulus $\bar{E}_x^{(1)}$ up to point 2, where the next ply or plies fail. Again, if the ply or plies fail instantly in a brittle manner, there will be a further sudden drop in the modulus to a value $\bar{E}_x^{(2)}$. This value corresponds to the laminate with all the failed plies to date discounted or reduced in stiffness. The drop in modulus will be manifested as a sudden

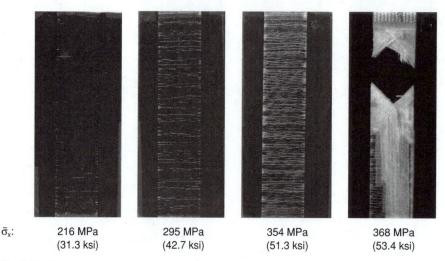

$\bar{\sigma}_x$:   216 MPa   295 MPa   354 MPa   368 MPa
                    (31.3 ksi)   (42.7 ksi)   (51.3 ksi)   (53.4 ksi)

Fig. 9.20 Damage progression in a $[0/\pm45/90]_s$ glass/epoxy laminate under increasing uniaxial tensile loading.

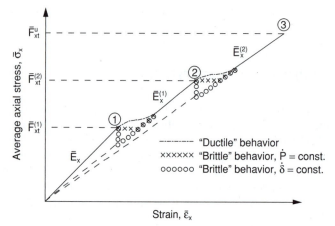

Fig. 9.21 Stress-strain curve of multidirectional laminate under uniaxial tension showing progressive failure ($\dot{P}$ = constant, load rate control; $\dot{\delta}$ = constant, strain rate control).

horizontal or vertical shift in the stress-strain curve, depending on the test control mode as discussed before. This progressive failure continues up to a point (say, point 3) where ultimate failure takes place. If the plies fail gradually due to ductile behavior or statistical variation in local strengths and failures, the stress-strain curve will appear smoother with gradual transitions in slope at the points of partial ply failures (Fig. 9.21).

At each stage of failure there is a corresponding strength. Thus, one can define the initial stage (or FPF) and last stage (ULF) tensile strengths, $\bar{F}_{xt}^{(1)}$ and $\bar{F}_{xt}^{u}$ or $\bar{F}_{FPF}$ and $\bar{F}_{ULF}$, respectively. The ratio of these two strengths, φ_L, is a measure of laminate efficiency and indicates the level of fiber strength utilization at FPF:

$$\varphi_L = \frac{\bar{F}_{xt}^{(1)}}{\bar{F}_{xt}^{u}} = \frac{\bar{F}_{FPF}}{\bar{F}_{ULF}} \tag{9.40}$$

This ratio depends on both the material system and the laminate layup.

A comparison of FPF and ULF of different multidirectional laminates under uniaxial tensile loading is shown in Tables 9.2 and 9.3 as done previously for a similar set of materials.[21] The laminate efficiency ratio for $[0/90]_s$ crossply laminates is fairly low for all material systems, ranging from 0.20 for glass/epoxy to 0.42 for boron/epoxy. In the case

TABLE 9.2 Comparison of Initial and Ultimate Failures of $[0/90]_s$ Laminates

Material System	First Ply Failure (FPF)			Ultimate Laminate Failure (ULF)			Laminate Efficiency Ratio $(\varphi_L = \bar{F}_{FPF}/\bar{F}_{ULF})$
	Strength $\bar{F}_{FPF}$, MPa (ksi)	Strain (%)	Modulus GPa (Msi)	Strength $\bar{F}_{ULF}$, MPa (ksi)	Strain (%)	Modulus GPa (Msi)	
Glass/epoxy	111 (16.1)	0.46	24.1 (3.5)	552 (80)	2.75	20.0 (2.9)	0.20
Kevlar/epoxy	241 (34.9)	0.52	46.2 (6.7)	655 (95)	1.50	44.9 (6.5)	0.37
Carbon/epoxy	420 (60.9)	0.50	78.2 (11.3)	1152 (167)	1.50	82.8 (12.0)	0.36
Boron/epoxy	288 (41.7)	0.35	111.8 (16.2)	690 (100)	0.60	103.5 (15.0)	0.42
Boron/aluminum	163 (23.6)	0.09	187.0 (27.1)	683 (99)	0.58	113.9 (16.5)	0.24

TABLE 9.3 Comparison of Initial and Ultimate Failures of $[0_2/\pm45]_s$ Laminates

Material System	First Ply Failure (FPF)			Ultimate Laminate Failure (ULF)			Laminate Efficiency Ratio $(\varphi_L = \bar{F}_{FPF}/\bar{F}_{ULF})$
	Strength $\bar{F}_{FPF}$, MPa (ksi)	Strain (%)	Modulus GPa (Msi)	Strength $\bar{F}_{ULF}$, MPa (ksi)	Strain (%)	Modulus GPa (Msi)	
Glass/epoxy	265 (38.4)	1.04	25.5 (3.7)	607 (88)	2.75	22.1 (3.2)	0.44
Kevlar/epoxy	607 (88.0)	1.26	48.0 (7.0)	683 (99)	1.50	45.5 (6.6)	0.88
Carbon/epoxy	460 (66.6)	0.53	86.2 (12.5)	1208 (175)	1.50	80.5 (11.7)	0.38
Boron/epoxy	601 (87.1)	0.53	114.5 (16.6)	745 (108)	0.70	106.4 (15.4)	0.81
Boron/aluminum	259 (37.5)	0.14	185.6 (26.9)	780 (113)	0.60	130.0 (18.8)	0.33

of $[0_2/\pm45]_s$ laminates the laminate efficiency ratios are much higher, ranging from 0.33 to 0.88.

9.12 ANALYSIS OF PROGRESSIVE AND ULTIMATE LAMINATE FAILURE

The determination of ultimate strength of a laminate requires an iterative procedure taking into account the damage progression in the various plies. The general approach consists of the following stages or steps.

9.12.1 Determination of First Ply Failure (FPF)

A lamination analysis is conducted to determine the state of stress in every ply as a function of the loading. This analysis is usually linear and includes curing residual stresses. The lamina stresses as a function of the loading are referred first to the laminate coordinates (x, y) and then transformed to the lamina coordinates $(1, 2)$. These stresses are checked against a selected lamina failure criterion (for example, one of those discussed in Chapter 6) to determine the load at FPF. Failure initiation in a given lamina is usually based on strength properties obtained by testing isolated lamina specimens, however, damage progression and saturation (CDS level) are related to enhanced in situ lamina properties due to constraints from the adjacent plies.[22] First-ply-failure envelopes predicted by various failure theories for a $[0/\pm45/90]_s$ quasi-isotropic carbon/epoxy laminate were shown in Figs. 9.10 and 9.11.

9.12.2 Discounting of Damaged Plies

The failed lamina or laminae and their failure modes are identified. Failure modes are directly identifiable when using limit or noninteractive theories (maximum stress, maximum strain) or failure-mode-based theories (Hashin-Rotem). However, this is not the case with fully interactive theories (Tsai-Wu, Tsai-Hill). For this case the following failure mode discrimination rule is proposed:

Matrix or interfiber failure

$$\left(\frac{\sigma_2}{F_2}\right)^2 + \left(\frac{\tau_6}{F_6}\right)^2 > \left(\frac{\sigma_1}{F_1}\right)^2$$

$$(9.41)$$

Fiber failure

$$\left(\frac{\sigma_2}{F_2}\right)^2 + \left(\frac{\tau_6}{F_6}\right)^2 < \left(\frac{\sigma_1}{F_1}\right)^2$$

When matrix or interfiber failures are identified, the matrix-dominated stiffnesses, such as E_2 and G_{12}, are reduced. Stiffness reduction factors can be selected based on analysis as discussed before (see Eq. (9.35)), or the affected stiffnesses may be discounted altogether (total ply discount method). If the first ply failure is a fiber failure, the corresponding fiber-dominated stiffness E_1 is discounted. After discounting the appropriate stiffnesses in the damaged plies, new laminate stiffnesses [A], [B], and [D] are calculated.

9.12.3 Stress Analysis of the Damaged Laminate

Lamina stresses are recalculated and checked against the selected failure criterion to verify that the undamaged laminae do not fail immediately under their increased share of stress following the FPF above. In this analysis the strengths of the previously failed laminae (with reduced or totally discounted stiffnesses) are assumed to be fictitiously very high to avoid repeated failure indication in the same plies.

9.12.4 Second Ply Failure

The load is increased until the next ply (or group of plies) fails. This could be a failure in a previously undamaged ply or a new failure in a previously damaged one. Then, steps 1, 2, and 3 above are repeated.

9.12.5 Ultimate Laminate Failure

The above process continues until the criterion for ultimate laminate failure (ULF) is met. Criteria such as maximum load, last ply failure, limit laminate strain, and first fiber failure have been proposed. Theoretical predictions of ULF vary widely depending on the definition of ULF. In the case of fiber-dominated failures, the criterion of first fiber failure

(FFF) seems to be the most reasonable and yields consistent results. In the case of matrix-dominated failures, such as that of angle-ply laminates, the last-ply-failure (LPF) definition yields more realistic results than the FFF criterion. A modification of the LPF criterion allows for a higher loading up to a maximum allowable relative rotation of adjacent plies (or a maximum interlaminar shear strain). According to the maximum load criterion of ULF, failure occurs when the laminate, at any stage of the progressive ply failures, cannot sustain the stresses.

9.12.6 Computational Procedure

A computational procedure for the multiple-ply-failure process is illustrated in the flowchart of Fig. 9.22 for a symmetric laminate under in-plane loading using the Tsai-Wu failure criterion. It consists of the following steps:

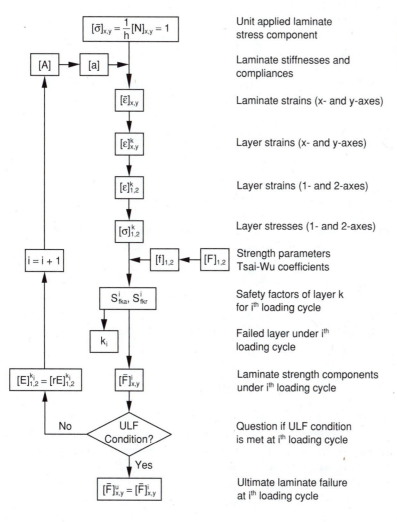

$[\bar{\sigma}]_{x,y} = \frac{1}{h}[N]_{x,y} = 1$	Unit applied laminate stress component
$[A] \rightarrow [a]$	Laminate stiffnesses and compliances
$[\bar{\varepsilon}]_{x,y}$	Laminate strains (x- and y-axes)
$[\varepsilon]^k_{x,y}$	Layer strains (x- and y-axes)
$[\varepsilon]^k_{1,2}$	Layer strains (1- and 2-axes)
$[\sigma]^k_{1,2}$	Layer stresses (1- and 2-axes)
$[f]_{1,2} \leftarrow [F]_{1,2}$	Strength parameters Tsai-Wu coefficients
S^i_{fka}, S^i_{fkr}	Safety factors of layer k for i^{th} loading cycle
k_i	Failed layer under i^{th} loading cycle
$[\bar{F}]^i_{x,y}$	Laminate strength components under i^{th} loading cycle
ULF Condition?	Question if ULF condition is met at i^{th} loading cycle
$[\bar{F}]^u_{x,y} = [\bar{F}]^i_{x,y}$	Ultimate laminate failure at i^{th} loading cycle

$i = i + 1$

$[E]^{k_i}_{1,2} = [rE]^{k_i}_{1,2}$

No

Yes

Fig. 9.22 Flowchart for determination of load-deformation curve and failure levels of symmetric laminates under in-plane loading.

Step 1   Enter a unit average laminate stress (or load) in the desired direction, that is, $\bar{\sigma}_x = 1$, $\bar{\sigma}_y = 1$, or $\bar{\tau}_s = 1$.

Step 2   Enter the laminate extensional stiffnesses $[A]$ and compliances $[a]$ obtained from basic lamina properties and laminate layup (steps 1–7 in the flowchart of Fig. 9.3).

Step 3   Calculate the average laminate strains $\bar{\varepsilon}_x$, $\bar{\varepsilon}_y$, and $\bar{\gamma}_s$ referred to the x- and y-axes using Eq. (7.80).

Step 4   Calculate the layer strains $(\varepsilon_x,\ \varepsilon_y,\ \gamma_s)_k$, which are equal to the laminate strains.

Step 5   Calculate the layer strains $(\varepsilon_1,\ \varepsilon_2,\ \gamma_6)_k$ referred to layer principal axes $(1, 2)$, using Eq. (4.58).

Step 6   Calculate the layer stresses $(\sigma_1,\ \sigma_2,\ \tau_6)_k$ referred to layer principal axes $(1, 2)$ using Eq. (4.31).

Step 7   Enter the lamina strengths, $[F]_{1,2}$, and calculate Tsai-Wu coefficients f_i and f_{ij}, using relations in Eqs. (6.32), (6.33), (6.35), and (6.40).

Step 8   Calculate the layer safety factors, S_{fka} and S_{fkr}, using Eqs. (9.7) and (9.8).

Step 9a   Identify the failed layer, k_i, under the ith loading cycle from $S_{fki} = (S_{fk})_{min}$.

Step 9b   Determine the laminate strength components $[\bar{F}]_{x,y}^i$ for the ith loading cycle.

Step 10   Question whether the conditions of ULF are met at the ith loading cycle.

Step 11a   If the answer is "no," that is, if loading can be increased, the damaged lamina k_i is replaced with one having the properties

$$E_1^{k_i} = r_1 E_1$$
$$E_2^{k_i} = r_2 E_2$$
$$G_{12}^{k_i} = r_{12} G_{12} \qquad (9.42)$$
$$\nu_{12}^{k_i} = r_1 \nu_{12}$$

where r_1, r_2, and r_{12} are stiffness reduction factors, obtained from analysis or prior experiments. Typical values of these factors are

$$r_1 \cong 1$$
$$r_2 \cong r_{12} \cong 0.25$$

In the conservative approach of complete ply discount,

$$r_1 \cong 1$$
$$r_2 \cong r_{12} \cong 0$$

Step 11b   Go to the $i + 1$ loading cycle and recalculate modified laminate stiffnesses $[A]$ and compliances $[a]$. Repeat steps 2 through 10 above. In step 7, the lamina strengths of the failed layer k_i should be made fictitiously very high to avoid repeated failure indication in the same layer.

Step 12   If the answer to the question of step 10 is "yes," then ultimate failure occurs at the ith loading cycle, that is, $\bar{F}^u = \bar{F}^i$.

Hygrothermal loading can be included in the above procedure as in the case of the flowchart for FPF (Fig. 9.9). However, in subsequent loading cycles, the coefficients of thermal and hygric expansion of the failed plies must be modified. The presence of hygrothermal (or residual) stresses also affects the local stresses in the cracked laminate, the extent of layer cracking, and the modulus reduction ratios, Eqs. (9.34) and (9.35). The effect of hygrothermal stresses on the laminate strength decreases with failure level, that is, it is much less pronounced on ultimate failure than on first ply failure. A similar procedure can be followed for the general case of an asymmetric laminate under general in-plane and moment loading by combining the flowcharts of Figs. 9.9 and 9.22.

9.13 LAMINATE FAILURE THEORIES—OVERVIEW, EVALUATION, AND APPLICABILITY

Failure theories applicable to a composite lamina were described and discussed in Chapter 6. They were classified into three groups: *limit or noninteractive theories* (maximum stress, maximum strain), *interactive theories* (Tsai-Hill, Tsai-Wu), and *partially interactive or failure-mode-based theories* (Hashin-Rotem, Puck). As discussed before, numerous theories have been proposed, but it is difficult to evaluate their applicability and to select one over the others, due to the lack of adequate and reliable experimental data. The difficulty is much greater in the case of a multidirectional laminate.

The scope of a laminate failure theory comprises the following:

1. a lamina failure theory for prediction of failure initiation, that is, first ply failure (FPF) in the laminate
2. a scheme of ply discounting and failure progression in the laminate after FPF
3. a criterion or definition of ultimate laminate failure (ULF)

Two main efforts recently aimed at evaluation of laminate failure theories are those by C. T. Sun[23,24] and the "Worldwide Failure Exercise" by Hinton, Soden, and Kaddour.[25,26] In these works, failure theories were reviewed, compared with each other, and with experimental results. The scope of these studies comprised several material systems, laminate configurations, and biaxial loading conditions. Results were expressed in the form of failure envelopes for first ply failure (FPF) and ultimate laminate failure (ULF). Progressive failure mechanisms and models were discussed extensively.

C. T. Sun[24] reviewed six failure theories and showed comparisons of theoretical predictions of laminate strengths with experimental results for six composite material systems and various loading conditions. He showed comparisons between theoretical predictions based on the ply-by-ply discount method and biaxial experimental data for quasi-isotropic carbon/epoxy laminates and off-axis data for various laminates. In the case of fiber-dominated laminates, where matrix-dominated strengths F_{2t}, F_{2c}, and F_6 play a small role, the limit theories (maximum stress and maximum strain) and the failure-mode-based theory (Hashin-Rotem) gave the best predictions. The interactive theories (Tsai-Hill, Tsai-Wu), which give weight to matrix-dominated strengths, require accurate determination of such properties in situ. In addition, it was found that residual stresses must be taken into account in all cases.

The "Worldwide Failure Exercise" initiated by Hinton, Soden, and Kaddour had a wider scope as described in three special issues of *Composites Science and Technology*[25] and a book.[26] The "exercise" covered nineteen theories, four composite material systems, six laminate configurations, and fourteen loading conditions. Biaxial test results were provided for a range of composite laminates as benchmark data for comparison with theoretical predictions.[27] Comparisons were presented in the form of failure envelopes and stress-strain curves to failure. Overall evaluation and conclusions were presented by the initiators of the exercise.[28]

In general, a wide variation was observed in the prediction of laminate failure by the various theories as noted in the "failure exercise." The divergence of the predictions is greater for FPF than for ULF; also it is greater for matrix-dominated failures than for fiber-dominated ones. The divergence observed may be attributed primarily to the following factors:

1. the different ways in which curing residual stresses were introduced in the predictions, especially in the case of first ply failure
2. the concept of in situ behavior of a lamina within the laminate, which is still debated
3. the different methods of modeling the progressive failure process and the definition of ultimate laminate failure
4. the nonlinear behavior of matrix-dominated laminates, for example, angle-ply laminates

Under uniaxial loading of a laminate the deciding factor in predicting ultimate failure is whether it is fiber or matrix dominated. Table 9.4 shows a comparison between measured

TABLE 9.4 Measured and Predicted Ultimate Axial Strengths of Carbon/Epoxy Laminates (AS4/3501-6)

Laminate	Tensile Strength, $\bar{F}_{xt}$, MPa						ULF Definition
	Experimental	Tsai-Wu	Tsai-Hill	Max. Stress	Max. Strain	Hashin-Rotem	
$[0/90_2]_s$	780	791	794	791	798	791	FFF
$[0_2/90_2]_s$	1112	1163	1168	1163	1172	1163	FFF
$[0/90_4]_s$	444	494	496	494	497	494	FFF
$[0/\pm45]_s$	875	816	806	851	844	850	FFF
$[0/(\pm45)_2]_s$	565	527	521	553	548	553	FFF
$[\pm20]_{2s}$	809	763	631	679	679	667	LPF
$[\pm45]_s$	151	105	119	152	152	119	LPF
$[\pm70]_{2s}$	—	65	67	73	73	67	LPF

Laminate	Compressive Strength, $\bar{F}_{xc}$, MPa						ULF Definition
	Experimental	Tsai-Wu	Tsai-Hill	Max. Stress	Max. Strain	Hashin-Rotem	
$[\pm20]_{2s}$	670	425	507	667	519	540	LPF
$[\pm70]_{2s}$	200	180	132	148	148	132	LPF

and predicted axial ultimate strengths of a number of laminates. Residual stresses were taken into account in the predictions by assuming a temperature difference of $\Delta T = -100\ °C$ $(-180\ °F)$ between the stress-free and current states. Lamina strengths used were those obtained by direct testing of isolated laminae without consideration of in situ effects. In the case of the first five laminates listed, it is clear that they are fiber dominated because the 0° plies carry a substantial portion of the total load. Thus, the first fiber failure (FFF) was used as a criterion of ultimate laminate failure (ULF). The predictions by the five failure theories used are in substantial agreement with each other and with the experimental measurements.

In the case of matrix-dominated angle-ply laminates, predictions by the limit or interactive theories are not usually in agreement with each other and with experimental results. Failure is governed by the lamina transverse normal stress σ_2 and the in-plane shear stress τ_6. When $\sigma_2 > 0$, as in the case of $[\pm 45]_s$ and $[\pm 70]_{2s}$ laminates under tension, the limit theories predict higher strengths in agreement with the experiment. When $\sigma_2 < 0$, as in the case of the $[\pm 20]_{2s}$ laminate under tension, the Tsai-Wu criterion comes closer to the experimental results. Although none of the theories used predicts fiber failure, such a failure mode does occur as a result of constraints and stress concentrations at the cracks of the adjacent plies. Figure 9.23 shows that after in-plane shear failure in the 20° plies, fiber bundle failures occurred through the thickness of the −20° layers.

Under axial compression, the transverse lamina normal stress σ_2 is tensile in the $[\pm 20]_{2s}$ laminate and compressive in the $[\pm 70]_{2s}$ one. The results in Table 9.4 show that for the $[\pm 20]_{2s}$ laminate, only the maximum stress theory prediction agrees with the experiment, whereas the other theories, especially the Tsai-Wu theory, underestimate the compressive strength. The failure mode consists of transverse tension and shear in the 20° plies followed by fiber bundle failures in the −20° layers (Fig. 9.24). For the $[\pm 70]_{2s}$ laminate under compression, only the Tsai-Wu prediction comes close to the experimental measurement, whereas all other theories underestimate the strength.

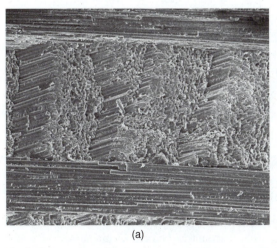
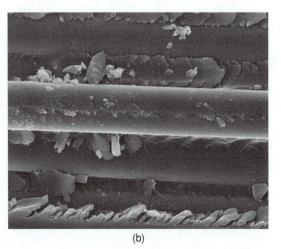

(a)   (b)

Fig. 9.23 Fracture surfaces of $[\pm 20]_{2s}$ carbon/epoxy laminate under axial tensile loading: (a) shear failure of 20° plies and fiber bundle failures in −20° layers and (b) clean debonding of fibers in the 20° ply.

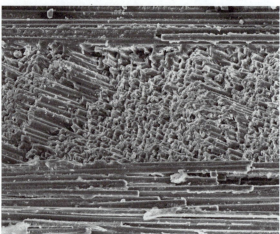

Fig. 9.24 Fracture surfaces of $[\pm 20]_{2s}$ carbon/epoxy laminate under axial compression.

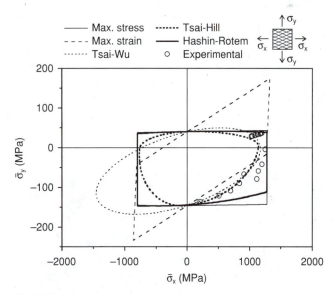

Fig. 9.25 Comparison between predicted ultimate failure envelopes and experimental results for glass/epoxy (E-glass/MY750) $[\pm 5]_s$ laminate under biaxial normal loading (data from Al-Khalil et al.[29] as quoted by Soden et al.[27]).

In the more general cases of biaxial loading, it is not easy to establish fiber or matrix dominance in failure, as that varies with the loading biaxiality. The case of a $[\pm 5]_s$ glass/epoxy angle-ply laminate under biaxial normal loading is illustrated in Fig. 9.25. In the region near the uniaxial $\bar{\sigma}_x$ loading, where the ultimate failure is fiber dominated, the data are closer to the maximum stress and the Hashin-Rotem predictions. However, in the second quadrant, characterized by high compression in the y-direction and decreasing tension in the x-direction, the behavior tends to be more matrix dominated and agrees well with the predictions of the Tsai-Hill and Tsai-Wu theories. No data are available in the compression-compression third quadrant, where the predictions differ the most from each other.

Theoretical predictions were compared with experimental results for a $[90/\pm 30]_s$ quasi-isotropic glass/epoxy laminate under biaxial shear, $\bar{\tau}_s$, and normal stress, $\bar{\sigma}_x$. The progressive failure scheme discussed in Section 9.12 was adopted. Ultimate laminate failure was defined as the first fiber failure (FFF). Residual stresses were taken into account as before. A comparison of predicted ultimate failure envelopes and experimental results for the laminate is shown in Fig. 9.26. All five theories tested are in reasonable agreement with the experimental results in the

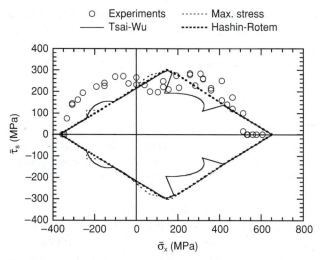

Fig. 9.26 Comparison between predicted failure envelopes and experimental results for [90/±30]$_s$ E-glass/LY556 epoxy laminate under biaxial normal and shear loading (data from Hütter et al.[30]).

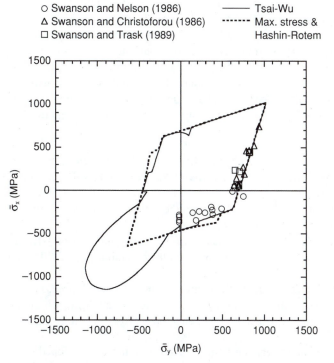

Fig. 9.27 Comparison between predicted failure envelopes and experimental results for [90/±45/0]$_s$ carbon/epoxy (AS4/3501-6) laminate under biaxial normal loading (data from Swanson et al.[31–33]).

shear-tension quadrant ($\bar{\sigma}_x > 0$), but they deviate noticeably from the experimental results in the shear-compression quadrant ($\bar{\sigma}_x < 0$), where the influence of the matrix is greater. The localized deviations of the Tsai-Wu predictions from the other theoretical predictions are due to the fiber failure mode discrimination rule used in this case for FFF, Eq. (9.41). A more restrictive fiber failure mode rule where $(\sigma_1/F_1)^2$ is much greater than $(\sigma_2/F_2)^2 + (\tau_6/F_6)^2$ would minimize the deviations.

Similar comparisons were made for the case of a [90/±45/0]$_s$ quasi-isotropic carbon/epoxy (AS4/3501-6) laminate under biaxial normal loading. The progressive failure scheme described before (Fig. 9.22) was used with the first fiber failure (FFF) defining ultimate laminate failure (ULF). Residual stresses were accounted for by assuming the difference between stress-free and current temperatures as $\Delta T = -100\ °C\ (-180\ °F)$. Predicted ultimate failure envelopes are compared with experimental results in Fig. 9.27. The agreement between predictions of the theories tested and experimental results is reasonable in the tension-tension ($\bar{\sigma}_x > 0$, $\bar{\sigma}_y > 0$) and tension-compression ($\bar{\sigma}_x < 0$, $\bar{\sigma}_y > 0$) quadrants. In the compression-compression regime, where the behavior is matrix dominated, no reliable experimental data exist and the predictions differ the most from each other. The interactive Tsai-Wu criterion in conjunction with the progressive failure scheme predicts very high compressive strengths.

It is difficult to reach definitive conclusions on the applicability of the various theories based on comparison with the limited experimental data available, especially in the cases of FPF and under biaxial compression or compression and shear. Theories based on the maximum stress criterion, a partly interactive approach (Puck et al.), or a totally interactive criterion (Tsai-Wu), give reasonable predictions of ultimate failure in fiber-dominated laminates if the first fiber failure (FFF) is used as the definition of ULF. In the case of matrix-dominated failure, a distinction must be made between cases of transverse tensile stress ($\sigma_2 > 0$) and transverse compressive stress ($\sigma_2 < 0$) in the lamina. In the former case, limit theories (maximum stress) agree better with experimental results, whereas interactive theories (Tsai-Wu) underestimate the strength. In the second case ($\sigma_2 < 0$), the opposite is true.

9.14 DESIGN CONSIDERATIONS

Structural composites are designed with a substantial fraction of layers oriented along major loading directions, that is, they are designed as fiber-dominated structures. For engineering design, only two failure levels are important, FPF and ULF. First ply failure, even in fiber-dominated laminates, is highly dependent on in-plane transverse and interfiber shear strengths (F_2, F_6) and the interaction of the respective stresses (σ_2, τ_6). The analysis of FPF must take into account residual stresses due to processing and hygrothermal service conditions and may also consider the in situ behavior of a lamina within the laminate.[22] On the other hand, analysis of ULF is mainly dependent on longitudinal stresses (σ_1) and strengths (F_{1t}, F_{1c}) and is less sensitive to interfiber stresses (σ_2, τ_6), stiffnesses (E_2, G_{12}), and strengths (F_{2t}, F_{2c}, F_6) of the already damaged layers, and to the mostly relieved residual stresses.

In some applications where FPF is crucial to the function of the structure (e.g., pressure vessels containing toxic chemicals under pressure), some simplifications are appropriate for a conservative approach. Crack initiation based on isolated lamina strengths (F_{2t}, F_6) is taken as an indication of FPF with no accounting for the in situ effect. The assumption of linear lamina stress-strain behavior contributes to a more conservative FPF prediction. The above assumptions simplify the analysis by limiting (to 13) the number of necessary mechanical and hygrothermal parameters for a two-dimensional analysis.

In view of the multitude of failure theories, the divergence of their predictions and the lack of definitive general conclusions regarding their applicability, a practical approach is recommended as follows:

Step 1 Select a classical representative theory from each category, that is, noninteractive (maximum stress), fully interactive (Tsai-Wu), and partly interactive (Hashin-Rotem).

Step 2 Compute and plot stress-strain relations of the laminate under representative mechanical and hygrothermal loading.

Step 3 Compute safety factors for FPF and ULF and compute and plot failure envelopes for the selected failure theories for the two failure levels (FPF and ULF).

Step 4 Select a prediction according to the degree of conservatism desired. For the most conservative approach, limit the state of stress (loading) to within the common domain of the selected failure envelopes (see Fig. 6.20).

All of the above computations and plots can be performed by a newly developed computer program.[34] The approach above is adequate for conservative structural design. More sophisticated theories and approaches exist as discussed before, incorporating nonlinear behavior and in situ effects.

9.15 INTERLAMINAR STRESSES AND STRENGTH OF MULTIDIRECTIONAL LAMINATES: EDGE EFFECTS

9.15.1 Introduction

One of the assumptions of the classical lamination theory discussed in Section 7.1 is that the laminate and all its layers are in a state of plane stress, that is, all out-of-plane stress components are zero, $\sigma_z = \tau_{xz} = \tau_{yz} = 0$. This assumption is justified away from geometric discontinuities, such as free edges. Mechanical and hygrothermal loadings can produce interlaminar stresses, both shear and normal, especially near free edges. Normal tensile interlaminar stresses, or peel stresses, tend to separate the laminae from each other. Interlaminar shear stresses tend to slide one lamina over adjacent ones. In both cases interlaminar stresses can cause interlaminar separation or *delamination*. Delamination can interact with transverse cracking in the failure process.[35] Interlaminar stresses are mainly a function of the laminate stacking sequence; thus, they can be controlled by proper design of the stacking sequence. For example, in some cases a tensile interlaminar normal stress, σ_z, can be transformed into a compressive one by simply rearranging the layer sequence.

There are three types of interlaminar stress problems associated with three types of laminates: [±θ] angle-ply laminates, [0/90] crossply laminates, and laminates combining both angle-ply and crossply configurations.

9.15.2 Angle-Ply Laminates

Consider for example a $[\pm\theta]_s$ angle-ply laminate under average axial tensile stress $\bar{\sigma}_x$ (Fig. 9.28). Each layer is subjected to the same axial stress, $\sigma_x = \bar{\sigma}_x$. Because of the off-axis orientation (shear coupling), layers θ and −θ when considered independently will undergo shear deformations of opposite signs as shown in Fig. 9.28. When bonded together in the laminate, the layers must have zero shear strain. This is achieved through interlaminar shear stresses τ_{zx} transmitted from one layer to the other. These stresses vary across the width of the specimen, being zero over most of the central region and peaking near the edges (Fig. 9.29). The moment produced by these stresses is equilibrated by intralaminar shear stresses τ_{xy} acting on the transverse cross section

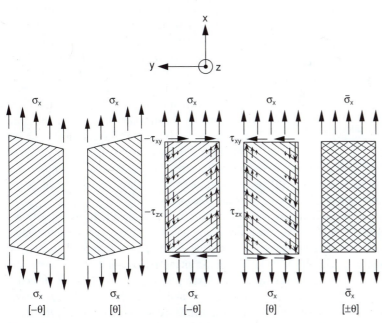

Fig. 9.28 Illustration of generation of interlaminar and intralaminar shear stresses in angle-ply laminate under axial tension.

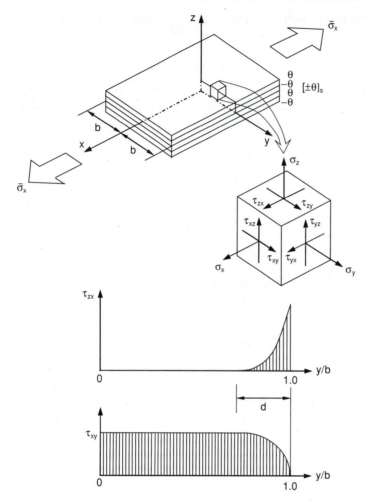

Fig. 9.29 Distribution of interlaminar (τ_{zx}) and intralaminar (τ_{xy}) shear stresses in θ-layer of [±θ]$_s$ angle-ply laminate under axial tension.

of the layer. These stresses are constant over most of the central region and drop to zero at the (stress-free) edges (Fig. 9.29). The interlaminar shear stress (τ_{zx}) is a function of fiber orientation in the [±θ]$_s$ laminate as illustrated in Fig. 9.30 for a carbon/epoxy [±θ]$_s$ laminate.[36] The τ_{zx} stress (as close to the interface and as near the edge as can be determined numerically) is zero at θ = 60° as well as at θ = 0° and θ = 90°. It reaches a peak value (for this carbon/epoxy) at θ ≅ 35°. The physical existence of interlaminar stresses was demonstrated by determining axial displacements across the width of a [±25]$_s$ carbon/epoxy specimen by means of the moiré method.[37]

9.15.3 Crossply Laminates

A different state of interlaminar stresses arises in the case of crossply laminates. Consider, for example, a [0/90]$_s$ crossply laminate under average axial tensile stress $\bar{\sigma}_x$ (Fig. 9.31). The load sharing is such that each layer undergoes the same axial deformation. Because of the different Poisson's ratios, the 0° and 90° layers will undergo different transverse deformations when acting independently. When bonded together in the laminate the 0° and 90° layers must have the same transverse strain. This is achieved through interlaminar shear stresses τ_{zy}, which tend to expand the 0° layer and compress the 90° layer in the y-direction. These stresses vary across the width of the specimen, being zero over the central region and peaking near the free edges.

The development of interlaminar stresses in a crossply laminate is further illustrated in Fig. 9.32. Considering a free body diagram of an element of the 0° ply near the edge, one can see that the interlaminar stresses τ_{zy} must be equilibrated by normal stresses σ_y acting on the layer. Moment equilibrium in the y-z plane requires interlaminar normal stresses σ_z with a distribution producing a zero force resultant in the z-direction and a moment equal and opposite to that produced by the τ_{zy} and σ_y stresses. For the case under discussion this means high tensile interlaminar normal stresses near the edge, that is, a tendency for delamination. Of course the sign of all interlaminar stresses is reversed when the applied stress $\bar{\sigma}_x$ is compressive.

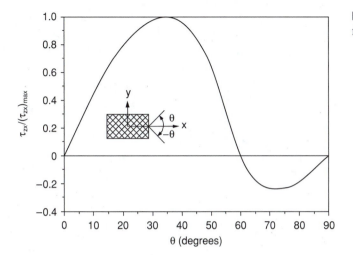

Fig. 9.30 Interlaminar shear stress as a function of fiber orientation.[36]

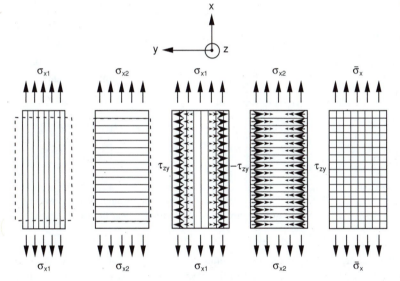

Fig. 9.31 Illustration of generation of interlaminar stresses in crossply laminate under axial tension.

9.15.4 **Effects of Stacking Sequence**

In the case of a general multidirectional laminate, all three types of interlaminar stresses are generated near free edges, that is, σ_z, τ_{zx}, and τ_{zy}. In all cases the magnitude and distribution of interlaminar stresses depend greatly on the stacking sequence of the laminate. The effect of stacking sequence on the interlaminar normal stress σ_z is illustrated in Fig. 9.33 for a laminate consisting of $\pm 15°$ and $\pm 45°$ layers under axial tension.[38] The distribution of stress σ_z through the thickness is given for three stacking sequences, $[\pm 15/\pm 45]_s$, $[15/\pm 45/-15]_s$, and $[\pm 45/\pm 15]_s$. It is seen that both the magnitude and sign of the stress can change drastically with stacking sequence. It is obvious that, for design considerations, stacking sequences resulting in low tensile or compressive σ_z stresses should be selected.

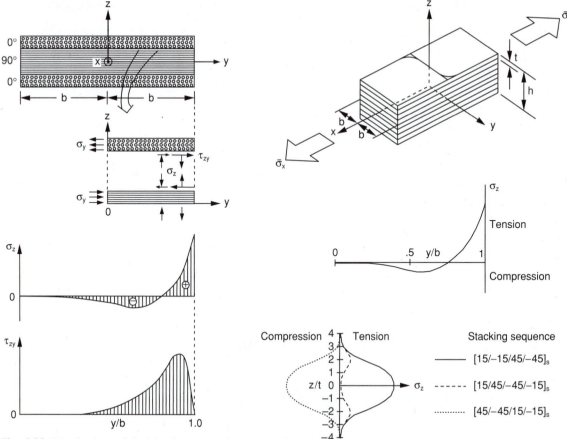

Fig. 9.32 Distribution of interlaminar normal stress σ_z and interlaminar shear stress τ_{zy} in $[0/90]_s$ laminate under axial tension.

Fig. 9.33 Effect of stacking sequence on through-the-thickness distribution of interlaminar normal stress σ_z near the free edge.[38]

The effect of stacking sequence on interlaminar edge stresses, for example, σ_z, and thereby on strength is dramatically illustrated in the case of laminates with circular holes. In this case, edge effects are accentuated by the stress concentration on the edge of the hole. Two boron/epoxy panels of $[0_2/\pm45/\overline{0}]_s$ and $[\pm45/0_2/\overline{0}]_s$ stacking sequences with circular holes were loaded in axial tension to failure.[39] These two stacking sequences result, respectively, in tensile and compressive interlaminar normal stresses near the edge of the hole at the point of maximum stress concentration. Figure 9.34 shows fringe patterns in a photoelastic coating around the hole near failure. The pattern for the $[0_2/\pm45/\overline{0}]_s$ specimen is fairly symmetric with lower stress concentration; the pattern for the $[\pm45/0_2/\overline{0}]_s$ specimen is skewed with higher stress concentration. The failure modes of the two specimens were dramatically different (Fig. 9.35). The $[\pm45/0_2/\overline{0}]_s$ specimen failed horizontally in a catastrophic manner at an average applied axial stress of 426 MPa (61.7 ksi). The $[0_2/\pm45/\overline{0}]_s$ specimen failed by vertical cracking in a noncatastrophic manner at an applied axial stress of 527 MPa (76.4 ksi). The specimen then split into two strips that carried a

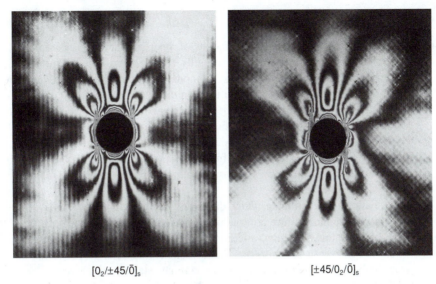

$[0_2/\pm45/\bar{0}]_s$ $[\pm45/0_2/\bar{0}]_s$

Fig. 9.34 Isochromatic fringe patterns in photoelastic coating around hole in boron/epoxy specimens of two different stacking sequences ($\bar{\sigma}_x$ = 392 MPa [56.8 ksi]).[39]

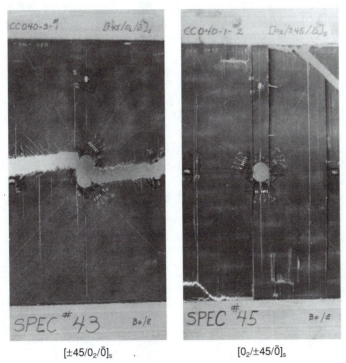

$[\pm45/0_2/\bar{0}]_s$. $[0_2/\pm45/\bar{0}]_s$

Fig. 9.35 Failure patterns of boron/epoxy tensile panels with holes of two different stacking sequences.[39]

much higher ultimate stress of 725 MPa (105 ksi). Thus, stacking sequence can influence, through interlaminar edge effects, the strength and mode of failure. In summary, there are three types of interlaminar stress problems associated with three types of laminates:

1. $[\pm\theta]$ angle-ply laminates exhibit effects of shear coupling mismatch, and thus only τ_{zx} interlaminar shear stresses are generated.
2. $[0/90]_s$ crossply laminates exhibit effects of Poisson's ratio mismatch, and thus only interlaminar shear stresses, τ_{zy}, and interlaminar normal stresses, σ_z, are generated.
3. General multidirectional laminates, combining angle-ply and crossply sublaminates, exhibit effects of both shear coupling and Poisson's ratio mismatch. Thus, all three types of interlaminar stresses, σ_z, τ_{zx}, and τ_{zy}, can be generated.

9.15.5 Interlaminar Strength

The analysis and exact determination of interlaminar stresses, which is essential for interlaminar strength evaluation, is highly complex and requires numerical methods that are beyond the scope of this book. Computer programs have been developed that aim at providing an engineering solution to this problem.[40] In practice the effect of interlaminar stresses can be controlled, since they are confined to a narrow zone near free edges or free hole boundaries. This is usually done by means of edge fastening or stitching that constrains the effect of interlaminar tensile (peel) stresses and prevents delamination propagation.[41] As long as the laminate is free of severe delaminations (which can be revealed by nondestructive evaluation), the basic assumptions of Section 7.1, including that of plane stress, are valid.

Failure of a laminate under the action of interlaminar stresses cannot be analyzed easily. The lamina failure criteria discussed in Chapter 6 are not applicable. In addition to the basic lamina strengths, interlaminar shear and interlaminar tensile strengths must be determined. These strengths may not be constant material properties but may also depend on the layer (fiber) orientation and the laminate stacking sequence.

Interlaminar strength is a matrix-dominated property, and thus it depends on factors such as moisture and temperature, which affect matrix and interfacial performance. Several test methods are available for determination of interlaminar strength, as will be discussed in Chapter 10. Results from such tests cannot be treated as design allowables, but may be used for comparative parametric investigations or for qualitative evaluation of interlaminar performance. Simplified tests exist that provide a good measure of interlaminar quality and serve as a means of quality control of the fabrication process.

SAMPLE PROBLEM 9.3
Interlaminar Shear Stresses Under Flexure

Consider a cantilever beam of unit width made of a multidirectional laminate and loaded by a concentrated force P at the free end (Fig. 9.36). The beam is subjected only to a bending moment M_x and a transverse shear force V_z such that

$$M_x = Px$$

$$V_z = P \tag{9.43}$$

All other forces and moments are zero.

The axial strains at any point of section *a-a* at a distance z from the reference plane are

$$\begin{bmatrix} \varepsilon_x \\ \varepsilon_y \\ \gamma_s \end{bmatrix}_a = z \begin{bmatrix} \kappa_x \\ \kappa_y \\ \kappa_s \end{bmatrix}_a \tag{9.44}$$

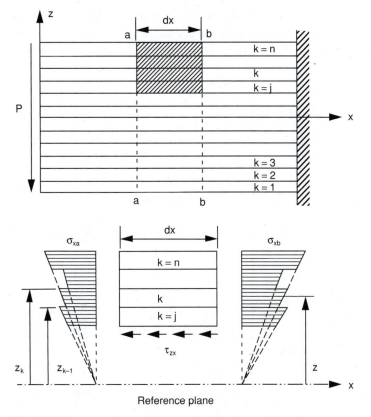

Fig. 9.36 Stresses acting on element of composite cantilever beam.

since the reference plane strains are zero. Referring to Eq. (7.25) and noting that $N_x = N_y = N_s = 0$ and $M_y = M_s = 0$, we can rewrite Eq. (9.44) as

$$
\begin{bmatrix} \varepsilon_x \\ \varepsilon_y \\ \gamma_s \end{bmatrix}_a = z \begin{bmatrix} d_{xx} \\ d_{yx} \\ d_{sx} \end{bmatrix} M_{xa}
\tag{9.45}
$$

where d_{ij} are the laminate compliances and M_{xa} the moment at section a-a.

Using the constitutive relations for layer k referred to the x-y coordinate system, we obtain the layer axial stress at section a-a and at location z as

$$
\sigma_{xa} = Q_{xx}^k \varepsilon_{xa} + Q_{xy}^k \varepsilon_{ya} + Q_{xs}^k \gamma_{sa} = zM_{xa}[Q_{xx}^k d_{xx} + Q_{xy}^k d_{yx} + Q_{xs}^k d_{sx}] = zM_{xa}[Qd]_k \tag{9.46}
$$

Similarly, we obtain the axial stress in layer k at section b-b and at location z as

$$
\sigma_{xb} = zM_{xb}[Qd]_k \tag{9.47}
$$

The crosshatched element of the beam in Fig. 9.36 is in equilibrium under the action of the axial normal stresses σ_{xa} and σ_{xb} and the interlaminar shear stress τ_{zx}, acting between layers $j-1$ and j. Equilibrium of forces in the x-direction,

$$N_x = 0$$

yields

$$\tau_{zx}dx = dM \sum_{k=j}^{n} [Q\,d]_k \int_{z_{k-1}}^{z_k} z\,dz \tag{9.48}$$

from which we obtain

$$\tau_{zx} = \frac{1}{2}\frac{dM}{dx} \sum_{k=j}^{n} [Q\,d]_k (z_k^2 - z_{k-1}^2) \tag{9.49}$$

or

$$\tau_{zx} = V_z \sum_{k=j}^{n} [Q\,d]_k \bar{z}_k t_k \tag{9.50}$$

where

$$V_z = \frac{dM}{dx} \qquad \text{(shear force)}$$

$\bar{z}_k =$ coordinate to center of layer k

$t_k =$ thickness of layer k

9.16 INTERLAMINAR FRACTURE TOUGHNESS

Interlaminar cracking or delamination can occur under three basic modes, opening or peel mode (mode I), forward sliding shear mode (mode II), or tearing mode (mode III), or under combinations thereof (Fig. 9.37). The resistance to delamination growth is expressed in terms of the interlaminar fracture toughness, which has three forms corresponding to the three basic delamination modes. The interlaminar fracture toughness is measured by the strain energy release rate (G_I, G_{II}, or G_{III}), which is the energy dissipated per unit area of delamination growth.

Consider a composite beam delaminated along its midplane over a length, a, and loaded at the ends by loads P as shown in Fig. 9.38. The total energy balance is expressed as[42]

$$W = U + T + D \tag{9.51}$$

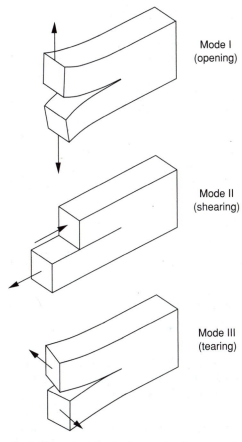

Mode I
(opening)

Mode II
(shearing)

Mode III
(tearing)

Fig. 9.37 Basic delamination modes in composite material.

where

W = external work

U = elastic strain energy

T = kinetic energy

D = dissipative energy associated with fracture

The energy released per unit area of crack extension is expressed as

$$G_I = \frac{1}{b}\frac{dD}{da} = \frac{1}{b}\left[\frac{dW}{da} - \frac{dU}{da} - \frac{dT}{da}\right] \qquad (9.52)$$

where b is the width of the beam. Noting that

$$\frac{dW}{da} = P\frac{d\delta}{da} \qquad (9.53)$$

and that, for linear elastic behavior,

$$U = \frac{1}{2}P\delta \qquad (9.54)$$

we can rewrite Eq. (9.52) as

$$G_I = \frac{1}{b}\frac{dD}{da} = \frac{1}{2b}\left[P\frac{d\delta}{da} - \delta\frac{dP}{da} - 2\frac{dT}{da}\right] \qquad (9.55)$$

Methods of analysis and applications of the double cantilever beam (DCB) specimen for determination of interlaminar fracture toughness have been discussed in the literature.[41,43–45] In the compliance method, the strain energy release rate is expressed in terms of the compliance

$$C = \frac{\delta}{P} \qquad (9.56)$$

Substituting $\delta = PC$ in Eq. (9.55) and neglecting kinetic energy, we obtain

$$G_I = \frac{P^2}{2b}\frac{dC}{da} \qquad (9.57)$$

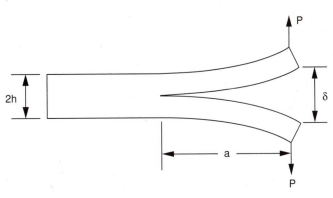

2h

δ

a

P

P

Fig. 9.38 Double cantilever beam (DCB) for measurement of mode I delamination fracture toughness.

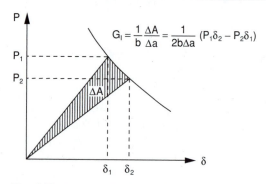

$$G_I = \frac{1}{b}\frac{\Delta A}{\Delta a} = \frac{1}{2b\Delta a}(P_1\delta_2 - P_2\delta_1)$$

Fig. 9.39 Area method for calculation of strain energy release rate using a double cantilever beam specimen.

The compliance C is calculated by considering the DCB specimen as two cantilever beams of length a joined at the crack tip. Thus, for a unidirectional composite DCB specimen with the fiber direction along the longitudinal axis,

$$C = \frac{\delta}{P} = \frac{24}{E_1 b}\left[\frac{1}{3}\left(\frac{a}{h}\right)^3 + \frac{1}{10}\left(\frac{E_1}{G_{31}}\right)\left(\frac{a}{h}\right)\right] \quad (9.58)$$

and

$$G_I = \frac{12P^2}{E_1 b^2 h}\left[\left(\frac{a}{h}\right)^2 + \frac{1}{10}\left(\frac{E_1}{G_{31}}\right)\right] \quad (9.59)$$

Another commonly used method of analysis is the so-called *area method*. In this approach the specimen is not modeled as a beam. The energy released per unit area of crack extension is simply calculated as

$$G_I = \frac{1}{2b\Delta a}(P_1\delta_2 - P_2\delta_1) \quad (9.60)$$

where load P_1 corresponding to opening deflection δ_1 drops to load P_2 corresponding to deflection δ_2 after a finite increment Δa in crack length (Fig. 9.39). The quantity $\frac{1}{2}(P_1\delta_2 - P_1\delta_1)$ in Eq. (9.60) is equal to the shaded area ΔA in Fig. 9.39. It should be noted that the relation above is valid only for linear load-deformation response. Variations of the DCB specimen above and other experimental and analytical procedures for determination of strain energy release rates G_I, G_{II} and G_{III} and combinations thereof are discussed in Chapter 10.

9.17 DESIGN METHODOLOGY FOR STRUCTURAL COMPOSITE MATERIALS

The design of composite structures is an integrated process involving material selection, process specification, optimization of laminate configuration, and design of the structural components. Design objectives vary according to the structural application. Specific application requirements define one or a combination of two or more of the following basic design objectives:

1. design for stiffness
2. design for strength (static and fatigue)
3. design for dynamic stability
4. design for environmental stability
5. design for damage tolerance

The design process is guided by certain considerations and optimization criteria. A major consideration in aerospace applications is weight savings. In commercial applications, such as the automotive and sports industries, an additional consideration is cost competitiveness with more conventional materials and processes. For critical applications of all types, in addition to the above requirements, the need is added for assurance of quality, durability, and reliability over the lifetime of the structure. The various design objectives, structural and material requirements, and typical materials and applications are summarized in Table 9.5.

Applications such as aircraft control surfaces, underground and underwater vessels, thin skins in compression, and sport products such as bicycles and tennis rackets require small deflections under working loads, high buckling loads, and low weight. The design objective in these cases is high stiffness and low weight, that is, high specific stiffness. This requires selection of high-stiffness fibers in general, such as boron, carbon, graphite, and aramid. For high flexural stiffness, a sandwich construction with composite skins and low-density aluminum honeycomb or foam cores, or hybrid laminates with high-stiffness outer layers and low-stiffness (and low-cost) inner layers, are recommended.

Pressure vessels, truss members in tension, thin composite skins in sandwich panels, ribs, and joints require high load-carrying capacity (static and dynamic) combined with low weight. A design for strength is indicated. This requires the selection of high-strength fibers, such as carbon, aramid (in tension only), and S-glass. The optimum laminate is one with a high efficiency ratio, φ_L, and a high degree of fiber strength utilization.[46] The first one, as defined in Eq. (9.40) is the ratio of the FPF to the ULF strength and depends primarily on the material system. The second one is measured by the ratio of the longitudinal stress reached in the lamina at FPF to its longitudinal strength and depends primarily on the laminate layup for a given loading. Both of these are related to the principal modulus and principal strength ratios of the lamina,[46]

$$\rho_E = \frac{E_1}{E_2} \tag{9.61}$$

$$\rho_F = \frac{F_1}{F_2} \tag{9.62}$$

and their ratio, the lamina stiffness to strength ratio,

$$\rho_{EF} = \frac{\rho_E}{\rho_F} = \frac{E_1/E_2}{F_1/F_2} \tag{9.63}$$

The ideal laminate, optimized for strength and minimum weight, would be one with high fiber strength utilization and with all its layers failing simultaneously in their fiber direction.

In addition to in-plane stresses, interlaminar stresses must be taken into account. Thus, in addition to meeting the requirements for in-plane loading, selected laminates must maintain high interlaminar normal and shear strengths and high interlaminar fracture toughness. These properties are primarily dominated by the matrix characteristics.

Rotating structural components such as turbine blades, rotor blades, and flywheels, as well as components subjected to vibration and flutter, must have long fatigue life, low mass, high stiffness, high damping, and high resonance frequency and allow for better vibration control. A design for dynamic control and stability is required. This is achieved

TABLE 9.5 Design Methodology for Structural Composite Materials

Design Objective	Structural Requirements	Material Requirements	Typical Materials	Typical Applications
Design for stiffness	Small deflections High buckling loads Low weight	High-stiffness fibers in sandwich or hybrid laminates for high flexural rigidity	Carbon, graphite, boron, and Kevlar fiber composites	Aircraft control surfaces Underground, underwater vessels Thin skins in compression Sporting goods Marine structures
Design for strength	High load capacity (static, dynamic) Low weight	High lamina strength with high degree of fiber utilization High stiffness to strength ratio (ρ_{EF}) High interlaminar strength	Carbon, Kevlar (in tension), and S-glass fiber composites	Pressure vessels Trusses (tension members) Thin skins in sandwich panels, ribs, joints Marine structures
Design for dynamic control and stability	Long fatigue life High resonance frequency Vibration control Low centrifugal forces	High-strength fibers Fibers with high specific stiffness (E/ρ) Ductile matrices or hybrids with high-damping layers	Carbon, graphite fibers Thermoplastic matrices Interleaving with thermoplastic layers	Engine components Aircraft components Helicopter rotor blades Flywheels
Design for environmental stability and durability	High dimensional stability under extreme environmental fluctuations High corrosion resistance	Low coefficients of thermal and moisture expansion Laminate design for hygrothermal isotropy High-stiffness anisotropic fibers	Carbon, graphite fiber composites Fabric composite coatings	Radar and space antennae Space mirrors, telescopes Solar reflectors Marine structures
Design for damage tolerance	High impact resistance High compressive strength after impact damage Resistance to damage growth	High fracture toughness (intra- and interlaminar) Energy-absorbent interlayers Textile laminates	Kevlar fiber composites Tough epoxy matrices, thermoplastic matrices Interleaving	Ballistic armor Bulletproof vests, helmets Impact-resistant structures Rotor blades (helicopters, wind tunnels, wind turbines)

by using high-strength fibers (carbon), fibers with high specific modulus (E/ρ), tough matrices, and hybrids with soft and high-damping interlayers.

Space structures (radar and space antennae, telescopes, mirrors, and solar reflectors) require high dimensional stability under extreme environmental fluctuations. A design for environmental stability is indicated, requiring high-modulus thermally anisotropic fibers, such as carbon and graphite (GY-70) in laminates designed for thermoelastic and hygroelastic isotropic behavior (see Section 8.11). Graphite fabric composite coatings are recommended for protection of structures in corrosive and erosive environments.

Impact-resistant structures such as helmets, ballistic armor, and bulletproof vests require materials and laminates with high impact resistance and high compressive strength after impact. A design for damage tolerance is recommended. Aramid (Kevlar) fibers and tough matrices such as thermoplastics and laminates with selective interleaving can provide the necessary impact and damage propagation resistance.[47]

Many structures have more than one requirement. For example, marine structures in the form of composite sandwich panels must have high flexural stiffness, high impact strength, must endure exposure to seawater, and they must be affordable. E-glass/vinylester or carbon/vinylester facesheet and balsa wood core materials meet these design requirements.

9.18 ILLUSTRATION OF DESIGN PROCESS: DESIGN OF A PRESSURE VESSEL

A thin-wall cylindrical pressure vessel is loaded by internal pressure, p, and an external torque, T, as shown in Fig. 9.40. It is also given that the vessel operates at room temperature and dry conditions and that curing residual stresses can be neglected. It is required to find the optimum composite material system and layup to achieve minimum weight and to compare it with an aluminum reference vessel. The allowable safety factor is $S_{all} = 2.0$. The design of the aluminum vessel is based on the von Mises criterion with a material yield strength $\sigma_{yp} = 242$ MPa (35 ksi). The density of aluminum is given as $\rho = 2.8$ g/cm³ (0.101 lb/in³). The design of the composite laminate is based on the Tsai-Wu failure criterion for FPF. Balanced symmetric laminates are to be investigated for three candidate composite materials, S-glass/epoxy, Kevlar/epoxy, and carbon/epoxy.

The unit loads acting on an element of the cylindrical shell along the axial and hoop directions (x and y) are obtained as follows:

$$N_x = \bar{\sigma}_x h = \frac{pD}{4}$$

$$N_y = \bar{\sigma}_y h = \frac{pD}{2} \qquad (9.64)$$

$$N_s = \bar{\tau}_s h \cong \frac{2T}{\pi D^2}$$

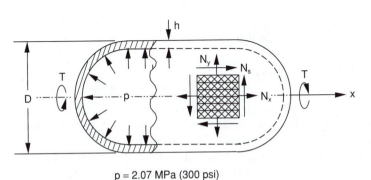

p = 2.07 MPa (300 psi)
T = 283 kN-m (2.5 × 10⁶ lb-in)
D = 89 cm (35 in)

Fig. 9.40 Thin-wall cylindrical pressure vessel under internal pressure and torque loading.

Substituting the data given, we obtain

$$N_x = 460 \text{ kN/m}$$

$$N_y = 920 \text{ kN/m} \tag{9.65}$$

$$N_s = 228 \text{ kN/m}$$

The principal stresses for the above state of stress are

$$\bar{\sigma}_1 = \frac{1014}{h} \quad \text{(in kPa)}$$

$$\bar{\sigma}_2 = \frac{366}{h} \quad \text{(in kPa)} \tag{9.66}$$

$$\bar{\sigma}_3 = 0$$

9.18.1 Aluminum Reference Vessel

According to the von Mises yield criterion,

$$\left[(\bar{\sigma}_1 - \bar{\sigma}_2)^2 + (\bar{\sigma}_2 - \bar{\sigma}_3)^2 + (\bar{\sigma}_3 - \bar{\sigma}_1)^2 \right]^{1/2} = \frac{\sqrt{2}\sigma_{yp}}{S_{all}} \tag{9.67}$$

Substituting the numerical results of Eq. (9.66) and the given data in Eq. (9.67), we obtain

$$\frac{1257}{h_a} = 170{,}766 \text{ kPa}$$

which yields

$$h_a = 7.36 \text{ mm } (0.290 \text{ in})$$

9.18.2 Crossply $[0_m/90_n]_s$ Laminates

Since the ratio of hoop stress to axial stress is 2:1, a similar ratio of the number of 90° and 0° layers, or n:m, is selected initially. The process of optimization for a given type of layup is best carried out by using one of several available computer programs.[34,48]

Initially the safety factor S_f is obtained for a $[0/90_2]_s$ laminate of the material investigated, the thickness of which is $h_o = 6t$, that is, six ply thicknesses. The multiples m_i and n_i for the initial trial are obtained as

$$m_i = \frac{n_i}{2} \cong \frac{S_{all}}{S_f} = \frac{2}{S_f} \tag{9.68}$$

TABLE 9.6 Optimum $[0_m/90_n]_s$ Layup for Three Composite Materials

	S-Glass/Epoxy	Kevlar/Epoxy	Carbon/Epoxy
Ply thickness (t, mm)	0.165	0.127	0.127
m	10	12	10
n	28	29	22
Safety factor, S_f	2.017	2.029	2.043
Optimum layup*	$[0_{10}/90_{28}]_s$	$[0_{12}/90_{29}]_s$	$[0_{10}/90_{22}]_s$
Laminate thickness (h, mm)	12.54	10.41	8.13

*To reduce interlaminar stresses, it is recommended to intersperse the plies and minimize layer thicknesses, as discussed in Section 7.11.

and the allowable laminate thickness is $h_a = 6mt = mh_o$. The optimum choice from the point of view of weight is reached by trying different values of m and n around the initial guess until the sum $(m + n)$ is minimized. Results for the three materials investigated are tabulated in Table 9.6.

9.18.3 Angle-Ply $[\pm\theta]_{ns}$ Laminates

Optimization of this type of laminate involves only one variable, θ. This is accomplished by selecting the angle θ for the basic laminate unit $[\pm\theta]_s$ that maximizes the safety factor S_f. The basic laminate unit has a thickness $h_o = 4t$, that is, four ply thicknesses. Then the allowable laminate thickness is

$$h_a = \frac{S_{all}}{S_f}h_o = \frac{8t}{S_f} \tag{9.69}$$

To find the optimum θ, the allowable (required) thickness h_a was computed and plotted versus θ for the three materials considered in Fig. 9.41. It is interesting to note that the optimum angle θ corresponding to the minimum allowable thickness is almost the same for all three materials, 55° for S-glass/epoxy and carbon/epoxy and 54° for Kevlar/epoxy. Although the three materials considered have comparable strength properties, the variation of the required minimum laminate thickness with angle θ is very different for each material (Fig. 9.41). The curve for Kevlar/epoxy shows the sharpest variation with angle, but it has roughly the same minimum ($h_a = 4.24$ mm) as that of carbon/epoxy ($h_a = 4.14$ mm). Although the strength properties of S-glass/epoxy are

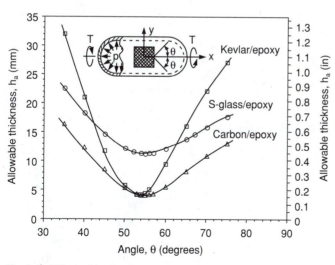

Fig. 9.41 Effect of lamination angle on allowable thickness of $[\pm\theta]_{ns}$ angle-ply laminate in pressure vessel.

higher than those of Kevlar/epoxy, the required minimum thickness for S-glass/epoxy is much higher ($h_a = 11.37$ mm). The results above are tabulated in Table 9.7.

The results obtained illustrate the important fact that the structural efficiency of a laminate is not only a function of the lamina strength properties but also of its lamina stiffnesses and their ratios (degree of anisotropy) as discussed before in Eqs. (9.61) to (9.63).[46]

9.18.4 $[90/\pm\theta]_{ns}$ Laminates

Optimization of the $[90/\pm\theta]_{ns}$ laminate again involves only one variable, θ. Safety factors are computed for the basic laminate unit $[90/\pm\theta]_s$ for the three materials investigated for various values of θ. The minimum allowable thickness for each laminate is obtained as

$$h_a = \frac{S_{all}}{S_f}h_o = \frac{12t}{S_f} \qquad (9.70)$$

Results are tabulated in Table 9.8. The optimum angle θ was found to be 48°, 45°, and 45° for the S-glass/epoxy, Kevlar/epoxy, and carbon/epoxy materials, respectively. Again, as in the previous case of the $[\pm\theta]_{ns}$ laminates, the Kevlar/epoxy material appears much better than the S-glass/epoxy because of its higher laminate efficiency and fiber utilization ratio.

TABLE 9.7 Optimum $[\pm\theta]_{ns}$ Layup for Three Composite Materials

	S-Glass/Epoxy	Kevlar/Epoxy	Carbon/Epoxy
Ply thickness (t, mm)	0.165	0.127	0.127
Optimum θ (degrees)	55	54	55
Safety factor (S_f, $n = 1$)	0.116	0.240	0.246
Minimum allowable thickness (h_a, mm)	11.37	4.24	4.14
Optimum layup	$[\pm55]_{18s}$	$[\pm54]_{9s}$	$[\pm55]_{9s}$
Safety factor (S_f, optimum layup)	2.091	2.125	2.209
Laminate thickness (h, mm, optimum layup)	11.89	4.57	4.57

TABLE 9.8 Optimum $[90/\pm\theta]_{ns}$ Layup for Three Composite Materials

	S-Glass/Epoxy	Kevlar/Epoxy	Carbon/Epoxy
Ply thickness (t, mm)	0.165	0.127	0.127
Optimum θ (degrees)	48	45	45
Safety factor (S_f, $n = 1$)	0.155	0.240	0.335
Minimum allowable thickness (h_a, mm)	12.76	6.35	4.55
Optimum layup	$[90/\pm48]_{13s}$	$[90/\pm45]_{9s}$	$[90/\pm45]_{6s}$
Safety factor (S_f, optimum layup)	2.018	2.159	2.012
Laminate thickness (h, mm, optimum layup)	12.87	6.86	4.57

9.18.5 $[0/\pm\theta]_{ns}$ Laminates

Optimization of the $[0/\pm\theta]_{ns}$ laminate is similar to the previous one. Safety factors and minimum allowable laminate thicknesses are calculated as before in Eq. (9.70). Results are tabulated in Table 9.9. The optimum angle θ was found to be 75°, 67°, and 67° for the S-glass/epoxy, Kevlar/epoxy, and carbon/epoxy materials, respectively. As in the previous case the required laminate thickness for the S-glass/epoxy material was approximately double that for the other two materials.

9.18.6 Quasi-isotropic $[0/\pm45/90]_{ns}$ Laminates

Quasi-isotropic $[0/\pm45/90]_{ns}$ laminates are investigated for reference purposes. Safety factors are calculated for the basic unit ($n = 1$) and allowable thicknesses computed as before,

$$h_a = \frac{S_{all}}{S_f}h_o = \frac{16t}{S_f} \tag{9.71}$$

Results are tabulated in Table 9.10. As can be seen, this is the least efficient layup for all three materials.

TABLE 9.9 Optimum $[0/\pm\theta]_{ns}$ Layup for Three Composite Materials

	S-Glass/Epoxy	Kevlar/Epoxy	Carbon/Epoxy
Ply thickness (t, mm)	0.165	0.127	0.127
Optimum θ (degrees)	75	67	67
Safety factor (S_f, $n = 1$)	0.157	0.254	0.293
Minimum allowable thickness (h_a, mm)	12.61	6.01	5.21
Optimum layup	$[0/\pm75]_{13s}$	$[0/\pm67]_{8s}$	$[0/\pm67]_{7s}$
Safety factor (S_f, optimum layup)	2.041	2.029	2.047
Laminate thickness (h, mm, optimum layup)	12.87	6.10	5.33

TABLE 9.10 Optimum $[0/\pm45/90]_{ns}$ Layup for Three Composite Materials

	S-Glass/Epoxy	Kevlar/Epoxy	Carbon/Epoxy
Safety factor (S_f, $n = 1$)	0.181	0.257	0.361
Minimum allowable thickness (h_a, mm)	14.59	7.91	5.63
Minimum n	12	8	6
Optimum layup	$[0/\pm45/90]_{12s}$	$[0/\pm45/90]_{8s}$	$[0/\pm45/90]_{6s}$
Safety factor (S_f, optimum layup)	2.173	2.055	2.169
Laminate thickness (h, mm, optimum layup)	15.84	8.13	6.10

TABLE 9.11 Summary of Optimum Layups for Three Composite Materials

	S-Glass/Epoxy	Kevlar/Epoxy	Carbon/Epoxy
Density (ρ, g/cm^3)	2.0	1.4	1.6
Ply thickness (t, mm)	0.165	0.127	0.127
Optimum layup	$[\pm55]_{18s}$	$[\pm54]_{9s}$	$[\pm55]_{9s}$
Safety factor (S_f, optimum layup)	2.091	2.125	2.209
Laminate thickness (h, mm, optimum layup)	11.89	4.57	4.57
Weight savings compared to aluminum ($\Delta W/W$, %)	−15.4	69.0	64.5

9.18.7 Summary and Comparison of Results

Results of the optimum layups for the three composite materials considered and the relative weight savings compared with an aluminum pressure vessel are summarized in Table 9.11. The optimum layup for all three materials is the angle-ply $[\pm\theta]_{ns}$ layup. Given the fixed ply thicknesses for the materials, total laminate thicknesses were obtained that resulted in safety factors slightly higher than the allowable one ($S_{all} = 2.0$). Both, the Kevlar/epoxy and carbon/epoxy materials resulted in the same laminate thickness, which is less than half of the required one for the S-glass/epoxy material. The relative weight savings compared with the aluminum reference pressure vessel were calculated as follows:

$$\frac{\Delta W}{W} = \frac{W_{al} - W_{comp}}{W_{al}}$$

or

$$\frac{\Delta W}{W} = 1 - \frac{\rho_{comp} h_{comp}}{\rho_{al} h_{al}} \tag{9.72}$$

As shown in Table 9.11 there are weight savings of 69% and 64.5% in the Kevlar/epoxy and carbon/epoxy designs, but a weight increase of 15.4% in the S-glass/epoxy design.

9.19 RANKING OF COMPOSITE LAMINATES

The results above are summarized in a bar graph in Fig. 9.42. Here the different composite laminate options are ranked according to their weight per unit wall area. A clear trend is observed that is common for the three material systems considered, that is, minimum weight for $[\pm\theta]_{ns}$ angle-ply configurations with $\theta \cong 55°$ and significantly higher weights for crossply $[0_m/90_n]_s$ and quasi-isotropic $[0/\pm45/90]_{ns}$ layups.

The design procedure illustrated before can be very time consuming if all potential layups are examined for each material system considered. Based on this illustration and prior experience, a shortened ranking procedure is recommended:

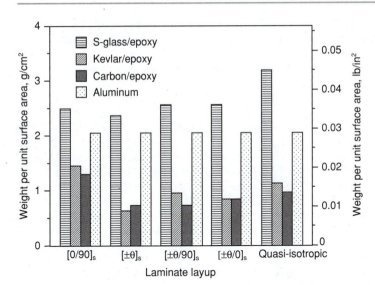

Fig. 9.42 Ranking of different material systems and laminate layups according to weight for pressure vessel design example.

Step 1 Select a material system and determine the best layup for this system to achieve minimum weight.

Step 2 Compare different material systems for this layup.

Step 3 Select the material system giving the lowest weight in step 2 and repeat step 1 for this material system.

It should be pointed out that the above sizing and ranking procedure is based on FPF, which is considered a conservative approach for many applications. A less conservative approach may be based on ULF, using higher allowable safety factors. In this approach, the carbon/epoxy system with its high ultimate strength would rank much more favorably than the other candidate material systems. This is related to the lower structural efficiency ratio, φ_L, of this carbon/epoxy, which means that its fiber strength is not utilized efficiently at FPF (see Tables 9.2 and 9.3). At ULF, which in many cases is related to fiber failure, the carbon/epoxy laminate is significantly stronger.

REFERENCES

1. K. L. Reifsnider and J. E. Masters, "An Investigation of Cumulative Damage Development in Quasi-Isotropic Graphite/Epoxy Laminates," in *Damage in Composite Materials*, ASTM STP 775, K. L. Reifsnider, Ed., American Society for Testing and Materials, West Conshohocken, PA, 1982, pp. 40–62.

2. A. L. Highsmith and K. L. Reifsnider, "Stiffness Reduction Mechanisms in Composite Laminates," in *Damage in Composite Materials*, ASTM STP 775, K. L. Reifsnider, Ed., American Society for Testing and Materials, West Conshohocken, PA, 1982, pp. 103–117.

3. A. S. D. Wang, P. C. Chou, and S. C. Lei, "A Stochastic Model for the Growth of Matrix Cracks in Composite Laminate," *Journal of Composite Materials*, Vol. 18, 1984, pp. 239–254.

4. H. Fukunaga, T. W. Chou, P. W. M. Peters, and K. Schulte, "Probabilistic Failure Strength Analysis of Graphite/Epoxy Cross-Ply Laminates," *Journal of Composite Materials*, Vol. 18, 1984, pp. 339–356.

5. R. Talreja, "Transverse Cracking and Stiffness Reduction in Composite Laminates," *J. Composite Materials*, Vol. 19, 1985, pp. 355–375.

6. Z. Hashin, "Analysis of Cracked Laminates: A Variational Approach," *Mech. Materials*, Vol. 4, 1985, pp. 121–136.

7. S. L. Ogin, P. A. Smith, and P. W. R. Beaumont, "Matrix Cracking and Stiffness Reduction During the Fatigue of [0/90] GFRP Laminates," *Composites Sci. Technol.*, Vol. 22, 1985, pp. 23–31.

8. I. M. Daniel, J. W. Lee, and G. Yaniv, "Damage Mechanisms and Stiffness Degradation in Graphite/Epoxy Composite," in *Proceedings of ICCM-6 and ECCM-2*, 1987, Vol. 4, pp. 129–138.

9. N. Laws and G. J. Dvorak, "Progressive Transverse Cracking in Composite Laminates," *J. Composite Materials*, Vol. 22, 1988, pp. 900–916.

10. Y. M. Han, H. T. Hahn, and R. B. Croman, "A Simplified Analysis of Transverse Ply Cracking in Cross-Ply Laminate," *Composites Science and Technology*, Vol. 31, No. 3, 1988, pp. 165–177.

11. J. A. Nairn, "The Strain Energy Release Rate of Composite Microcracking: A Variational Approach," *Journal of Composite Materials*, Vol. 23, 1989, pp. 1106–1129.

12. S. G. Lim and C. S. Hong, "Prediction of Transverse Cracking and Stiffness Reduction in Cross-Ply Laminated Composites," *J. Composite Materials*, Vol. 23, 1989, pp. 695–713.

13. I. M. Daniel and J.-W. Lee, "Damage Development in Composite Laminates Under Monotonic Loading," *Journal of Composites Technology & Research*, JCTR, 1990, Vol. 12, No. 2, pp. 98–102.

14. J.-W. Lee and I. M. Daniel, "Progressive Transverse Cracking of Crossply Composite Laminates," *J. Composite Materials*, Vol. 24, 1990, pp. 1225–1243.

15. J. M. Berthelot and J. F. LeCorre, "Statistical Analysis of the Progression of Transverse Cracking and Delamination in Cross-Ply Laminates," *Composites Science and Technology*, Vol. 60, No. 14, 2000, pp. 2659–2669.

16. S. Ogihara, A. Kobayashi, N. Takeda, and S. Kobayashi, "Damage Mechanics Analysis of Transverse Cracking Behavior in Composite Laminates," *International Journal of Damage Mechanics*, Vol. 9. No. 2, 2000, pp. 113–129.

17. Z. Sun, I. M. Daniel, and J.-J. Luo, "Statistical Damage Analysis of Transverse Cracking in High Temperature Composite Laminates," *Materials Science and Engineering A*, Vol. 341, 2003, pp. 49–56.

18. C.-L. Tsai, I. M. Daniel, and J.-W. Lee, "Progressive Matrix Cracking of Crossply Composite Laminates Under Biaxial Loading," in *Microcracking-Induced Damage in Composites*, G. J. Dvorak and D. C. Lagoudas, Eds., Proc. of ASME 1990 Winter Annual Meeting, AMD-Vol. 111, MD-Vol. 22, American Soc. of Mechanical Engineers, New York, 1990, pp. 9–18.

19. I. M. Daniel and C.-L. Tsai, "Analytical/Experimental Study of Cracking in Composite Laminates Under Biaxial Loading," *Composites Engineering*, Vol. 1, 1991, pp. 355–362.

20. C.-L. Tsai and I. M. Daniel, "The Behavior of Cracked Crossply Laminates Under Shear Loading," *Int. J. Solids Structures*, Vol. 29, 1992, pp. 3251–3267.

21. J. C. Halpin, "Structure-Property Relations and Reliability Concepts," *J. Composite Materials*, Vol. 6, 1972, pp. 208–231.

22. D. L. Flaggs and M. H. Kural, "Experimental Determination of the In-situ Transverse Lamina Strength in Graphite/Epoxy Laminates," *Journal of Comp. Materials*, Vol. 16, 1982, pp. 103–116.

23. C. T. Sun and B. J. Quinn, "Evaluation of Failure Criteria Using Off-Axis Laminate Specimens," *Proc. American Society for Composites*, Ninth Technical Conf., Sept. 1994, pp. 97–105.

24. C. T. Sun, "Strength Analysis of Unidirectional Composites and Laminates," in *Comprehensive Composite Materials*, A. Kelly and C. Zweben, Eds., Ch. 1.20, Elsevier Science, Ltd., Oxford, 2000.

25. *Composites Science and Technology*, Vol. 58, 1998, pp. 999–1254; Vol. 62, 2002, pp. 1479–1797; Vol. 64, 2004, pp. 309–605.

26. M. J. Hinton, P. D. Soden, and A. S. Kaddour, *Failure Criteria in Fibre-Reinforced-Polymer Composites*, Elsevier, Oxford, 2004.

27. P. D. Soden, M. J. Hinton, and A. S. Kaddour, "Biaxial Test Results for Strength and Deformation of a Range of E-glass and Carbon Fibre Reinforced Composite Laminates: Failure Exercise Benchmark Data," *Composites Science and Technology*, Vol. 62, 2002, pp. 1489–1514.

28. M. J. Hinton, A. S. Kaddour, and P. D. Soden, "A Comparison of the Predictive Capabilities of Current Failure Theories for Composite Laminates, Judged Against Experimental Evidence," *Composites Science and Technology*, Vol. 62, 2002, pp. 1725–1798.

29. M. F. S. Al-Khalil, P. D. Soden, R. Kitching, and M. J. Hinton, "The Effects of Radial Stresses on the Strength of Thin Walled Filament Wound GRP Composite Pressure Cylinders," *Int. J. of Mech. Scie.*, Vol. 38, 1996, pp. 97–120.

30. U. Hütter, H. Schelling, and H. Krauss, "An Experimental Study to Determine Failure Envelope of Composite Materials with Tubular Specimens Under Combined Loads and Comparison with Several Classical Criteria," in *Failure Modes of Composite Materials with Organic Matrices and Their Consequences on Design*, NATO, AGARD-CP-163, 1975, pp. 3-1–3-11.

31. S. R. Swanson and M. Nelson, "Failure Properties of Carbon/Epoxy Laminates Under Tension-Compression Biaxial Stress," *Recent Advances in Japan and United States*, Proc. Japan-US CCM-III, Tokyo, 1986, pp. 279–286.

32. S. R. Swanson and A. P. Christoforou, "Response of Quasi-isotropic Carbon/Epoxy Laminates to Biaxial Stress," *J. Composite Materials*, Vol. 20, 1986, pp. 457–471.

33. S. R. Swanson and B. C. Trask, "Strength of Quasi-isotropic Laminates Under Off-Axis Loading," *Composites Science and Technology*, Vol. 34, 1989, pp. 19–34.

34. J. J. Luo, "Webcomp: Stress and Failure Analysis of Laminate Composites," http:www.composites.northwestern.edu/~webcomp, 2004.

35. E. Altus and O. Ishai, "Transverse Cracking and Delamination Interaction in the Failure Process of Composite Laminates," *J. Comp. Science and Technology*, Vol. 26, 1986, pp. 59–77.

36. R. B. Pipes and N. J. Pagano, "Interlaminar Stresses in Composite Laminates Under Uniform Axial Extension," *J. Composite Materials*, Vol. 4, 1970, pp. 538–548.

37. R. B. Pipes and I. M. Daniel, "Moiré Analysis of the Interlaminar Shear Edge Effect in Laminated Composites," *J. Composite Materials*, Vol. 5, 1971, pp. 255–259.

38. N. J. Pagano and R. B. Pipes, "The Influence of Stacking Sequence on Laminate Strength," *J. Composite Materials*, Vol. 5, 1971, pp. 50–57.

39. I. M. Daniel, R. E. Rowlands, and J. B. Whiteside, "Effects of Material and Stacking Sequence on Behavior of Composite Plates with Holes," *Exp. Mech.*, Vol. 14, 1974, pp. 1–9.

40. N. J. Pagano, "Automated System for Composite Analysis (ASCA)," Software Package, Ad Tech Systems, Inc., Dayton, OH, 1991.

41. O. Ishai, "Interlaminar Fracture Toughness of Selectively Stitched Thick CFRP Fabric Composite Laminates," *Plastics, Rubber and Composites*, Vol. 29, No. 3, 2000, pp. 134–143.

42. K. Hellan, *Introduction to Fracture Mechanics*, McGraw-Hill Book Co., New York, 1984.

43. J. M. Whitney, C. E. Browning, and W. Hoogsteden, "A Double Cantilever Beam Test for Characterizing Mode I Delamination of Composite Materials," *J. Reinforced Plastics and Composites*, Vol. 1, 1982, pp. 297–313.

44. A. A. Aliyu and I. M. Daniel, "Effects of Strain Rate on Delamination Fracture Toughness of Graphite/Epoxy," in *Delamination and Debonding of Materials*, ASTM STP 876, W. S.

Johnson, Ed., American Society for Testing and Materials, West Conshohocken, PA, 1985, pp. 336–348.

45. N. Sela and O. Ishai, "Interlaminar Fracture Toughness and Toughening of Laminated Composite Materials: A Review," *Composites*, Vol. 20, 1989, pp. 423–435.

46. O. Ishai, S. Krishnamachari, and L. J. Broutman, "Structural Design Optimization of Composite Laminates," *J. Reinforced Plastics and Composites*, Vol. 7, 1988, pp. 459–474.

47. O. Ishai and C. Hiel, "Damage Tolerance of Composite Sandwich with Interleaved Foam Core," *J. Comp. Technology and Research*, Vol. 14, No. 3, 1992, pp. 155–168.

48. S. C. Wooh and I. M. Daniel, *ICAN-Interactive Composite Analysis*, Version 1.02, Northwestern University, Evanston, IL, 1992.

PROBLEMS

9.1 A $[0/90]_s$ laminate is subjected to uniaxial loading N_x in the 0° direction. Which of the following four answers is a correct approximation (for a high-stiffness composite) of the ratio of the σ_2 stress in the 90° layer to the σ_1 stress in the 0° layer, before FPF?

(a) $\dfrac{\sigma_2}{\sigma_1} \cong \dfrac{2E_2}{E_1}$

(b) $\dfrac{\sigma_2}{\sigma_1} \cong \dfrac{E_2}{E_1}$

(c) $\dfrac{\sigma_2}{\sigma_1} \cong \dfrac{v_{12}E_2}{E_1}$

(d) $\dfrac{\sigma_2}{\sigma_1} \cong \dfrac{E_2}{v_{12}E_1}$

9.2 Determine the FPF strength of a $[0/90]_s$ laminate under uniaxial tension or compression based on

(a) the maximum stress criterion

(b) the Tsai-Wu criterion (material: AS4/3501-6 carbon/epoxy, Table A.4)

9.3 Determine the axial tensile strength $\bar{F}_{xt}$ at FPF of a $[\pm45]_s$ laminate using the Tsai-Wu criterion. Obtain exact and approximate results for a high-stiffness, high-strength carbon/epoxy composite (AS4/3501-6, Table A.4).

9.4 Determine the axial tensile strength $\bar{F}_{xt}$ at FPF of a $[\pm45]_s$ laminate using the Tsai-Hill criterion for the following materials listed in Table A.4:

(a) E-glass/epoxy

(b) Kevlar/epoxy

(c) carbon/epoxy (AS4/3501-6)

9.5 Determine the axial tensile strength $\bar{F}_{xt}$ of a $[\pm45]_s$ carbon/epoxy laminate (AS4/3501-6, Table A.4) at FPF by the Tsai-Wu, maximum stress, and Hashin-

Rotem criteria. Include effect of residual stresses by assuming $\Delta T = -100$ °C (-180 °F), the difference between stress-free and service temperatures.

9.6 Determine the axial compressive strength $\bar{F}_{xc}$ at FPF of a $[\pm45]_s$ laminate using the Tsai-Hill criterion for the following materials listed in Table A.4:

(a) E-glass/epoxy

(b) Kevlar/epoxy

(c) carbon/epoxy (AS4/3501-6)

9.7 Determine the axial compressive strength $\bar{F}_{xc}$ at FPF of a $[\pm45]_s$ carbon/epoxy (AS4/3501-6, Table A.4) laminate at FPF by the Tsai-Wu, Tsai-Hill, and maximum stress criteria. Include effect of residual stresses by assuming $\Delta T = -100$ °C (-180 °F), the difference between the stress-free and service temperatures.

9.8 Determine the shear strength $\bar{F}_s$ at FPF of a $[\pm45]_s$ laminate using the Tsai-Hill criterion for the following materials listed in Table A.4:

(a) E-glass/epoxy

(b) Kevlar/epoxy

(c) carbon/epoxy (AS4/3501-6)

9.9 Determine the axial tensile and compressive strengths, $\bar{F}_{xt}$ and $\bar{F}_{xc}$, and the shear strength $\bar{F}_s$ at FPF of a $[\pm45]_s$ carbon fabric/epoxy (AGP370-5H/3501-6S, Table A.5) using

(a) the maximum stress criterion

(b) the Tsai-Hill criterion

9.10 Determine the shear strength $\bar{F}_s$ at FPF of a $[\pm45]_s$ laminate in terms of lamina properties using the maximum stress criterion.

9.11 A $[\pm45]_{ns}$ laminate is loaded in pure shear and uniaxial tension as shown in Fig. P9.11. Which ones of the four following statements are wrong?

(a) The tensile strength $\bar{F}_{xt}$ is primarily controlled by the lamina in-plane shear strength F_6.

(b) The laminate positive shear strength $\bar{F}_s^{(+)}$ is equal to the negative shear strength $\bar{F}_s^{(-)}$.

(c) The tensile strength $\bar{F}_{xt}$ is primarily controlled by the lamina transverse tensile strength F_{2t}.

(d) The laminate shear strength is primarily controlled by the fiber strength.

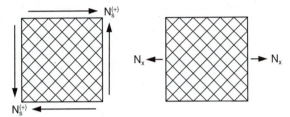

Fig. P9.11

9.12 A $[\pm45]_s$ laminate is loaded under biaxial tension $\bar{\sigma}_x = 2\bar{\sigma}_y$ (Fig. P9.12). Determine the ultimate value of $\bar{\sigma}_x = \bar{F}$ at first ply failure in terms of lamina strengths using the Tsai-Hill criterion and using high-stiffness and high-strength approximations. (Hint: Use results from Sample Problem 9.1 and superposition to determine stresses.)

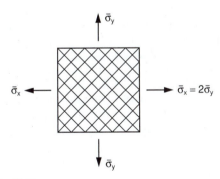

Fig. P9.12

9.13 A $[\pm45]_s$ laminate is loaded biaxially as shown in Fig. P9.13. Determine the magnitude of the biaxial stress $\bar{F}_o = \bar{\sigma}_x^u$ at FPF using both the maximum stress and Tsai-Wu criteria (material: AS4/3501-6 carbon/epoxy, Table A.4). (Hint: Transform the $[\pm45]_s$ into a $[0/90]_s$ layup.)

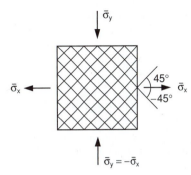

Fig. P9.13

9.14 For the laminate and loading of Problem 9.13, the biaxial stress at FPF based on the maximum stress criterion is $\bar{G}_x^U = \bar{F}_o$ of the following. Select the correct answer from the following list:

(a) $\bar{F}_o = F_6$

(b) $\bar{F}_o = F_{2c}$

(c) $\bar{F}_o = F_{2t}$

(d) $\bar{F}_o = F_{1c}$

9.15 A $[\pm30]_s$ laminate is loaded in uniaxial compression as shown in Fig. P9.15. Determine the compressive strength $\bar{F}_{xc}$ at FPF according to the maximum strain theory for the following given properties:

$$\varepsilon_{1t}^u = 0.015$$
$$\varepsilon_{2t}^u = 0.006$$
$$\varepsilon_{1c}^u = -0.015$$
$$\varepsilon_{2c}^u = -0.024$$
$$\gamma_6^u = 0.015$$
$$\bar{E}_x = 61.4 \text{ GPa (8.9 Msi)}$$
$$\bar{\nu}_{xy} = 1.2$$

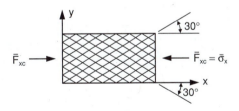

Fig. P9.15

9.16 For the laminate of Problem 9.15, determine the compressive strength $\bar{F}_{xc}$ according to the Tsai-Hill criterion for an E-glass/epoxy material (Table A.4) given the laminate properties $\bar{E}_x = 23.0$ GPa (3.33 Msi) and $\bar{\nu}_{xy} = 0.66$.

9.17 Determine the axial tensile and compressive strengths $\bar{F}_{xt}$ and $\bar{F}_{xc}$ of a $[\pm 20]_s$ carbon/epoxy (AS4/3501-6, Table A.4) by the Tsai-Wu and Hashin-Rotem criteria, including residual stresses for $\Delta T = -100\ °C\ (-180\ °F)$.

9.18 A beam of a symmetric crossply laminate $[0/90]_s$ is loaded under pure bending M_x as shown in Fig. P9.18. Determine the maximum σ_1 and σ_2 stresses in the top ply in terms of the lamina stiffnesses Q_{ij}, thickness t, laminate bending stiffnesses D_{ij}, and the applied moment M_x. (Note: $\varepsilon_x = z\kappa_x$, $\varepsilon_y = z\kappa_y$.)

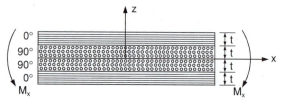

Fig. P9.18

9.19 Using the maximum stress criterion, determine the tensile strength $\bar{F}_{xt}$ at first ply failure of a $[0/90_4]_s$ laminate (x in the 0° direction) for the lamina and laminate properties.

$$E_1 = 185\ \text{GPa}\ (26.8\ \text{Msi})$$
$$E_2 = 10\ \text{GPa}\ (1.45\ \text{Msi})$$
$$\nu_{12} = 0.30$$
$$F_{2t} = 60\ \text{MPa}\ (8.7\ \text{ksi})$$
$$\bar{E}_x = 45\ \text{GPa}\ (6.5\ \text{Msi})$$
$$\bar{\nu}_{xy} = 0.015$$

(Note: Laminate properties are given to reduce the computation.)

9.20 Determine the axial tensile strength $\bar{F}_{xt}$ at FPF of a $[0/\pm 45]_s$ laminate by the Tsai-Hill criterion (Fig. P9.20). Obtain exact and approximate expressions for a high-strength, high-stiffness composite in terms of the lamina properties, the laminate modulus $\bar{E}_x$, and Poisson's ratio $\bar{\nu}_{xy}$.

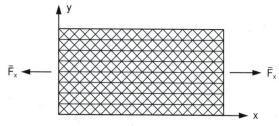

Fig. P9.20

9.21 Determine the strength $\bar{F}_{xt}$ in Problem 9.20 for the carbon/epoxy material AS4/3501-6 listed in Table A.4 using the maximum stress theory. The laminate modulus is $\bar{E}_x = 65.5\ \text{GPa}\ (7.49\ \text{Msi})$ and Poisson's ratio is $\bar{\nu}_{xy} = 0.67$. What is the prevailing failure mode?

9.22 Determine the uniaxial tensile and compressive strengths at FPF of the $[0/90]_s$ laminate of Problem 9.2, taking into account the residual stresses due to cooldown $\Delta T = -150\ °C\ (-270\ °F)$ during curing. Use both maximum stress and Tsai-Wu criteria. What are the failure modes in tension and compression? Thermal stresses are given as $\sigma_{2e} = -\sigma_{1e} = 39\ \text{MPa}\ (5.65\ \text{ksi})$ for both layers.

9.23 A $[0/90]_s$ laminate is cured at an elevated temperature and cooled down to room temperature. What is the effect of increasing the moisture concentration c on the FPF uniaxial laminate strength $\bar{F}_x$? Neglect the effect of moisture on unidirectional lamina properties. Select the correct answer from the following list:

(a) reduction in compressive strength and increase in tensile strength

(b) increase in compressive strength and reduction in tensile strength

(c) no effect on compressive strength and increase in tensile strength

(d) increase in both tensile and compressive strengths

9.24 Determine the shear strength $\bar{F}_s$ at FPF of a $[\pm 45]_{ns}$ carbon/epoxy (AS4/3501-6) laminate, taking into account the residual stresses due to the hygrothermal loading $\Delta T = -150\ °C\ (-270\ °F)$ and $\Delta c = 0.5\%$. Compare results based on the maximum stress and Tsai-Wu criteria and identify the failure mode.

9.25 Determine the biaxial stress $\bar{F}_o$ of Problem 9.13, taking into account the residual stresses due to the hygrothermal loading $\Delta T = -150\ °C\ (-270\ °F)$ and $\Delta c = 0.5\%$.

9.26 A $[\pm 30]_{ns}$ angle-ply laminate is loaded under a uniaxial stress $\bar{\sigma}_x$ and a hygrothermal loading $\Delta T = -150\ °C\ (-270\ °F)$ and $\Delta c = 0.5\%$ (Fig. P9.26). Determine

(a) the residual stresses due to hygrothermal loading alone

(b) the mechanical stresses due to applied stress $\bar{\sigma}_x$ as a function of this stress

(c) the axial tensile strength $\bar{F}_{xt}$ at FPF under the combined mechanical and hygrothermal loading

Use both the maximum stress and Tsai-Wu criteria and identify the prevailing failure mode (material: AS4/3501-6 carbon/epoxy, Table A.4).

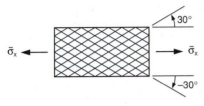

Fig. P9.26

9.27 Determine the axial compressive strength of the laminate of Problem 9.26 for the same hygro-thermal conditions. What is the failure mode in this case?

9.28 Determine the ultimate laminate failure tensile strength and the laminate efficiency ratio φ_L for the $[0/90]_s$ laminate of Problem 9.2, using the maximum stress criterion and total ply discount method.

9.29 The laminate efficiency ratio $\varphi_L = \bar{F}_{FPF}/\bar{F}_{ULF}$ (ratio of FPF to ultimate laminate failure strength) for a $[0/90]_s$ laminate is approximately equal to one of the following ratios. Select the correct one from the following list:

(a) $\varphi_L \cong \dfrac{F_{2t}}{F_{1t}}$

(b) $\varphi_L \cong \dfrac{E_2 F_{2t}}{E_1 F_{1t}}$

(c) $\varphi_L \cong \left(\dfrac{\nu_{12}}{\nu_{21}}\right)\left(\dfrac{F_{2t}}{F_{1t}}\right)$

(d) $\varphi_L \cong \left(\dfrac{E_2}{E_1}\right)\left(\dfrac{F_{1t}}{F_{2t}}\right)$

9.30 Determine the FPF and ULF uniaxial tensile strengths of a $[0/90_2]_s$ carbon/epoxy laminate using the maximum stress criterion and total ply discount method. Plot the stress-strain curve to failure (material: AS4/3501-6 carbon/epoxy, Table A.4).

9.31 For a $[0_m/90_n]_s$ carbon/epoxy (AS4/3501-6) laminate under axial tension compute the laminate efficiency ratio as a function of m and n. Use the maximum stress criterion for FPF.

9.32 What is the effect of temperature reduction on the uniaxial tensile strength $\bar{F}_{xt}$ of a $[\pm45]_s$ laminate? Based on the maximum stress criterion, select the correct answer from the following list:

(a) increase in FPF strength but no effect on ULF strength

(b) reduction in FPF strength but no effect on ULF strength

(c) increase in both FPF and ULF strengths

(d) no effect on either FPF or ULF strength

9.33 A cantilever beam of unit width made of a $[0/\pm45/90]_s$ carbon/epoxy laminate is subjected to a concentrated force P at the free end (see Fig. 9.36). Determine the force P at initiation of delamination for the properties of AS4/3501-6 carbon/epoxy listed in Table A.4, interlaminar shear strength $F_{zx} = 90$ MPa (13 ksi) and lamina thickness $t = 0.127$ mm (0.005 in).

9.34 A thin-wall cylindrical pressure vessel made of a $[\pm30]_{ns}$ carbon/epoxy laminate was cured at 180 °C (356 °F) and cooled down to 30 °C (86 °F) (Fig. P9.34). Subsequently, it absorbed 0.5% moisture by volume and was loaded by an internal pressure $p = 1$ MPa (145 psi). Calculate

(a) the stresses in each layer due to the pressure loading only

(b) the stresses due to the hygrothermal loading only

(c) the total stresses due to the combined mechanical and hygrothermal loading (material: AS4/3501-6 carbon/epoxy, Table A.4)

The $[\pm30]_s$ laminate properties are

$$\bar{E}_x = 58.1 \text{ GPa (8.42 Msi)}$$
$$\bar{E}_y = 13.8 \text{ GPa (2.00 Msi)}$$
$$\bar{G}_{xy} = 30.3 \text{ GPa (4.39 Msi)}$$
$$\bar{\nu}_{xy} = 1.22$$
$$\bar{\alpha}_x = -3.63 \times 10^{-6}/°C \ (-2.02 \times 10^{-6}/°F)$$
$$\bar{\alpha}_y = 13.50 \times 10^{-6}/°C \ (7.50 \times 10^{-6}/°F)$$
$$\bar{\beta}_x = -8.57 \times 10^{-3}$$
$$\bar{\beta}_y = 0.108$$

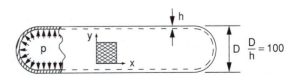

Fig. P9.34

9.35 For the pressure vessel of Problem 9.34, calculate the allowable pressure for FPF based on the maximum stress criterion and an allowable safety factor $S_{all} = 2.0$. What is the expected failure mode?

9.36 A thin-wall cylindrical pressure vessel made of a $[\pm 60]_{ns}$ S-glass/epoxy laminate was cured at 100 °C (212 °F), cooled down to 20 °C (68 °F), and stored in a dry environment (Fig. P9.36). For a design internal pressure of $p = 1.2$ MPa (174 psi), calculate the required wall thickness (h_a) for an allowable safety factor $S_{all} = 2.0$. Use both the maximum stress and Tsai-Wu failure criteria for first ply failure and compare results (material: S-glass/epoxy, Table A.4).

Measured laminate thermal strains due to cooldown:

$$\bar{\varepsilon}_x^T = -1.85 \times 10^{-3}$$
$$\bar{\varepsilon}_y^T = -0.49 \times 10^{-3}$$
$$\bar{\gamma}_s^T = 0$$

Mechanical (pressure) strains given for $h = 1.0$ mm (0.039 in):

$$\bar{\varepsilon}_x^M = 6.57 \times 10^{-3}$$
$$\bar{\varepsilon}_y^M = 9.59 \times 10^{-3}$$
$$\bar{\gamma}_s^M = 0$$

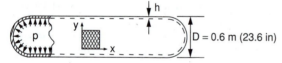

Fig. P9.36

9.37 A thin-wall cylindrical pressure vessel of the same overall dimensions as that of Problem 9.36 is made of a $[\pm 60]_{ns}$ carbon/epoxy laminate (AS4/3501-6, Table A.4). It is subjected to a hygrothermal loading $\Delta T = -150$ °C (−270 °F) and $\Delta c = 0.005$. For a design pressure $p = 1.2$ MPa (174 psi) and allowable safety factor $S_{all} = 2.0$,

(a) determine the required wall thickness and compare it with the corresponding thickness obtained for the S-glass/epoxy vessel of Problem 9.36, using both the maximum stress and Tsai-Wu failure criteria

(b) determine the required thickness for a pressure vessel of the same overall dimensions made of aluminum with a yield stress of $\sigma_{yp} = 200$ MPa (29 ksi), using the von Mises failure theory

(c) compare the required thicknesses of the aluminum and carbon/epoxy vessels and calculate the weight savings of the carbon/epoxy vessel with respect to the aluminum one

The laminate strains produced by cooldown and moisture absorption were measured as

$$\bar{\varepsilon}_x^{HT} = -1.48 \times 10^{-3}$$
$$\bar{\varepsilon}_y^{HT} = 0.501 \times 10^{-3}$$
$$\bar{\gamma}_s^{HT} = 0$$

Strains due to pressure loading only for a 1 mm (0.039 in) thick vessel are given as

$$\bar{\varepsilon}_x^M = 5.48 \times 10^{-3}$$
$$\bar{\varepsilon}_y^M = 2.41 \times 10^{-3}$$
$$\bar{\gamma}_s^M = 0$$

The densities of aluminum and carbon/epoxy are 2.8 g/cm³ and 1.60 g/cm³, respectively.

10 Experimental Methods for Characterization and Testing of Composite Materials

10.1 INTRODUCTION

The analysis and design of composite structures requires the input of reliable experimental data. As in the case of analysis, experimental characterization can be done on several scales: micromechanical, macromechanical, and structural. Testing of composite materials has three major objectives: determination of basic properties of the unidirectional lamina for use as an input in structural design and analysis; investigation and verification of analytical predictions of mechanical behavior; and independent experimental study of material and structural behavior for specific geometries and loading conditions. Under these general objectives, specific types and applications of testing include the following:

1. characterization of constituent materials (i.e., fiber, matrix, and interphase) for use in micromechanics analyses (Knowing these properties, one can predict, in principle, the behavior of the lamina and hence of laminates and structures.)
2. characterization of the basic unidirectional lamina, which forms the basic building block of all laminated structures
3. determination of interlaminar properties
4. material behavior under special conditions of loading (e.g., multiaxial, fatigue, creep, impact, and high-rate loading)
5. experimental stress and failure analysis of composite laminates and structures, especially those involving geometric discontinuities such as free edges, cutouts, joints, and ply drop-offs
6. assessment of structural integrity by means of nondestructive testing

A variety of experimental methods are used for the various applications above. Most of these deal with measurement of deformation or strains. Experimental methods for composite materials are more complex than for isotropic materials and require significant modifications. Test methods have been reviewed and extensive related references have appeared in the literature.[1–10]

10.2 CHARACTERIZATION OF CONSTITUENT MATERIALS

Constituent properties are important for understanding and predicting the macroscopic behavior of composite materials by means of micromechanics. These include physical and mechanical properties of fibers and matrices.

10.2.1 Mechanical Fiber Characterization

Various test methods are used for characterization of chemical, physical, and mechanical properties of reinforcing fibers for application to composite materials. It is recognized that the effective properties of the fibers within the composite are more relevant than those of isolated fibers. For example, the surface treatment (sizing) applied to the fibers affects their properties, especially their effective properties within the composite.

The most commonly measured mechanical properties are the longitudinal modulus, tensile strength, and ultimate tensile strain.[11–13] Fibers can be tested as single filaments, in impregnated tows, or in composite laminae. A single-filament test method is described in ASTM (American Society for Testing and Materials) specification D3379-89.[5] The method is recommended for fibers with an elastic modulus greater than 21 GPa (3 Msi). The filament is mounted along the centerline of a slotted paper tab, and axial alignment is accomplished without damaging the fiber (Fig. 10.1). After the specimen is mounted in the test machine, the paper tab is cut to allow for filament elongation. Specimens of various gage lengths are tested to failure at a constant crosshead rate, and the load-displacement curve is obtained.

To determine the elastic modulus of the fiber, the measured load-displacement curves must be corrected for the system compliance. The measured or "apparent" compliance is assumed to be the sum of the fiber and system compliances:

$$C_a = \frac{u}{P} = \frac{u_f}{P} + \frac{u_s}{P} = \frac{l}{AE_{1f}} + \frac{u_s}{P} \tag{10.1}$$

where

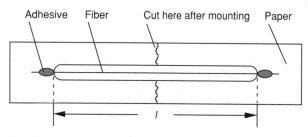

Fig. 10.1 Fiber specimen mounted on slotted paper tab (ASTM D3379-89).[5]

C_a = apparent compliance

u = crosshead displacement

u_f = actual fiber elongation

u_s = displacement due to system compliance

P = load

l = fiber gage length

E_{1f} = longitudinal fiber modulus

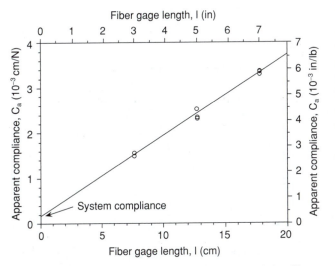

Fig. 10.2 Apparent compliance versus fiber gage length for silicon carbide fiber (SCS-2, Textron Specialty Materials).

The system compliance, $C_s = u_s/P$, is obtained as the zero gage length intercept by plotting the apparent compliance obtained from the various specimens versus fiber length. A typical compliance versus gage length curve is shown in Fig. 10.2 for a silicon carbide fiber. The fiber modulus is determined from Eq. (10.1). The cross-sectional area A is determined by measurements of representative fiber cross sections under the microscope.

The fiber strength is simply obtained from the maximum load as

$$F_{ft} = \frac{P_{max}}{A} \tag{10.2}$$

The ultimate strain is obtained from the maximum fiber elongation:

$$\varepsilon_{ft}^u = \frac{(u_f)_{max}}{l} \tag{10.3}$$

A newer mechanical test method suitable for measuring fiber stiffness at various temperatures has been described.[13] A length of fiber is fixed at the ends on a plate of very low-thermal-expansion material (Fig. 10.3). Titanium silicate, having a coefficient of thermal expansion of $0.3 \times 10^{-7}/°C$ ($1.7 \times 10^{-8}/°F$), can be used as the support plate. The fiber is fixed to the titanium silicate plate over a span length l with the plate held vertically inside a furnace. Small incremental weights, W_1, W_2, W_3, and so on, are suspended at the center of the fiber and the corresponding deflections, δ_1, δ_2, δ_3, and so on, are recorded photographically. The equilibrium condition yields the following relation:

$$\frac{W_i}{2} = \frac{\sigma A \delta_i}{\sqrt{(l/2)^2 + \delta_i^2}} \tag{10.4}$$

where

σ = stress in fiber

A = cross-sectional area of fiber

From geometric considerations we obtain the following relation for the strain in the fiber:

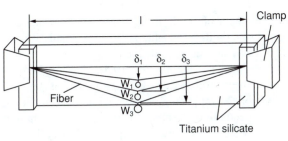

Fig. 10.3 Fixture for stiffness measurement of fiber.[13]

$$\varepsilon = \varepsilon_0 + \varepsilon_w = \varepsilon_0 + \frac{2\sqrt{(l/2)^2 + \delta_i^2} - l}{l} \tag{10.5}$$

where

$$\varepsilon_0 = \text{initial strain}$$

$$\varepsilon_w = \text{strain produced by deflection under weight}$$

The stress obtained from Eq. (10.4) is plotted versus the weight-induced strain obtained from Eq. (10.5), that is,

$$\sigma = \frac{W_i}{2A}\sqrt{(l/2\delta_i)^2 + 1} \tag{10.6}$$

versus

$$\varepsilon_w = \sqrt{1 + (2\delta_i/l)^2} - 1 \tag{10.7}$$

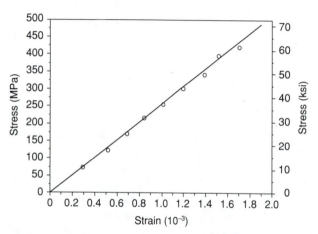

Fig. 10.4 Stress-strain curve of carbon fiber (IM6, Hercules, Inc.).[13]

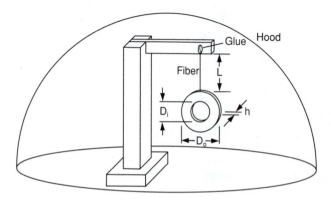

Fig. 10.5 Torsional pendulum for measuring shear modulus of fibers.[14]

A typical stress-strain curve for an intermediate-modulus carbon fiber (IM6, Hercules, Inc.) is shown in Fig. 10.4.

In the case of carbon or graphite fibers, it is also recommended that tensile properties be obtained of fiber strands or tows impregnated in resin (ASTM D4018). In this case a minimum amount of resin is used that is compatible with the fibers and has a higher ultimate strain. It is also important to ensure that the fibers in the strand are well collimated. Average fiber properties can also be obtained indirectly by testing unidirectional laminae and employing the micromechanics relations discussed in Chapters 3 and 5.

In addition to longitudinal properties, there is a need to measure other properties, such as transverse modulus, E_{2f}, longitudinal shear modulus, G_{12f}, and Poisson's ratio, ν_{12f}, especially for anisotropic fibers such as carbon and aramid (Kevlar) fibers. A technique has been described for determination of the longitudinal shear modulus G_{12f} of uniform-diameter single fibers.[14] A length of single fiber suspended from one end and with a weight, such as a small washer, attached at the other end was used as a torsional pendulum (Fig. 10.5). The longitudinal shear modulus can be expressed as a function of the frequency of oscillation of the torsional pendulum. Results were obtained for different types of fibers including carbon, Kevlar 49, silicon carbide, and glass.

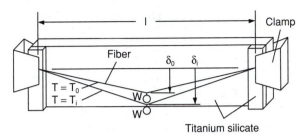

Fig. 10.6 Deflection of fiber with fixed ends under suspended load at various temperatures.[13]

10.2.2 Thermal Fiber Characterization

In addition to mechanical properties, thermal properties of fibers are very important. There are no standardized methods for measuring the coefficient of thermal expansion (CTE), and relatively few results are available.[13,15] Commercial instruments such as the thermomechanical analyzer (TMA) can be adapted for the purpose. The CTE of a fiber can also be derived from thermal expansion measurements made on unidirectional laminae using relations in Eqs. (8.1) and (8.2).

A different method for measuring CTE of the fiber utilizes a length of fiber fixed loosely at the ends to a titanium silicate plate (Fig. 10.6).[13] A constant weight is suspended at the center and the deflection δ_0 at room temperature T_0 is measured optically. The stretched length of the fiber under these conditions (T_0 and W) is

$$L_0 = 2\sqrt{(l/2)^2 + \delta_0^2} \tag{10.8}$$

The temperature is then raised to the next step T_i and the corresponding deflection δ_i is measured. The new fiber length is then

$$L_i = 2\sqrt{(l/2)^2(1 + \alpha_r \Delta T)^2 + \delta_i^2} \tag{10.9}$$

where

α_r = coefficient of thermal expansion of mounting plate material (titanium silicate)

$\Delta T = T_i - T_0$ = temperature difference

The difference in fiber length, $L_i - L_0$, is due primarily to the thermal expansion and in a smaller part to any possible changes in the fiber stiffness with temperature. For small loads and fibers of high modulus not varying much with temperature, the mechanical changes are negligible. Then, the thermal strain is given by

$$\varepsilon^T \cong \frac{L_i - L_0}{L_0} \cong \sqrt{\frac{(1 + \alpha_r \Delta T)^2 + (2\delta_i/l)^2}{1 + (2\delta_0/l)^2}} - 1 \tag{10.10}$$

The thermal strain obtained from Eq. (10.10) is plotted versus temperature, and the coefficient of thermal expansion at any temperature is the slope of the curve at that temperature. A typical thermal strain versus temperature curve for a silicon carbide fiber (Nicalon, Nippon Carbon Co.) is shown in Fig. 10.7. The technique described here was modified for stiff large-diameter fibers, such as boron and silicon carbide (SCS-type) filaments.[13]

There are no known techniques for measuring the transverse CTE directly. One indirect method is to measure the transverse CTE of a unidirectional lamina and obtain the transverse CTE of the fiber, α_{2f}, using Eq. (8.3).

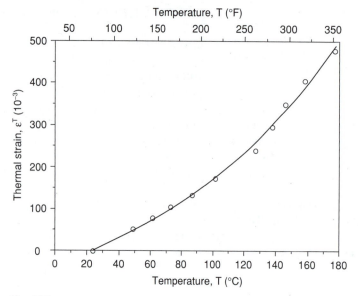

Fig. 10.7 Thermal strain versus temperature for silicon carbide fiber (Nicalon NLM-202, Nippon Carbon Co.).[13]

10.2.3 Matrix Characterization

Polymeric matrices are characterized by casting the material in sheet form and cutting and testing coupons from these sheets. The specimens are usually prismatic or dog bone shaped. The suggested geometry and dimensions depend on the sheet thickness and are described in ASTM specifications D638-02 and D882-02.[5,16] Strains are measured by means of strain gages or an extensometer. When the specimen thickness is small and/or the material stiffness is low, optical techniques are recommended. Uniaxial tensile tests to failure yield the following properties: Young's modulus, E_m; Poisson's ratio, v_m; tensile strength, F_{mt}; and ultimate tensile strain, ε_{mt}^u. In the above, it is assumed that the matrix is isotropic and that its bulk properties are the same as the in situ properties in the composite.

Metallic matrices are characterized in a similar way by testing coupons of the matrix metal. However, the bulk properties of the metal or metallic alloy may be appreciably different from the in situ properties of the matrix within the composite due to the heat treatment during the fabrication process and interactions with the fibers. In such cases, it is important to identify the condition of the matrix within the composite and to characterize a bulk material of similar properties. For example, in the fabrication of silicon carbide/aluminum composites (such as SCS-2/6061-Al), the matrix within the composite after processing is considered equivalent to bulk aluminum of T4 temper.

Ceramic matrices can also be tested in bulk form to obtain elastic properties. However, strength and failure characteristics are not the same as those of the ceramic matrix within the composite. These properties can be obtained indirectly by testing the composite material.

10.2.4 Interface/Interphase Characterization

The nature of the bond between fiber and matrix, whether occurring through a zero-thickness interface or an interphase region, plays a profound role in the failure mechanisms, toughness, and the overall deformation behavior of the composite. Characterization of the interface/interphase is a very challenging problem.

Many methods and techniques have been developed for determination of properties of the interface region. Two of the most commonly used tests are the pushout and pullout tests, which examine the response of a single fiber embedded in a matrix material.[17–20] In the pushout or indentation test, a microindenter is pressed against a fiber end in the direction of the fiber axis.[17–19]

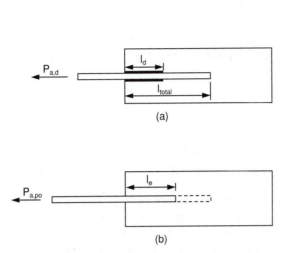

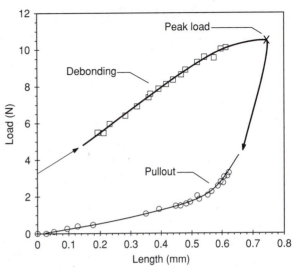

Fig. 10.8 Single-fiber pullout test: (a) debonding, (b) pullout. (The peak load occurs when $l_d = l_{total}$, just as the fiber debonds completely before pullout.)[21]

Fig. 10.9 Load versus debond length during debonding and load versus embedded length during pullout.[21]

In the fiber pullout test, a fiber partially embedded in a matrix material block is loaded in tension (Fig. 10.8).[21] If the fiber is bonded at the interface, and if the load required for interfacial crack growth is less than the load for fiber failure, then, fiber debonding will occur. Fiber debonding starts at the free surface of the block and progresses along the embedded length until the fiber is completely debonded or until the fiber failure load is reached. Thereafter, the fiber starts sliding, resisted by interfacial friction, until complete pullout. Figure 10.9 shows the progressive debonding followed by pullout of a silicon carbide (SCS-2 SiC) fiber embedded in barium borosilicate glass (7059, Corning Glass Works). The slopes of the linear portions of the debonding and pullout curves in Fig. 10.9 yield the interfacial shear strength and frictional shear stress as follows:

$$F_{is} = \frac{m_d}{2\pi r} \qquad (10.11)$$

$$\tau_f = \frac{m_p}{2\pi r} \qquad (10.12)$$

where

$$F_{is} = \text{interfacial shear strength}$$

$$\tau_f = \text{frictional shear stress}$$

$$m_d, m_p = \text{slopes of the curves (lines) of load versus debonded length and}$$
$$\text{load versus embedded length during pullout, respectively}$$

$$r = \text{fiber radius}$$

10.3 PHYSICAL CHARACTERIZATION OF COMPOSITE MATERIALS

Physical characterization of composites includes the determination of density, fiber volume ratio, void volume ratio, coefficients of thermal expansion, coefficients of moisture expansion, and heat conduction coefficients.

10.3.1 Density

The procedure for measuring the density of a composite material is the same as that used for any other solid and is based on ASTM specification D792-00.[22] The procedure consists of the following steps:

Step 1 Weigh the specimen in air to the nearest 0.1 mg.

Step 2 Attach the specimen to an analytical balance with a thin wire and weigh while the specimen and a portion of the wire are immersed in distilled water.

Step 3 Weigh the wire alone, partially immersed up to the same point as in the previous step.

The density of the material at 23 °C (73.4 °F) is determined as follows:

$$\rho = \frac{a}{a + w - b}(0.9975) \tag{10.13}$$

where

ρ = density (in g/cm^3)

a = weight of specimen in air

b = apparent weight of fully immersed specimen and partially immersed wire

w = apparent weight of partially immersed wire

0.9975 = density of distilled water at 23 °C (in g/cm^3)

10.3.2 Fiber Volume Ratio

A variety of methods exist for determination of fiber volume ratio, an important property of a composite. When it can be confirmed that the composite material has zero or negligible (less than 1%) porosity, the fiber volume ratio can be obtained from the densities of the composite and the constituents by the following gravimetric relation:

$$V_f = \frac{\rho_c - \rho_m}{\rho_f - \rho_m} \tag{10.14}$$

where

ρ_c, ρ_m, ρ_f = densities of composite, matrix, and fiber, respectively

The ignition or burnout method, based on ASTM specification D2584-02, can be applied to composites having inorganic fibers in an organic matrix, such as glass/epoxy and boron/epoxy composites.[5] A sample of the composite material is oven dried, weighed, and then heated in a crucible until the matrix is completely burnt. The residue is washed of the ashes, dried, and weighed. The fiber volume ratio is obtained as

$$V_f = \frac{W_f/\rho_f}{W_c/\rho_c} \tag{10.15}$$

where

$$W_c, W_f = \text{weight of composite and fibers, respectively}$$

The acid digestion method, described in ASTM specification D3171-99, is used with composites having a matrix that is soluble in some acid that does not attack the fiber.[5] A sample of the composite material is dried and weighed. Then it is immersed in an acid solution to dissolve the matrix. The type of acid used is one that dissolves the matrix without attacking the fibers. The residue is filtered, washed, dried, and weighed, and the fiber volume ratio is determined by Eq. (10.15).

The fiber volume ratio can also be determined reliably by optical techniques based on image analysis of photomicrographs of transverse (to the fibers) cross sections of the composite. An elementary approach consists of counting the number of fiber cross sections and fractions thereof within the frame of the photomicrograph, calculating the total area of the fiber cross sections, and dividing it by the total area photographed. More sophisticated image analysis techniques are also used and can determine both fiber volume ratio and void volume ratio in one operation, as will be discussed below.

10.3.3 Void Volume Ratio (Porosity)

The void volume ratio (or porosity) is obtained as described in ASTM specification D2734-94 (2003).[5] It is expressed in terms of the quantities measured in the acid digestion or ignition methods as follows:

$$V_v = 1 - \frac{W_f/\rho_f + (W_c - W_f)/\rho_m}{W_c/\rho_c} \tag{10.16}$$

In the above relation the void volume ratio, which is usually a small number, is expressed as the difference of two much larger numbers; therefore, the result is very sensitive to the measurement accuracy involved.

A preferred method for determination of porosity in a composite is the image analysis method mentioned before. A photomicrograph of a cross section of a carbon/epoxy composite, for example, shows the fibers, matrix, and voids as light gray, dark gray, and black, respectively (Fig. 10.10). In the image analysis procedure, the specimen cross section is viewed by a CCD camera, which transmits the image in digital form to the "image grabber" of a computer. The image is converted into a rectangular array of integers, corresponding to the digitized gray levels of the picture elements (pixels). The image

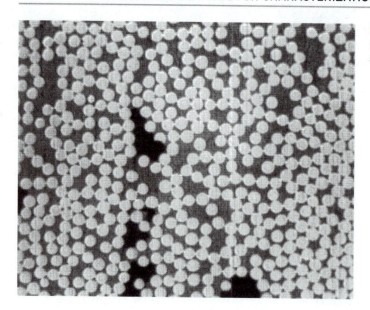

Fig. 10.10 Digitized image of carbon/epoxy cross section used for determination of fiber volume and void volume ratios.

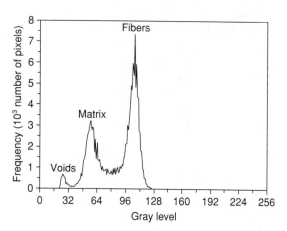

Fig. 10.11 Trimodal gray-level histogram of the image of Fig. 10.10.

processing board in the computer processes this digital image information and represents it in the form of a gray-level histogram, such as the one shown in Fig. 10.11. This histogram summarizes the gray-level content of the image. It shows for each gray level the number of pixels in the image that have that gray level. The three peaks of the histogram of Fig. 10.11 correspond to the porosity, matrix, and fibers. The pixels closest to the three peaks are separated by a thresholding process into three groups corresponding to porosity, matrix, and fibers. The void volume, matrix volume, and fiber volume ratios are obtained as follows:

$$V_v = \frac{n_1}{N}$$

$$V_m = \frac{n_2}{N} \qquad (10.17)$$

$$V_f = \frac{n_3}{N}$$

where

n_1, n_2, n_3 = number of pixels corresponding to gray levels associated with porosity, matrix, and fibers, respectively

N = total number of pixels

10.3.4 Coefficients of Thermal Expansion

As mentioned in Chapter 8, the thermal behavior of a unidirectional lamina can be fully characterized by two principal coefficients of thermal expansion, α_1 and α_2. Determination of these coefficients consists of measuring the corresponding thermal strains in a unidirectional composite specimen as a function of temperature. These measurements can be made by means of interferometric, dilatometric, optical noninterferometric, or strain gage methods.[23–32] The interferometric method is the most sensitive one and can give results for the coefficients α_1 and α_2 with a resolution of $10^{-8}/°C$. The thermomechanical analyzer (TMA) measures dimensional changes in a sample with high precision. The sample rests on a quartz base inside the furnace. A measuring quartz probe rests on the specimen and senses changes in length by means of a sensitive position transducer, normally a linear variable differential transformer (LVDT). This instrument gives results with high precision, but it requires one specimen for each coefficient.

Strain gages have been shown to be a practical and adequate means of measuring thermal strain in composites.[24–26] However, they must be properly compensated for the purely thermal output. One method of temperature compensation employs an identical gage bonded to a reference material of known thermal expansion exposed to the same temperature as the composite specimen.[23] The true thermal strain in the composite is given by

$$\varepsilon_{tc} = \varepsilon_{ac} - \varepsilon_{ar} + \varepsilon_{tr} \qquad (10.18)$$

where

ε_{tc} = true thermal strain in composite specimen

ε_{ac} = apparent strain in composite specimen

ε_{tr} = true thermal strain in reference specimen

ε_{ar} = apparent strain in reference specimen

Reference materials used are usually ceramics of low and stable coefficient of thermal expansion, such as fuzed quartz ($\alpha_r = 0.7 \times 10^{-6}/°C$) and titanium silicate ($\alpha_r = 0.03 \times 10^{-6}/°C$). For a temperature change of ΔT, the true thermal strain in the reference material is $\alpha_r \Delta T$.

A unidirectional composite specimen is usually instrumented with two-gage rosettes to record thermal strains along the fiber (1) and transverse to the fiber (2) directions (Fig. 10.12). For better results it is preferable to use gages on both surfaces of the specimen, or even embedded gages, to correct for any possible bending of the specimen due to asymmetries or small thermal gradients through the thickness. Measured thermal strains are plotted

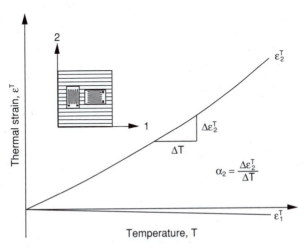

Fig. 10.12 Measurement of coefficients of thermal expansion by means of strain gages.

versus temperature as shown in Fig. 10.12. The slopes of these curves at any point give the coefficients of thermal expansion. Thermal strain curves for typical composites were shown in Fig. 8.9 and typical results in Table 8.2.

10.3.5 Coefficients of Hygric (Moisture) Expansion

The hygric (moisture) behavior of a unidirectional lamina, like the thermal behavior, can be fully characterized in terms of two principal coefficients of moisture expansion, β_1 and β_2. Determination of these coefficients consists of measuring the principal direction strains in a unidirectional composite as a function of moisture concentration. Specimens are pre-conditioned by drying them in an oven at a moderately high temperature (65 °C; 150 °F) for approximately two hours. Subsequently they are exposed to the moisture conditioning environment. Moisture absorption can be accomplished by immersing the specimens in a water bath inside an oven at a moderately high temperature, for example, 50 °C (120 °F).

Moisture expansion or swelling has been measured by means of a micrometer or a caliper gage.[33,34] The application of strain gages has been difficult because conventional strain gage adhesives are attacked by moisture.[1] Furthermore, the presence of the gage on the surface of the specimen may inhibit locally the process of moisture diffusion. A method utilizing embedded strain gages has been shown to be more reliable and consistent than previously used techniques.[35] The method consists of embedding encapsulated strain gages in the midplane of the specimen. The technique results in good adhesion without the need for additional adhesive and does not cause any local disturbance in moisture diffusion, since the gage is located at a plane of symmetry.

Unidirectional specimens with and without embedded strain gages are dried (precondi-tioned) and then immersed in a 50 °C (120 °F) water bath inside an oven. The embedded gages in the immersed specimen are connected to a data logger and monitored continu-ously throughout the duration of conditioning. Specimens without gages exposed to the same environment are removed periodically from the water bath and weighed on an ana-lytical balance to determine the relative weight gain, M. The average moisture concentra-tion c representing the relative volume occupied by water is related to the weight gain as follows:

$$c = \frac{V_w}{V_c} = \frac{W_w/\rho_w}{W_c/\rho_c} = \frac{\rho_c}{\rho_w}\frac{W_w}{W_c} = \frac{\rho_c}{\rho_w}M \tag{10.19}$$

where

$$V_w, V_c = \text{volumes of water and composite, respectively}$$

$$W_w, W_c = \text{weights of water and composite, respectively}$$

$$\rho_w, \rho_c = \text{densities of water and composite, respectively}$$

The measured hygric strains are plotted versus average moisture concentration c as illustrated in Fig. 10.13 for a carbon/epoxy composite. The slopes of these curves yield the coefficients of moisture expansion β_1 and β_2.

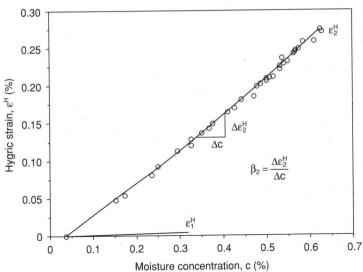

Fig. 10.13 Hygric strains in unidirectional AS4/3501-6 carbon/epoxy composite as a function of moisture concentration.

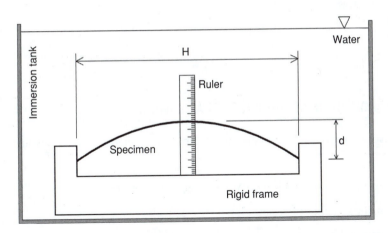

Fig. 10.14 Setup for measuring the coefficient of hygric expansion of a composite specimen.[36]

   Two new experimental methods were described for measuring hygric properties of a composite material.[36,37] The first of these is related to the method discussed before for measuring the CTE of fibers.[13] It is based on measuring the transverse deflection of an axially constrained specimen (Fig. 10.14). A dry specimen is placed between the walls of the fixture with enough interference to cause a slight transverse deflection. The specimen with the fixture is immersed in a water bath allowing the specimen to absorb moisture. The outward deflection is measured at various time intervals corresponding to various average moisture concentrations. The corresponding moisture concentrations are obtained by prior moisture diffusion experiments. The deflection changes are related to changes in specimen length and, thereby, hygric strain, ε^H. The coefficient of hygric expansion along the axis of the specimen is obtained as

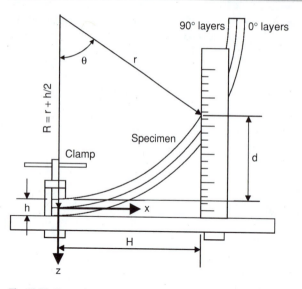

Fig. 10.15 Setup for measuring the curvature of an asymmetric specimen to determine the coefficient of hygric expansion.[37]

$$\beta = \frac{\Delta\varepsilon^H}{\Delta\bar{c}} \qquad (10.20)$$

where $\bar{c}$ is the average moisture concentration.

The second method yielding results with improved sensitivity is based on measurement of the curvature of an antisymmetric crossply laminate (Fig. 10.15).[37] The initial curvature of the as-cured dry specimen is measured first. Thereafter, the specimen is immersed in water for varying periods of time. After each immersion, the specimen is weighed and then mounted in the fixture of Fig. 10.15 to measure the curvature. The results are analyzed by means of hygrothermomechanical lamination theory discussed in Chapter 8, to yield β_1, β_2, and the stress-free temperature. Results obtained for the AS4/3501-6 carbon/epoxy are

$$\beta_1 = 0.01$$

$$\beta_2 = 0.20$$

$$T_f = 173\ °C\ (345\ °F) \qquad \text{(stress-free temperature)}$$

10.4 DETERMINATION OF TENSILE PROPERTIES OF UNIDIRECTIONAL LAMINAE

Uniaxial tensile tests are conducted on unidirectional laminae to determine the following properties:

E_1, E_2 = longitudinal and transverse Young's moduli, respectively

ν_{12}, ν_{21} = major and minor Poisson's ratios, respectively

F_{1t}, F_{2t} = longitudinal and transverse tensile strengths, respectively

$\varepsilon_{1t}^u, \varepsilon_{2t}^u$ = longitudinal and transverse ultimate tensile strains, respectively

Tensile specimens are straight-sided coupons of constant cross section with adhesively bonded beveled glass/epoxy tabs (Fig. 10.16). More details are given in ASTM specification D3039/D3039M-00.[5] The longitudinal (0°) coupon is usually 1.27 cm (0.50 in) wide, while the transverse (90°) coupon is 2.54 cm (1.0 in) wide. Recommended thicknesses are 0.5 to 2.5 mm (0.020 to 0.100 in), usually six plies for the longitudinal specimen and at least eight plies for the transverse one. Both specimens have an overall length of 22.9 cm (9.0 in) and a gage length of 15.2 cm (6.0 in). The specimens are loaded to failure under uniaxial tensile loading. A continuous record of load and deformation is obtained by an appropriate digital data acquisition system. Axial and transverse strains are obtained by means of a pair of two-gage rosettes mounted on both sides of the specimen. In some

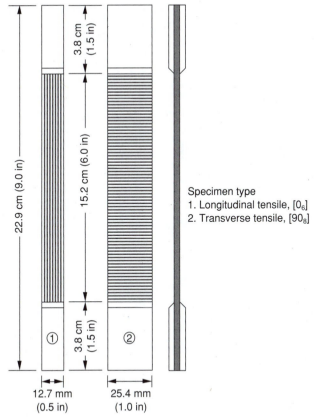

Specimen type
1. Longitudinal tensile, $[0_6]$
2. Transverse tensile, $[90_8]$

Fig. 10.16 Specimen geometries for determination of tensile properties of unidirectional lamina.

cases when the transverse strain is not needed for determination of Poisson's ratio, the axial deformation alone is recorded with an extensometer mounted on the specimen. Typical stress-strain curves for 0° and 90° carbon/epoxy specimens are shown in Figs. 10.17 and 10.18. Results obtained from these curves are shown in the figures as well as in Table A.4. Typical fractures of unidirectional tensile coupons were shown in Figs. 5.6 and 5.22. Failure of the 0° specimens consists of fiber fractures, matrix splitting, and fiber pullout. The latter mechanism is much more pronounced in the "brooming" failure pattern of glass/epoxy. Transverse (90°) specimens fail in a brittle manner by matrix tensile failure between fibers.

Sometimes the unidirectional material to be characterized is prepared in thin-wall tubular form, following the same fabrication procedure used for tubular structures. Rings approximately 2.54 cm (1.0 in) wide are machined from a thin-wall composite cylinder. They are instrumented on the outer surface with strain gages along the axial and hoop directions. They are mounted in a specially designed fixture and loaded to failure under internal pressure (Fig. 10.19). The rings are thus subjected to uniaxial hoop stress equivalent to the axial stress in a flat coupon. The hoop stress σ_θ is obtained from the internal pressure p as

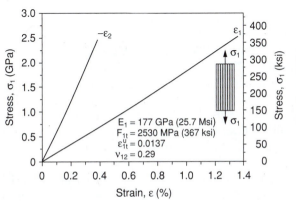

Fig. 10.17 Stress-strain curves for $[0_6]$ carbon/epoxy specimen under uniaxial tensile loading (IM6G/3501-6).

$E_1 = 177$ GPa (25.7 Msi)
$F_{1t} = 2530$ MPa (367 ksi)
$\varepsilon_{1t}^u = 0.0137$
$\nu_{12} = 0.29$

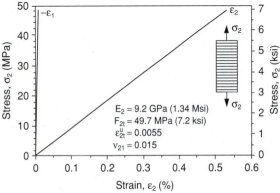

Fig. 10.18 Stress-strain curves for $[90_8]$ carbon/epoxy specimen under uniaxial tensile loading (IM6G/3501-6).

$E_2 = 9.2$ GPa (1.34 Msi)
$F_{2t} = 49.7$ MPa (7.2 ksi)
$\varepsilon_{2t}^u = 0.0055$
$\nu_{21} = 0.015$

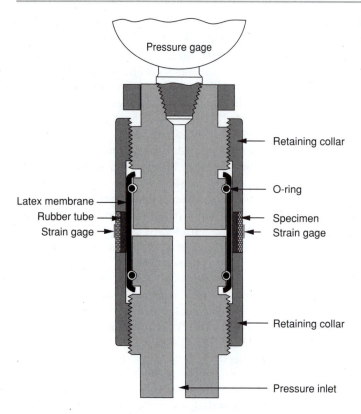

Fig. 10.19 Sketch of fixture for subjecting composite ring specimen to uniform tensile hoop stress.

$$\sigma_\theta = \frac{p\bar{r}}{h} \tag{10.21}$$

where $\bar{r}$ is the mean radius and h the wall thickness. A record is obtained of the pressure and strain gage signals up to failure, and stress-strain curves similar to those of Figs. 10.17 and 10.18 are obtained.

10.5 DETERMINATION OF COMPRESSIVE PROPERTIES OF UNIDIRECTIONAL LAMINAE

Compression testing of composite materials is one of the most difficult types of testing because of the tendency for premature failure due to global buckling or end crushing. The test is sensitive to many experimental parameters, such as alignment, specimen geometry, load introduction scheme, and stability. Over the years many test methods have been developed and used, primarily for thin laminates incorporating a variety of specimen designs and loading fixtures. These methods have been reviewed before.[1]

One of the most commonly used test methods today is the one employing the IITRI (IIT Research Institute) fixture or its modifications.[38] The method makes use of coupon specimens, 14.1 cm (5.5 in) long, 15 to 20 plies thick, and 0.64 cm (0.25 in) wide as described in ASTM D3410M-03.[5] The coupons are tabbed with long glass/epoxy tabs,

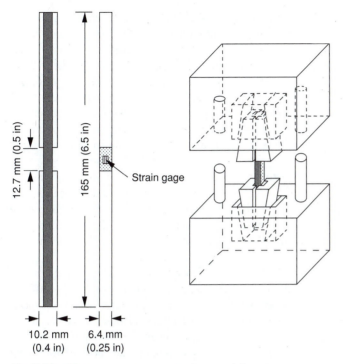

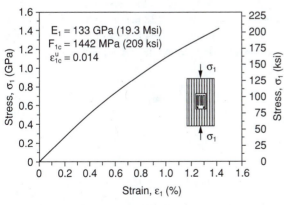

Fig. 10.20 IITRI compression test specimen and fixture.

leaving a gage length section 1.27 cm (0.5 in) long. Load is introduced through shear by trapezoidal wedge grips (Fig. 10.20). The fixture permits precompression of the specimen tabs to prevent slippage in the early stages of loading. The lateral alignment of the top and bottom halves of the fixture is assured by two parallel roller bushings. Strains are measured by means of strain gages mounted on both sides of the specimen to confirm that it fails in compression and not by buckling. Typical results obtained with this fixture for a carbon/epoxy material are plotted in Figs. 10.21 and 10.22. The limitation of the IITRI fixture is the maximum load that can be transmitted through shear by the tabs, which is controlled by the adhesive shear strength, the interlaminar shear strength of the tab and specimen materials, and the total tabbed area. The only practical means for increasing the testing load capacity is to introduce end loading in addition to shear loading. Tests based on end loading include a modified ASTM D695-02 test, known as SACMA (Suppliers of Advanced Composite Materials Association) SRM 1R-94, and a European version ICSTM (Imperial College of Science, Technology, and Medicine).[39]

A new test method and fixture, the NU (Northwestern University) fixture, incorporating both shear and end-loading concepts was developed.[40] The NU fixture consists of two steel blocks with machined cavities into which trapezoidal wedge grips are inserted (Fig. 10.23). Two steel adapters for the hydraulic grips of the testing machine are attached

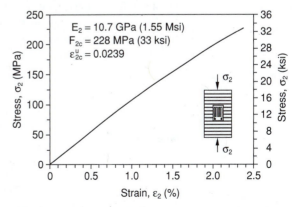

Fig. 10.21 Stress-strain curve of $[0_{16}]$ carbon/epoxy specimen under uniaxial compressive loading (AS4/3501-6).

Fig. 10.22 Stress-strain curve for $[90_{16}]$ carbon/epoxy specimen under uniaxial compressive loading (AS4/3501-6).

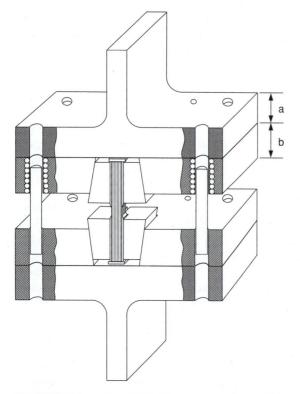

Fig. 10.23 Schematic of NU (Northwestern University) compression test fixture (dimensions: adapter $a = 25$ mm, base $b = 38$ mm).

to the outer ends of the blocks. Lateral alignment of the top and bottom halves of the fixture is controlled by means of two hardened steel pins and linear bearings. Specimens used in this fixture are similar to those used in the IITRI fixture, with more flexibility in varying the width and thickness of the specimen. Steel endcaps are attached to the ends of the specimen to prevent premature crushing or longitudinal splitting at the ends. A schematic of the specimen and wedge grip assembly is shown in Fig. 10.24. The concept of the testing procedure is to transmit the first part of the load through the tabs by shear loading and thereafter, just before shear failure of the tabs or specimen, engage the ends to apply the additional load to failure by end loading. The NU test fixture has been used successfully in testing thick composite materials.

Another approach to compression testing consists of bonding the composite laminates to a low-stiffness honeycomb or foam core to produce a sandwich specimen. Sandwich specimens can be tested in direct edgewise compression or in pure bending. In the first case, two composite coupons are bonded to a core block as described in ASTM C364-99.[5] The recommended specimen width is on the order of 50 mm (2 in). The unsupported length should be limited to prevent global buckling and should not exceed 12 times the total specimen thickness.

The sandwich flexure specimen recommended by ASTM C393-00 consists of a honeycomb or foam core with a composite facesheet bonded to the top (compressive) side and a metal or composite sheet bonded to the bottom (tensile) side (Fig. 10.25). The beam is loaded under four-point bending at the quarter points as shown, which subjects the top composite facesheet to nearly uniform compression. The overall beam is 56 cm (22 in) long and 2.54 cm (1 in) wide; the honeycomb core is 3.8 cm (1.5 in) deep, and the thicknesses of the top and bottom facesheets are adjusted to ensure compressive failure in the top facesheet. This is ensured when

$$\frac{L}{h} \geq \frac{4F_{xc}}{F_{cs}} \tag{10.22}$$

where

L = span length

h = composite facesheet thickness

F_{xc} = compressive strength of composite facesheet

F_{cs} = allowable core shear stress

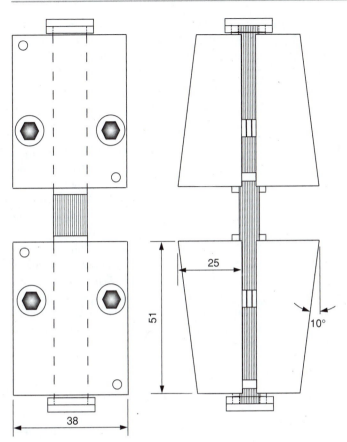

Fig. 10.24 Schematic of specimen and wedge grip assembly (with dimensions given in mm).

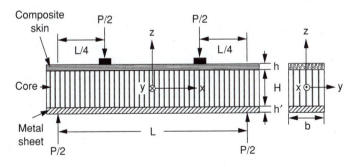

Fig. 10.25 Sandwich beam specimen for compression testing of composites (ASTM C393-00).

If the beam is not long enough, core failure or indentation may occur before composite compressive failure.

The beam is loaded through hard rubber pads on the top face or loading blocks embedded in the honeycomb core next to the metal sheet to prevent local crushing. Strains are measured by means of strain gages mounted on the composite facesheet. The center deflection can also be monitored with a deflectometer. The stress in the composite

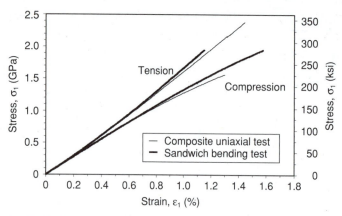

Fig. 10.26 Tensile and compressive stress-strain curves of a unidirectional carbon/epoxy composite obtained from a composite sandwich beam under pure bending (AS4/3501-6).[41]

facesheet is determined by assuming uniform deformation in the facesheets and neglecting the bending stresses in the core, that is,

$$\bar{\sigma}_x = \frac{N_x}{h} = \frac{PL}{4bh(2H + h + h')} \quad (10.23)$$

This stress can be plotted versus the measured strain ε_x on the facesheet to obtain the compressive stress-strain behavior. The ultimate value of the stress is the compressive strength of the facesheet material. Figure 10.26 shows tensile and compressive stress-strain curves of a unidirectional carbon/epoxy composite obtained from a sandwich beam with a foam core and composite facesheets loaded in pure bending.[41]

Results obtained from sandwich beam tests tend to be higher than those obtained by the other methods discussed, probably because of the constraint produced by the core and the biaxial state of stress induced in the composite facesheet.

10.6 DETERMINATION OF SHEAR PROPERTIES OF UNIDIRECTIONAL LAMINAE

Full characterization of a unidirectional composite in two dimensions requires the determination of lamina properties under in-plane shear parallel to the fibers, that is, shear modulus, G_{12}, shear strength, F_6, and ultimate shear strain, γ_6^u. There are four generally accepted test methods for determination of these properties: (1) the $[\pm 45]_{ns}$ coupon test, (2) the 10° off-axis test, (3) the rail shear test, and (4) the torsion test.

The first test method utilizes an eight-ply $[\pm 45]_{2s}$ coupon of the same dimensions as the 90° unidirectional tensile coupon discussed before (Fig. 10.27).[42] The test procedure is described in ASTM standard D3518M-94 (2001).[5] When this coupon is subjected to a uniaxial tensile stress, $\bar{\sigma}_x$, the stresses acting on the lamina element shown are

$$\sigma_1 = \frac{\bar{\sigma}_x}{2} + \tau_{xy}$$

$$\sigma_2 = \frac{\bar{\sigma}_x}{2} - \tau_{xy} \quad (10.24)$$

$$\tau_6 = \frac{\bar{\sigma}_x}{2}$$

where τ_{xy} is the in-plane shear stress generated by the shear coupling mismatch (see Figs. 9.28 and 9.29). The in-plane lamina strains are

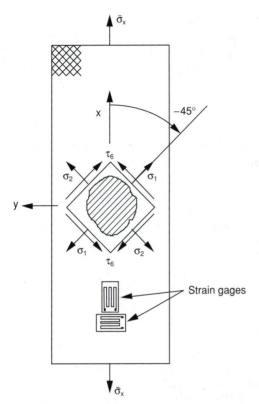

Fig. 10.27 $[\pm 45]_{ns}$ angle-ply specimen under uniaxial tension for determination of in-plane lamina shear properties.

$$\varepsilon_1 = \varepsilon_2 = \frac{\bar{\varepsilon}_x + \bar{\varepsilon}_y}{2} \tag{10.25}$$

$$\gamma_6 = \bar{\varepsilon}_x - \bar{\varepsilon}_y \tag{10.26}$$

where $\bar{\varepsilon}_x$ and $\bar{\varepsilon}_y$ are the axial and transverse strains in the coupon measured with two-gage rosettes. This in-plane (or intralaminar) shear modulus of the unidirectional lamina is obtained from the initial slope of the τ_6 versus γ_6 curve as

$$G_{12} = \frac{\bar{\sigma}_x}{2(\bar{\varepsilon}_x - \bar{\varepsilon}_y)} \tag{10.27}$$

This value of the modulus is not affected by the edge effects present in this specimen or by the biaxial state of stress existing in the lamina.

Equation (10.27) can be rewritten in the following form (by dividing numerator and denominator by $\bar{\varepsilon}_x$):

$$G_{12} = \frac{\bar{E}_x}{2(1 + \bar{\nu}_{xy})} \tag{10.28}$$

Thus, the lamina shear modulus G_{12} can be obtained in terms of the axial modulus $\bar{E}_x$ and Poisson's ratio $\bar{\nu}_{xy}$ of the $[\pm 45]_{ns}$ laminate (see Sample Problems 7.3 and 7.5 and Problem 7.23).

The above method tends to overestimate the in-plane shear strength of the lamina because of the constraint imposed on the lamina by the adjacent plies. In estimating this strength the method does not take into account edge effects or the influence of the other stress components σ_1 and σ_2 on the lamina (see Sample Problem 9.1). A typical shear stress versus shear strain curve obtained from a $[\pm 45]_{2s}$ carbon/epoxy coupon is shown in Fig. 10.28.

The second test method is the 10° off-axis test.[24,43] The 10° angle is chosen to minimize the effects of longitudinal and transverse stress components σ_1 and σ_2 on the shear response. The specimen is a six-ply unidirectional coupon with the fibers oriented at 10° to the loading axis, 1.27 cm (0.5 in) wide and approximately 23 cm (9 in) long. It is tabbed with beveled tabs and instrumented with a two-gage rosette on each side of the test section. The two gages

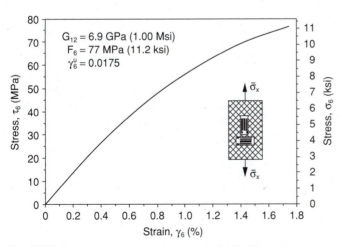

Fig. 10.28 Shear stress versus shear strain in $[\pm 45]_{2s}$ carbon/epoxy specimen under uniaxial tensile loading (AS4/3501-6).

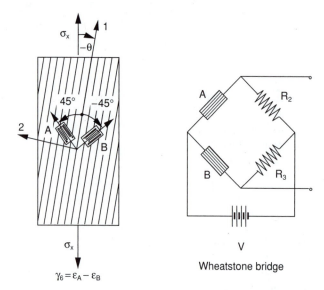

$$\gamma_6 = \varepsilon_A - \varepsilon_B$$

Fig. 10.29 Arrangement of strain gages on an off-axis composite specimen for measurement of in-plane shear strain.

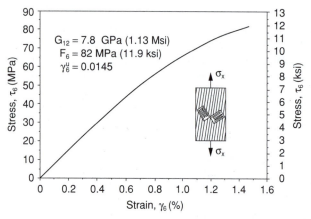

Fig. 10.30 Shear stress versus shear strain in $[10]_6$ carbon/epoxy specimen under uniaxial tensile loading (AS4/3501-6).

A and B of the rosette are oriented at $45°$ and $-45°$ to the fiber direction as shown in Fig. 10.29. The algebraic difference of the strain readings of gages A and B gives the in-plane shear strain directly:

$$\gamma_6 = \varepsilon_A - \varepsilon_B \tag{10.29}$$

This difference is read directly by the gage instrumentation when the two gages are connected to adjacent arms of the Wheatstone bridge.

The specimen is subjected to a uniaxial tensile stress σ_x up to failure. The intralaminar shear stress referred to the fiber coordinate system is given by

$$\tau_6 = -\sigma_x \sin\theta \cos\theta = 0.171\sigma_x \quad (10.30)$$

where

$$\theta = -10°$$

The in-plane shear modulus is obtained by plotting τ_6 versus γ_6 and taking the initial slope of the curve. The ultimate values of τ_6 and γ_6 define the shear strength and ultimate shear strain. A typical shear stress versus shear strain curve obtained from a $10°$ off-axis carbon/epoxy specimen is shown in Fig. 10.30. This method tends to underestimate the ultimate properties due to interaction of the transverse tensile stress across the fibers.

In both methods above the estimation of shear strength as the ultimate value of τ_6 is based on the implicit assumption of the maximum stress criterion. Different values of shear strength would be obtained if interaction failure criteria were used. The Tsai-Hill failure criterion of Eq. (6.16) can be solved for the shear strength F_6 as

$$F_6 = \frac{\tau_6}{(1 - \sigma_1^2/F_1^2 - \sigma_2^2/F_2^2 + \sigma_1\sigma_2/F_1^2)^{1/2}} \tag{10.31}$$

Using the Tsai-Wu criterion of Eq. (6.29) we obtain

$$F_6 = \frac{1}{\sqrt{f_{66}}} = \frac{\tau_6}{(1 - f_1\sigma_1 - f_2\sigma_2 - f_{11}\sigma_1^2 - f_{22}\sigma_2^2 - 2f_{12}\sigma_1\sigma_2)^{1/2}} \tag{10.32}$$

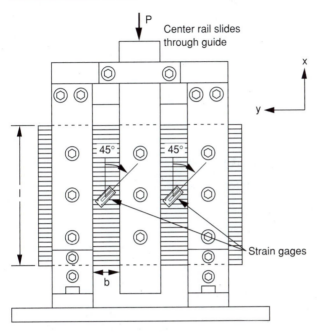

Fig. 10.31 Three-rail shear test fixture.

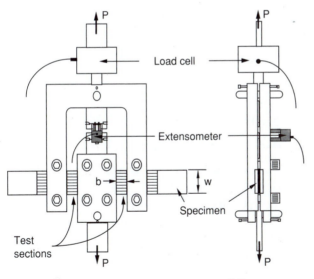

Fig. 10.32 Modified three-rail shear test fixture.[44,45]

In the above relations, τ_6 is the ultimate value of the shear stress in the test, and σ_1 and σ_2 are the corresponding normal stresses at failure.

The third method of determining shear properties is the rail shear test, the two-rail or the three-rail test as described in ASTM standard D4255M-01.[5] In the two-rail test, a rectangular composite coupon is gripped along its long edges by two pairs of rails that are loaded in a direction nearly parallel to the edges. In the three-rail test, a rectangular composite coupon is clamped between three parallel pairs of rails (Fig. 10.31). The load is applied to one end of the middle rails and reacted at the opposite ends of the two outer pairs of rails. The average shear stress applied to the specimen is

$$\tau_6 = \frac{P}{2lh} \tag{10.33}$$

where

$$P = \text{load}$$

$$l = \text{specimen length along rails}$$

$$h = \text{specimen thickness}$$

The shear strain is obtained from a single gage placed at the center of the exposed specimen at 45° to the rail axes,

$$\gamma_6 = 2(\varepsilon_x)_{\theta=45°} \tag{10.34}$$

Sometimes a three-gage rectangular rosette, with additional gage elements in the x- and y-directions, is used to ensure that the state of stress at the center of the specimen is pure shear. This condition is best approximated when the aspect ratio of length to width of the exposed specimen section is large, usually greater than 8:1. The state of stress near the ends is not pure shear, and the large normal stress concentrations at the ends may result in premature failures.

Modifications of the three-rail shear test have been proposed.[44,45] The modified fixture allows the use of standard tensile coupons (Fig. 10.32). The aspect ratio of the two test sections is 2:1. The shear deformation is determined by means of an extensometer, which measures the relative motion of the central rails with respect to the outer ones. The Timoshenko beam theory is used to account for bending effects in the determination of the shear modulus:[45]

$$G_{12} = \frac{6PbE_1w^2}{108E_1hw^3 - 5Pb^3}$$

(10.35)

where

$$\delta = \text{deflection measured by extensometer}$$

$$w = \text{specimen height (or coupon width)}$$

$$h = \text{specimen thickness}$$

$$b = \text{test section width}$$

This fixture can be applied to 90° and $[0/90]_{ns}$ crossply coupons as well. In the latter case it has been used to monitor shear modulus degradation due to matrix cracking, and the results showed excellent agreement with theoretical predictions.[44,45]

The fourth method is the torsion method utilizing a solid rod or a hollow tubular specimen subjected to torque. For a unidirectional tubular specimen with the fibers in the axial direction, the maximum shear stress and shear strain are

$$(\tau_6)_{\text{max}} = \frac{2Tr_o}{\pi(r_o^4 - r_i^4)}$$

(10.36)

$$(\gamma_6)_{\text{max}} = \psi r_o = (\varepsilon_x)_{\theta=45°} - (\varepsilon_x)_{\theta=-45°} = 2(\varepsilon_x)_{\theta=45°}$$

(10.37)

where

$$r_i, r_o = \text{inner and outer radii}$$

$$\psi = \text{angle of twist per unit length}$$

$$(\varepsilon_x)_{\theta=45°}, (\varepsilon_x)_{\theta=-45°} = \text{surface strains at 45° and } -45° \text{ to tube axis}$$

For a thin-wall tube Eq. (10.36) can be approximated as follows:

$$(\tau_6)_{\text{max}} \cong \frac{T}{2\pi\bar{r}^2h}$$

(10.38)

where

$$\bar{r} = \tfrac{1}{2}(r_o + r_i)$$

is the mean radius. For a solid rod, $r_i = 0$ in Eq. (10.36). The shear strain can be obtained by measuring the angle of twist or the strains at 45° and/or −45° with strain gages.

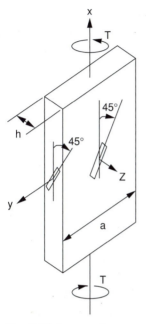

Fig. 10.33 Rectangular coupon under torsion for determination of shear moduli.[47]

Although the tube torsion test seems very desirable from the mechanics point of view, tubular specimens are difficult to make and load. The solid rod torsion test is less desirable because of the shear stress gradient across the section.

Another torsion method has been developed that utilizes thin rectangular coupons of the type used in tensile testing (Fig. 10.33).[46,47] A closed-form solution was obtained for such a specimen under torsion relating the applied torque, T, with the angle of twist, ψ, in terms of the shear moduli of the lamina. For a unidirectional laminate twisted about the longitudinal (fiber) axis, the torque-twist relationship is

$$T = \frac{1}{3} a \psi G_{12} h^3 \left(1 - \frac{\tanh \beta}{\beta} \right) \tag{10.39}$$

where

$$a = \text{specimen width}$$

$$\psi = \text{angle of twist per unit length}$$

$$h = \text{specimen thickness}$$

$$G_{12}, G_{13} = \text{in-plane and out-of-plane shear moduli}$$

and

$$\beta = \frac{a}{2h} \sqrt{10 G_{13}/G_{12}} \tag{10.40}$$

For a unidirectional laminate twisted about the transverse axis,

$$\beta = \frac{a}{2h} \sqrt{10 G_{23}/G_{12}} \tag{10.41}$$

where

$$G_{23} = \text{out-of-plane shear modulus in the transverse 2-3 plane}$$

The three shear moduli of a unidirectional composite, G_{12}, G_{13}, and G_{23}, can be determined by conducting selected tests on unidirectional prismatic coupons and measuring the torque and angle of twist. A minimum of three tests would be needed, for example, two tests with 0° specimens of different cross-sectional dimensions (h, a) and one test with a 90° specimen. However, direct measurement of the overall angle of twist does not yield accurate results because of end effects from the specimen tabs and grips. In the method described by Tsai and Daniel,[47] all three shear moduli can be obtained from two tests of a unidirectional material twisted about the fiber and transverse to the fiber directions. In each test the strains are measured on the face and on the edge of the coupon at 45° to the torque axis. Typical curves of applied torque versus surface and edge strains at 45° for a $[0_{32}]$

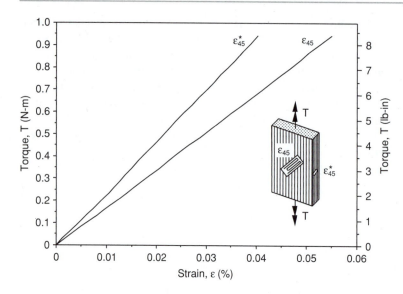

Fig. 10.34 Torque versus surface and edge strains at 45° to torque axis for [0₃₂] carbon/epoxy specimen (AS4/3501-6).[47]

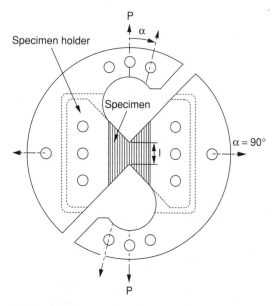

Fig. 10.35 Arcan loading fixture and specimen holder for pure shear and mixed-mode loading.[48]

carbon/epoxy specimen are shown in Fig. 10.34. Results of the complete shear modulus characterization for a carbon/epoxy material (AS4/3501-6) are

$$G_{12} = 6.90 \text{ GPa } (1.00 \text{ Msi})$$

$$G_{13} = 6.97 \text{ GPa } (1.01 \text{ Msi})$$

$$G_{23} = 3.73 \text{ GPa } (0.54 \text{ Msi})$$

Another type of shear test is based on the fact that a shear force transmitted through a section between two edge notches produces a nearly uniform shear stress along the section. Two test methods and fixtures are based on this principle, the Arcan and Iosipescu tests. The Arcan fixture is illustrated in Fig. 10.35.[48] The specimen is a short coupon with two 90° notches. The coupon is mounted on the fixture through a bolted specimen holder. The load can be applied at various orientations with respect to the section through the notches. This allows the application of any biaxial state of stress from pure shear to transverse tension or any combination thereof. Pure shear loading is obtained for $\alpha = 0°$. In the Iosipescu test the specimen is a beam with two 90° notches loaded in the fixture shown in Fig. 10.36.[49] In both cases the average shear stress applied through the notched section of the specimen is

$$\tau_6 = \frac{P}{lh} \tag{10.42}$$

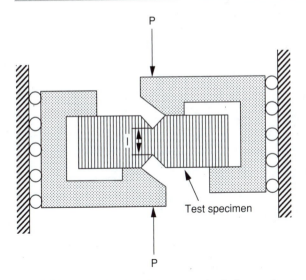

Fig. 10.36 Schematic of loading fixture for Iosipescu shear test.[49]

where

l = specimen height at notch location

h = specimen thickness

The shear strain, γ_6, can be measured with a strain gage located at the center of the notched section at 45° to the loading direction, as in the case of the three-rail shear specimen, Eq. (10.34). The in-plane shear modulus, G_{12}, is obtained as the slope of the τ_6 versus γ_6 curve. The ultimate value of τ_6 yields the in-plane shear strength F_6 of the lamina.

The Arcan test can also be used to measure the out-of-plane (interlaminar) shear modulus, G_{31}, by bonding the faces of a thick unidirectional lamina to the specimen holder in Fig. 10.35. The relative motion of the two parts of the specimen holder must then be measured as a function of applied load.

10.7 DETERMINATION OF THROUGH-THICKNESS PROPERTIES

Through-thickness elastic and strength properties are required to study the behavior of thick composite sections (20 mm or greater), to analyze thick laminates under a three-dimensional state of stress, and to evaluate the effect of various types of through-thickness reinforcement. Through-thickness testing is more problematic than in-plane testing because it is difficult to fabricate material of sufficient thickness and uniform quality. It is also difficult to introduce the loading without the deleterious influence of end effects and stress concentrations. Different tests have been proposed over the years, but the amount of data is limited. An overview of through-thickness test methods was given by Lodeiro et al.[50] A number of test methods were described with results for textile composites.[51] Through-thickness mechanical characterization is performed primarily by three types of tests, tensile, compressive, and interlaminar shear tests.

10.7.1 Through-Thickness Tensile Properties

Determination of through-thickness (interlaminar) tensile properties has the following objectives:

1. assessment of laminate integrity in the out-of-plane direction (3- or z-direction) for quality control of the lamination process
2. generation of design allowables for thick composite laminates subjected to three-dimensional states of stress
3. evaluation of laminate resistance to delamination caused by interlaminar tensile stresses near free edges under in-plane loading

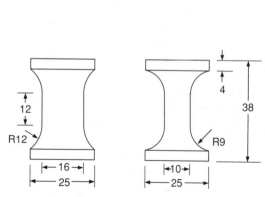

Fig. 10.37 Geometry of through-thickness waisted block specimen (Ferguson et al.[52]; dimensions in mm).

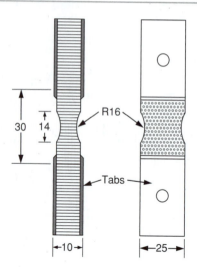

Fig. 10.38 Specimen for determination of interlaminar tensile strength.[53]

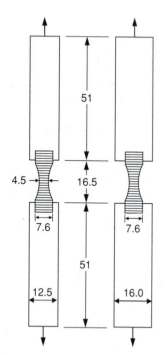

Fig. 10.39 Through-thickness tensile specimen used at Northwestern University[51] (dimensions in mm).

The direct approach is similar to that used for determination of in-plane transverse tensile properties. Coupons of rectangular or square cross section are machined in the thickness direction. These blocks are usually short, as they are limited by the thickness of the material (usually less than 40 mm). Various configurations of short tensile specimens have been used.[50–53] The specimen proposed by DERA (Defense Evaluation and Research Agency, UK) is a waisted block with rectangular cross section and large radii at the ends of the gage section (Fig. 10.37). The specimen used at the Technion Research and Development Foundation (Israel) consists of three bonded pieces, each of which is machined from a laminate less than 30 mm thick. It is circularly waisted at the center and tabbed as shown in Fig. 10.38. The specimen used at Northwestern University (NU) is made from a laminate less than 27 mm thick and is waisted along its length by circular arcs to a 4.5 × 4.5 mm (0.177 × 0.177 in) square cross section (Fig. 10.39). The NU specimen is bonded into specially machined wells in aluminum shanks used for load introduction. The adhesive works under both tension and shear to ensure failure in the specimen gage section. The specimen is instrumented with strain gages on all four sides to monitor any possible misalignment. A through-thickness tensile stress-strain curve is shown in Fig. 10.40 for a woven carbon/epoxy composite (AGP370-5H/3501-6).

An indirect test for determination of through-thickness tensile strength, the so-called *split ring test*, was originally suggested for characterization of filament-wound composite elements.[54] This test was later modified and improved.[55] It consists of a curved laminated beam specimen loaded in tension, as shown in Fig. 10.41. Under this loading, a pure interlaminar radial tensile stress, σ_z, is induced in the apex of the specimen test section. For semicircular geometry this stress can be closely estimated by the following relation:

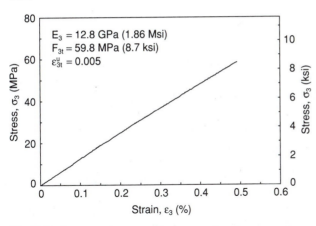

Fig. 10.40 Stress-strain curve of woven carbon/epoxy specimen under through-thickness tensile loading (AGP370-5H/3501-6S).[51]

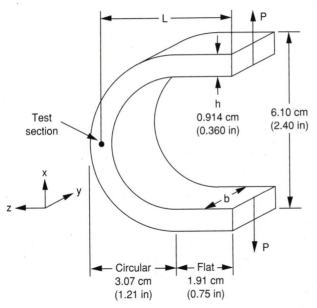

Fig. 10.41 Semicircular curved beam specimen for determination of interlaminar tensile strength.[55]

$$\sigma_z = \frac{3PL}{2bhR} \qquad (10.43)$$

where

P = applied tensile load

L = distance between load and test section

b = beam width

h = beam thickness

R = mean radius of circular section

Test results obtained with this specimen are characterized by high scatter and low mean value of F_{3t}. Better results were obtained by using a "scarfed circular specimen" with significantly reduced test section and also by using specimens with elliptical geometry.[55] Regardless of the experimental approach used, the quality of the results can be assessed by how close the values of F_{3t} are to those of F_{2t}.

10.7.2 Through-Thickness Compressive Properties

The specimens for determination of through-thickness compressive properties can be straight-sided blocks with square or rectangular cross section or waisted blocks similar to those used in tensile testing. The specimens are bonded to steel blocks and loaded directly. It is very important to ensure that the bonded ends of the specimen and loaded ends of the steel blocks are parallel to prevent buckling.[51]

10.7.3 Interlaminar Shear Strength

Interlaminar shear strength is a measure of the in situ shear strength of the matrix layer between plies. There is no method available for exact determination of this property. Approximate values of the interlaminar shear strength, or apparent interlaminar shear strength, can be obtained by various tests.

The most commonly used test is the short beam under three-point bending (Fig. 10.42). The beam is machined from a relatively thick (at least 16 plies thick) unidirectional laminate with the fibers in the axial direction and is loaded normally to the plies, in the 3-direction. (ASTM D2344M-00).[5] Some doubts have been raised about the validity of results

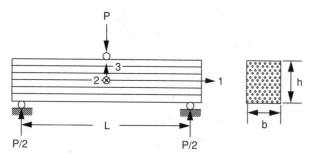

Fig. 10.42 Short-beam shear test for measurement of interlaminar shear strength.

obtained from thin laminates (less than 16 plies thick) because of local compressive failure near the loaded points. Better results are obtained with thicker (approximately 50 plies thick) laminates.[56,57] Because of its simplicity, the short-beam shear (SBS) test is used as a quality control (qualification) test of the lamination process and related matrix-dominated properties of the composite.

If the beam is sufficiently short compared to its depth, shear failure will take place at the mid-plane in the form of delamination. This is true only sufficiently far away from the load and reaction points, where a parabolic shear stress distribution through the thickness can be assumed. Finite element analyses have shown that the shear stress distribution is skewed near the load and reaction points and that interlaminar shear stresses larger than those predicted by classical beam theory exist.[58] The apparent interlaminar shear strength obtained from classical beam theory is given by

$$F_{31} \cong \frac{3P}{4bh} \tag{10.44}$$

where

$$P = \text{load at failure initiation}$$

$$b = \text{width of beam}$$

$$h = \text{depth of beam (laminate thickness)}$$

If the beam is too long compared to its depth, flexural failure (tensile or compressive) may take place at the outer plies of the beam. To ensure interlaminar shear failure prior to flexural failure, the span to depth ratio must satisfy the relationship

$$\frac{2L}{h} < \frac{F_1}{F_{31}} \tag{10.45}$$

where

$$L = \text{beam span}$$

$$F_1 = \text{flexural strength of beam in the fiber direction (smaller of } F_{1t} \text{ or } F_{1c})$$

For a typical 15- to 20-ply carbon/epoxy laminate, a span length $L = 1$ cm (0.4 in), width $b = 0.64$ cm (0.25 in), and thickness $h = 1.9$ to 2.5 mm (0.075 to 0.100 in) are used.

The short-beam shear test method is problematic in the case of materials with low flexural to interlaminar shear strength ratio, F_1/F_{31}, such as Kevlar, textile, and carbon/carbon composites.[59] Modifications of the method for this case have been proposed.[59-61] In one of the approaches, the direct concentrated loading in the SBS test is replaced by a

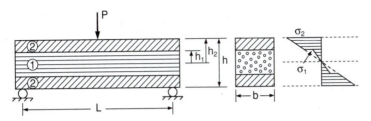

Fig. 10.43 Short sandwich beam specimen for measurement of interlaminar shear strength of weak composites.[59]

distributed patch loading through an aluminum plate and a rubber pad.[61] In another approach, the test lamina or laminate is sandwiched between two strips of steel or other tougher composite material and tested under three- or four-point bending.[59,60] The short sandwich beam (SSB) test is illustrated in Fig. 10.43. The maximum normal stresses in the core (test) and facesheet material are

$$\sigma_1 = \frac{PLh_1}{4I_e} \tag{10.46}$$

$$\sigma_2 = \frac{\sigma_1}{kn} \tag{10.47}$$

where

$$I_e = \frac{2b}{3n}\left[h_1^3(n-1) + h_2^3\right] = \text{equivalent moment of inertia of beam cross section} \tag{10.48}$$

and

$$k = \frac{h_1}{h_2}$$

$$n = \frac{E_1}{E_2} = \text{ratio of axial moduli of core and facesheet materials}$$

The interlaminar shear stress on the midplane of the specimen is given by[59]

$$\tau = \frac{3P}{4bh}\beta \tag{10.49}$$

where

$$\beta = \frac{1 + k^2(n-1)}{1 + k^3(n-1)} \tag{10.50}$$

The factor β can be considered as a "correction factor" for the heterogeneous beam.

Another test proposed for measurement of interlaminar shear strength is the double-notch shear test described in ASTM specification D3846-02.[5] The specimen is a unidirectional coupon 79.5 mm (3.13 in) long, 12.7 mm (0.50 in) wide, and 2.54 to 6.60 mm (0.100 to 0.260 in) thick (Fig. 10.44). Two parallel notches or grooves are machined, one on each face of the specimen, 6.4 mm (0.25 in) apart and of depth equal to half the specimen

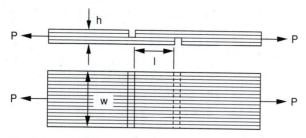

Fig. 10.44 Double-notch specimen for determination of interlaminar shear strength.

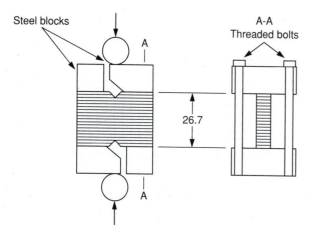

Fig. 10.45 Modified Iosipescu fixture for through-thickness shear testing (with dimension given in mm).[51]

thickness. When this specimen is loaded in uniaxial tension or compression, shear failure results along the midplane of the specimen between the notches. In the case of compressive loading, a supporting fixture is recommended to prevent buckling as shown in ASTM D3846-02.[5] The interlaminar shear strength is then given by

$$F_{31} \cong \frac{P}{wl} \qquad (10.51)$$

where P is the failure load, l is the distance between notches, and w the width of the specimen.

A stress analysis of the notched specimen showed that the interlaminar shear stress along the midplane between the notches is nonuniform, but becomes more uniform as the distance between the notches decreases.[62] For shear failure to take place between the notches, the following condition must be satisfied (for tensile or compressive loading):

$$\frac{F_{31}}{F_{1t}} < \frac{h}{2l} \qquad (10.52)$$

or

$$\frac{F_{31}}{F_{1c}} < \frac{h}{2l}$$

where

F_{1t}, F_{1c} = longitudinal tensile and compressive strength of the lamina

h = specimen thickness

In a modification of the above test method, the notched coupon is incorporated as a compression facesheet of a sandwich beam loaded in pure bending.[59]

The interlaminar shear strength can also be measured by means of the Arcan test discussed before. The faces of a unidirectional coupon are bonded to the specimen holder of Fig. 10.35 and the failure load recorded. The interlaminar shear strength is obtained by an expression similar to Eq. (10.51) where now l is the distance between notches and w is the width of the coupon. Interlaminar shear tests have also been performed using the Iosipescu fixture discussed before (Fig. 10.36). The shear strain is measured with strain gages mounted on the line connecting the V-notches oriented at 45° to the line or loading direction. A modified Iosipescu fixture has been used successfully to determine the through-thickness shear behavior of a woven fabric/epoxy composite (Fig. 10.45).[51]

10.8 DETERMINATION OF INTERLAMINAR FRACTURE TOUGHNESS

Most high-performance composites are designed to have superior in-plane stiffness and strength. However, one of the most common failures in composites is delamination, separation of layers or plies under shear and normal stresses. Two measures of the resistance of the composite to delamination are the through-thickness tensile strength and the interlaminar shear strength discussed in Section 10.7. Another measure of resistance to delamination is based on a fracture mechanics approach and is expressed as the energy release rate associated with a propagating delamination crack. As discussed previously (see Section 9.16), interlaminar crack propagation can occur under opening, forward shearing, tearing, or a combination thereof; therefore, delamination fracture toughness can be characterized by stress intensity factors or strain energy release rates corresponding to modes I, II, and III. Several test methods have been developed for these modes and combinations thereof. Reviews of some of these methods have been published.[8,10,63-66]

10.8.1 Mode I Testing

The most commonly used specimen for mode I characterization is the double cantilever beam (DCB) specimen (ASTM D5528-01). The DCB specimen shown in Fig. 10.46 was introduced and described in 1982.[63,67]

Methods of analysis of the DCB specimen were discussed in Section 9.16. In the beam analysis or compliance method, the specimen is assumed to consist of two identical cantilever beams with built-in ends and length equal to the length of the crack. For quasi-static loading the critical energy release rate is given by

$$G_{Ic} = \frac{12P^2}{E_1 b^2 h}\left[\left(\frac{a}{h}\right)^2 + \frac{E_1}{10G_{31}}\right] \qquad (9.59 \text{ bis})$$

where

$P = $ maximum applied load at crack extension

$b = $ specimen width

$h = $ cantilever beam thickness

$E_1 = $ longitudinal modulus (in the fiber direction)

$G_{31} = $ transverse shear modulus

$a = $ crack length

Another commonly used method of analysis is the area method. The energy released per unit area of crack extension is simply calculated as

$$G_{Ic} = \frac{1}{2b\Delta a}(P_1\delta_2 - P_2\delta_1) \qquad (9.60 \text{ bis})$$

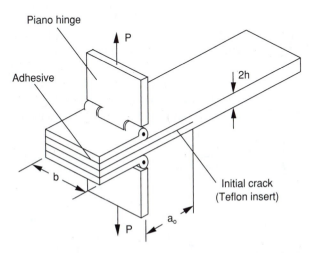

Fig. 10.46 Double cantilever beam specimen with initial crack and attached hinges for loading.

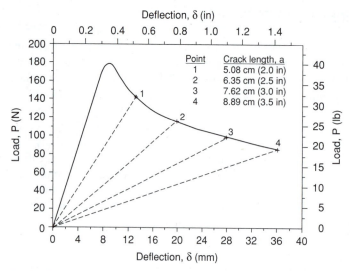

Point	Crack length, a
1	5.08 cm (2.0 in)
2	6.35 cm (2.5 in)
3	7.62 cm (3.0 in)
4	8.89 cm (3.5 in)

Fig. 10.47 Load versus crack opening deflection for uniform DCB specimen with an initial crack of $a_o = 3.81$ cm (1.5 in) loaded at a crosshead rate of 0.0085 mm/s (0.02 in/min) (AS4/3501-6 carbon/epoxy).[69]

where the load P_1 corresponding to opening deflection δ_1 drops to load P_2 corresponding to deflection δ_2 after an increment Δa in crack length (see Fig. 9.39).

The uniform DCB specimen is usually 22.9 cm (9 in) long, 2.54 cm (1 in) wide, and 3 to 3.6 mm (0.120 to 0.140 in) thick, with an initial artificial crack of 3.81 cm (1.5 in) length at one end. This crack is produced by inserting a 0.025 mm (0.001 in) thick folded polytetrafluoroethylene (PTFE, Teflon) film at the midplane of the laminate near one end. Metallic piano hinges are bonded to the cracked end of the specimen as shown in Fig. 10.46 to allow for unrestrained rotation at that end during load introduction.

The specimen is loaded in a testing machine at a low crosshead rate on the order of 0.5 to 1.3 mm/min (0.02 to 0.05 in/min) in order to produce stable crack growth. The opening deflection is determined by measuring the crosshead displacement or by means of a linear variable differential transformer (LVDT) extensometer. A continuous load-deflection curve is obtained as shown in Fig. 10.47. Incremental crack lengths are marked on this curve during the test. Crack extension is monitored in several different ways. At low crosshead rates, crack extension can be monitored visually. At higher rates, the crack can be monitored either by means of strain gages mounted on the top face of the specimen or by means of a conductive paint circuit applied to the edge of the DCB specimen.[68,69]

The use of the uniform DCB specimen requires monitoring of the crack length a, which becomes more difficult at high loading rates. This problem is alleviated with a different type of specimen, the width-tapered double cantilever beam (WTDCB) specimen.[70] The WTDCB specimen has the property of constant rate of change of compliance with crack length, that is, $dC/da = $ constant. For this reason this specimen does not require exact monitoring of the crack length and yields a constant crack velocity for a constant opening deflection rate.[71] The critical strain energy release rate based on the compliance method is

$$G_{Ic} = \frac{12P^2k^2}{E_1h^3}\left[1 + \frac{1}{10}\frac{E_1}{G_{31}}\left(\frac{h}{a}\right)^2\right] \tag{10.53}$$

where

$$k = \frac{a}{b}$$

$b = $ beam width at crack length a

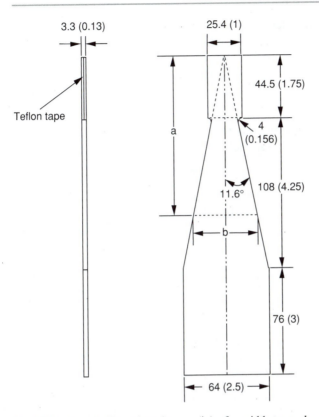

Fig. 10.48 Typical dimensions in mm (in) of a width-tapered double cantilever beam (WTDCB) specimen.[71]

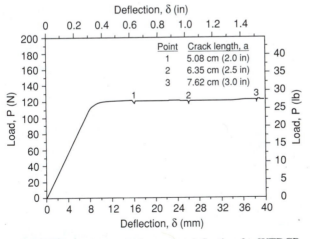

Fig. 10.49 Load versus crack opening deflection for WTDCB specimen loaded at a crosshead rate of 0.85 mm/s (2 in/min) (T300/F-185 carbon/epoxy).[69]

An alternative expression based on the area method is

$$G_{Ic} = \frac{P_1\delta_2 - P_2\delta_1}{a_2^2 - a_1^2}k \qquad (10.54)$$

where a_1 and a_2 are crack lengths corresponding to loads P_1 and P_2 and deflections δ_1 and δ_2, respectively.

Typical dimensions of a WTDCB specimen are given in Fig. 10.48. A rectangular section is provided near the end to facilitate the attachment of hinges for load introduction. A continuous load-deflection curve obtained with a WTDCB carbon/epoxy specimen is shown in Fig. 10.49.

Another type of specimen proposed and used for the study of loading rate effects on interlaminar fracture toughness is the height-tapered double cantilever beam (HTDCB) specimen.[72]

Numerous studies have been reported on the effects of material, layup, temperature, viscoelastic behavior, and loading rate on delamination fracture toughness.[68,69,71–78] Typical results of critical strain energy release rates for various types of carbon fiber composites are shown in Table 10.1.[63,67–69,71,74–76] These results demonstrate the effectiveness of the DCB test in manifesting the significantly higher (approximately one order of magnitude higher) G_{1c} values of composites with tough thermoplastic matrices like PEEK compared to those with standard epoxy matrices.

10.8.2 Mode II Testing

The significance and measurement of interlaminar shear fracture toughness was discussed at length.[79] The most popular method for measuring mode II delamination fracture toughness is the end-notched flexure (ENF) test.[80] The width of the specimen is usually 2.54 cm (1 in), the crack length also 2.54 cm (1 in), and the total span 10.16 cm (4 in). The beam is first loaded as a DCB specimen in mode I up to the point of crack initiation. Then it is loaded in flexure under three-point or four-point (4ENF) bending until further crack growth occurs at the maximum load. A load-deflection curve is recorded. Independently, a compliance calibration is performed by testing similar specimens with various crack lengths.

For three-point bending shown in Fig. 10.50, the strain energy release rate based on linear beam theory with linear elastic behavior, including effects of shear deformation, is given by

TABLE 10.1 Critical Strain Energy Release Rates for Various Carbon Fiber Composite Materials

| Material | Test* | Strain Energy Release Rate, Jm⁻² (lb/in) | |
		G_{Ic}	G_{IIc}
T300/5208	DCB	88–103 (0.50–0.59)	
	CLS		154–433 (0.88–2.47)
T300/914	DCB	185 (1.06)	
	ENF		520–600 (2.97–3.43)
	CBEN		496 (2.83)
T300/F-185	WTDCB	1880 (10.73)	
AS4/3501-6	DCB	130–190 (0.74–1.08)	
	HTDCB	189 (1.08)	
	ENF		570–810 (3.25–4.62)
AS1/3501-6	DCB	110–130 (0.62–0.74)	
	ENF		460–670 (2.62–3.82)
AS4/3502	DCB	160–227 (0.91–1.29)	
	ELS		560 (3.20)
	ENF		587 (3.35)
AS4/PEEK	DCB	1260–1680 (7.19–9.59)	
	ENF		1200–1765 (6.84–10.07)
	CBEN		1860 (10.62)
	ENCB		1780 (10.16)
AS4 Fabric/LY564	DCB	720 (4.11)	
	ENF		3500 (19.98)
IM7/8552	DCB	200 (1.14)	
	4ENF		1100–1670 (6.28–9.53)
IM7/E7T1	DCB	180 (1.03)	
	ENF		1100–2450 (6.28–13.98)

*DCB = double cantilever beam; CLS = cracked-lap shear; ENF = end-notched flexure; CBEN = cantilever beam with embedded notch; WTDCB = width tapered double cantilever beam; HTDCB = height-tapered double cantilever beam; ELS = end-loaded split laminate; ENCB = end-notched cantilever beam; 4ENF = four-point end-notched flexure.

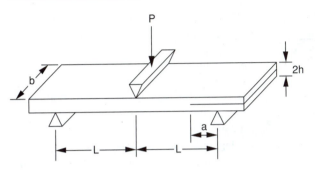

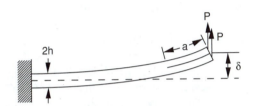

Fig. 10.50 End-notched flexure (ENF) specimen for determination of mode II interlaminar fracture toughness.

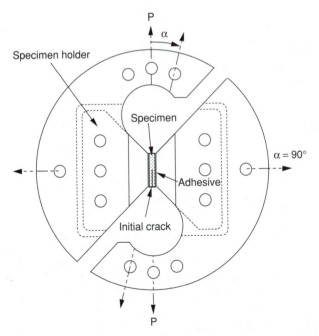

Fig. 10.51 End-loaded split laminate (ELS) specimen for determination of mode II interlaminar fracture toughness.

Fig. 10.52 Specimen holder and loading fixture for pure and mixed-mode interlaminar fracture testing in Arcan fixture.[87]

$$G_{IIc} = \frac{9P^2a^2}{16E_1b^2h^3}\left[1 + 0.2\frac{E_1}{G_{31}}\left(\frac{h}{a}\right)^2\right] \quad (10.55)$$

Neglecting shear deformation, the above is expressed in terms of the measured compliance C as

$$G_{IIc} = \frac{9P^2a^2C}{2b(2L^3 + 3a^3)} \quad (10.56)$$

Many applications of the ENF specimen have been reported, including effects of fatigue, adhesive interleaving, and loading rate.[76,81–86]

In general, mode II interlaminar fracture toughness may be obtained from mixed-mode tests. One such test is the end-loaded split laminate (ELS) test (Fig. 10.51).[82] In this case, the expression for G_{IIc}, neglecting shear deformation, is similar to the one for the ENF test.

$$G_{IIc} = \frac{9P^2a^2}{4E_1b^2h^3} \quad (10.57)$$

The Arcan test configuration can also be used in testing for pure mode II as well as any combination of modes I and II by attaching an end-notched coupon in the specimen holder (Fig. 10.52).[87] Additional mixed-mode tests used for computation of G_{IIc} include the end-notched cantilever beam (ENCB) and the cantilever beam with enclosed notch (CBEN).[76]

Typical results for mode II critical strain energy release rates are summarized in Table 10.1 for various types of carbon/epoxy composites.[64,67,74,76,80,82] These results show how the ENF test manifests the higher (approximately three times higher) G_{IIc} values of composites with tough thermoplastic matrices like PEEK, compared to composites with standard epoxy matrices.

10.8.3 Mixed-Mode Testing

Mixed-mode (I and II) fracture toughness has been measured by a variety of test methods such as the cracked-lap shear (CLS), mixed-mode bending (MMB), edge delamination tension

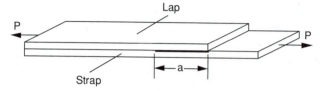

Fig. 10.53 Cracked-lap shear specimen.

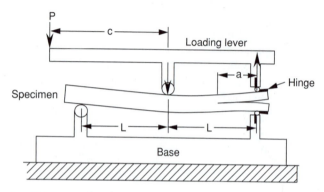

Fig. 10.54 Mixed-mode bending specimen and test apparatus.[89]

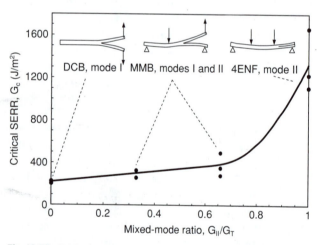

Fig. 10.55 Critical strain energy release rate (SERR) as a function of mixed-mode ratio for carbon/epoxy composite (IM7/8552). (Courtesy of T. K. O'Brien.[91])

(EDT), and the Arcan specimen. The CLS specimen shown in Fig. 10.53 has been used for composite materials as well as adhesive bond evaluation.[67,70] Uniaxial loading is applied to one arm (strap) of a split unidirectional laminate. The load transfer to the other arm (lap) produces both shear (mode II) and peel (mode I) stresses along the interface between the lap and strap arms. The relative magnitudes of G_I and G_{II} can be modified by adjusting the relative thicknesses of the strap and lap parts. The exact determination of the two fracture toughness components is usually done by means of finite element analysis.[88]

The mixed-mode bending test is a combination of the DCB and ENF tests described before.[89] A sketch of the specimen and loading are shown in Fig. 10.54. Detailed drawings and the test procedure are given in ASTM D6671M-04. The individual components of fracture toughness are determined as follows:

$$G_{Ic} = \frac{3P^2a^2}{4b^2h^3L^2E_1}(3c - L)^2 \quad (10.58)$$

$$G_{IIc} = \frac{9P^2a^2}{16b^2h^3L^2E_1}(c + L)^2 \quad (10.59)$$

Their ratio

$$G_{Ic}/G_{IIc} = \frac{4}{3}\frac{(3c - L)^2}{(c + L)^2} \quad c \geq \frac{L}{3} \quad (10.60)$$

is only a function of the load position c and half span length L. Results can also be presented in terms of the ratio G_{II}/G_T where $G_T = G_I + G_{II}$.[90] The variation of the total critical strain energy release rate G_c with mixed-mode ratio G_{II}/G_T is illustrated in Fig. 10.55 for a carbon/epoxy composite (IM7/8552).[91]

The EDT test utilizes $[(\pm\theta)_2/90/\overline{90}]_s$ and $[\pm\theta/0/90]_s$ laminates designed to delaminate at the edges under tensile loading.[92,93] The orientation θ is usually 30° in the first laminate and 35° in the second. In these laminates, a noticeable change in the load-deflection curve occurs at the onset of edge delamination. The total critical strain energy release rate associated with edge delamination growth in an unnotched composite laminate is given by[92]

$$G_c = \frac{\varepsilon_c^2 h}{2}(\bar{E}_x - \bar{E}_x^*)$$
(10.61)

where

ε_c = tensile strain at delamination onset

h = specimen thickness

$\bar{E}_x$ = laminate modulus before delamination

$\bar{E}_x^*$ = laminate modulus after total delamination along one or more interfaces

The value of G_c is independent of delamination size, but it depends on the laminate layup, which determines $\bar{E}_x$, and on the location of the delaminated surfaces, which determines $\bar{E}_x^*$. The two moduli are determined by both laminate plate theory and the rule of mixtures. The total strain energy release rate above consists of components G_I, G_{II}, and G_{III}. As in the case of the CLS specimen, numerical analyses are required to determine the individual components.

The Arcan test configuration can be used in principle to apply any desired combination of mode I and mode II interlaminar loading (Fig. 10.52).[87]

10.8.4 Mode III Testing

Compared with mode I and mode II testing, relatively little work has been reported on mode III testing. A split DCB specimen made by bonding two aluminum bars to the faces of the laminate at the split end is loaded in the direction parallel to the crack plane and normal to the beam axis.[94] As in the case of the mode I DCB test, the load drops suddenly as the crack extends, causing some uncertainty in the crack length corresponding to the critical load.

Another test method proposed for determination of G_{IIIc} uses a doubly split DCB specimen, as shown in Fig. 10.56.[95] The symmetry of the specimen ensures self-balancing and prevents twisting. The uncertainty in identifying and measuring the correct crack length at the critical load and the so-called *stick-slip phenomenon* are eliminated by adding a support to the doubly split DCB specimen, as shown in Fig. 10.56. The strain energy release rate for this specimen is given by

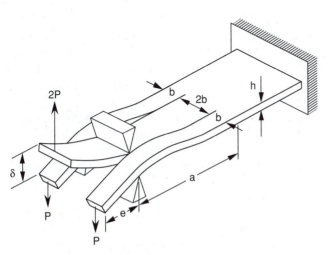

Fig. 10.56 Doubly split double cantilever beam specimen for determination of mode III fracture toughness.[95]

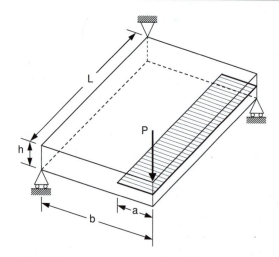

Fig. 10.57 Edge-cracked torsion (ECT) specimen for determination of mode III delamination fracture toughness.

$$G_{\mathrm{III}c} = \frac{3P^2 e^2}{Ebh^4} \qquad (10.62)$$

where

e = distance between the end load and the support

E = modulus of laminate along beam axis

b = width of outer split beams

h = laminate thickness

The fracture toughness in this case is independent of crack length $(a + e)$, and the critical tearing load P remains constant as the crack propagates.

Another test method introduced for the determination of mode III delamination toughness is the so-called *edge-cracked torsion* (ECT) test.[96] The test specimen is a rectangular laminate containing an edge delamination at the midplane (Fig. 10.57). The recommended layup for carbon/epoxy is $[90/(\pm45)_3/(\mp45)_3/90]_s$. A precrack of length a is obtained by inserting a 0.013 mm (0.0005 in) thick Teflon film between the 90° plies at the midplane. A detailed analysis of the specimen and a simplified test procedure have been published.[97,98] At the time of this writing the ECT specimen is being considered for standardization by ASTM.

10.9 BIAXIAL TESTING

10.9.1 Introduction

Failure theories, as discussed in the previous chapter, can predict FPF of multidirectional laminates under any state of in-plane stress. Ultimate failure, however, is difficult to predict analytically on the basis of single lamina properties because of unknown in situ effects, nonlinear behavior, and not well-defined damage progression and ultimate laminate failure. To check or verify analytical predictions and to generate useful failure envelopes for design purposes, it is necessary to conduct extensive testing of composite laminates under biaxial states of stress. The application of a general in-plane biaxial state of stress, including normal tension and compression and shear components, poses a difficult problem in composites testing. Some of the basic requirements for a biaxial test specimen are as follows:

1. A significant volume of the material must be under a homogeneous state of stress.
2. Primary failure must occur in the test section.
3. The state of stress in the test section must be known a priori or easily determined without the need for secondary measurements or analysis.
4. It must be possible to vary the three in-plane stress components ($\bar{\sigma}_x$, $\bar{\sigma}_y$, $\bar{\tau}_s$) independently.

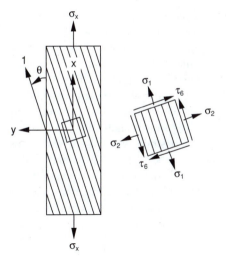

Fig. 10.58 Off-axis specimen for biaxial testing of unidirectional lamina.

A variety of specimen types and techniques have been proposed and used for biaxial testing of laminates. They include the off-axis coupon or ring, the crossbeam sandwich specimen, bulge plate, rectangular plate under biaxial tension, and the thin-wall tubular specimen.

10.9.2 Off-Axis Uniaxial Test

Uniaxial loading of a unidirectional lamina along a direction other than one of the principal axes produces a biaxial state of stress (Fig. 10.58). The state of stress referred to the principal material axes under uniaxial stress σ_x is obtained from the transformation relations in Eq. (4.57):

$$\sigma_1 = m^2 \sigma_x$$
$$\sigma_2 = n^2 \sigma_x \qquad (10.63)$$
$$\tau_6 = -mn\sigma_x$$

where

$$m = \cos\theta$$
$$n = \sin\theta$$

The off-axis specimen has been used successfully in coupon and ring form.[99–102] In the latter case thin-wall rings with the principal material axes at an angle to the circumferential direction are subjected to internal pressure loading. Some of the limitations of the off-axis specimen are as follows:

1. The biaxial normal stresses are always of the same sign.
2. There is no possibility for independent variation of the three stress components (nonproportional loading).
3. Erroneous stiffness and strength results may be obtained when using tensile coupon specimens of dimensions and with clamping conditions customarily used in testing along principal axes.

If the applied stress σ_x were uniform throughout, that is, if the specimen were free to deform, the specimen strains would be

$$\varepsilon_x = \frac{1}{E_x}\sigma_x$$

$$\varepsilon_y = -\frac{\nu_{xy}}{E_x}\sigma_x \qquad (4.74 \text{ bis})$$

$$\gamma_s = \frac{\eta_{xs}}{E_x}\sigma_x$$

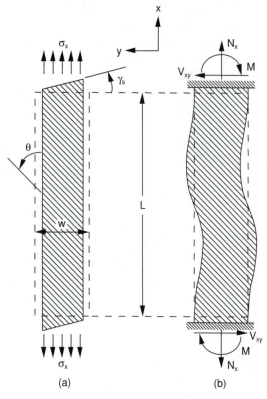

Fig. 10.59 Effect of end constraint on off-axis tensile specimen: (a) free ends and (b) clamped ends.[103]

which means that a rectangular coupon would deform into a parallelogram-shaped one as shown in Fig. 10.59a, due to shear coupling. If the specimen is clamped at the ends during loading, the shear deformation is constrained and the specimen deforms, as shown in Fig. 10.59b.[103] This constraint induces shear and bending moments at the ends as shown that disturb the stress field uniformity in the specimen. It was shown analytically that a uniform state of stress and strain will exist at the center of the specimen if the aspect ratio L/w (length/width) is sufficiently large.[104] Furthermore, an expression was obtained for the error involved and a correction factor for determination of the true axial modulus as follows:[104]

$$E_x = E_x^*(1 - \zeta) \tag{10.64}$$

and

$$\zeta = \frac{6S_{xs}^2}{S_{xx}\left[6S_{ss} + S_{xx}\left(\dfrac{2L}{w}\right)^2\right]}$$

$$= \frac{\eta_{xs}^2}{\dfrac{E_x}{G_{xy}} + \dfrac{2}{3}\left(\dfrac{L}{w}\right)^2} \tag{10.65}$$

where

$$E_x = \text{true axial modulus}$$

$$E_x^* = \frac{\sigma_x}{\varepsilon_x} = \text{apparent (uncorrected) axial modulus}$$

$$\zeta = \text{correction factor}$$

$$S_{xx}, S_{xs}, S_{ss} = \text{compliance parameters of off-axis lamina (functions of } \theta)$$

$$E_x, G_{xy}, \eta_{xs} = \text{engineering parameters of off-axis lamina (see Chapter 4)}$$

$$L = \text{specimen length}$$

$$w = \text{specimen width}$$

The shear coupling effect in a uniaxially loaded off-axis specimen as characterized by the correction factor ζ in Eq. (10.65) depends on the following variables:

1. the clamping conditions
2. the degree of anisotropy of the composite, which affects the ratio E_x/G_{xy}

3. the shear coupling parameter η_{xs}, which depends primarily on the off-axis angle θ and becomes predominant in the range $10° < \theta < 45°$

4. the specimen aspect ratio or length to width ratio L/w

Of these variables, the first and last are the only ones that can be controlled in testing of a given material at a given off-axis angle. When the aspect ratio L/w is large enough, the shear coupling effect becomes independent of clamping conditions.[105] For an E-glass/epoxy specimen with a ratio $L/w = 24$, small differences were observed in strength and stiffness for clamped and hinged end conditions ($\zeta < 0.05$).

Off-axis testing is not limited to the unidirectional lamina. Multidirectional symmetric laminates can be tested under uniaxial loading at an angle to one of the principal laminate axes ($\bar{x}, \bar{y}$). The off-axis laminate can also be tested in compression by using it as a skin in a soft-core sandwich beam under pure bending or direct compression. Within the limitations stated before, the specimen is reliable and simple to use.

10.9.3 Flat Plate Specimen

The flat plate specimen is usually a square plate subjected to tension-tension loading on its sides through fiberglass tabs. A variety of biaxial states of stress (in the tension-tension shear space) can be achieved by rotating the principal material axes with respect to the loading directions. Nonproportional loading is possible to some degree. To ensure stress homogeneity within a reasonable test section and failure within this region, it is necessary to design the tabs and transition region very carefully.

A typical geometry of a plate specimen designed for equal biaxial tensile loading is shown in Fig. 10.60.[106,107] The composite specimen is a 40.6×40.6 cm (16×16 in) square plate with the corners cut off. It is tabbed with glass/epoxy tabs that have a 20.3 cm (8 in) diameter cutout. Loading is introduced by means of four whiffle-tree grip linkages designed to apply four equal loads to each side of the specimen. A photograph of such a biaxial specimen with the loading grip linkages is shown in Fig. 10.61.

There are two problems with this type of specimen. The state of stress in the test section is not easily determined from the applied loads because of the unknown load sharing between the tab and the specimen. This problem can be resolved by prior calibration of the system and by ensuring (or assuming) that the tab stiffness remains constant throughout the test. The other problem is premature failure at the corners due

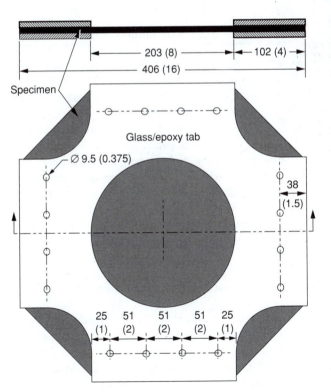

Fig. 10.60 Geometry of plate specimen for biaxial testing with dimensions given in mm (in).[106]

Fig. 10.61 Whiffle-tree linkage grips for load introduction in biaxial plate specimen.[106]

to stress concentrations. This problem can be alleviated by proper design of the tab geometry. This specimen is most suitable for the investigation of the influence of biaxial stress on notches.[106,107]

The uniformity or homogeneity of the state of stress in the test section has been verified experimentally by means of photoelastic coatings.[106,107] In the case of unequal biaxial loading, an elliptical reduced-thickness region (or tab cutout in this case) has been proposed to ensure uniform biaxial stresses throughout the test section.[108] The ellipse is selected to satisfy the relation

$$\frac{b}{a} = \sqrt{\frac{\bar{\sigma}_y}{\bar{\sigma}_x}} \qquad (10.66)$$

where $\bar{\sigma}_x$ and $\bar{\sigma}_y$ are the average stresses along the two loading directions, and a and b are the major and minor semiaxes of the ellipse along the corresponding directions.

10.9.4 Thin-Wall Tubular Specimen

Of the various biaxial test specimens mentioned, the tubular specimen appears to be the most versatile and offers the greatest potential. It offers the possibility of applying any desired biaxial state of stress with or without proportional loading. A state of generalized plane stress can be achieved by the independent application of axial loads, internal or external pressure, and torque. Tubular specimens have been used successfully with metals because stress concentrations in the load introduction region are relieved by plastic yielding. This is not the case with brittle materials such as composites. For this reason, no entirely satisfactory solution has been found to date. To achieve the full potential of the thin-wall tubular specimen, the following requirements must be met:

1. The tube must be loaded without constraints that would produce local extraneous or nonhomogeneous stresses.
2. Surface pressures on the laminate in the test section, used for producing circumferential or axial stresses, should be minimized to avoid adding a high radial stress component resulting in a triaxial state of stress.
3. Functional or material failures of the load introduction tabs must be avoided.
4. Undesirable buckling prior to material failure must be avoided.
5. The cost of specimen fabrication, equipment, and testing process must not be prohibitive.

A general biaxial state of stress is produced by means of internal and external pressures p_i and p_o, longitudinal load P_x (which can also be applied by means of p_i and p_o), and torque T_x about the longitudinal axis. The axial, circumferential (hoop), and shear stresses are obtained as

$$\bar{\sigma}_x = \frac{P_x}{2\pi\bar{r}h}$$

$$\bar{\sigma}_y = (p_i - p_o)\frac{\bar{r}}{h} - \frac{1}{2}(p_i + p_o) \qquad (10.67)$$

$$\tau_s = \frac{T_x}{2\pi\bar{r}^2h}$$

where $\bar{r}$ is the mean radius and h the tube thickness.

Testing of composite tubular specimens has been discussed by several investigators who analyzed the various problems arising in this type of specimen.[109–114] One of the most frequent and most critical problems encountered is that of introducing and maintaining a uniform biaxial state of stress in the specimen test section and inducing failure in the test region. To overcome or minimize end constraints and gripping problems, the concepts of tab and grip pressurization have been used.[99,115] These concepts, however, have not been implemented with full success because of the inherent difficulty of the problem.

A typical tubular specimen geometry is illustrated in Fig. 10.62. The specimen is designed to have end tabs for gripping and load introduction. These tabs are made of epoxy or glass/epoxy premachined and bonded to the specimen ends. The stiffness of these tabs in the axial and circumferential directions relative to those of the composite tube can be varied by varying the glass/epoxy layup and the tab thickness. This specimen geometry has been analyzed extensively using finite element methods.[116] The objective of these analyses was to minimize stress discontinuities in the transition between the test section and tabbed section of the specimen by varying the tab materials and geometry and the tab-compensating pressure.

The preparation of tubular specimens must be done very carefully to ensure that the specimen is of a quality similar to that of flat laminates fabricated and cured in an autoclave. Composite tubes can be fabricated in any desired laminate layup and stacking sequence. Fabrication techniques have been developed and are described in the technical literature.[117,118] The glass/epoxy tabs required for the specimen are fabricated as tubes in a similar manner. They are subsequently machined to size and bonded on the composite tube.

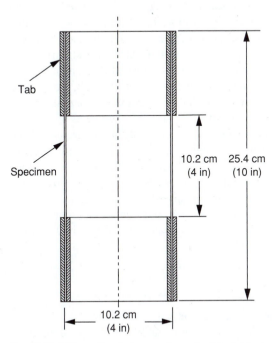

Tab

Specimen

10.2 cm
(4 in)   25.4 cm
          (10 in)

10.2 cm
(4 in)

Fig. 10.62 Thin-wall tubular specimen for biaxial testing of composites.[116]

The problem of biaxial testing of tubular specimens is inherently difficult and is not amenable to a complete solution. The analysis to date shows that it is possible to obtain valid tests in many cases for a variety of laminates over large portions of the failure envelope. Indeed a large part of the data used for comparison with theoretical predictions in the "failure exercise" discussed in Section 9.13 was obtained from tubular specimens.

10.10 CHARACTERIZATION OF COMPOSITES WITH STRESS CONCENTRATIONS

10.10.1 Introduction

The behavior of composite laminates with stress concentrations is of great interest in design because of the resulting strength reduction and life reduction due to damage growth around these stress concentrations. Stress distributions and stress concentrations around notches can be determined by linear elastic analysis, finite element methods, and experimental methods. The problem of failure of notched composite laminates has been dealt with by several approaches. One approach is based on concepts of linear elastic fracture mechanics carried over from homogeneous isotropic materials.[119,120] A second approach is based on actual stress distributions in the vicinity of the notch and makes use of simplified stress fracture criteria.[121] According to the average stress criterion proposed, failure occurs when the average stress over an assumed characteristic dimension (or volume) from the boundary of the notch equals the tensile strength of the unnotched material. Comparison with results from uniaxial tensile tests showed satisfactory agreement between predicted and measured strengths for a narrow range of values of the characteristic dimension. In a similar approach, lamina failure criteria are used, and a characteristic dimension (volume) is postulated.[122] Failure is said to occur if the average state of stress (or strain) on the boundary of this characteristic volume falls on the failure envelope of the lamina.

Experimental methods using strain gages, photoelastic coatings, and moiré have proven very useful in studying the deformation and failure of composite laminates with circular holes and through-the-thickness cracks of various sizes.[123] The effects of laminate layup, stacking sequence, notch size, and far-field stress biaxiality on failure have been investigated. The approach used was to load composite plate specimens with holes or cracks under uniaxial and biaxial tension, measure deformations by means of experimental strain analysis techniques, and determine strain distributions, failure modes, and strength reduction ratios. Experimental results are compared with predictions based on linear elastic fracture mechanics, an average stress criterion, and a progressive degradation model. A review of theoretical and experimental results on notched composite laminates was published by Awerbuch and Madhukar.[124]

10.10.2 Laminates with Holes

Stress distributions around a circular hole in an infinite plate can be obtained by anisotropic elasticity.[125] For an orthotropic laminate with a hole under uniaxial loading along a principal axis x, the maximum stress is the circumferential stress on the hole boundary at $\varphi = 90°$ (Fig. 10.63). The stress concentration at this location is obtained as

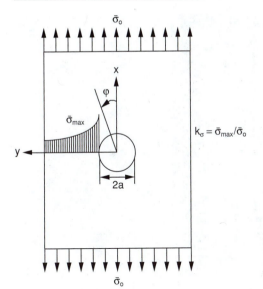

$$k_\sigma = \frac{\bar{\sigma}_{max}}{\bar{\sigma}_o} = 1 + \sqrt{2\left[\sqrt{\frac{\bar{E}_x}{\bar{E}_y}} - \bar{\nu}_{xy}\right] + \frac{\bar{E}_x}{\bar{G}_{xy}}} \qquad (10.68)$$

where

$\bar{\sigma}_o$ = applied far-field average stress

$\bar{\sigma}_{max}$ = maximum circumferential stress on hole boundary (at $\varphi = 90°$)

$\bar{E}_x, \bar{E}_y$ = average Young's moduli in the x- and y-directions

$\bar{G}_{xy}$ = average shear modulus

$\bar{\nu}_{xy}$ = average Poisson's ratio

The x- and y-axes are principal axes of the laminate.

The circumferential stress on the hole boundary at $\varphi = 0°$ is given by

Fig. 10.63 Composite laminate with circular hole under uniaxial tensile loading.

$$(\bar{\sigma}_\varphi)_{\varphi=0°} = -\bar{\sigma}_x\sqrt{\frac{\bar{E}_y}{\bar{E}_x}} \qquad (10.69)$$

Although the exact stress distribution along the transverse axis through the hole is known, the following approximate expression can be more useful:[126]

$$\frac{\bar{\sigma}_x(0, y)}{\bar{\sigma}_o} \cong 1 + \frac{1}{2}\rho^{-2} + \frac{3}{2}\rho^{-4} - \frac{k_\sigma - 3}{2}(5\rho^{-6} - 7\rho^{-8}) \qquad (10.70)$$

where

$\bar{\sigma}_x(0, y)$ = axial stress along the y-axis

$\bar{\sigma}_o$ = applied far-field axial stress

$\rho = \dfrac{y}{a}$

a = hole radius

k_σ = anisotropic stress concentration factor (obtained by Eq. (10.68) for large plate)

The influence of laminate layup on stress distribution and stress concentration is clearly illustrated by the fringe patterns of a photoelastic coating around the hole of boron/epoxy plates. These were 66 cm (26 in) long and 25.4 cm (10 in) wide laminates of various layups with a 2.54 cm (1 in) diameter central hole. They were loaded in longitudinal tension to failure. Figure 10.64a shows that, for a $[0/\pm45/0/\overline{90}]_s$ laminate, the stress distribution is similar to that of an isotropic plate, and the stress concentration factor is close to 3. Figure 10.64b shows that, for a $[0/90]_{2s}$ crossply laminate, the stress gradient at the hole boundary is sharp and the stress concentration factor is high (approximately 5). The influence of the hole along the transverse axis extends over a distance of approximately

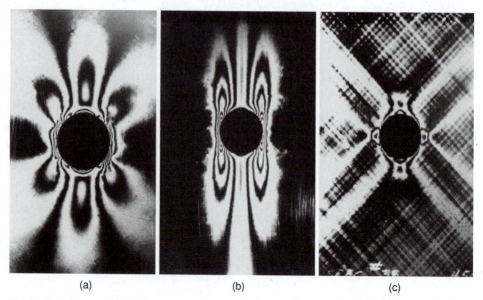

(a)   (b)   (c)

Fig. 10.64 Isochromatic fringe patterns in photoelastic coating around hole in boron/epoxy plates of different layups under uniaxial tensile loading: (a) $[0/\pm45/0/\overline{90}]_s$, $\bar{\sigma}_0 = 293$ MPa (42.4 ksi), (b) $[0/90]_{2s}$, $\bar{\sigma}_0 = 170$ MPa (24.6 ksi), and (c) $[\pm45]_{2s}$, $\bar{\sigma}_0 = 77$ M Pa (11.1 ksi).[127]

half the radius. On the other hand, the fringe pattern for the $[\pm45]_{2s}$ angle-ply laminate in Fig. 10.64c shows that the gradient is very mild, the stress concentration factor is moderate (approximately 2), and the influence of the hole extends over the entire plate.

The influence of material and stacking sequence on failure was studied for a variety of layups.[127] Some results are listed in Table 10.2. The ultimate strength is a function of both stress concentration and percentage of 0° plies. Laminates with a high percentage of 0° plies, but with a sufficient number of 45° plies to mollify the stress concentration factor, are the strongest. The $[0/90]_{2s}$ layup with a high percentage (50%) of 0° plies is not strong because of the high stress concentration factor. The $[\pm45]_{2s}$ lay-up is the weakest because of the absence of 0° plies, although the stress concentration factor is the lowest. Stacking sequence was also found to have a noticeable influence on strength and failure patterns. Stacking sequences resulting in tensile interlaminar normal stresses near the boundary of the hole reduce the strength of the laminate. In some cases, stacking sequence variations can cause drastic differences in strength related to changes in failure modes from catastrophic to noncatastrophic (see Fig. 9.35).

The influence of hole diameter in uniaxially loaded plates can be described by using the average stress criterion.[121] According to this criterion, failure occurs when the axial stress, averaged over a characteristic distance a_o from the hole boundary, equals the strength $\bar{F}_o$ of the unnotched laminate (Fig. 10.65). The strength reduction ratio, ratio of notched to unnotched strength, is expressed as

$$k_F = \frac{\bar{F}_N}{\bar{F}_o} = \frac{2}{(1 + \xi)[2 + \xi^2 + (k_\sigma - 3)\xi^6]} \qquad (10.71)$$

TABLE 10.2 Effect of Laminate Layup and Stacking Sequence on Stress Concentration and Strength of Boron/Epoxy Plates with Circular Holes Under Uniaxial Tensile Loading

Layup	Axial Modulus $\bar{E}_x$, GPa (Msi)	Measured Stress Concentration Factor k_σ	Predicted Stress Concentration Factor k_σ	Notched Strength $\bar{F}_N$, MPa (ksi)	Unnotched Strength $\bar{F}_o$, MPa (ksi)	Strength Ratio $k_F = \bar{F}_N/\bar{F}_o$
$[0/90/0/90]_s$	115.2 (16.70)	4.82	5.80	194 (28.1)	617 (89.5)	0.314
$[0_2/\pm45/\overline{0}]_s$	133.9 (19.40)	3.58	3.68	498 (72.2)	807 (117.0)	0.617
$[\pm45/0_2/\overline{0}]_s$	127.3 (18.45)	4.02	3.68	426 (61.7)	807 (117.0)	0.529
$[0/\pm45/0/\overline{90}]_s$	115.2 (16.70)	3.34	3.45	291 (42.2)	669 (97.0)	0.435
$[0_2/\pm45/\overline{90}]_s$	116.3 (16.85)	3.15	3.45	291 (42.2)	669 (97.0)	0.435
$[0/\pm45/90]_s$	79.5 (11.52)	3.08	3.00	180 (26.1)	457 (66.2)	0.394
$[45/90/0/-45]_s$	81.4 (11.80)	3.10	3.00	213 (30.9)	459 (66.5)	0.465
$[\pm45/0/\pm45]_s$	59.3 (8.59)	2.46	2.45	206 (29.9)	378 (54.8)	0.546
$[\pm45/\pm45]_s$	19.9 (2.88)	2.06	1.84	125 (18.1)	137 (19.9)	0.909
$[45_2/-45_2]_s$	20.2 (2.93)	2.55	1.84	115 (16.6)	137 (19.9)	0.833

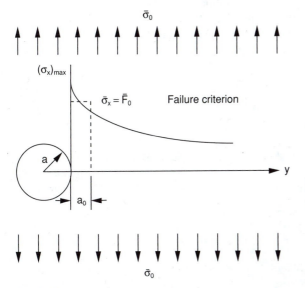

Fig. 10.65 Stress distribution and average stress criterion for uniaxially loaded composite plate with hole.

where

$$\xi = \frac{a}{a + a_o}$$

Experimental results obtained for uniaxially loaded $[0_2/\pm45]_s$ carbon/epoxy laminates with holes of various diameters are in good agreement with predictions based on the average stress criterion for a characteristic length $a_o = 5$ mm, as shown in Fig. 10.66. The laminates were 12.7 cm (5 in) wide and 56 cm (22 in) long with central holes of diameters ranging from 0.64 cm (0.25 in) to 2.54 cm (1 in). One interesting result observed in this and other similar cases is the existence of a threshold hole diameter, approximately 1.5 mm in this case, below which the laminate becomes notch insensitive.

The behavior of biaxially loaded composite plates with holes has been studied experimentally.[106,128] A typical failure pattern of a $[0/\pm45/90]_s$ carbon/epoxy plate with a hole under equal biaxial tensile loading is shown in Fig. 10.67. The strength reduction ratio, ratio of notched biaxial strength to unnotched uniaxial strength, is plotted versus hole radius in Fig. 10.68. These ratios are higher than corresponding values for uniaxial loading by approximately 30%. The variation of strength reduction ratio with hole diameter was satisfactorily described by using an average biaxial stress criterion. Radial and circumferential stresses around the hole were

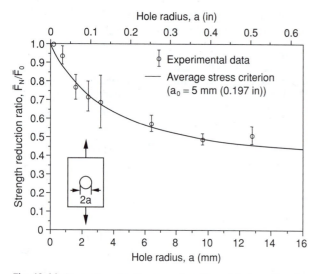

Fig. 10.66 Strength reduction as a function of hole radius for $[0_2/\pm45]_s$ carbon/epoxy plates with circular holes under uniaxial tensile loading.[123]

averaged over an annulus of 3 mm width and compared with the biaxial strength envelope for the unnotched quasi-isotropic laminate. Results were also in good agreement with predictions based on the Tsai-Wu failure criterion for the individual lamina and a progressive degradation model.[129] The strength reduction ratios for uniaxial and equal biaxial tensile loading represent lower and upper bounds for any biaxial tensile loading of this laminate. The strength reduction for any other tensile biaxiality ratio would fall between these two bounds and could be estimated approximately by interpolation.

10.10.3 Laminates with Cracks

Through-the-thickness cracks introduce much more severe stress concentrations in composite laminates. Stress distributions near a crack tip in an orthotropic material have been obtained in terms of mode I and mode II stress intensity factors and the laminate compliances.[130] An expression for the axial stress ahead of the crack, obtained as a limiting case of the solution for an elliptical hole (Fig. 10.69),[131] is

$$\bar{\sigma}_x = \frac{\bar{\sigma}_o y}{\sqrt{y^2 - a^2}} \tag{10.72}$$

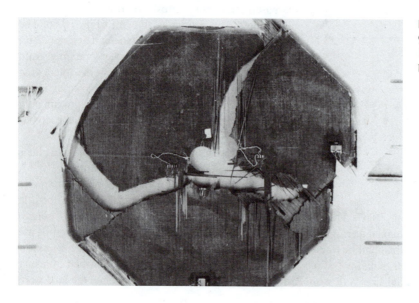

Fig. 10.67 Failure pattern in $[0/\pm45/90]_s$ carbon/epoxy specimen with 1.91 cm (0.75 in) hole under equal biaxial tensile loading.[106]

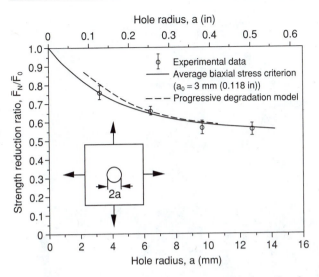

Fig. 10.68 Strength reduction as a function of hole radius for [0/±45/90]$_s$ carbon/epoxy plates with circular holes under equal biaxial tensile loading.[106]

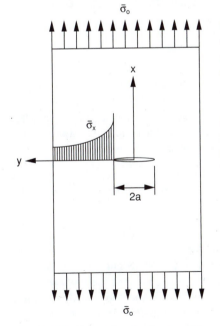

Fig. 10.69 Composite laminate with through-the-thickness transverse crack under uniaxial tensile loading.

where

$\bar{\sigma}_x$ = axial stress along transverse axis

$\bar{\sigma}_o$ = applied far-field stress

a = half crack length

The failure of composite laminates with through-the-thickness cracks can also be described by the semiempirical average stress criterion.[121] According to this criterion, failure occurs when the axial stress averaged over a characteristic distance a_o from the crack tip equals the unnotched strength of the material. Using the stress distribution of Eq. (10.72), the average stress criterion predicts the following strength reduction ratio:

$$k_F = \frac{\bar{F}_N}{\bar{F}_o} = \sqrt{\frac{a_o}{a_o + 2a}} \qquad (10.73)$$

where $\bar{F}_N$ and $\bar{F}_o$ are the notched and unnotched laminate strengths, respectively, and a_o is the characteristic length dimension.

The deformation and failure of carbon/epoxy plates with cracks of different lengths have been investigated by experimental techniques.[132, 133] In general, failure at the tip of the crack takes the form of a damage zone consisting of ply cracking along fiber directions, local delaminations, and fiber breakage in adjacent plies along the initial cracks. The strain distribution around the crack tip and the phenomenon of damage zone formation and growth for a uniaxially loaded plate are vividly illustrated by the isochromatic fringe patterns in the photoelastic coating (Fig. 10.70). A noticeable characteristic is the apparent extension of the damage zone at a 45° angle to the crack direction, which is related to initial cracking of the 45° ply. The size of this zone increases with applied stress up to some critical value, at which point the specimen fails catastrophically. Far-field strains and the crack opening displacement were obtained from moiré fringe patterns around the crack.

Experimental results for the strength reduction ratio agree well with predictions based on the average stress criterion using a characteristic dimension $a_o = 5$ mm (Fig. 10.71). Comparison of these results with those from similar specimens with circular holes shows that strength reduction in this case is nearly independent of notch geometry, that is, specimens with holes and cracks of the same size have nearly the same strength. Experimental results for specimens with holes along with predictions based on an average stress criterion for holes are also shown in Fig. 10.71.

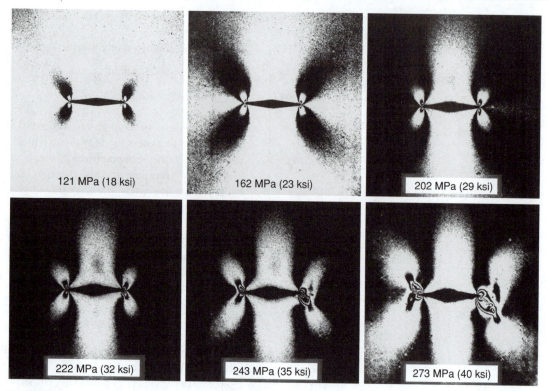

Fig. 10.70 Isochromatic fringe patterns in photoelastic coating around 1.27 cm (0.50 in) crack of $[0/\pm45/90]_s$ carbon/epoxy specimen at various levels of applied stress.[132]

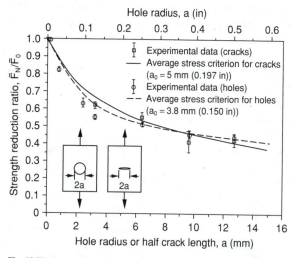

Fig. 10.71 Strength reduction as a function of hole radius or half crack length for $[0/\pm45/90]_s$ carbon/epoxy plates with circular holes and transverse cracks under uniaxial tensile loading.[132]

The result above can also be analyzed by linear elastic fracture mechanics. If the half crack length a is adjusted to include the length a_o of the critical damage zone near the crack tip, the critical stress intensity factor is given by

$$K_{1c} = \bar{F}_N \sqrt{\pi(a + a_o)} \qquad (10.74)$$

The length of the critical damage zone can be taken equal to the characteristic dimension a_o in the average stress criterion (5 mm). It can also be approximated by the diameter of the damage zone near failure as detected by the photoelastic coating. It is shown in Fig. 10.72 that the critical stress intensity factor, as modified to account for the critical damage zone at the crack tip, is nearly constant with initial crack length.

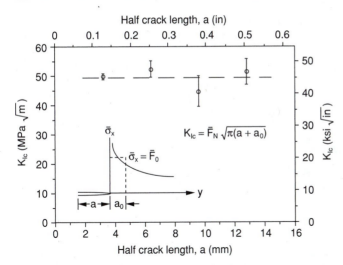

Fig. 10.72 Critical stress intensity factor as a function of half crack length for $[0/\pm45/90]_s$ carbon/epoxy plates with transverse cracks under uniaxial tensile loading.[132]

Similar studies have been conducted with quasi-isotropic plates under biaxial loading having cracks of various lengths.[107,134] Under these loading conditions it is not possible to use simplified failure criteria as discussed before. However, experimental results can be compared satisfactorily with predictions based on a maximum stress criterion for the individual lamina and a progressive degradation model.

The strength of notched laminates in general is a measure of notch sensitivity and toughness. This has prompted the recommendation of simple test methods for qualitative evaluation of laminate toughness by ASTM, NASA, and SACMA (Suppliers of Advanced Composite Materials Association). In the NASA specification, quasi-isotropic laminates with holes are tested in tension and compression.[135] The ASTM specification D5766M-02 describes the method for determining the open-hole tensile strength of multidirectional composite laminates. The laminates are symmetric and balanced and can be tape based or fabric. The SACMA recommendation describes open-hole tension and compression testing of the unidirectional material.[136]

10.11 TEST METHODS FOR TEXTILE COMPOSITES

Fabric-reinforced or textile composites are increasingly used in aerospace, automotive, naval, and other applications. The microstructure of composite laminates reinforced with woven, braided, or stitched networks is significantly different from that of unidirectional tape laminates. This leads to a higher degree of inhomogeneity. Furthermore, the relative magnitudes of in-plane and through-thickness elastic and strength properties of textile composites are different from those of tape-based composites. Consequently, testing of textile composites requires some special considerations. Specimen dimensions, instrumentation, and loading methods developed for tape laminates, may not be applicable to textile composites. A thorough review of test methods for textile composites was given in a NASA report.[137] A new ASTM guide was issued recently (D6856-03).[138] Test methods for 3-D textile composites have also been reviewed.[139]

Strains during testing can be measured with extensometers and strain gages. Special care must be taken to select gages of sufficiently large size compared to the fabric unit cell and to prepare the specimen surface to ensure that it is flat and smooth.

10.11.1 In-Plane Tensile Testing

The method for determination of in-plane tensile properties is essentially the same as described before (Section 10.4, ASTM D3039). The geometry is also similar to that shown

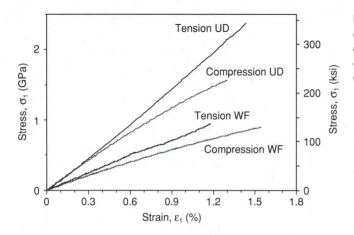

Fig. 10.73 Tensile and compressive stress-strain curves of unidirectional and woven fabric carbon/epoxy composites in the longitudinal and warp directions (AS4/3501-6 and AGP370-5H/3501-6S).

in Fig. 10.16, except that the width of the specimen should be large enough compared to the size of the textile unit cell. The recommended width is typically 25.4 mm (1.00 in), and the minimum gage length is 127 mm (5.00 in). Tests must be conducted along the principal directions of the fabric, the warp and fill directions, to determine

E_1, E_2 = Young's moduli in warp and fill directions, respectively

ν_{12}, ν_{21} = major and minor Poisson's ratios, respectively

F_{1t}, F_{2t} = tensile strengths in warp and fill directions, respectively

ε_{1t}^u, ε_{2t}^u, = ultimate tensile strains in warp and fill directions, respectively

Typical tensile stress-strain curves for a woven carbon/epoxy material and a unidirectional lamina with the same fiber type and matrix are shown in Fig. 10.73.[140] It is noted that the modulus and strength of the fabric composite are roughly half those of the corresponding unidirectional lamina. This is due to the fact that the five-harness satin-weave-reinforced composite behaves approximately like a $[0/90]_s$ crossply laminate made of unidirectional laminae.

10.11.2 In-Plane Compressive Testing

Compression testing of textile composites can be performed as in the case of unidirectional tape-based composites (Section 10.5, ASTM D3410) with special considerations for the size of the gages and specimen width. Special guidelines for compression testing are given in a NASA report.[141] A prismatic specimen of 305 mm (12 in) length, 38 mm (1.50 in) width, and 3.18 mm (0.125 in) thickness is recommended for two-dimensional braids with large unit cells. A special fixture was proposed to provide lateral support to the specimen during compressive loading.[141] Compression testing of satin-weave fabric composites has been conducted successfully with the NU fixture (Fig. 10.23) using an IITRI-type specimen of 114 mm (4.5 in) length with long glass/epoxy tabs leaving an unsupported gage length of 12.7 mm (0.50 in).[140] Typical compressive stress-strain curves for a woven carbon/epoxy and a unidirectional carbon/epoxy with the same type of fiber and matrix

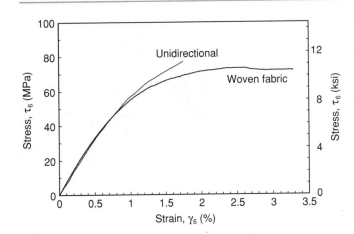

Fig. 10.74 In-plane shear stress-strain curves of unidirectional and woven fabric carbon/epoxy composites (AS4/3501-6 and AGP370-5H/3501-6S).

are shown in Fig. 10.73, together with the tensile stress-strain curves. The same observation can be made here, that is, that the modulus and compressive strength of the fabric composite are approximately half those of the corresponding unidirectional composite.

10.11.3 In-Plane Shear Testing

The ASTM-recommended test methods for in-plane shear characterization are shear of a Iosipescu-type beam with V-notches (ASTM D5379M-98) or the standard two-rail or three-rail shear tests (D4255M-01) discussed before (see Fig. 10.31). The $[\pm45]_{ns}$ coupon test discussed before (Section 10.6, Fig. 10.27) can be used only for woven or orthogonally braided composites with equal yarn sizes and spacings in the warp and fill (or $\pm\theta$ braiding) directions, or for uniweaves of $[\pm45]$ layup. In the cases above it is also possible to use off-axis tensile testing as discussed previously for unidirectional composites (see Fig. 10.29). In all testing above, strain measurements should be made using strain gages of sufficient size relative to the textile unit cell. A compact Iosipescu-type specimen instrumented with a special strain gage that yields the average shear strain over the test section has been used successfully for a variety of textile composites.[142] Typical shear stress-strain curves for a woven carbon/epoxy and a unidirectional carbon/epoxy with the same type of fiber and matrix are shown in Figure 10.74. It can be observed that the two stress-strain curves nearly coincide in the initial quasi-linear region, up to a shear strain of $\gamma_6 \cong 0.8\%$. Beyond this strain level, the woven fabric composite shows much more ductility, with an ultimate strain exceeding 3%.

10.11.4 Through-Thickness Testing

Through-thickness testing of textile composites is more challenging than testing of unidirectional composites, because of material thickness limitations and the requirement for a sufficiently large test section compared to the textile unit cell size. Methods and results were published recently for a woven carbon/epoxy composite.[51]

Through-thickness tensile testing was conducted by using a short circularly waisted block specimen (Fig. 10.39). A through-thickness tensile stress-strain curve is shown in Fig. 10.40. The specimens for through-thickness compressive testing are prismatic blocks

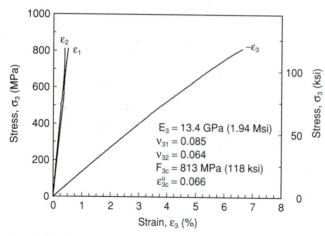

Fig. 10.75 Stress-strain curves of carbon fabric/epoxy under through-thickness compression (AGP370-5H/3501-6S).[51]

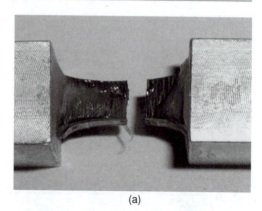

(a)

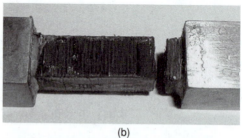

(b)

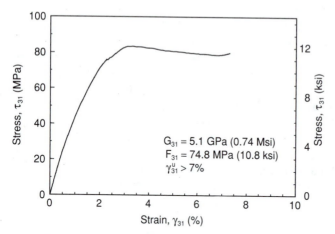

Fig. 10.76 Shear stress-strain curve of carbon fabric/epoxy under through-thickness shear (3-1 plane; AGP370-5H/3501-6S).[51]

(c)

Fig. 10.77 Failure patterns of carbon fabric/epoxy composite under through-thickness loading: (a) tension, (b) compression, (c) shear (AGP370-5H/3501-6S).[51]

bonded to steel end blocks. A typical compressive stress-strain curve to failure is shown in Fig. 10.75.

The ASTM-recommended through-thickness shear test method is the same as for in-plane shear, that is, a Iosipescu-type V-notch beam under shear (ASTM D5379M-98). A compact version of this test specimen was shown in Fig. 10.45. A typical through-thickness shear stress-strain curve for a woven carbon/epoxy is shown in Fig. 10.76. Failure patterns of carbon fabric/epoxy specimens tested in through-thickness tension, compression, and shear are shown in Fig. 10.77.

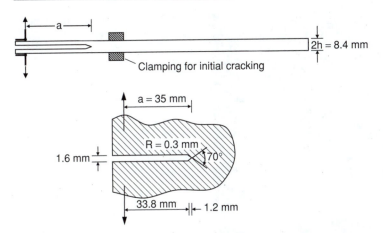

Fig. 10.78 DCB specimen and notch geometry for evaluation of G_{Ic} for textile composites.[144]

The above methods are not suitable for textile composites with through-thickness yarns. These composites have high through-thickness tensile and shear strengths compared to their in-plane strengths.

10.11.5 Interlaminar Fracture Toughness

Interlaminar cracks in textile composites propagate along the interlaminar matrix phase as in the case of tape-based laminates. However, there is more resistance to interlaminar crack propagation due to the undulations of the fabric weave. It has been observed that the interlaminar fracture toughness (IFT) of textile composites is usually two to eight times that of corresponding tape-based laminates. The IFT of textile composites is determined essentially by the same methods as those described previously for unidirectional composites, with special considerations for the expected higher IFT values and the increased specimen thickness required as a result.[143,144]

The most commonly used tests are the double cantilever beam (DCB) test for mode I (ASTM D5528-01) and the end-notched flexure (ENF) test for mode II. In the case of mode I testing using the DCB specimen, the recommended specimen thickness is large enough to prevent premature bending failures of the cantilever beams before crack propagation. A DCB specimen and its notch geometry used for textile composites are shown in Fig. 10.78.[144] In this case the specimen was machined from a 8.4 mm (0.331 in) thick laminate made of A193 plain-weave AS4 carbon and an LY564/HY2954 epoxy. A central notch was machined at the end as shown in Fig. 10.78. The specimen is preloaded to form and extend a natural precrack, the length of which is controlled by a clamp.

10.12 STRUCTURAL TESTING

The increasing applications of composites in larger structures and the infrastructure create the need for additional testing on a larger scale. The structural behavior of large-scale structures manufactured by industrial methods, such as RTM or VARTM, cannot be predicted reliably based on small-scale coupon tests using specimens fabricated under laboratory conditions. Structural components representative of larger structures, such as joints, pipe sections, sandwich panels, and stiffened sections, must be tested under conditions similar to those encountered in service.

Sandwich structures consisting of a lightweight core and composite facesheets are very desirable and widely used because of their light weight and high in-plane and flexural stiffness. The overall behavior of sandwich structures depends on the material properties of the constituents (facesheets, adhesive, and core), geometric dimensions, and type of loading.

Fig. 10.79 Simply supported composite sandwich panel under central loading.

Analysis of this behavior is difficult because of the nonlinear and inelastic behavior of the constituent materials and the complex interactions of failure modes.[41,145] For these reasons it is important to investigate experimentally the behavior of sandwich beams and panels under various loadings. Sandwich beams under four-point bending are used for determination of compressive strength of laminates as discussed before (Figs. 10.25 and 10.26). Sandwich panels are tested under monotonic and impact loadings (Fig. 10.79). In the case of impact loading, the damage is assessed by nondestructive evaluation and the residual strength is measured by means of compressive testing (compression after impact).

One major application of composites to infrastructure is in the form of pipes for transportation of water or oil.[146] Large-diameter pipes are made of glass/polyester composites because of their corrosion resistance and easy maintenance. However, they are less stiff than metal pipes and more expensive to fabricate. Underground pipes are subjected mainly to internal fluid pressure producing hoop tensile stresses, external soil pressure inducing compressive and flexural stresses and possible global buckling, and longitudinal flexure due to pipeline curvature and structural constraints. Five basic tests are recommended in ASTM specification D3517-04. Of these the hoop stiffness is measured directly by loading a pipe segment between flat plates under diametral compression (ASTM D2412-02, Fig. 10.80). Another important test is the measurement of hoop tensile strength by the split disk method (ASTM D2290-00, Fig. 10.81).

10.13 SUMMARY AND DISCUSSION

Methods of physical and mechanical characterization of constituents and composite materials were reviewed. The physical and mechanical characterization of the fiber and matrix constituents and the interface is necessary for the application of micromechanics in

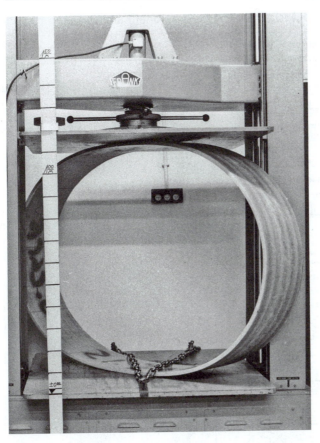

Fig. 10.80 Segment of 1.52 m (60 in) diameter glass/polyester pipe under 25% diametral compression.[146]

Fig. 10.81 Split disk tensile test and failure of ring specimen cut from 51 cm (20 in) diameter glass/polyester pipe.[146]

understanding and predicting the macroscopic behavior of composite materials. In the macromechanical approach, the unidirectional lamina is considered the basic building block, and its full characterization is essential for any subsequent theoretical, numerical, or experimental analysis of composite structures. Many of the methods discussed have been standardized, and the relevant ASTM specifications were mentioned in the text. A summary of test methods for characterization of the unidirectional lamina is given in Table 10.3.

In the case of compressive testing, variations of the IITRI test fixture are widely used, with modifications being introduced by many investigators. An extension of this test method to thick composites involving combined shear and end loading was described (NU fixture).

There is a variety of methods used for in-plane shear testing, with the $[\pm 45]_{2s}$ and the off-axis lamina tests being the most commonly used. Many types of tests may yield reasonable values for the in-plane shear modulus, but the ultimate properties are not always determined reliably. In the case of off-axis testing, it was shown that the determination of the shear strength is linked to the assumed failure criterion.

TABLE 10.3 Test Methods for Mechanical Characterization of Unidirectional Lamina

Test (ASTM Standard)	Specimen Configuration	Elastic Properties	Strength Parameters
Longitudinal tension (D3039M-00)	Ring	E_1, ν_{12}	$F_{1t}, \varepsilon_{1t}^u$
Transverse tension (D3039M-00)		E_2, ν_{21}	$F_{2t}, \varepsilon_{2t}^u$
Longitudinal compression (D3410M-03) (C364-99) (C393-00)	IITRI — Sandwich column — Sandwich beam	E_1	$F_{1c}, \varepsilon_{1c}^u$
Transverse compression (D3410M-03) (C364-99) (C393-00)	IITRI (or sandwich construction)	E_2	$F_{2c}, \varepsilon_{2c}^u$
In-plane shear D3518M-94 (2001) D4255M-01	[±45] 10° off-axis Three-rail shear Tube in torsion Coupon in torsion Iosipescu Arcan	G_{12}	F_6, γ_6^u
Interlaminar shear (for quality assessment) (D2344M-00) (D3846-02)	Short beam Double-notch coupon		F_5
Through-thickness tension		E_3 ν_{31} ν_{32}	F_{3t} ε_{3t}^u
Through-thickness compression		E_3	F_{3c} ε_{3c}^u
Through-thickness shear		G_{13} G_{23}	F_5 F_4

Through-thickness mechanical characterization is a particularly challenging problem because of practical limitations on laminate thickness. Direct tensile or compressive testing in the thickness direction makes use of small coupons loaded through end blocks. The short-beam interlaminar shear test is primarily used as a quality control test. A modified Iosipescu test can be used with short notched blocks to determine the through-thickness shear moduli and strengths.

Interlaminar fracture toughness is an important property characterizing the resistance to delamination onset and propagation. It is well characterized in mode I by using the double cantilever beam (DCB) specimen. The end-notched flexure (ENF) specimen under three- or four-point bending is well accepted for determination of mode II fracture toughness. A combination of the two tests above, standardized by ASTM, yields mixed-mode (I and II) results. Of the various mode III tests proposed, the edge-cracked torsion (ECT) test is on its way to standardization.

Limited biaxial results can be obtained by off-axis testing. Valid results can be obtained by careful testing of thin-wall cylindrical specimens under internal pressure, axial tension or compression, and torque.

Stress and failure analysis around stress concentrations was discussed at some length. However, most of the work to date deals with well-defined flaws, such as holes and through-the-thickness cracks. There is need to extend this work to study more realistic defects, such as partial delaminations, scratches, and impact-damaged areas. The strength of notched composites serves as a measure of notch sensitivity and toughness and has prompted the recommendation of standard open-hole tension and compression test methods by ASTM, NASA, and SACMA.

Testing of textile composites received attention because of the increasing applications of textile composites. Test methods are essentially the same as for tape-based composites but with special considerations for gage instrumentation and specimen size. The strain gages and the width of the specimen should be sufficiently larger than the fabric unit cell. In-plane, through-thickness, and interlaminar fracture toughness test methods were discussed.

A brief discussion was included of large-scale structural testing because of the increasing applications of composites to larger structures and infrastructure.

Several areas of mechanical characterization were not addressed in this review. For example, fatigue testing is very important in studying damage mechanisms, damage accumulation, and the attendant degradation of stiffness and residual strength. Some specifications for fatigue testing are given in ASTM specifications D3479M-96 (2002) and D4762-04. Fatigue studies must be extended to include spectrum fatigue loading, environmental effects, and delamination crack growth. Future efforts will be directed toward methods of accelerated testing for determining the reliability and life of structures.

Environmental effects on mechanical properties are important and must be evaluated by special testing. The extent of property degradation as a function of temperature and moisture concentration must be measured for the various materials. Special testing procedures are needed to evaluate time effects, both long-term (creep) and short-term (dynamic) effects, and physical aging. Creep testing is useful in determining the viscoelastic characteristics of the composite, that is, time-dependent stiffnesses and compliances. It can be conducted at various temperatures to obtain master curves covering many decades of time. A variety of methods have been developed for high-rate testing to evaluate short-term effects. Closely related to this is impact testing and evaluation of impact damage tolerance.

Another important area of experimental characterization not covered in this chapter is nondestructive evaluation (NDE). A variety of NDE techniques are used for evaluating the integrity of composite materials and structures. They include radiographic (X-ray and neutron), optical (moiré, birefringent coatings, holographic interferometry, and speckle shearing interferometry), thermographic, acoustic (acoustic wave, acoustic emission), embedded sensors, ultrasonic, and electromagnetic techniques. The most widely used methods, especially for polymer-matrix composites, are ultrasonic and X-radiographic techniques. They are used to detect and characterize flaws introduced during fabrication and in service, such as delaminations, porosity, matrix cracking, fiber/matrix debonding, fiber misalignment, inclusions, and fiber fractures. The discussion of NDE methods and their application to composites is beyond the scope of this book.

REFERENCES

1. J. M. Whitney, I. M. Daniel, and R. Byron Pipes, *Experimental Mechanics of Fiber Reinforced Composite Materials*, Monograph No. 4, Society for Experimental Mechanics, Bethel, CT, Prentice-Hall, Englewood Cliffs, NJ, Revised Edition, 1985.
2. I. M. Daniel, "Methods of Testing Composite Materials," in *Fracture Mechanics and Methods of Testing*, G. C. Sih and A. M. Skudra, Vol. Eds., in *Handbook of Fibrous Composites*, A. Kelly and Y. N. Rabotnov, Eds., North Holland Publishing Co., Amsterdam, 1985, pp. 277–373.
3. Y. M. Tarnopolskii and T. Kincis, *Static Test Methods for Composites*, G. Lubin, Trans. Ed., Van Nostrand Reinhold Co., New York, 1985 (originally published in Russian, 1981).
4. I. M. Daniel, "Composite Materials," in *Handbook on Experimental Mechanics*, A. S. Kobayashi, Ed., VCH Publishers, New York, 1993, pp. 829–904.
5. "Space Simulation; Aerospace and Aircraft; Composite Materials," in *Annual Book of ASTM Standards*, Vol. 15.03, American Society for Testing and Materials, West Conshohocken, PA, 2004.
6. I. M. Daniel, "Testing, Mechanical Characterization," in *International Encyclopedia of Composites*, S. M. Lee, Ed., Vol. 5, VCH Publishers, New York, 1991, pp. 436–464.
7. *The Composite Materials Handbook*, MIL-HDBK-17, Materials Sciences Corp., University of Delaware, Army Research Laboratory, http://www.mil17.org, 2000.
8. J. M. Hodgkinson, Ed., *Mechanical Testing of Advanced Fibre Composites*, CRC Press, Boca Raton, FL, 2000.
9. L. A. Carlsson, R. L. Crane, and K. Uchino, "Test Methods, Nondestructive Evaluation, and Smart Materials," in *Comprehensive Composite Materials*, A. Kelly and C. Zweben, Eds., Elsevier Science, Ltd., Oxford, UK, 2000.
10. D. F. Adams, L. A. Carlsson, and R. Byron Pipes, *Experimental Characterization of Advanced Composite Materials*, Third Edition, CRC Press, Boca Raton, FL, 2003.
11. H. D. Wagner, S. L. Phoenix, and P. Schwartz, "A Study of Statistical Variability in the Strength of Single Aramid Filaments," *J. Composite Materials*, Vol. 18, 1984, pp. 312–338.
12. J. A. DiCarlo, "Creep of Chemically Vapor Deposited SiC Fibers," *J. Materials Sci.*, Vol. 21, 1986, pp. 217–224.
13. C.-L. Tsai and I. M. Daniel, "Method for Thermomechanical Characterization of Single Fibers," *Composites Sci. Technol.*, Vol. 50, 1994, pp. 7–12.
14. C.-L. Tsai and I. M. Daniel, "Determination of Shear Modulus of Single Fibers," *Exper. Mech.*, Vol. 39, No. 4, 1999, pp. 284–286.
15. N. J. Hillmer, "Thermal Expansion of Chemically Vapor Deposited Silicon Carbide Fibers," in *Symposium on High Temperature Composites*, Proceedings of the American Society for Composites, Technomic Publishing Co., Lancaster, PA, June 13–15, 1989, pp. 206–213.

16. "Standard Test Method for Tensile Properties of Thin Plastic Sheeting," ASTM Standard D882-02, *Annual Book of ASTM Standards*, Section 8, Plastics, Vol. 08.01, American Society for Testing and Materials, West Conshohocken PA, 2004.

17. R. J. Kerans and T. A. Parthasarathy, "Theoretical Analysis of the Fiber Pullout and Pushout Tests," *J. Amer. Ceramic Soc.*, Vol. 74, 1991, pp. 1585–1596.

18. R. N. Singh and M. Sutcu, "Determination of Fibre-Matrix Interfacial Properties in Ceramic-Matrix Composites by a Fibre-Push-Out Technique," *J. Materials Science*, Vol. 26, 1991, pp. 2547–2556.

19. P. D. Warren, T. J. Mackin, and A. G. Evans, "Design, Analysis and Application of an Improved Push-Through Test for the Measurement of Interface Properties in Composites," *Acta Metall. Mater.*, Vol. 40, 1992, pp. 1243–1249.

20. J. W. Hutchinson and H. M. Jensen, "Models of Fiber Debonding and Pullout in Brittle Composites with Friction," *Mech. of Materials*, Vol. 9, 1990, pp. 139–163.

21. R. D. Cordes and I. M. Daniel, "Determination of Interfacial Properties from Observations of Progressive Fiber Debonding and Pullout," *Composites Engineering*, Vol. 5, No. 6, 1995, pp. 633–648.

22. "Test Methods for Density and Specific Gravity (Relative Density) of Plastics by Displacement," ASTM Standard D792-00, Vol. 08.01, American Society for Testing and Materials, West Conshohocken, PA, 2004.

23. W. Freeman and M. D. Campbell, "Thermal Expansion Characteristics of Graphite Reinforced Composite Materials," *Composite Materials: Testing and Design* (Second Conference), ASTM STP 497, American Society for Testing and Materials, West Conshohocken, PA, 1972, pp. 121–142.

24. I. M. Daniel and T. Liber, *Lamination Residual Stresses in Fiber Composites*, NASA CR-134826, NASA-Glen Research Center, Cleveland, OH, 1975.

25. I. M. Daniel, T. Liber, and C. C. Chamis, "Measurement of Residual Strains in Boron/Epoxy and Glass/Epoxy Laminates," in *Composite Reliability*, ASTM STP 580, American Society for Testing and Materials, West Conshohocken, PA, 1975, pp. 340–351.

26. I. M. Daniel, "Thermal Deformations and Stresses in Composite Materials," in *Thermal Stresses in Severe Environments*, D. P. H. Hasselman and R. A. Heller, Eds., Plenum Press, New York, 1980, pp. 607–625.

27. A. S. D. Wang, R. B. Pipes, and A. Ahmadi, "Thermoelastic Expansion of Graphite-Epoxy Unidirectional and Angle-Ply Composites," in *Composite Reliability*, ASTM STP 580, American Society for Testing and Materials, West Conshohocken, PA, 1975, pp. 574–585.

28. E. Wolff, *Measurement Techniques for Low Expansion Materials*, Vol. 9, National SAMPE Technical Conference Series, Society for the Advancement of Material and Process Eng., Covina, CA, Oct. 1977.

29. R. R. Johnson, M. H. Kural, and G. B. Mackey, *Thermal Expansion Properties of Composite Materials*, NASA CR-165632, NASA-Glen Research Center, Cleveland, OH, 1981.

30. D. E. Bowles, D. Post, C. T. Herakovich, and D. R. Tenney, "Moiré Interferometry for Thermal Expansion of Composites," *Exp. Mechanics*, Vol. 21, 1981, pp. 441–448.

31. S. S. Tompkins, D. E. Bowles, and W. R. Kennedy, "A Laser-Interferometric Dilatometer for Thermal-Expansion Measurements of Composites," *Exp. Mechanics*, Vol. 26, 1986, pp. 1–6.

32. S. S. Tompkins, "Techniques for Measurement of the Thermal Expansion of Advanced Composite Materials," in *Metal Matrix Composites: Testing, Analysis and Failure Modes*, ASTM STP 1032, W. S. Johnson, Ed., American Society for Testing and Materials, West Conshohocken, PA, 1989, pp. 54–67.

33. S. Gazit and O. Ishai, "Hygroelastic Behavior of Glass-Reinforced Plastics Exposed to Different Relative Humidity Levels," in *Proceedings of Conference on Environmental Degradation of Engineering Materials*, Virginia Polytechnic Inst., Blacksburg, VA, Oct. 1977, pp. 383–392.

34. H. T. Hahn and R. Y. Kim, "Swelling of Composite Laminates," in *Environmental Effects on Composite Materials*, ASTM STP 658, J. R. Vinson, Ed., American Society for Testing and Materials, West Conshohocken, PA, 1978, pp. 98–120.

35. G. Yaniv, G. Peimanidis, and I. M. Daniel, "Method for Hygrothermal Characterization of Graphite/Epoxy Composite," *J. Composites Technol. Res.*, Vol. 9, 1987, pp. 21–25.

36. C.-L. Tsai and S.-C. Wooh, "Hygric Characterization of Woven Glass/Epoxy Composites," *Exper. Mech.*, Vol. 41, No. 1, 2001, pp. 70–76.

37. C.-L. Tsai, S.-C. Wooh, S.-F. Hwang, and Y. Du, "Hygric Characterization of Composites Using an Antisymmetric Cross-Ply Specimen," *Exper. Mech.*, Vol. 41, No. 3, 2001, pp. 270–276.

38. K. E. Hofer and P. N. Rao, "A New Static Compression Fixture for Advanced Composite Materials," *J. Test Eval.*, Vol. 5, 1977, pp. 278–283.

39. J. G. Häberle and F. L. Matthews, "An Improved Technique for Compression Testing of Unidirectional Fibre-Reinforced Plastics; Developments and Results," *Composites*, Vol. 25, 1994, pp. 358–371.

40. H.-M. Hsiao, I. M. Daniel, and S.-C. Wooh, "A New Compression Test Method for Thick Composites," *J. Comp. Materials*, Vol. 29, No. 13, 1995, pp. 1789–1806.

41. I. M. Daniel and J. L. Abot, "Fabrication, Testing and Analysis of Composite Sandwich Beams," *J. Fiber Science and Tech.*, Vol. 60, 2000, pp. 2455–2463.

42. B. W. Rosen, "A Simple Procedure for Experimental Determination of the Longitudinal Shear Modulus of Unidirectional Composites," *J. Composite Materials*, Vol. 6, 1972, pp. 552–554.

43. C. C. Chamis and J. H. Sinclair, "Ten-Degree Off-Axis Test for Shear Properties in Fiber Composites," *Exp. Mech.*, Vol. 17, 1977, pp. 339–346.

44. G. Yaniv, I. M. Daniel, and J.-W. Lee, "Method for Monitoring In-Plane Shear Modulus in Fatigue Testing of Composites," *Test Methods and Design Allowables for Fibrous Composites*, Second Volume, ASTM STP 1003, C. C. Chamis, Ed., American Society for Testing and Materials, West Conshohocken, PA, 1989, pp. 276–284.

45. C.-L. Tsai and I. M. Daniel, "The Behavior of Cracked Crossply Composite Laminates Under Simple Shear Loading," *Composites Engineering*, Vol. 1, 1991, pp. 3–11.

46. C.-L. Tsai, I. M. Daniel, and G. Yaniv, "Torsional Response of Rectangular Composite Laminates," *J. of Appl. Mech.*, Vol. 112, 1990, pp. 383–387.

47. C.-L. Tsai and I. M. Daniel, "Determination of In-Plane and Out-of-Plane Shear Moduli of Composite Materials," *Exp. Mech.*, Vol. 30, 1990, pp. 295–299.

48. M. Arcan, Z. Hashin, A. Voloshin, "A Method to Produce Uniform Plane Stress States with Applications to Fiber Reinforced Materials," *Exp. Mech.*, Vol. 18, 1978, pp. 141–146.

49. D. E. Walrath and D. F. Adams, "The Iosipescu Shear Test as Applied to Composite Materials," *Exp. Mech.*, Vol. 23, 1983, pp. 105–110.

50. M. J. Lodeiro, W. R. Broughton, and G. D. Sims, "Understanding the Limitations of Through-Thickness Test Methods," *Proc. 4th European Conf. on Composites: Testing and Standardisation*, IOM Communications Ltd., London, 1998, pp. 80–90.

51. J. L. Abot and I. M. Daniel, "Through-Thickness Mechanical Characterization of Woven Fabric Composites," *J. Comp. Materials*, Vol. 38, No. 7, 2004, pp. 543–553.

52. R. F. Ferguson, M. J. Hinton, and M. J. Hiley, "Determining the Through-Thickness Properties of FRP Materials," *Comp. Science and Tech.*, Vol. 58, 1998, pp. 1411–1420.

53. O. Ishai, "Strengthening of Composite Materials in the Third Dimension," *Annual Report of Technion Research and Development Foundation*, Technion, Haifa, Israel, 1995.

54. V. P. Nikolaev, "A Method of Testing the Transverse Tensile Strength of Wound Articles," *Mekhanica Polymerov (Polymer Mechanics*; in Russian), Vol. 9, 1973, pp. 675–677.

55. C. C. Hiel, M. Sumich, and D. P. Chappell, "A Curved Beam Test for Determination of the Interlaminar Tensile Strength of a Laminated Composite," *J. Composite Materials*, Vol. 25, 1991, pp. 854–868.

56. J. M. Whitney and C. E. Browning, "On Interlaminar Beam Experiments for Composite Materials," in *Proceedings of the V International Congress on Experimental Mechanics*, Society for Experimental Mechanics, Bethel, CT, 1984, pp. 97–101.

57. C. E. Browning, F. L. Abrams, and J. M. Whitney, "A Four-Point Shear Test for Graphite/Epoxy Composites," in *Composite Materials: Quality Assurance and Processing*, ASTM STP 797, C. E. Browning, Ed., American Society for Testing and Materials, West Conshohocken, PA, 1983, pp. 54–74.

58. C. A. Berg, J. Tirosh, and M. Israeli, "Analysis of Short Beam Bending of Fiber Reinforced Composites," *Composite Materials: Testing and Design* (Second Conference), ASTM STP 497, American Society for Testing and Materials, West Conshohocken, PA, 1972, pp. 206–218.

59. H. Weisshaus and O. Ishai, "A Comparative Study of the Interlaminar Shear Strength of C/C Composites," *Proc. of 32nd International SAMPE Technical Conference*, Nov. 5–9, 2000, pp. 854–872.

60. S. R. Short, "Characterization of Interlaminar Shear Failures of Graphite-Epoxy Composite Materials," *Composites*, Vol. 26, No. 6, 1995, pp. 431–449.

61. F. Abali, A. Pora, and K. Shivakumar, "Modified Short Beam Test for Measurement of Interlaminar Shear Strength of Composites," *J. Composite Materials*, Vol. 37, No. 5, 2003, pp. 453–464.

62. G. Menges and R. Kleinholz, "Comparison of Different Methods for the Determination of Interlaminar Shear Strength" (in German), *Kunststoffe*, Vol. 59, 1969, pp. 959–966.

63. J. M. Whitney, C. E. Browning, and W. Hoogsteden, "A Double Cantilever Beam Test for Characterizing Mode I Delamination of Composite Materials," *J. Reinforced Plastics and Composites*, Vol. 1, 1982, pp. 297–313.

64. N. Sela and O. Ishai, "Interlaminar Fracture Toughness and Toughening of Laminated Composite Materials: A Review," *Composites*, Vol. 20, 1989, pp. 423–435.

65. T. K. O'Brien and R. H. Martin, "Round Robin Testing for Mode I Interlaminar Fracture Toughness of Composite Materials," *J. Composites Tech. and Research*, Vol. 15, No. 4, 1993, pp. 269–281.

66. P. Davies, B. R. K. Blackman, and A. J. Brunner, "Standard Test Methods for Delamination Resistance of Composite Materials: Current Status," *Applied Comp. Materials*, Vol. 5, No. 6, 1998, pp. 345–364.

67. D. J. Wilkins, J. R. Eisenmann, R. A. Camin, W. S. Margolis, and R. A. Benson, "Characterizing Delamination Growth in Graphite-Epoxy," in *Damage in Composite Materials: Basic Mechanisms, Accumulation, Tolerance, Characterization*, ASTM STP 775, K. L. Reifsnider, Ed., American Society for Testing and Materials, West Conshohocken, PA, 1982, pp. 168–183.

68. A. A. Aliyu and I. M. Daniel, "Effects of Strain Rate on Delamination Fracture Toughness of Graphite/Epoxy," in *Delamination and Debonding of Materials*, ASTM STP 876, W. S. Johnson, Ed., American Society for Testing and Materials, West Conshohocken, PA, 1985, pp. 336–348.

69. I. M. Daniel, I. Shareef, and A. A. Aliyu, "Rate Effects on Delamination Fracture-Toughness of a Toughened Graphite/Epoxy," in *Toughened Composites*, ASTM STP 937, N. J. Johnston, Ed., American Society for Testing and Materials, West Conshohocken, PA, 1987, pp. 260–274.

70. T. R. Brussat S. T. Chin, and S. Mostovoy, *Fracture Mechanics for Structural Adhesive Bonds, Phase II*, AFML-TR-77-163, Wright Aeronautical Labs, Dayton, OH, 1978.

71. I. M. Daniel, G. Yaniv, and J. W. Auser, "Rate Effects on Delamination Fracture Toughness of Graphite/Epoxy Composites," in *Composite Structures*-4 (Proceedings of Fourth International Conference on Composite Structures, Paisley, Scotland, 1978), I. H. Marshall, Ed., Elsevier Applied Science, New York, 1987, pp. 2.258–2.272.

72. G. Yaniv and I. M. Daniel, "Height-Tapered Double Cantilever Beam Specimen for Study of Rate Effects on Fracture Toughness of Composites," in *Composite Materials: Testing and Design*, ASTM STP 972, J. D. Whitcomb, Ed., American Society for Testing and Materials, West Conshohocken, PA, 1988, pp. 241–258.

73. D. F. Devitt, R. A. Schapery, and W. L. Bradley, "A Method for Determining Mode I Delamination Fracture Toughness of Elastic and Viscoelastic Composite Materials," *J. Composite Materials*, Vol. 14, 1980, pp. 270–285.

74. R. L. Ramkumar and J. D. Whitcomb, "Characterization of Mode I and Mixed Mode Delamination Growth in T300/5208 Graphite/Epoxy," in *Delamination and Debonding of Materials*, ASTM STP 876, W. S. Johnson, Ed., American Society for Testing and Materials, West Conshohocken, PA, 1985, pp. 315–335.

75. J. W. Gillespie, L. A. Carlsson, and A. J. Smiley, "Rate Dependent Mode I Interlaminar Crack Growth Mechanisms in Graphite/Epoxy and Graphite/PEEK," *Composites Sci. Technol.*, Vol. 28, 1987, pp. 1–15.

76. Y. J. Prel, P. Davies, M. L. Benzeggagh, and F. X. de Charentenay, "Mode I and Mode II Delamination of Thermosetting and Thermoplastic Composites," in *Composite Materials: Fatigue and Fracture*, Second Volume, ASTM STP 1012, Paul A. Lagace, Ed., American Society for Testing and Materials, West Conshohocken, PA, 1989, pp. 251–269.

77. Z. Yang and C. T. Sun, "Interlaminar Fracture Toughness of a Graphite/Epoxy Multidirectional Composite," *J. Eng. Materials and Technology*, Vol. 122, No. 4, 2000, pp. 428–433.

78. C. Guo and C. T. Sun, "Dynamic Mode I Crack Propagation in a Carbon/Epoxy Composite," *Comp. Science and Tech.*, Vol. 58, No. 9, 1998, pp. 1405–1410.

79. T. K. O'Brien, "Composite Interlaminar Shear Fracture Toughness, G_{IIc}: Shear Measurement or Sheer Myth?" *Composite Materials: Fatigue and Fracture*, Seventh Volume, ASTM STP 1330, R. B. Bucinell, Ed., American Society for Testing and Materials, West Conshohocken, PA, 1998, pp. 3–18.

80. A. J. Russell and K. N. Street, "Moisture and Temperature Effects on the Mixed-Mode Delamination Fracture of Unidirectional Graphite/Epoxy," in *Delamination and Debonding of Materials*, ASTM STP 876, W. S. Johnson, Ed., American Society for Testing and Materials, West Conshohocken, PA, 1985, pp. 349–370.

81. L. A. Carlsson, J. W. Gillespie, and R. B. Pipes, "Analysis and Design of End-Notched Flexure (ENF) Specimen for Mode II Testing," *J. Composite Materials*, Vol. 30, 1986, pp. 594–604.

82. C. R. Corletto and W. L. Bradley, "Mode II Delamination Fracture Toughness of Unidirectional Graphite/Epoxy Composites," in *Composite Materials: Fatigue and Fracture*, Second Volume, ASTM STP 1012, Paul A. Lagace, Ed., American Society for Testing and Materials, West Conshohocken, PA, 1989, pp. 201–221.

83. T. K. O'Brien, G. B. Murri, and S. A. Salpekar, "Interlaminar Shear Fracture Toughness and Fatigue Thresholds for Composite Materials," in *Composite Materials: Fatigue and Fracture*, Second Volume, ASTM STP 1012, Paul A. Lagace, Ed., American Society for Testing and Materials, West Conshohocken, PA, 1989, pp. 222–250.

84. O. Ishai, H. Rosenthal, N. Sela, and E. Drucker, "Effect of Selective Adhesive Interleaving on Interlaminar Fracture Toughness of Graphite/Epoxy Composite Laminate," *Composites*, Vol. 19, 1988, pp. 49–54.

85. N. Sela, O. Ishai, and L. Banks-Sills, "The Effect of Adhesive Thickness on Interlaminar Fracture Toughness of Interleaved CFRP Specimens," *Composites*, Vol. 20, 1989, pp. 257–264.

86. J. L. Tsai, C. Guo, and C. T. Sun, "Dynamic Delamination Fracture Toughness in Unidirectional Polymeric Composites," *Comp. Science and Tech.*, Vol. 61, No. 1, 2001, pp. 87–94.

87. L. Arcan, M. Arcan, and I. M. Daniel, "SEM Fractography of Pure and Mixed-Mode Interlaminar Fractures in Graphite/Epoxy Composites," in *Fractography of Modern*

Engineering Materials, ASTM STP 948, J. Masters and J. Au, Eds., American Society for Testing and Materials, West Conshohocken, PA, 1987, pp. 41–67.

88. W. S. Johnson, "Stress Analysis of the Cracked-Lap-Shear Specimen: An ASTM Round Robin," *J. Test. Eval.*, Vol. 15, 1987, pp. 303–324.

89. J. H. Crews, Jr., and J. R. Reeder, *A Mixed-Mode Bending Apparatus for Delamination Testing*, NASA Tech-Memorandum 100662, NASA-Langley Research Center, Hampton, VA, 1988.

90. M. L. Benzeggagh and M. Kenane, "Measurement of Mixed-Mode Delamination Fracture Toughness of Unidirectional Glass/Epoxy Composites with Mixed-Mode Bending Apparatus," *Comp. Science and Technology*, Vol. 56, No. 4, 1996, pp. 439–449.

91. R. Krueger, I. L. Paris, T. K. O'Brien, and P. J. Miguet, "Fatigue Life Methodology for Bonded Composite Skin/Stringer Configurations," *J. Composites Technology and Research*, Vol. 24, No. 2, 2002, pp. 56–79.

92. T. K. O'Brien, "Characterization of Delamination Onset and Growth in a Composite Laminate," in *Damage in Composite Materials*, ASTM STP 775, K. L. Reifsnider, Ed., American Society for Testing and Materials, West Conshohocken, PA, 1982, pp. 140–167.

93. T. K. O'Brien, "Mixed-Mode Strain-Energy-Release Rate Effects on Edge Delamination of Composites," in *Effects of Defects in Composite Materials*, ASTM STP 836, American Society for Testing and Materials, West Conshohocken, PA, 1984, pp. 125–142.

94. S. L. Donaldson, "Mode III Interlaminar Fracture Characterization of Composite Materials," *Composites Sci. Technol.*, Vol. 32, 1988, pp. 225–249.

95. G. Yaniv and I. M. Daniel, "New Test Method and Monitoring Technique to Evaluate Mode III and Mixed Mode Fracture Toughness in Composite Laminates," in *Composites: Design, Manufacture and Application*, Proceedings of Eighth International Conference on Composite Materials (ICCM/VIII), S. W. Tsai and G. S. Springer, Eds., Honolulu, 1991, pp. 36-A-1–15.

96. S. M. Lee, "An Edge Crack Torsion Method for Mode III Delamination Fracture Testing," *J. Comp. Technol. and Research*, Vol. 15, No. 3, 1993, pp. 193–201.

97. J. A. Li and Y. J. Wang, "Analysis of a Symmetrical Laminate with Mid-Plane Free-Edge Delamination Under Torsion-Theory and Application to the Edge Crack Torsion (ECT) Specimen for Mode III Toughness Characterization," *Eng. Fract. Mech.*, Vol. 49, No. 2, 1994, pp. 179–194.

98. J. A. Li and T. K. O'Brien, "Simplified Data Reduction Methods for the ECT Test for Mode III Interlaminar Fracture Toughness," *J. Comp. Technol. and Research*, Vol. 18, No. 2, 1996, pp. 96–101.

99. B. W. Cole and R. B. Pipes, *Utilization of the Tubular and Off-Axis Specimens for Composite Biaxial Characterization*, AFFDL-TR-72-130, Wright Aeronautical Labs, Dayton, OH, 1972.

100. R. E. Rowlands, "Analytical-Experimental Correlation of Polyaxial States of Stress in Thornel-Epoxy Laminates," *Exp. Mechanics*, Vol. 18, 1978, pp. 253–260.

101. V. D. Azzi and S. W. Tsai, "Anisotropic Strength of Composites," *Exp. Mech.*, Vol. 5, 1965, pp. 283–288.

102. O. Ishai, "Failure of Unidirectional Composites in Tension," *J. Eng. Mech. Div.*, Proc. of the ASCE, Vol. 97, No. EM2, 1971, pp. 205–221.

103. J. C. Halpin, *Primer on Composite Materials: Analysis*, Revised Edition, Technomic Publishing Co., Inc., Lancaster, PA, 1984.

104. N. J. Pagano and J. C. Halpin, "Influence of End Constraint in the Testing of Anisotropic Bodies," *J. Composite Materials*, Vol. 2, 1968, pp. 18–31.

105. O. Ishai and R. E. Lavengood, "Characterizing Strength of Unidirectional Composites," *Composite Materials: Testing and Design*, ASTM STP 460, American Society for Testing and Materials, West Conshohocken, PA, 1970, pp. 271–281.

106. I. M. Daniel, "Behavior of Graphite/Epoxy Plates with Holes Under Biaxial Loading," *Exp. Mech.*, Vol. 20, 1980, pp. 1–8.

107. I. M. Daniel, "Biaxial Testing of Graphite/Epoxy Laminates with Cracks," in *Test Methods and Design Allowables for Fibrous Composites*, ASTM STP 734, C. C. Chamis, Ed., American Society for Testing and Materials, West Conshohocken, PA, 1981, pp. 109–128.

108. C. W. Bert, B. L. Mayberry, and J. D. Ray, "Behavior of Fiber-Reinforced Plastic Laminates Under Biaxial Loading," in *Composite Materials: Testing and Design*, ASTM STP 460, American Society for Testing and Materials, West Conshohocken, PA, 1969, pp. 362–380.

109. N. J. Pagano and J. M. Whitney, "Geometric Design of Composite Cylindrical Characterization Specimens," *J. Composite Materials*, Vol. 4, 1970, pp. 360–378.

110. J. M. Whitney, G. C. Grimes, and P. H. Francis, "Effect of End Attachment on the Strength of Fiber-Reinforced Composite Cylinders," *Exp. Mechanics*, Vol. 13, 1973, pp. 185–192.

111. T. L. Sullivan and C. C. Chamis, "Some Important Aspects in Testing High Modulus Fiber Composite Tubes in Axial Tension," in *Analysis of the Test Methods for High Modulus Fibers and Composites*, ASTM STP 521, American Society for Testing and Materials, West Conshohocken, PA, 1973, pp. 277–292.

112. T. R. Guess and F. P. Gerstle, Jr., "Deformation and Fracture of Resin Matrix Composites in Combined Stress States," *J. Composite Materials*, Vol. 7, 1977, pp. 448–464.

113. U. Hütter, H. Schelling, and H. Krauss, "An Experimental Study to Determine Failure Envelope of Composite Materials with Tubular Specimens Under Combined Loads and Comparison with Several Classical Criteria," in *Failure Modes of Composite Materials with Organic Matrices and Their Consequences on Design*, North Atlantic Treaty Organization, AGARD-CP-163, 1975, pp. 3-1–3-11.

114. M. F. S. Al-Khalil, P. D. Soden, R. Kitching, and M. J. Hinton, "The Effects of Radial Stresses on the Strength of Thin Walled Filament Wound GRP Composite Pressure Cylinders," *Int. J. of Mech. Scie.*, Vol. 38, 1996, pp. 97–120.

115. U. S. Lindholm, A. Nagy, L. M. Yeakley, and W. L. Ko, *Design of a Test Machine for Biaxial Testing of Composite Laminate Cylinders*, AFFDL-TR-75-83, Wright Aeronautical Labs, Dayton, OH, 1975.

116. I. M. Daniel, T. Liber, R. Vanderby, and G. M. Koller, "Analysis of Tubular Specimen for Biaxial Testing of Composite Laminates," in *Advances in Composite Materials* (Proc. Third International Conference on Composite Materials, ICCM 3), Vol. 1, A. R. Bunsell, C. Bathias, A. Martrenchar, D. Menkes, G. Verchery, Eds., Pergamon Press, New York, 1980, pp. 840–855.

117. D. N. Weed and P. H. Francis, "Process Development for the Fabrication of High Quality Composite Tubes," *Fibre Sci. Tech.*, Vol. 10, 1977, pp. 89–100.

118. T. Liber, I. M. Daniel, R. H. Labedz, and T. Niiro, "Fabrication and Testing of Composite Ring Specimens," *Proc. 34th Annual Tech. Conf., Reinf. Plastics/Composites Institute*, The Society of the Plastics Industry, New York, 1979, Sect. 22-B.

119. M. E. Waddoups, J. R. Eisenmann, and B. E. Kaminski, "Microscopic Fracture Mechanisms of Advanced Composite Materials," *J. Composite Materials*, Vol. 5, 1971, pp. 446–454.

120. T. A. Cruse, "Tensile Strength of Notched Laminates," *J. Composite Materials*, Vol. 7, 1973, pp. 218–229.

121. J. M. Whitney and R. J. Nuismer, "Stress Fracture Criteria for Laminated Composites Containing Stress Concentrations," *J. Composite Materials*, Vol. 8, 1974, pp. 253–265.

122. E. M. Wu, "Failure Criteria to Fracture Mode Analysis of Composite Laminates," in *Failure Modes of Composite Materials and Their Consequences on Design*, North Atlantic Treaty Organization, AGARD CP-163, 1975, pp. 2-1–2-11.

123. I. M. Daniel, "Failure Mechanisms and Fracture of Composite Laminates with Stress Concentrations," *Proceedings of the Seventh International Conference on Experimental Stress Analysis*, Technion, Israel Institute of Technology, Haifa, Israel, 1982, pp. 1–20.

124. J. Awerbuch and M. S. Madhukar, "Notched Strength of Composite Laminates; Pedictions and Experiments—A Review," *J. Reinforced Plastics and Composites*, Vol. 4, 1985, pp. 3–159.

125. S. G. Lekhnitskii, *Theory of Anisotropic Elastic Body*, P. Fern, Trans., J. J. Brandstatter, Ed., Holden-Day, San Francisco, 1963.

126. H. J. Konish and J. M. Whitney, "Approximate Stresses in an Orthotropic Plate Containing a Circular Hole," *J. Composite Materials*, Vol. 9, 1975, pp. 157–166.

127. I. M. Daniel, R. E. Rowlands, and J. B. Whiteside, "Effects of Material and Stacking Sequence on Behavior of Composite Plates with Holes," *Exp. Mech.*, Vol. 14, 1974, pp. 1–9.

128. I. M. Daniel, "Biaxial Testing of [0₂/±45]ₛ Graphite/Epoxy Plates with Holes," *Exp. Mech.*, Vol. 22, 1982, pp. 188–195.

129. K. H. Lo and E. M. Wu, "Failure Strength of Notched Composite Laminates," in *Advanced Composite Serviceability Program*, J. Altman, B. Burroughs, R. Hunziker, and D. Konishi, Eds., AFWAL-TR-80-4092, Wright Aeronautical Labs, Dayton, OH, 1980.

130. G. C. Sih, P. C. Paris, and G. R. Irwin, "On Cracks in Rectilinearly Anisotropic Bodies," *Int. J. Fracture Mech.*, Vol. 1, 1965, pp. 189–203.

131. S. G. Lekhnitskii, *Anisotropic Plates* (translated from the Second Russian Edition by S. W. Tsai and T. Cheron), Gordon and Breach, Science Publishers, Inc., New York, 1968.

132. I. M. Daniel, "Strain and Failure Analysis in Graphite/Epoxy Laminates with Cracks," *Exp. Mech.*, Vol. 18, 1978, pp. 246–252.

133. Y. T. Yeow, D. H. Morris, and H. F. Brinson, "The Fracture Behavior of Grapite/Epoxy Laminates," *Exp. Mech.*, Vol. 19, 1979, pp. 1–8.

134. I. M. Daniel, "Mixed Mode Failure of Composite Laminates with Cracks," *Exp. Mech.*, Vol. 25, 1985, pp. 413–420.

135. *Standard Tests for Toughened Resin Composites*, Revised Edition, NASA Reference Publication 1092, July 1983.

136. "Open-Hole Compression Properties of Oriented Fiber-Resin Composites," SACMA SRM 3-88, in *SACMA Recommended Methods*, Suppliers of Advanced Composite Materials Association (SACMA), Composites Fabricators Association (CFA), 1999.

137. J. E. Masters and M. Portanova, *Standard Test Methods for Textile Composites*, NASA CR 4751, 1996.

138. *Standard Guide for Testing Fabric-Reinforced "Textile" Composite Materials*, ASTM, D6856-03, 2003.

139. Y. M. Tarnopolskii, V. L. Kulakov, and A. K. Arnautov, "Experimental Determination of Mechanical Characteristics of 3-D Textile Composites," *Proc. International Conf. on Composites in Material and Structural Engineering*, CMSE/1, M. Cerny, Ed., Prague, 2001.

140. J. L. Abot, A. Yasmin, A. J. Jacobsen, and I. M. Daniel, "In-Plane Mechanical, Thermal and Viscoelastic Properties of a Satin Fabric Carbon/Epoxy Composite," *Comp. Science and Technology*, Vol. 64, 2004, pp. 263–268.

141. J. E. Masters, *Compression Testing of Textile Composites*, NASA CR 198285, 1996.

142. P. G. Ifju, "Shear Testing of Textile Composite Materials," *J. Composites Technology and Research*, Vol. 17, 1995, pp. 199–204.

143. W. D. Bascom, R. J. Bitner, R. J. Moulton, and A. R. Siebert, "The Interlaminar Fracture of Organic-Matrix Woven Reinforced Composites," *Composites*, Vol. 11, 1980, pp. 9–18.

144. O. Ishai, "Interlaminar Fracture Toughness of Selectively Stitched Thick Carbon Fibre Reinforced Polymer Fabric Composite Laminates," *Plastics, Rubber and Composites*, Vol. 29, No. 3, 2000, pp. 134–143.

145. I. M. Daniel, E. E. Gdoutos, K.-A. Wang, and J. L. Abot, "Failure Modes of Composite Sandwich Beams," *Intern. Journal of Damage Mechanics*, Vol. 11, 2002, pp. 309–334.

146. O. Ishai and J. M. Lifshitz, "Quality Assurance of GFRP Pipes for Water Cooling of Power Plants," *J. of Composites for Construction*, 1999, pp. 27–37.

Material Properties

The data given in this appendix were obtained experimentally and collected from different sources without complete verification. It is recommended that these data be used primarily for educational and preliminary design purposes and not for final structural design before verifying them experimentally or checking with other sources.

TABLE A.1 Properties of Typical Fibers

Type	Diameter μm (10^{-3} in)	Density g/cm³ (lb/in³)	Modulus GPa (Msi)	Tensile Strength MPa (ksi)
Glass				
E-glass	8–14 (0.30–0.55)	2.54 (0.092)	73 (10.5)	3450 (500)
S-glass	10 (0.40)	2.49 (0.090)	86 (12.4)	4500 (650)
Carbon				
AS4	7 (0.28)	1.81 (0.065)	235 (34)	3700 (535)
T300		1.76 (0.063)	230 (33)	3100 (450)
T-400H		1.80 (0.065)	250 (36)	4500 (650)
IM-6	4 (0.16)	1.80 (0.065)	255 (37)	4500 (650)
IM-7	4 (0.16)	1.80 (0.065)	290 (42)	5170 (750)
Graphite				
T-50		1.67 (0.060)	390 (57)	2070 (300)
GY-70		1.86 (0.067)	520 (75)	1725 (250)
P100 S	10 (0.4)	2.02 (0.073)	720 (105)	1725 (250)
Boron	140 (5.6)	2.50 (0.090)	395 (57)	3450 (500)
Aramid				
Kevlar 49	12 (0.47)	1.45 (0.052)	131 (19)	3800 (550)
Kevlar 149	12 (0.47)	1.45 (0.052)	186 (27)	3400 (490)
Silicon carbide				
(SCS-2)	140 (5.6)	3.10 (0.112)	400 (58)	4140 (600)
Nicalon	15 (0.6)	2.60 (0.094)	172 (25)	2070 (300)
Alumina				
FP-2		3.70 (0.133)	380 (55)	1725 (250)
Nextel 610		3.75 (0.135)	370 (54)	1900 (275)
Saphikon		3.80 (0.137)	380 (55)	3100 (450)
Sapphire whiskers		3.96 (0.143)	410 (60)	21,000 (3000)
Silica	—	2.19 (0.079)	73 (10.5)	5800 (840)
Tungsten	—	19.3 (0.696)	410 (60)	4140 (600)

TABLE A.2 Mechanical and Thermal Properties of Representative Fibers

Property	E-Glass	S-Glass	AS-4 Carbon	T-300 Carbon	IM7 Carbon	Boron	Kevlar 49 Aramid	Silicon Carbide (Nicalon)
Longitudinal modulus, E_{1f}, GPa (Msi)	73 (10.5)	86 (12.4)	235 (34)	230 (33)	290 (42)	395 (57)	131 (19)	172 (25)
Transverse modulus, E_{2f}, GPa (Msi)	73 (10.5)	86 (12.4)	15 (2.2)	15 (2.2)	21 (3)	395 (57)	7 (1.0)	172 (25)
Axial shear modulus, G_{12f}, GPa (Msi)	30 (4.3)	35 (5.0)	27 (4.0)	27 (4.0)	14 (2)	165 (24)	21 (3.1)	73 (10.6)
Transverse shear modulus, G_{23f}, GPa (Msi)	30 (4.3)	35 (5.0)	7 (1.0)	7 (1.0)	—	—	—	—
Poisson's ratio, v_{12f}	0.23	0.23	0.20	0.20	0.20	0.13	0.33	0.20
Longitudinal tensile strength, F_{1f}, MPa (ksi)	3450 (500)	4500 (650)	3700 (535)	3100 (450)	5170 (750)	3450 (500)	3800 (550)	2070 (300)
Longitudinal coefficient of thermal expansion, α_{1f}, 10^{-6}/°C (10^{-6}/°F)	5.0 (2.8)	5.6 (3.1)	−0.5 (−0.3)	−0.7 (−0.4)	−0.2 (−0.1)	16 (8.9)	−2 (−1.1)	3.2 (1.8)
Transverse coefficient of thermal expansion, α_{2f}, 10^{-6}/°C (10^{-6}/°F)	5.0 (2.8)	5.6 (3.1)	15 (8.3)	12 (6.7)	10 (5.6)	16 (8.9)	60 (33)	3.2 (1.8)

TABLE A.3 Properties of Typical Polymer Matrix Materials

Property	Epoxy (3501-6)	Epoxy (977-3)	Epoxy (HY6010/HT917/DY070)	Polyesters	Vinylester (Derakane)	Polyimides	Poly-ether-ether-ketone (PEEK)
Density, ρ, g/cm^3 (lb/in^3)	1.27 (0.046)	1.28 (0.046)	1.17 (0.043)	1.1–1.5 (0.040–0.054)	1.15 (0.042)	1.4–1.9 (0.050–0.069)	1.32 (0.049)
Young's modulus, E_m, GPa (Msi)	4.3 (0.62)	3.7 (0.54)	3.4 (0.49)	3.2–3.5 (0.46–0.51)	3–4 (0.43–0.58)	3.1–4.9 (0.45–0.71)	3.7 (0.53)
Shear modulus, G_m, GPa (Msi)	1.60 (0.24)	1.37 (0.20)	1.26 (0.18)	0.7–2.0 (0.10–0.30)	1.1–1.5 (0.16–0.21)		
Poisson's ratio, v_m	0.35	0.35	0.36	0.35	0.35		
Tensile strength, F_{mt}, MPa (ksi)	69 (10)	90 (13)	80 (11.6)	40–90 (5.8–13.0)	65–90 (9.4–13.0)	70–120 (10.1–17.4)	96 (14)
Compressive strength, F_{mc}, MPa (ksi)	200 (30)	175 (25)	104 (15.1)	90–250 (13–35)	127 (18.4)		
Shear strength, F_{ms}, MPa (ksi)	100 (15)	52 (7.5)	40 (5.8)	45 (6.5)	53 (29)		
Coefficient of thermal expansion, α_m, 10^{-6}/°C (10^{-6}/°F)	45 (25)		62 (3.4)	60–200 (33–110)	100–150 (212–514)	90 (50)	
Glass transition temperature, T_g, °C (°F)	200 (390)	200 (390)	152 (305)	50–110 (120–230)		280–320 (540–610)	143 (290)
Maximum use temperature, T_{max}, °C (°F)	150 (300)	177 (350)				300–370 (570–700)	250 (480)
Ultimate tensile strain, ε_{mt}^u (%)	2–5			2–5	1–5	1.5–3.0	

TABLE A.4 Properties of Typical Unidirectional Composites (Two-Dimensional)

Property	E-Glass/Epoxy	S-Glass/Epoxy	Kevlar/Epoxy (Aramid 49/Epoxy)	Carbon/Epoxy (AS4/3501-6)	Carbon/Epoxy (IM6G/3501-6)
Fiber volume ratio, V_f	0.55	0.50	0.60	0.63	0.66
Density, ρ, g/cm^3 (lb/in^3)	1.97 (0.071)	2.00 (0.072)	1.38 (0.050)	1.60 (0.058)	1.62 (0.059)
Longitudinal modulus, E_1, GPa (Msi)	41 (6.0)	45 (6.5)	80 (11.6)	147 (21.3)	169 (24.5)
Transverse modulus, E_2, GPa (Msi)	10.4 (1.50)	11.0 (1.60)	5.5 (0.80)	10.3 (1.50)	9.0 (1.30)
In-plane shear modulus, G_{12}, GPa (Msi)	4.3 (0.62)	4.5 (0.66)	2.2 (0.31)	7.0 (1.00)	6.5 (0.94)
Major Poisson's ratio, v_{12}	0.28	0.29	0.34	0.27	0.31
Minor Poisson's ratio, v_{21}	0.06	0.06	0.02	0.02	0.02
Longitudinal tensile strength, F_{1t}, MPa (ksi)	1140 (165)	1725 (250)	1400 (205)	2280 (330)	2240 (325)
Transverse tensile strength, F_{2t}, MPa (ksi)	39 (5.7)	49 (7.1)	30 (4.2)	57 (8.3)	46 (6.7)
In-plane shear strength, F_6, MPa (ksi)	89 (12.9)	70 (10.0)	49 (7.1)	76 (11.0)	73 (10.6)
Ultimate longitudinal tensile strain, ε_{1t}^u	0.028	0.029	0.015	0.015	0.013
Ultimate transverse tensile strain, ε_{2t}^u	0.005	0.006	0.005	0.006	0.005
Longitudinal compressive strength, F_{1c}, MPa (ksi)	620 (90)	690 (100)	335 (49)	1725 (250)	1680 (245)
Transverse compressive strength, F_{2c}, MPa (ksi)	128 (18.6)	158 (22.9)	158 (22.9)	228 (33)	215 (31)
Longitudinal thermal expansion coefficient, α_1, 10^{-6}/°C (10^{-6}/°F)	7.0 (3.9)	7.1 (3.9)	−2.0 (−1.1)	−0.9 (−0.5)	−0.9 (−0.5)
Transverse thermal expansion coefficient, α_2, 10^{-6}/°C (10^{-6}/°F)	26 (14.4)	30 (16.7)	60 (33)	27 (15)	25 (13.9)
Longitudinal moisture expansion coefficient, β_1	0	0	0	0.01	0
Transverse moisture expansion coefficient, β_2	0.2	0.2	0.3	0.2	—

TABLE A.4 Properties of Typical Unidirectional Composites (Two-Dimensional) (cont'd)

Property	Carbon/Epoxy (IM7/977-3)	Carbon/PEEK (AS4/APC2)	Carbon/Polyimide (Mod I/WRD9371)	Graphite/Epoxy (GY-70/934)	Boron/Epoxy (B5.6/5505)
Fiber volume ratio, V_f	0.65	0.58	0.45	0.57	0.50
Density, ρ, g/cm^3 (lb/in^3)	1.61 (0.058)	1.57 (0.057)	1.54 (0.056)	1.59 (0.058)	2.03 (0.073)
Longitudinal modulus, E_1, GPa (Msi)	190 (27.7)	138 (19.9)	216 (31.3)	294 (42.7)	201 (29.2)
Transverse modulus, E_2, GPa (Msi)	9.9 (1.44)	8.7 (1.27)	5.0 (0.72)	6.4 (0.92)	21.7 (3.15)
In-plane shear modulus, G_{12}, GPa (Msi)	7.8 (1.13)	5.0 (0.73)	4.5 (0.65)	4.9 (0.71)	5.4 (0.78)
Major Poisson's ratio, ν_{12}	0.35	0.28	0.25	0.23	0.17
Minor Poisson's ratio, ν_{21}	0.02	0.02	0.01	0.01	0.02
Longitudinal tensile strength, F_{1t}, MPa (ksi)	3250 (470)	2060 (300)	807 (117)	985 (145)	1380 (200)
Transverse tensile strength, F_{2t}, MPa (ksi)	62 (8.9)	78 (11.4)	15 (2.2)	29 (4.3)	56 (8.1)
In-plane shear strength, F_6, MPa (ksi)	75 (10.9)	157 (22.8)	22 (3.2)	49 (7.1)	62 (9.1)
Ultimate longitudinal tensile strain, ε_{1t}^u	0.016	0.016	0.004	0.002	0.007
Ultimate transverse tensile strain, ε_{2t}^u	0.006	0.009	0.003	0.005	0.003
Longitudinal compressive strength, F_{1c}, MPa (ksi)	1590 (230)	1100 (160)	655 (95)	690 (100)	1600 (232)
Transverse compressive strength, F_{2c}, MPa (ksi)	200 (29)	196 (28.4)	71 (10.2)	98 (14.2)	125 (18)
Longitudinal thermal expansion coefficient, α_1, 10^{-6}/°C (10^{-6}/°F)	−0.9 (−0.5)	−0.2 (−0.1)	0	−0.1 (−0.06)	6.1 (3.4)
Transverse thermal expansion coefficient, α_2, 10^{-6}/°C (10^{-6}/°F)	22 (12.4)	24 (13.3)	25 (14.1)	26 (14.4)	30 (17)
Longitudinal moisture expansion coefficient, β_1	0	0	0	0	0
Transverse moisture expansion coefficient, β_2	—	0.4	0.3	0.4	0.2

TABLE A.5 Properties of Typical Fabric Composites (Two-Dimensional)

Property	Woven Glass/ Epoxy (7781/5245C)	Woven Glass/ Epoxy (120/3501-6)	Woven Glass/ Epoxy (M10E/3783)	Kevlar 49 Fabric/Epoxy (K120/M10.2)	Carbon Fabric/ Epoxy (AGP370-5H/3501-6S)
Fiber volume ratio, V_f	0.45	0.55	0.50	—	0.62
Density, ρ, g/cm³ (lb/in³)	2.20 (0.080)	1.97 (0.071)	1.90 (0.068)	—	1.60 (0.58)
Longitudinal modulus, E_1, GPa (Msi)	29.7 (4.31)	27.5 (3.98)	24.5 (3.55)	29 (4.2)	77 (11.2)
Transverse modulus, E_2, GPa (Msi)	29.7 (4.31)	26.7 (3.87)	23.8 (3.45)	29 (4.2)	75 (10.9)
In-plane shear modulus, G_{12}, GPa (Msi)	5.3 (0.77)	5.5 (0.80)	4.7 (0.68)	18 (2.6)	6.5 (0.94)
Major Poisson's ratio, ν_{12}	0.17	0.14	0.11	0.05	0.06
Minor Poisson's ratio, ν_{21}	0.17	0.13	0.10	0.05	0.06
Longitudinal tensile strength, F_{1t}, MPa (ksi)	367 (53)	435 (63)	433 (62.8)	369 (53.5)	963 (140)
Transverse tensile strength, F_{2t}, MPa (ksi)	367 (53)	386 (56)	386 (55.9)	369 (53.5)	856 (124)
In-plane shear strength, F_6, MPa (ksi)	97.1 (14.1)	55 (7.9)	84 (12.2)	113 (16.4)	71 (10.3)
Ultimate longitudinal tensile strain, ε_{1t}^u	0.025	0.019	0.022	—	0.013
Ultimate transverse tensile strain, ε_{2t}^u	0.025	0.018	0.021	—	0.012
Longitudinal compressive strength, F_{1c}, MPa (ksi)	549 (80)	—	377 (54.6)	129 (18.7)	900 (130)
Transverse compressive strength, F_{2c}, MPa (ksi)	549 (80)	—	335 (48.6)	129 (18.7)	900 (130)
Longitudinal thermal expansion coefficient, α_1, $10^{-6}/°C$ ($10^{-6}/°F$)	10.0 (5.6)	—	—	—	3.4 (1.9)
Transverse thermal expansion coefficient, α_2, $10^{-6}/°C$ ($10^{-6}/°F$)	10.0 (5.6)	—	—	—	3.7 (2.1)
Longitudinal moisture expansion coefficient, β_1	0.06	—	—	—	0.05
Transverse moisture expansion coefficient, β_2	0.06	—	—	—	0.05

TABLE A.6 Properties of Typical Unidirectional and Fabric Composite Materials (Three-Dimensional)

Property	E-Glass/Epoxy		Kevlar 49/Epoxy		Carbon/Epoxy	
	Unidirectional	Fabric (M10E/3783)	Unidirectional	Fabric (K120/M10.2)	Unidirectional (AS4/3501-6)	Woven Fabric (AGP370-5H/3501-6S)
Fiber volume ratio, V_f	0.55	0.50	0.60	—	0.63	0.62
Density, ρ, g/cm³ (lb/in³)	1.97 (0.071)	1.90 (0.068)	1.38 (0.050)	—	1.60 (0.058)	1.60 (0.058)
Longitudinal modulus, E_1, GPa (Msi)	41 (6.00)	24.5 (3.55)	80 (11.6)	29 (4.2)	147 (21.3)	77 (11.2)
Transverse in-plane modulus, E_2, GPa (Msi)	10.4 (1.50)	23.8 (3.45)	5.5 (0.80)	29 (4.2)	10.3 (1.50)	75 (10.9)
Transverse out-of-plane modulus, E_3, GPa (Msi)	10.4 (1.50)	11.6 (1.68)	5.5 (0.80)	—	10.3 (1.50)	13.8 (2.0)
In-plane shear modulus, G_{12}, GPa (Msi)	4.3 (0.62)	4.7 (0.68)	2.2 (0.31)	18 (2.6)	7.0 (1.00)	6.5 (0.94)
Out-of-plane shear modulus, G_{23}, GPa (Msi)	3.5 (0.50)	3.6 (0.52)	1.8 (0.26)	—	3.7 (0.54)	4.1 (0.59)
Out-of-plane shear modulus, G_{13}, GPa (Msi)	4.3 (0.62)	2.6 (0.38)	2.2 (0.31)	—	7.0 (1.00)	5.1 (0.74)
Major in-plane Poisson's ratio, ν_{12}	0.28	0.11	0.34	0.05	0.27	0.06
Out-of-plane Poisson's ratio, ν_{23}	0.50	0.20	0.40	—	0.54	0.37
Out-of-plane Poisson's ratio, ν_{13}	0.28	0.15	0.34	0.05	0.27	0.50
Longitudinal tensile strength, F_{1t}, MPa (ksi)	1140 (165)	433 (62.8)	1400 (205)	369 (53.5)	2280 (330)	963 (140)
Transverse tensile strength, F_{2t}, MPa (ksi)	39 (5.7)	386 (55.9)	30 (4.2)	369 (53.5)	57 (8.3)	856 (124)
Out-of-plane tensile strength, F_{3t}, MPa (ksi)	39 (5.7)	27 (3.9)	30 (4.2)	—	57 (8.3)	60 (8.7)
Longitudinal compressive strength, F_{1c}, MPa (ksi)	620 (90)	377 (54.6)	335 (49)	129 (18.7)	1725 (250)	900 (130)
Transverse compressive strength, F_{2c}, MPa (ksi)	128 (18.6)	335 (48.6)	158 (22.9)	129 (18.7)	228 (33)	900 (130)
Out-of-plane compressive strength, F_{3c}, MPa (ksi)	128 (18.6)	237 (34.4)	158 (22.9)	—	228 (33)	813 (118)
In-plane shear strength, F_6, MPa (ksi)	89 (12.9)	84 (12.2)	49 (7.1)	113 (16.4)	76 (11.0)	71 (10.3)
Out-of-plane shear strength, F_4, MPa (ksi)	—	44 (6.3)	—	33 (4.8)	—	65 (9.5)
Out-of-plane shear strength, F_5, MPa (ksi)	—	41 (5.9)	37 (5.4)	33 (4.8)	—	75 (10.8)
Longitudinal thermal expansion coefficient, α_1, 10⁻⁶/°C (10⁻⁶/°F)	7.0 (3.9)	—	-2.0 (-1.1)	—	-0.9 (-0.5)	3.4 (1.9)
Transverse thermal expansion coefficient, α_2, 10⁻⁶/°C (10⁻⁶/°F)	26 (14.4)	—	60 (33)	—	27 (15)	3.7 (2.1)
Out-of-plane thermal expansion coefficient, α_3, 10⁻⁶/°C (10⁻⁶/°F)	26 (14.4)	—	60 (33)	—	27 (15)	52 (29)
Longitudinal moisture expansion coefficient, β_1	0	—	0	—	0.01	0.05
Transverse moisture expansion coefficient, β_2	0.2	—	0.3	—	0.20	0.05
Out-of-plane moisture expansion coefficient, β_3	0.2	—	0.3	—	0.20	0.27

TABLE A.7 Properties of Typical High-Temperature Composite Materials

Property	Carbon/ Polyimide IM7/PETI-5	Woven Carbon/ Phenolic (8H Fabric)	Woven Graphite/ Phenolic (Plain Weave)	Carbon/ Carbon (T300 Fabric) 1K DV 8367	Boron/ Aluminum (B4/6061-Al)	Silicon Carbide/Al (SCS2/6061-Al)	Silicon Carbide/ Ceramic (SiC/CAS)
Fiber volume ratio, V_f	0.52	—	—	—	0.50	0.43	0.39
Density, ρ, g/cm³ (lb/in³)	1.65	—	—	—	2.65 (0.096)	2.85 (0.103)	2.72 (0.10)
Longitudinal modulus, E_1, GPa (Msi)	151 (21.9)	20 (2.9)	15 (2.2)	70.7 (10.3)	235 (34.1)	204 (29.6)	121 (17.6)
Transverse in-plane modulus, E_2, GPa (Msi)	9.65 (1.40)	19 (2.8)	13 (1.9)	73.4 (10.6)	137 (19.9)	118 (17.1)	112 (16.2)
Out-of-plane modulus, E_3, GPa (Msi)	9.65 (1.40)	14.6 (2.1)	7 (1.0)	—	—	—	—
In-plane shear modulus, G_{12}, GPa (Msi)	6.34 (0.92)	6.8 (0.99)	3.9 (0.57)	—	47 (6.8)	41 (5.9)	44 (6.4)
Major Poisson's ratio, ν_{12}	0.34	0.23	0.20	0.04	0.30	0.27	0.20
Minor Poisson's ratio, ν_{21}	0.02	0.22	0.17	0.04	0.17	0.12	0.18
Out-of-plane Poisson's ratio, ν_{13}	0.34	0.25	0.13	—	—	—	—
Long. tensile strength, F_{1t}, MPa (ksi)	2120 (307)	136 (19.7)	87 (12.6)	59.7 (8.6)	1373 (199)	1462 (212)	393 (57)
Transv. tensile strength, F_{2t}, MPa (ksi)	62 (9.0)	108 (15.7)	72 (10.4)	40.3 (5.8)	118 (17.1)	86 (12.5)	22 (3.2)
Out-of-plane tensile strength, F_{3t}, MPa (ksi)	31 (4.5)	10.9 (1.6)	8.7 (1.3)	—	—	—	—
Ultimate longitudinal tensile strain, ε_{1t}^u	—	—	—	—	0.006	0.009	0.008
Ultimate transverse tensile strain, ε_{2t}^u	—	—	—	—	0.001	0.001	0.0002
Long. compressive strength, F_{1c}, MPa (ksi)	1550 (225)	—	—	230 (33)	1573 (228)	2990 (434)	—
Trans. compressive strength, F_{2c}, MPa (ksi)	255 (37)	—	—	193 (28)	157 (22.8)	285 (41.4)	—
In-plane shear strength, F_6, MPa (ksi)	100 (14.5)	—	—	—	128 (18.5)	113 (16.4)	—
Out-of-plane shear strength, F_4, MPa (ksi)	—	21.4 (3.1)	18.7 (2.7)	10.4 (1.5)	—	—	—
Out-of-plane shear strength, F_5, MPa (ksi)	119 (17.2)	21.4 (3.1)	18.7 (2.7)	14.6 (2.1)	—	—	—
Longitudinal CTE, α_1, 10⁻⁶/°C (10⁻⁶/°F)	−0.16 (−0.09)	—	—	—	6.0 (3.3)	9.1 (5.0)	4.1 (2.3)
Transverse CTE, α_2, 10⁻⁶/°C (10⁻⁶/°F)	28 (15.6)	—	—	—	20 (11)	17.8 (9.9)	4.2 (2.4)

TABLE A.8 Properties of Typical Structural Metals

Property	Aluminum (2024 T3)	Steel (AISI 1025)	Titanium (MILT)
Density, ρ, g/cm^3 (lb/in^3)	2.80 (0.101)	7.80 (0.282)	4.40 (0.159)
Young's modulus, E, GPa (Msi)	73 (10.4)	207 (30)	108 (15.7)
Shear modulus, G, GPa (Msi)	26.6 (3.86)	79 (11.4)	42.4 (6.1)
Poisson's ratio, ν	0.33	0.30	0.30
Tensile strength, F_t, MPa (ksi)	414 (60)	394 (57)	550 (80)
Compressive strength, F_c, MPa (ksi)	414 (60)	394 (57)	475 (69)
Shear strength, F_s, MPa (ksi)	248 (36)	248 (36)	295 (43)
Coefficient of thermal expansion, α, 10^{-6}/°C (10^{-6}/°F)	23 (13)	11 (6)	11 (6)

TABLE A.9 Characteristic Properties of Sandwich Core Materials

Property	Divinycell H80	Divinycell H100	Divinycell H160	Divinycell H250	Balsa Wood CK57	Aluminum Honeycomb PAMG 5052	Foam-Filled Honeycomb Style 20	Polyurethane FR-3708
Density, ρ, kg/m³ (lb/ft³)	80 (5)	100 (6.2)	160 (9.9)	250 (15.5)	150 (9.4)	130 (8.1)	128 (8)	128 (8)
In-plane modulus, E_1, MPa (ksi)	77 (11.2)	95 (13.8)	150 (21.7)	240 (34.8)	65 (9.4)	8.3 (1.2)	25 (3.7)	35 (5.1)
In-plane modulus, E_2, MPa (ksi)	77 (11.2)	95 (13.8)	150 (21.7)	230 (33.3)	65 (9.4)	6.0 (0.87)	7.6 (1.1)	35 (5.1)
Out-of-plane modulus, E_3, MPa (ksi)	115 (16.7)	140 (20.3)	250 (36.2)	403 (58.4)	5200 (750)	2415 (350)	270 (39)	110 (15.9)
Transverse shear modulus, G_{13}, MPa (ksi)	40 (5.8)	50 (7.2)	75 (10.9)	115 (16.7)	58.7 (8.5)	580 (84)	8.5 (1.23)	10 (1.45)
In-plane compressive strength, F_{1c}, MPa (ksi)	1.2 (0.17)	1.7 (0.25)	2.8 (0.41)	4.6 (0.67)	0.78 (0.11)	0.2 (0.03)	0.4 (0.06)	1.2 (0.17)
In-plane tensile strength, F_{1t}, MPa (ksi)	2.3 (0.33)	2.7 (0.39)	4.5 (0.65)	7.2 (1.04)	1.13 (0.16)	1.6 (0.24)	0.5 (0.07)	1.1 (0.16)
In-plane compressive strength, F_{2c}, MPa (ksi)	1.2 (0.17)	1.7 (0.25)	2.8 (0.41)	4.6 (0.67)	0.78 (0.11)	0.2 (0.03)	0.3 (0.05)	1.2 (0.17)
Out-of-plane compressive strength, F_{3c}, MPa (ksi)	1.5 (0.22)	2.2 (0.32)	3.5 (0.51)	5.7 (0.83)	9.6 (1.39)	11.8 (1.7)	1.4 (0.20)	1.7 (0.25)
Transverse shear strength, F_5, MPa (ksi)	1.5 (0.22)	1.8 (0.26)	3.0 (0.43)	5.0 (0.72)	3.7 (0.54)	3.5 (0.50)	0.75 (0.11)	1.4 (0.20)

Three-Dimensional Transformations of Elastic Properties of Composite Lamina

TABLE B.1   Three-Dimensional Transformations for Elastic Parameters of Unidirectional Lamina

	$S_{11}(C_{11})$	$S_{12}(C_{12})$	$S_{13}(C_{13})$	$S_{22}(C_{22})$	$S_{23}(C_{23})$	$S_{33}(C_{33})$	$S_{44}(4C_{44})$	$S_{55}(4C_{55})$	$S_{66}(4C_{66})$
$S_{xx}(C_{xx})$	m_1^4	$2m_1^2 m_2^2$	$2m_1^2 m_3^2$	m_2^4	$2m_2^2 m_3^2$	m_3^4	$m_2^2 m_3^2$	$m_1^2 m_3^2$	$m_1^2 m_2^2$
$S_{xy}(C_{xy})$	$m_1^2 n_1^2$	$m_1^2 n_2^2 + m_2^2 n_1^2$	$m_1^2 n_3^2 + m_3^2 n_1^2$	$m_2^2 n_2^2$	$m_2^2 n_3^2 + m_3^2 n_2^2$	$m_3^2 n_3^2$	$m_2 m_3 n_2 n_3$	$m_1 m_3 n_1 n_3$	$m_1 m_2 n_1 n_2$
$S_{xz}(C_{xz})$	$m_1^2 p_1^2$	$m_1^2 p_2^2 + m_2^2 p_1^2$	$m_1^2 p_3^2 + m_3^2 p_1^2$	$m_2^2 p_2^2$	$m_2^2 p_3^2 + m_3^2 p_2^2$	$m_3^2 p_3^2$	$m_2 m_3 p_2 p_3$	$m_1 m_3 p_1 p_3$	$m_1 m_2 p_1 p_2$
$S_{xq}(2C_{xq})$	$2m_1^2 n_1 p_1$	$2m_1^2 n_2 p_2 + 2m_2^2 n_1 p_1$	$2m_1^2 n_3 p_3 + 2m_3^2 n_1 p_1$	$2m_2^2 n_2 p_2$	$2m_2^2 n_3 p_3 + 2m_3^2 n_2 p_2$	$2m_3^2 n_3 p_3$	$m_2 m_3 \times (n_2 p_3 + n_3 p_2)$	$m_1 m_3 \times (n_1 p_3 + n_3 p_1)$	$m_1 m_2 \times (n_1 p_2 + n_2 p_1)$
$S_{xr}(2C_{xr})$	$2m_1^3 p_1$	$2m_1 m_2 \times (m_1 p_2 + m_2 p_1)$	$2m_1 m_3 \times (m_1 p_3 + m_3 p_1)$	$2m_2^3 p_2$	$2m_2 m_3 \times (m_2 p_3 + m_3 p_2)$	$2m_3^3 p_3$	$m_2 m_3 \times (m_2 p_3 + m_3 p_2)$	$m_1 m_3 \times (m_1 p_3 + m_3 p_1)$	$m_1 m_2 \times (m_1 p_2 + m_2 p_1)$
$S_{xs}(2C_{xs})$	$2m_1^3 n_1$	$2m_1 m_2 \times (m_1 n_2 + m_2 n_1)$	$2m_1 m_3 \times (m_1 n_3 + m_3 n_1)$	$2m_2^3 n_2$	$2m_2 m_3 \times (m_2 n_3 + m_3 n_2)$	$2m_3^3 n_3$	$m_2 m_3 \times (m_2 n_3 + m_3 n_2)$	$m_1 m_3 \times (m_1 n_3 + m_3 n_1)$	$m_1 m_2 \times (m_2 n_1 + m_1 n_2)$
$S_{yy}(C_{yy})$	n_1^4	$2n_1^2 n_2^2$	$2n_1^2 n_3^2$	n_2^4	$2n_2^2 n_3^2$	n_3^4	$n_2^2 n_3^2$	$n_1^2 n_3^2$	$n_1^2 n_2^2$
$S_{yz}(C_{yz})$	$n_1^2 p_1^2$	$n_1^2 p_2^2 + n_2^2 p_1^2$	$n_1^2 p_3^2 + n_3^2 p_1^2$	$n_2^2 p_2^2$	$n_2^2 p_3^2 + n_3^2 p_2^2$	$n_3^2 p_3^2$	$n_2 n_3 p_2 p_3$	$n_1 n_3 p_1 p_3$	$n_1 n_2 p_1 p_2$
$S_{yq}(2C_{yq})$	$2n_1^3 p_1$	$2n_1 n_2 \times (n_1 p_2 + n_2 p_1)$	$2n_1 n_3 \times (n_1 p_3 + n_3 p_1)$	$2n_2^3 p_2$	$2n_2 n_3 \times (n_2 p_3 + n_3 p_2)$	$2n_3^3 p_3$	$n_2 n_3 p_2 p_3$	$n_1 n_3 p_1 p_3$	$n_1 n_2 p_1 p_2$
$S_{yr}(2C_{yr})$	$2m_1 n_1^2 p_1$	$2m_1 n_2^2 p_1 + 2m_2 n_1^2 p_2$	$2m_1 n_3^2 p_1 + 2m_3 n_1^2 p_3$	$2m_2 n_2^2 p_2$	$2m_2 n_3^2 p_2 + 2m_3 n_2^2 p_3$	$2m_3 n_3^2 p_3$	$n_2 n_3 \times (m_2 p_3 + m_3 p_2)$	$n_1 n_3 \times (m_1 p_3 + m_3 p_1)$	$n_1 n_2 \times (m_1 p_2 + m_2 p_1)$
$S_{ys}(2C_{ys})$	$2m_1 n_1^3$	$2n_1 n_2 \times (m_1 n_2 + m_2 n_1)$	$2n_1 n_3 \times (m_1 n_3 + m_3 n_1)$	$2m_2 n_2^3$	$2n_2 n_3 \times (m_2 n_3 + m_3 n_2)$	$2m_3 n_3^3$	$n_2 n_3 \times (m_2 n_3 + m_3 n_2)$	$n_1 n_3 \times (m_1 n_3 + m_3 n_1)$	$n_1 n_2 \times (m_1 n_2 + m_2 n_1)$

$S_{zz}(C_{zz})$	p_1^4	$2p_1^2p_2^2$	$2p_1^2p_3^2$	p_2^4	$2p_2^2p_3^2$	p_3^4	$p_2^2p_3^2$	$p_1^2p_3^2$	$p_1^2p_2^2$
$S_{zq}(2C_{zq})$	$2n_1p_1^3$	$2p_1p_2 \times (n_1p_2+n_2p_1)$	$2p_1p_3 \times (n_1p_3+n_3p_1)$	$2n_2p_2^3$	$2p_2p_3 \times (n_2p_3+n_3p_2)$	$2n_3p_3^3$	$p_2p_3 \times (n_2p_3+n_3p_2)$	$p_1p_3 \times (n_1p_3+n_3p_1)$	$p_1p_2 \times (n_1p_2+n_2p_1)$
$S_{zr}(2C_{zr})$	$2m_1p_1^3$	$2p_1p_2 \times (m_1p_2+m_2p_1)$	$2p_1p_3 \times (m_1p_3+m_3p_1)$	$2m_2p_2^3$	$2p_2p_3 \times (m_2p_3+m_3p_2)$	$2m_3p_3^3$	$p_2p_3 \times (m_2p_3+m_3p_2)$	$p_1p_3 \times (m_1p_3+m_3p_1)$	$p_1p_2 \times (m_1p_2+m_2p_1)$
$S_{zs}(2C_{zs})$	$2m_1n_1p_1^2$	$2m_1n_1p_2^2 + 2m_2n_2p_1^2$	$2m_1n_1p_3^2 + 2m_3n_3p_1^2$	$2m_2n_2p_2^2$	$2m_2n_2p_3^2 + 2m_3n_3p_2^2$	$2m_3n_3p_3^2$	$p_2p_3 \times (m_2n_3+m_3n_2)$	$p_1p_3 \times (m_1n_3+m_3n_1)$	$p_1p_2 \times (m_1n_2+m_2n_1)$
$S_{qq}(4C_{qq})$	$4n_1^2p_1^2$	$8n_1n_2p_1p_2$	$8n_1n_3p_1p_3$	$4n_2^2p_2^2$	$8n_2n_3p_2p_3$	$4n_3^2p_3^2$	$(n_2p_3+n_3p_2)^2$	$(n_1p_3+n_3p_1)^2$	$(n_1p_2+n_2p_1)^2$
$S_{qr}(4C_{qr})$	$4m_1n_1p_1^2$	$4p_1p_2 \times (m_1n_2+m_2n_1)$	$4p_1p_3 \times (m_1n_3+m_3n_1)$	$4m_2n_2p_2^2$	$4p_2p_3 \times (m_2n_3+m_3n_2)$	$4m_3n_3p_3^2$	$(n_2p_3+n_3p_2) \times (m_2p_3+m_3p_2)$	$(n_1p_3+n_3p_1) \times (m_1p_3+m_3p_1)$	$(n_1p_2+n_2p_1) \times (m_1p_2+m_2p_1)$
$S_{qs}(4C_{qs})$	$4m_1n_1^2p_1$	$4n_1n_2 \times (m_1p_2+m_2p_1)$	$4n_1n_3 \times (m_1p_3+m_3p_1)$	$4m_2n_2^2p_2$	$4n_2n_3 \times (m_2p_3+m_3p_2)$	$4m_3n_3^2p_3$	$(n_2p_3+n_3p_2) \times (m_2n_3+m_3n_2)$	$(n_1p_3+n_3p_1) \times (m_1n_3+m_3n_1)$	$(n_1p_2+n_2p_1) \times (m_1n_2+m_2n_1)$
$S_{rr}(4C_{rr})$	$4m_1^2p_1^2$	$8m_1m_2p_1p_2$	$8m_1m_3p_1p_3$	$4m_2^2p_2^2$	$8m_2m_3p_2p_3$	$4m_3^2p_3^2$	$(m_2p_3+m_3p_2)^2$	$(m_1p_3+m_3p_1)^2$	$(m_1p_2+m_2p_1)^2$
$S_{rs}(4C_{rs})$	$4m_1^2n_1p_1$	$4m_1m_2 \times (n_1p_2+n_2p_1)$	$4m_1m_3 \times (n_1p_3+n_3p_1)$	$4m_2^2n_2p_2$	$4m_2m_3 \times (n_2p_3+n_3p_2)$	$4m_3^2n_3p_3$	$(m_2p_3+m_3p_2) \times (m_2n_3+m_3n_2)$	$(m_1p_3+m_3p_1) \times (m_1n_3+m_3n_1)$	$(m_1p_2+m_2p_1) \times (m_1n_2+m_2n_1)$
$S_{ss}(4C_{ss})$	$4m_1^2n_1^2$	$8m_1m_2n_1n_2$	$8m_1m_3n_1n_3$	$4m_2^2n_2^2$	$8m_2m_3n_2n_3$	$4m_3^2n_3^2$	$(m_2n_3+m_3n_2)^2$	$(m_1n_3+m_3n_1)^2$	$(m_1n_2+m_2n_1)^2$

Answers to Selected Problems

CHAPTER 3

3.3 $V_{max} = 0.5804$

3.6 $\eta = 1,\ P^* = P_m \dfrac{1 + \xi V_f}{V_m}$

      $\eta = 0,\ P^* = P_m$

      $\eta = -\dfrac{1}{\xi},\ P^* = P_m \dfrac{\xi V_m}{\xi + V_f}$

3.11   Mechanics of materials: $E_2 = 7.56$ GPa
       Halpin-Tsai: $E_2 = 8.13$ GPa

3.14 $E_2 = 21.86$ GPa

3.17   Mechanics of materials: $G_{12} = 2.68$ GPa
       Halpin-Tsai: $G_{12} = 3.84$ GPa

3.20 $\xi = 13.29$

3.22 $\xi = 2.425$
       $G_{12} = 10.59$ GPa (1.535 Msi)

3.24   Cox: $1/r = 58.4$
       Halpin: $1/r = 75.45$

3.26   Kevlar/epoxy: $V_f = 0.237$
       Carbon nanotube/epoxy: $V_f = 0.035$

CHAPTER 4

4.12 (c)

4.14 $(E_x)_{\theta=45°} = 16.42$ GPa
$(E_x)_{\theta=45°} \cong 16.62$ GPa

4.16 $(\nu_{xy})_{\theta=45°} = 0.189$
$(\nu_{xy})_{\theta=45°} \cong 0.205$

4.18 $(\eta_{sx})_{\theta=45°} = -0.42$
$(\eta_{sx})_{\theta=45°} \cong -0.50$

4.20 $(\nu_{xy})_{\theta=45°} = 0.443$

4.22 $(\nu_{xy})_{\theta=60°} \cong 0.111$

4.24 $\dfrac{1}{(E_x)_{\theta=45°}} = \dfrac{4}{3(E_x)_{\theta=30°}} - \dfrac{1}{2E_1} + \dfrac{1}{6E_2}$

4.26 $(\nu_{xy})_{max} = \dfrac{E - 2G_{12}}{E + 2G_{12}}$ at $\theta = 45°$

4.28 $(\nu_{xy})_{max} = \dfrac{E - 2(1 - \nu)G_{12}}{E + 2(1 - \nu)G_{12}}$ at $\theta = 45°$

$(\eta_{sx})_{max} = \dfrac{1}{4}\dfrac{E - 2(1 + \nu)G_{12}}{\sqrt{2(H\nu)EG_{12}}}$ at $\theta = \dfrac{\pi}{2} - \dfrac{1}{2}\tan^{-1}\left[\dfrac{E}{2(1 + \nu)G_{12}}\right]^{1/2}$

4.29 $\eta_{sx} = -0.241$ at $\theta = 29.4°$
$\eta_{sx} = 0.189$ at $\theta = 62.8°$

4.33 $\varepsilon_2' = \varepsilon_2 + k_E\varepsilon_1$

4.36 $k \cong -\dfrac{m^2}{n^2} = -\cot^2\theta$

4.38 (a) $\eta_{sx} = \dfrac{G_{xy}}{\tau_s}\left(\varepsilon_x - \dfrac{\sigma_x}{E_x}\right)$

(b) $\gamma_s = 0.167 \times 10^{-3}$.

4.40 (a) $k = -\eta_{sx}$
(b) $\nu_{xy} = -\eta_{xs}\eta_{sy}$
(c) $\eta_{xs}\eta_{sx} = 1$

4.43 $E_1 = 61.82$ GPa
$E_2 = 11.15$ GPa
$G_{12} = 5.32$ GPa

CHAPTER 5

5.2 $E_1 = 154.2$ GPa (22.3 Msi)
$F_{1t} = 2264$ MPa (328 ksi)

5.4 $E_1 = 127$ GPa (18.4 Msi)
$F_{1t} = 180.7$ MPa (26.2 ksi)
The strength F_{1t} increases linearly with E_{1f} up to the point where $\varepsilon_{ft}^u = \varepsilon_{mt}^u$, thereafter, it decreases asymptotically to the value of 772 MPa.

5.6 $(V_f)_{min} = 0.049$

5.8 $E_1^0 = 142.4$ GPa (20.6 Msi)
$E_1' = 81.88$ GPa (11.87 Msi)

5.10 $(\tau_i)_{max} = 259$ MPa (37.5 ksi)

5.12 $E_1 = 12.16$ GPa (1.76 Msi)
$\sigma_1 = 352$ MPa (51.0 ksi)

5.15 Maximum stress criterion: $F_{2t} = 54.7$ MPa (7.9 ksi)
Maximum strain criterion: $F_{2t} = 87.8$ MPa (12.7 ksi)

5.17 $F_{xt} = 427$ MPa (61.9 ksi)

CHAPTER 6

6.1 $\sigma_0^u = 80$ MPa; in-plane shear failure mode

6.4 $\theta_{sc} = 26.6°$

6.6 $(\sigma_2)_{max} = 103$ MPa

6.8 $F_{xt} = 144$ MPa, $\theta = 38.3°$

6.10 $F_s^{(+)} \cong 2F_{2t}$, $F_s^{(-)} \cong \dfrac{2F_{2t}}{\sqrt{3}(1 + \nu_{21})}$

6.12 $F_{(xy)} = 40$ MPa

6.14 $\theta_1 = 31.4°$, $\theta_2 = 58.6°$

6.17 $F_s = 131$ MPa

6.20 $F_0 = 252$ MPa (Tsai-Hill); $F_0 = 352$ MPa (max. stress)

6.23 $F_0 = 84$ MPa

6.24 $\sigma_0'' = 98.3$ MPa

6.27 $F_s^{(+)} \cong F_{2c}, F_s^{(-)} \cong F_{2t}$

6.32 Tsai-Wu: $S_f = 1.121$
Max. stress: $S_f = 1.014$

6.34 $S_f = 1.075$ (for both Tsai-Wu and Tsai-Hill)

6.38 $F_0 = 253$ MPa

6.41 $p = 3.14$ MPa

6.43 $F_s^{(+)} = 192$ MPa (Tsai-Wu)
$= 164$ MPa (Hashin-Rotem)
$= 222$ MPa (Max. stress)

6.45 $F_s^{(+)} = 900$ MPa (Max. stress and Hashin-Rotem)
$= 850$ MPa (Max. strain)
$= 540$ MPa (Tsai-Wu)
$= 543$ MPa (Tsai-Hill)

6.47 $F_0 = 152$ MPa (Max. stress)
$= 122$ MPa (Tsai-Wu)
$= 106$ MPa (Tsai-Hill)

CHAPTER 7

7.2 (a) Symmetric, balanced, $B_{ij} = 0$, $A_{xs} = A_{ys} = 0$
(b) Asymmetric, balanced, $A_{xs} = A_{ys} = 0$
(c) Symmetric crossply, $B_{ij} = 0$, $A_{xs} = A_{ys} = D_{xs} = D_{ys} = 0$
(f) Antisymmetric crossply, $B_{xx} = -B_{yy}$ all other $B_{ij} = 0$, $A_{xs} = A_{ys} = D_{xs} = D_{ys} = 0$
(h) Antisymmetric, $A_{xs} = A_{ys} = D_{xs} = D_{ys} = 0$, $B_{xx} = B_{yy} = B_{xy} = B_{ss} = 0$
(i) Symmetric angle-ply, balanced $B_{ij} = 0$, $A_{xs} = A_{ys} = 0$

7.4 (a) Symmetric angle-ply, $n =$ odd, $B_{ij} = 0$
(b) Symmetric balanced angle-ply, $B_{ij} = 0$, $A_{xs} = A_{ys} = 0$
(c) Antisymmetric angle-ply, $A_{xs} = A_{ys} = D_{xs} = D_{ys} = 0$, $B_{xx} = B_{yy} = B_{xy} = B_{ss} = 0$
(d) Same as (c)

7.6 $B_{xs} = \dfrac{h^2}{2n} Q_{xs}(\theta)$ (see Eq. 4.67 for transformation of Q_{xs})

7.10 $B_{xs} = B_{ys} = B_{sx} = B_{sy} = \dfrac{(E_1 - E_2)t^2}{4(1 - \nu_{12}\nu_{21})}$

$B_{xx} = B_{yy} = B_{xy} = B_{ss} = 0$

7.11 $A_{xx} = 60.24$ MN/m
$A_{xy} = 17.53$ MN/m
$A_{yy} = 26.43$ MN/m
$A_{ss} = 20.56$ MN/m
$A_{xs} = A_{ys} = A_{sx} = A_{sy} = 0$

7.13 $A_{xx} = A_{yy} = 9.55E_o t \cong 7.50E_o t$
$A_{xy} = 7.15E_o t \cong 7.50E_o t$
$A_{ss} = 7.75E_o t \cong 7.50E_o t$
$A_{xs} = A_{ys} = A_{sx} = A_{sy} = 0$

$B_{xs} = B_{ys} = B_{sx} = B_{sy} = 3.52E_o t^2 \cong 3.75E_o t^2$
$B_{xx} = B_{xy} = B_{yy} = B_{ss} = 0$

$D_{xx} = D_{yy} = 3.18E_o t^3 \cong 2.50E_o t^3$
$D_{xy} = 2.38E_o t^3 \cong 2.50E_o t^3$
$D_{ss} = 2.58E_o t^3 \cong 2.50E_o t^3$
$D_{xs} = D_{ys} = D_{sx} = D_{sy} = 0$

7.19 $r = 7.67$

7.20 $E_1 = 160$ GPa, $G_{12} = 7.06$ GPa

7.23 $\bar{E}_x = 23.9$ GPa, $\bar{G}_{xy} = 37.6$ GPa, $\bar{\nu}_{xy} = 0.706$

7.25 $\bar{\nu}_{xy} = 1.210$

7.27 $\bar{\nu}_{xy} = 0.663$

7.29 $\bar{E}_x = 58.4$ GPa, $\bar{G}_{xy} = 23$ GPa, $\bar{\nu}_{xy} = 0.27$

7.33 $\bar{\nu}_{xy} = 0.281$

7.35 $N_o = 1,056$ N/m

7.37 (a) $\gamma_o = -2\varepsilon_x^s$
       (b) $\bar{\nu}_{xy} = \bar{\nu}_{yx} = 1$

7.41 $M_x = 1,955\ \varepsilon_{max}$ RN $\cdot$ m/m
$M_x \cong 1,926\ \varepsilon_{max}$ RN $\cdot$ m/m

CHAPTER 8

8.1 $\alpha_1 = 8.19 \times 10^{-6}/°C$
$\alpha_2 = 55.8 \times 10^{-6}/°C$

8.3 $\alpha_1 = 4.05 \times 10^{-6}/°C$
$\alpha_2 = 4.35 \times 10^{-6}/°C$

8.5 $N_x^T = N_y^T = 24.62t$ MPa
$N_s^T = 0$

8.8 $\bar{\alpha}_x = -3.63 \times 10^{-6}/°C$

8.10 $r = 3.66$

8.12 $\bar{\alpha}_x = -0.85 \times 10^{-6}/°C$
$\bar{\alpha}_y = 5.24 \times 10^{-6}/°C$

8.16 $\Delta c = 0.856\%$

8.18 $\sigma_{1e} = 109.7$ MPa, $\sigma_{2e} = 16.5$ MPa

8.20 $\sigma_1 = -98.8$ MPa
$\sigma_2 = 20.1$ MPa
$\tau_6 = 0$

8.22 $\sigma_x = -\dfrac{2E_x}{\eta_{xs}}(\alpha_1 - \alpha_2)mn\Delta T$

8.24 $\tau_s = (\alpha_2 - \alpha_1)\Delta T\left[\dfrac{1 + \nu_{12}}{E_1} + \dfrac{1 + \nu_{21}}{E_2}\right]^{-1}$

8.26 $N_x = 10.91$ kN/m

8.27 $N_x = 493$ kN/m

8.31   (b) $N_x = N_y = 3.51$ kN/m
$M_x = -M_y = -0.343$ N $\cdot$ m/m

CHAPTER 9

9.2   Maximum stress criterion:
$\bar{F}_{xt} = 438$ MPa
$\bar{F}_{xc} = 922$ MPa
Tsai-Wu criterion:
$\bar{F}_{xt} = 436$ MPa
$\bar{F}_{xc} = 1,282$ MPa

9.4 (a) $\bar{F}_{xt} = 122$ MPa, $\bar{F}_{xt} \cong 100$ MPa
     (b) $\bar{F}_{xt} = 94$ MPa, $\bar{F}_{xt} \cong 98$ MPa
     (c) $\bar{F}_{xt} = 148$ MPa, $\bar{F}_{xt} \cong 152$ MPa

9.6 (a) $\bar{F}_{xc} = 167$ MPa, $\bar{F}_{xc} \cong 178$ MPa
     (b) $\bar{F}_{xc} = 95$ MPa, $\bar{F}_{xc} \cong 98$ MPa
     (c) $\bar{F}_{xc} = 151$ MPa, $\bar{F}_{xc} \cong 152$ MPa

9.8 (a) $\bar{F}_{s} = 112$ MPa, $\bar{F}_{s} \cong 99$ MPa
     (b) $\bar{F}_{s} = 153$ MPa, $\bar{F}_{s} \cong 149$ MPa
     (c) $\bar{F}_{s} = 482$ MPa, $\bar{F}_{s} \cong 468$ MPa

9.12 $\bar{F} \cong 3.25 F_{6}$

9.13 $\bar{F}_{o} = 76$ MPa (by both Tsai-Wu and maximum stress criteria)

9.15 $\bar{F}_{xc} = 483$ MPa

9.19 $\bar{F}_{xt} = 270$ MPa

9.21 $\bar{F}_{xt} = 427$ MPa, in-plane shear mode

9.22 Maximum stress criterion:
      $\bar{F}_{xt} = 138$ MPa, transverse tensile failure in 90° –ply
      $\bar{F}_{xc} = 901$ MPa, longitudinal compressive failure in 0° –ply

      Tsai-Wu criterion
      $\bar{F}_{xt} = 132$ MPa
      $\bar{F}_{xc} = 694$ MPa

9.24 $\bar{F}_{s} = 189$ MPa (Tsai-Wu)
      $\bar{F}_{s} = 273$ MPa (maximum stress)
      Transverse tensile failure

9.26 (a) $\sigma_{1e} = -\sigma_{2e} = -20.9$ MPa, $\tau_{6e} = -12.0$ MPa
      (b) $\sigma_{1} = 1.098\ \bar{\sigma}_{x}$, $\sigma_{2} = -0.098\ \bar{\sigma}_{x}$, $\tau_{6} = -0.232\ \bar{\sigma}_{x}$
      (c) $\bar{F}_{xt} = 276$ MPa (maximum stress criterion)
           $\bar{F}_{xt} = 293$ MPa (Tsai-Wu criterion)
                In-plane shear failure

9.28 $\bar{F}_{FPF} = 438$ MPa, $\bar{F}_{ULF} = \bar{F}_{xt}^{u} = 1140$ MPa
      $\varphi_{L} = 0.384$

9.30 $\bar{F}_{FPF} = 311$ MPa
      $\bar{F}_{ULF} = 760$ MPa
      $\varphi_{L} = 0.409$

9.32 (d)

9.34 (a) $\sigma_1 = 51.9$ MPa, $\sigma_2 = 23.3$ MPa, $\tau_6 = 22.4$ MPa
(b) $\sigma_{1e} = -\sigma_{2e} = -20.9$ MPa, $\tau_{6e} = 12.0$ MPa
(c) $\sigma_1 = 31$ MPa, $\sigma_2 = 44$ MPa, $\tau_6 = 10.6$ MPa

9.36 $h_a = 5.59$ mm by maximum stress criterion
$h_a = 4.70$ mm by Tsai-Wu criterion
Transverse tensile failure

Author Index

Subject Index